SEMICONDUCTOR SENSORS

SEMICONDUCTOR SENSORS

Edited by

S. M. Sze
UMC Chair Professor
Department of Electronics Engineering and
Microelectronics and Information-Systems Research Center
National Chiao Tung University
Hsinchu, Taiwan, R.O.C.

A Wiley-Interscience Publication
JOHN WILEY & SONS, INC.

New York · Chichester · Brisbane · Toronto · Singapore

This text is printed on acid-free paper.

Copyright © 1994 by John Wiley & Sons, Inc.

All rights reserved. Published simultaneously in Canada.

Reproduction or translation of any part of this work beyond that permitted by Section 107 or 108 of the 1976 United States Copyright Act without the permission of the copyright owner is unlawful. Requests for permission or further information should be addressed to the Permissions Department, John Wiley & Sons, Inc., 605 Third Avenue, New York, NY 10158-0012.

Library of Congress Cataloging in Publication Data:
Semiconductor sensors / edited by S. M. Sze.
      p.   cm.
   Includes bibliographical references and index.
   ISBN 0-471-54609-7
    1. Detectors.   2. Semiconductors.   I. Sze, S. M., 1936– .
  TA185.S447   1994
  001'.2—0020   94-22271
                                                                    CIP

Printed in the United States of America

10 9 8 7 6 5 4 3 2

CONTRIBUTORS

S. AUDET,   Princeton Gamma-Tech, Inc., Princeton, New Jersey, USA

H. BALTES,   Swiss Federal Institute of Technology Zürich, Zürich, Switzerland

R. CASTAGNETTI,   Swiss Federal Institute of Technology Zürich, Zürich, Switzerland

A. DEWA,   Micropump Corporation, Vancouver, Washington, USA

S. VAN HERWAARDEN,   Xensor Integration, Delft, The Netherlands

B. KLOECK,   University of Neuchâtel, Neuchâtel, Switzerland

W. H. KO,   Case Western Reserve University, Cleveland, Ohio, USA

C. H. MASTRANGELO,   University of Michigan, Ann Arbor, Michigan, USA

G. C. M. MEIJER,   Delft University, Delft, The Netherlands

S. R. MORRISON,   Simon Fraser University, Burnaby, B.C., Canada

M. E. MOTAMEDI,   Rockwell Science Center, Thousand Oaks, California, USA

K. NAJAFI,   University of Michigan, Ann Arbor, Michigan, USA

N. NAJAFI,   IBM Microelectronics Division, Essex Junction, Vermont, USA

N. F. DE ROOIJ,   University of Neuchâtel, Neuchâtel, Switzerland

J. STEIGERWALD,   Analog Devices, Wilmington, Massachusetts, USA

S. M. SZE,   National Chiao Tung University, Hsinchu, Taiwan, ROC

W. C. TANG,   Ford Microelectronics, Inc., Colorado Springs, Colorado, USA

R. M. WHITE,   University of California, Berkeley, California, USA

K. D. WISE,   University of Michigan, Ann Arbor, Michigan, USA

CONTENTS

Preface　　　　　　　　　　　　　　　　　　　　　　　　　　　　xi

1 Classification and Terminology of Sensors　　　　　　　　　　1
   S. M. Sze

   1.1 Semiconductor Sensors   1
   1.2 Classification of Semiconductor Sensors   3
   1.3 Sensor Characterization   7
   1.4 Evolution of Semiconductor Sensors   10
   1.5 Organization of the Book   13
        References   15

2 Semiconductor Sensor Technologies　　　　　　　　　　　　17
   C. H. Mastrangelo and W. C. Tang

   2.1 Introduction   17
   2.2 Basic Fabrication Processes   20
   2.3 Micromechanical Process Design   31
   2.4 Bulk Micromachining   42
   2.5 Surface Micromachining   55
   2.6 Other Micromachining Techniques   75
   2.7 Summary and Future Trends   80
        Problems   81
        References   84

3 Acoustic Sensors　　　　　　　　　　　　　　　　　　　　　97
   M. E. Motamedi and R. M. White

   3.1 Introduction   97
   3.2 Acoustic Waves   99
   3.3 Piezoelectric Materials   104
   3.4 Acoustic Sensing   110
   3.5 SAW Sensors   117
   3.6 Sensor Applications   126
   3.7 Summary and Future Trends   143
        Problems   144
        References   146

4 Mechanical Sensors — 153
B. Kloeck and N. F. de Rooij

- 4.1 Introduction 153
- 4.2 Piezoresistivity 160
- 4.3 Piezoresistive Sensors 174
- 4.4 Capacitive Sensors 185
- 4.5 Applications 187
- 4.6 Summary and Future Trends 194
    - Problems 197
    - References 199

5 Magnetic Sensors — 205
H. Baltes and R. Castagnetti

- 5.1 Introduction 205
- 5.2 Effects and Materials 211
- 5.3 Integrated Hall Sensors 231
- 5.4 Magnetotransistors 246
- 5.5 Other Magnetic Sensors 258
- 5.6 Summary and Future Trends 264
    - Problems 265
    - References 266

6 Radiation Sensors — 271
S. Audet and J. Steigerwald

- 6.1 Introduction 271
- 6.2 Common Physics 273
- 6.3 HgCdTe Infrared Sensors 298
- 6.4 Visible-Light Color Sensors 305
- 6.5 High-Energy Photodiodes 308
- 6.6 Silicon Drift Chamber X-Ray Sensors 314
- 6.7 Summary and Future Trends 322
    - Problems 324
    - References 326

7 Thermal Sensors — 331
S. Van Herwaarden and G. C. M. Meijer

- 7.1 Introduction 331
- 7.2 Heat Transfer 332
- 7.3 Thermal Structures 340
- 7.4 Thermal-Sensing Elements 350
- 7.5 Thermal and Temperature Sensors 357
- 7.6 Summary and Future Trends 373
    - Problems 375
    - References 380

8 Chemical Sensors 383
S. R. Morrison

- 8.1 Introduction 383
- 8.2 Interaction of Gaseous Species at Semiconductor Surfaces 384
- 8.3 Catalysis, the Acceleration of Chemical Reactions 393
- 8.4 The Electrical Properties of Compressed Powders 396
- 8.5 Thin-Film Sensors 397
- 8.6 Thick-Film and Pressed-Pellet Sensors 399
- 8.7 FET Devices for Gas and Ion Sensing 404
- 8.8 Summary and Future Trends 409
  - Problems 409
  - References 410

9 Biosensors 415
A. Dewa and W. H. Ko

- 9.1 Introduction 415
- 9.2 Immobilization of Biological Elements 423
- 9.3 Mass Transport in Biosensors 427
- 9.4 Transduction Principles 437
- 9.5 Packaging of Biosensors 455
- 9.6 Summary and Future Trends 463
  - Problems 465
  - References 469

10 Integrated Sensors 473
K. Najafi, K. D. Wise, and N. Najafi

- 10.1 Introduction 473
- 10.2 System Organization and Functions 477
- 10.3 Interface Electronics 493
- 10.4 Fabrication Techniques 502
- 10.5 Examples of Integrated Sensors 511
- 10.6 Summary and Future Trends 523
  - Problems 526
  - References 527

Appendix A List of Symbols 531
Appendix B International System of Units 533
Appendix C Physical Constants 534
Appendix D Properties of Si and GaAs in 300K 535

Index 537

PREFACE

Since the early 1980s, there has been tremendous growth and progress in the development of a variety of semiconductor sensors. These sensors, with their ever-improving performance–cost ratio, will be key components in the further penetration of microelectronics into new products and new applications. The purpose of this book is to provide an introduction to the basic principles and operational characteristics of these sensors.

The text is interdisciplinary in nature and is intended for use by senior undergraduate or first-year graduate students in applied physics, chemical engineering, electrical engineering, mechanical engineering, and materials science; and as a reference for engineers and scientists involved in sensor research and development. It is assumed that the reader has already acquired a basic understanding of semiconductor device operations, such as those given in *Semiconductor Devices: Physics and Technology* (Wiley, 1985).

Chapter 1 provides a classification of semiconductor sensors and defines the basic sensor terminology. Chapter 2 is concerned with sensor technology, emphasizing bulk and surface micromachining. This information will be used extensively in subsequent chapters. In Chapters 3 through 9, each chapter considers a sensor family related to a special physical, chemical, or biological input signal. Lastly, Chapter 10 addresses integrated sensors (i.e., a monolithic integration of sensors and specially designed electronic circuits on the same semiconductor substrate). Each chapter has a brief historical review, as well as a general outline in its introduction. The operational principles, as well as the fabrication and characterization of the sensors are then presented. Each chapter ends with a summary and a discussion of future trends for sensors. The problem set at the end of each chapter plays an integral part in the development of the topics. More than 20,000 references provided a background for this text.

In the course of writing this text, many people have assisted us and offered their support. First we express our appreciation to the management of our industrial and academic institutions, without whose help this book could not have been written. We have benefited from suggestions made by our reviewers: Dr. S. Bouwstra of the Technical University of Denmark, Prof. W. N. Carr of the New Jersey Institute of Technology, Prof. C. Y. Chang of the National Chiao

Tung University, Prof. K. Colbow and Dr. B. Miremedi of Simon Fraser University, Dr. J. Friel and Dr. I. Taylor of Princeton Gamma-Tech Inc., Prof. C. C. Liu and Prof. J. E. Zull of Case Western Reserve University, Dr. G. Michalowicz of the National Semiconductor Company, Prof. A. Nathan of the University of Waterloo, Dr. R. S. Popovic of the Swiss Federal Institute of Technology Lausanne, and Dr. P. P. L. Regtien of Twente University.

We are further indebted to Mr. N. Erdos of AT&T Bell Laboratories for technical editing of the manuscript, and Dr. K. K. Ng and Mr. W. F. Wright for literature search and suggestions. At our publisher, John Wiley & Sons, we acknowledge Mr. G. Telecki who encouraged us to undertake this project. We also thank Ms. T. W. Sze for preparing the Appendices and Ms. P. L. Huang, Ms. L. J. Chang, Ms. Y. M. Chen, Ms. F. F. Fang, Ms. S. L. Hsiau, Ms. S. Y. Tseng, and Mr. S. Y. Wu for preparing the Index.

As editor, I would especially express my thanks to the United Microelectronics Corporation (UMC), Taiwan, ROC, for the UMC Chair Professorship grant that provided the environment to work on this book.

S. M. SZE

Hsinchu, Taiwan

SEMICONDUCTOR SENSORS

1 Classification and Terminology of Sensors

S. M. SZE
National Chiao Tung University
Hsinchu, Taiwan, ROC

1.1 SEMICONDUCTOR SENSORS

The word "sensor" is derived from the Latin *sentire* which means "to perceive". A sensor, therefore, suggests some connection with our human senses. It may provide us with informations about physical and chemical signals which could not otherwise be directly perceived by our senses. A dictionary definition (Webster's Collegiate Dictionary) of "sensor" is "a device that responds to a physical (or chemical) stimulus (such as heat, light, sound, pressure, magnetism, or a particular motion) and transmits a resulting impulse (as for measurement or operating a control)". Thus, a sensor can detect an input signal (or energy) and convert it to an appropriate output signal (or energy).

Semiconductor sensors are semiconductor devices in which the semiconductor materials are chiefly responsible for sensor operation. However, when semiconductors are not the optimum materials for a particular sensor type, alternative materials can be deposited on top of the semiconductor substrates to form the sensor (e.g., a surface-acoustic-wave sensor with ZnO deposited on silicon). Both approaches can lead to the possibility of integrating semiconductor sensors with microelectronics circuits. Therefore, we have two forms of semiconductor sensors: sensors *in* semiconductors, and sensors *on* semiconductors. The most important semiconductor for both forms is silicon.

The semiconductor sensor is differentiated from other solid-state sensors by its small size, and by the techniques used in its manufacture. Most semiconductor sensors are fabricated by processes that have been developed for integrated

Semiconductor Sensors, Edited by S. M. Sze.
ISBN 0-471-54609-7 © 1994 John Wiley & Sons, Inc.

2 CLASSIFICATION AND TERMINOLOGY OF SENSORS

circuits (ICs). By using standard batch processing, as in the IC industry, hundreds and thousands of idential semiconductor sensors can be produced in one run, thus substantially improving their performance/cost ratio.

The small size of semiconductor sensors not only contributes to their potentially low cost, but also allows them to be integrated with microelectronic circuits, the so-called integrated sensors; thus, further enhancing their performance. Figure 1 shows a comparison of the scale of semiconductor sensors with other quantities.[1] It is clear that semiconductor sensors are of the order of micrometers (microns) in dimension. Therefore they are sometimes referred to as microsensors. In such sensors, deposited films as thin as one-tenth of a micron with transversal dimensions ranging from a few microns up to a few millimeters are commonly used. Spacings between semiconductor sensors' parts can be one micron or less.

Another word closely related to sensor is "transducer", which is derived from the Latin *transducere*, meaning "to lead across". Because it is a device that converts energy from one system to another in the same or in different form, signals as well as energy can be "led across" a transducer. Sensors and transducers are sometimes used as synonymous terms. The differences between sensors and transducers are very slight. A sensor performs a transducing action, and the transducer must necessarily sense some physical or chemical signals. In this book, we shall reserve the word *sensor* for a device that detects or measures an input signal, and the word *transducer* for a device which performs *subsequent* transduction operations in a measurement or control system.

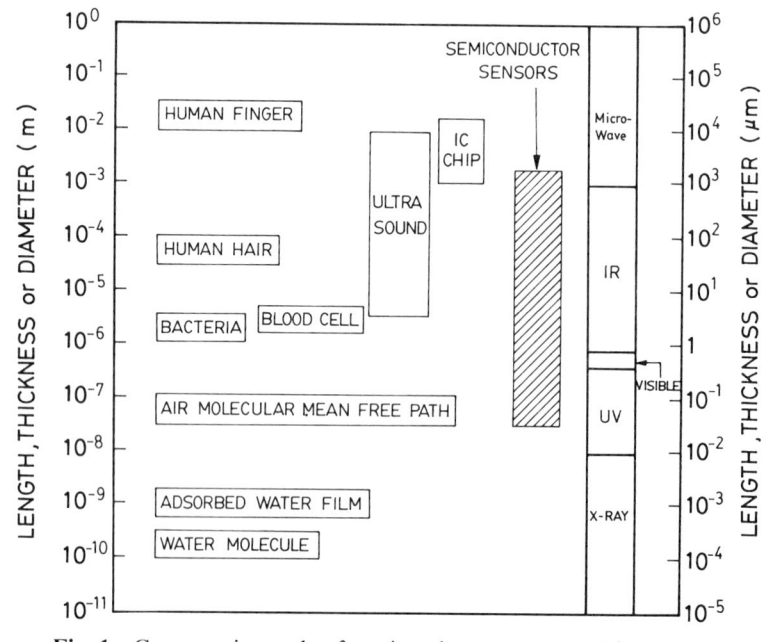

Fig. 1 Comparative scale of semiconductor sensors. (After Ref. 1)

1.2 CLASSIFICATION OF SEMICONDUCTOR SENSORS

1.2.1 Types of Signal

A key characteristic of a sensor is the conversion of energy from one form to another. It is, therefore, useful to consider the various forms of energy. From a physical point of view, we can distinguish the following ten forms of energy, listed in alphabetical order:[2,3]

1. Atomic energy—is related to the force between nuclei and electrons.
2. Electrical energy—pertains to electric field, current, voltage, etc.
3. Gravitational energy—is related to the gravitational attraction between a mass and the earth.
4. Magnetic energy—deals with magnetic field, etc.
5. Mass energy—is described by Einstein as part of his relativity theory, and is given by $E = mc^2$
6. Mechanical energy—pertains to the motion, displacement, force, etc.
7. Molecular energy—is the binding energy in molecules.
8. Nuclear energy—is the binding energy between nuclei.
9. Radiant energy—is related to electromagnetic radiowaves, microwaves, infrared, visible light, ultraviolet, X-rays, and gamma rays.
10. Thermal energy—is related to the kinetic energy of atoms and molecules.

Each form of energy has a corresponding signal associated with it. For practical sensors in measurement systems, we will not consider the nuclear and mass energy. Atomic energy and molecular energy can be brought together as the chemical signal. Gravitational and mechanical energy are both related to the mechanical signal. Therefore, for measurement purposes, we have six types of signal:[3]

1. Chemical signal
2. Electrical signal
3. Magnetic signal
4. Mechanical signal
5. Radiant signal
6. Thermal signal.

A general form of measurement systems is shown in Fig. 2.[2] The signal is fed to a sensor, which changes the form of energy, usually into electrical. In the modifier the signal is processed or modified, but the form of the signal is not changed. For example, when electrical energy is used, it is possible for an analog signal at the input of the modifier to be converted into a digital signal at the output. The modifier can also amplify the input electrical signal or can

4 CLASSIFICATION AND TERMINOLOGY OF SENSORS

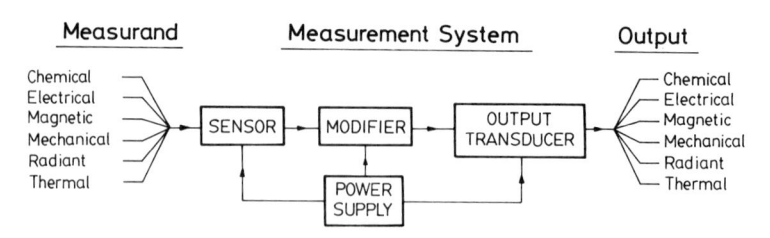

Fig. 2 A general measurement system. (After Ref. 2)

modulate/demodulate the input signal. Finally, the output transducer converts the energy into a form suitable for display, recording, or performing some action (called actuation).

For example, we consider a photodetection system. In the sensor, radiant energy is converted into electrical energy by means, for instance, of the photovoltaic effect. In the modifier, the small electronic signal is modified, e.g., amplified, and converted from an analog to a digital signal. In the subsequent transducer the digital signal is converted into a radiant signal, e.g., by means of a light-emitting-diode display panel. Thus, it is possible to read the wave amplitude, spectrum, or other properties of the incoming radiant signal, visually.

As shown in Fig. 2, the sensor is connected to a power supply. The modifier, as well as the transducer, is also connected to a power supply. The sensor, in this configuration, is referred to as a modulating sensor. In such sensors, an energy flow supplied by the power supply is modulated by the measurand (e.g., a photoconductive cell). Without the power supply, the cell will respond to the input light energy but does not produce a usable signal; a signal is obtained by applying an electrical voltage and monitoring the resulting current. If a sensor can function without a power supply, it is referred to as a self-generating sensor, e.g., a photodiode. The photodiode will itself generate a voltage due to the photovoltaic effect.

The output signal from a sensor may be of different types, but the preferred type is the electrical signal.[4] This is because electrical signals are used in most measurement systems. The advantages of electrical measurement systems include: (1) sensors can be designed for any nonelectrical signals by selecting an appropriate material. Any variation in a nonelectric parameter yields a variation in an electric parameter because of the electronic structure of matter; (2) a multitude of microelectronic circuits are available for electric signal conditioning and modification (e.g., amplification, filtering, modulation, etc); (3) there exist many options for information display or recording by electronic means, and (4) the electrical signal is better suited for signal transmission.

1.2.2 A Classification Scheme for Semiconductor Sensors

In order to compare sensors and obtain a comprehensive overview of them, a flexible classification scheme has been proposed by White.[5] Table 1 lists the

TABLE 1 Measurands[5]

1. **Acoustic**
    1.1 Wave amplitude, phase, polarization, spectrum
    1.2 Wave velocity
    1.3 Other (specify)
2. **Biological**
    2.1 Biomass (identities, concentrations, states)
    2.2 Other (specify)
3. **Chemical**
    3.1 Components (identities, concentrations, states)
    3.2 Other (specify)
4. **Electric**
    4.1 Charge, current
    4.2 Potential, potential difference
    4.3 Electric field (amplitude, phase, polarization, spectrum)
    4.4 Conductivity
    4.5 Permittivity
    4.6 Other (specify)
5. **Magnetic**
    5.1 Magnetic field (amplitude, phase, polarization, spectrum)
    5.2 Magnetic flux
    5.3 Permeability
    5.4 Other (specify)
6. **Mechanical**
    6.1 Position (linear, angular)
    6.2 Velocity
    6.3 Acceleration
    6.4 Force
    6.5 Stress, pressure
    6.6 Strain
    6.7 Mass, density
    6.8 Moment, torque
    6.9 Speed of flow, rate of mass transport
    6.10 Shape, roughness, orientation
    6.11 Stiffness, compliance
    6.12 Viscosity
    6.13 Crystallinity, structural integrity
    6.14 Other (specify)
7. **Optical**
    7.1 Wave amplitude, phase, polarization, spectrum
    7.2 Wave velocity
    7.3 Other (specify)
8. **Radiation**
    8.1 Type
    8.2 Energy
    8.3 Intensity
    8.4 Other (specify)
9. **Thermal**
    9.1 Temperature
    9.2 Flux
    9.3 Specific heat
    9.4 Thermal conductivity
    9.5 Other (specify)
10. **Other (specify)**

6 CLASSIFICATION AND TERMINOLOGY OF SENSORS

measurands which are defined as the input quantities, properties, or conditions that are detected or measured by sensors. For example, if the measurand is thermal, it is measured by a thermal sensor; if it is pressure, it is measured by a pressure sensor. The measurands, arranged in alphabetical order, are: acoustic, biological, chemical, electrical, magnetic, mechanical, optical, radiation, and thermal. Each entry in Table 1 not only represents the measurand itself but also its temporal or spatial distribution. For example, the entry "wave amplitude" under the heading "optical" can apply to a sensor that measures the intensity of steady infrared radiation at a point, or a fast photodiode detecting time-varying optical flux.

The measurands listed in Table 1 are closely related to the six types of signal. Some measurands are the same as the corresponding types of signal. However, some signals are subdivided into two measurand groups: the biological and chemical measurands pertain to the chemical signal; the acoustic and mechanical measurands are related to the mechanical signal, and the optical and radiation measurands deal with the radiant signal.

For a particular measurand, we are primarily interested in sensor characteristics, such as allowable ambient conditions, sensitivity, selectivity, etc. These are termed "technological aspects" and are listed in column 1 of Table 2. Since these terms will be used for sensor characterization in later chapters, we shall define them in Section 1.3.

The detection means used in sensors are listed in column 2 of Table 2. Entries in column 3 are for the sensor conversion phenomena, that is, the primary phenomena that are adopted to convert the measurand into a form suitable for producing the sensor output. The entries under "physical" are derived from the interactions among five physical variables: elastic, electric, electromagnetic, magnetic, and thermal. The physical phenomena listed are the ten possible combinations obtained by selecting only two of the physical variables to interact. The list of physical phenomena will, of course, increase substantially if we consider interactions from more than two physical variables (e.g., photomagnetoelectric effect).

To illustrate how we can characterize sensors using the terms in Tables 1 and 2, we consider an example of a "Diaphragm Pressure Sensor"[5] (considered in Chapter 4). An applied pressure distorts a thin semiconductor diaphragm. The deflection is inferred from the change of the values of resistors diffused into the diaphragm. Thus, the *measurand* is pressure (Table 1, item 6.5), the *technological aspects* include the sensitivity of the resistance change per unit pressure (Table 2, column 1, item 13), the primary *detection means* is mechanical (Table 2, column 2, item 5), and the sensor *conversion phenomenon* is elastoelectric (Table 2, column 3, item 3.7) due to piezoresistance, that is, a mechanical force acting on a semiconductor causes an appreciable change in the resistivity.

This classification scheme is useful for describing and comparing semiconductor sensors. For a solid-state or a general sensor, we must also consider other sensor materials such as inorganic, organic, conductor, insulator, liquid, gas, plasma, or biological substances. We may also include fields of application

TABLE 2 Technological Aspects, Detection Means and Conversion Phenomena of Sensors[5]

Technological Aspects	Detection Means	Conversion Phenomena

1. Ambient conditions allowed
2. Full-scale output
3. Hysteresis
4. Linearity
5. Measured range
6. Offset
7. Operating life
8. Output format
9. Overload characteristics
10. Repeatability
11. Resolution
12. Selectivity
13. Sensitivity
14. Speed of response
15. Stability
16. Others (specify)
	1. Biological
	2. Chemical
	3. Electric, magnetic, or electromagnetic wave
	4. Heat, temperature
	5. Mechanical displacement or wave
	6. Radioactivity, radiation
	7. Others (specify)
		1. Biological
		  1.1 Biochemical transformation
		  1.2 Physical transformation
		  1.3 Effects on test organism
		  1.4 Spectroscopy
		  1.5 Others (specify)
		2. Chemical
		  2.1 Chemical transformation
		  2.2 Physical transformation
		  2.3 Electrochemical process
		  2.4 Spectroscopy
		  2.5 Others (specify)
		3. Physical
		  3.1 Thermoelectric
		  3.2 Photoelectric
		  3.3 Photomagnetic
		  3.4 Magnetoelectric
		  3.5 Elastomagnetic
		  3.6 Thermoelastic
		  3.7 Elastoelectric
		  3.8 Thermomagnetic
		  3.9 Thermo-optic
		  3.10 Photoelastic
		  3.11 Others (specify)

e.g., health, telecommunication, manufacturing, transportation, etc) as a part of the classification scheme.[5]

1.3 SENSOR CHARACTERIZATION

The "technological aspects" listed in column 1 of Table 2 will be used in subsequent chapters to evaluate sensor characteristics and to compare sensor performances. We shall define them in this section.[6-8]

Ambient Conditions Allowed. Ambient conditions may have profound effects on sensor operation. These include temperature, acceleration, vibration, shock, ambient pressure (e.g., high altitudes), moisture, corrosive materials, and electromagnetic field. The allowed ambient conditions for a sensor should be specified so that the sensor can perform within its specified tolerance.

Full Scale Output (FSO). The algebraic difference between the end points of the output. The upper limit of sensor output over the measurand range is called the *full scale* (FS), see Fig. 3.

8   CLASSIFICATION AND TERMINOLOGY OF SENSORS

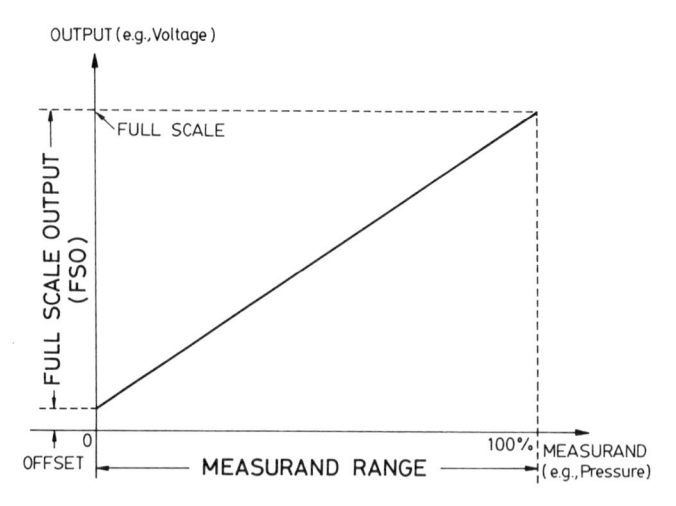

Fig. 3. Output–measurand relationship of a linear-output sensor with an offset. (After Ref. 6)

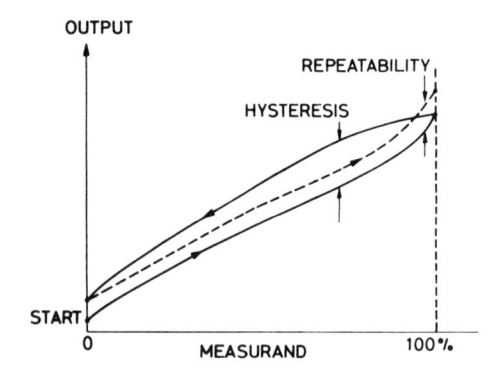

Fig. 4 Hysteresis and repeatability. (After Ref. 6)

Hysteresis. The maximum difference in output, at any measurand value, within the measurand range when the value is approached first with an increasing and then decreasing measurand (Fig. 4). Hysteresis is expressed in percent of FSO during one calibration cycle.

Linearity. The closeness between the calibration curve and a specified straight line. It is measured as the maximum deviation of any calibration point from a specified, straight line, during any one calibration cycle and is generally expressed as a percentage of FSO. If the specified straight line is the line connecting the two end point (at 0 and 100% measurand) then the result is called the end-point or terminal-based linearity.

1.3 SENSOR CHARACTERIZATION

Measurand Range. The value of the measurand over which the sensor is intended to measure, specified by upper and lower limits (Fig. 3).

Offset. The output of a sensor, under room-temperature condition unless otherwise specified, with zero measurand applied (Fig. 3).

Operating Life. The minimum length of time over which the sensor will operate, either continuously or over a number of on–off cycles whose duration is specified, without changing performance characteristics beyond specified tolerances.

Output Format. The output is usually the electrical quantity produced by a sensor and is a function of the measurand. The output format includes *analog output* (e.g., a continuous function of the measurand such as voltage amplitude, voltage ratio, and changes in capacitance). Frequency output (i.e., the number of cycles or pulses per second as a function of the measurand) and frequency-modulated output (i.e., frequency deviation from a center frequency) are also forms of analog output. Another output format is the *digital output*, which represents the measurand in the form of discrete quantities coded in some system of notation (e.g., binary code).

Overload Characteristics. Overload (or overrange) is the maximum magnitude of measurand that can be applied to a sensor without causing a change in performance beyond specified tolerance. A key parameter of the overload characteristics is the *recovery time*, which is the amount of time allowed to elapse after removal of an overload condition before the sensor again performs within the specified tolerance.

Repeatability. The ability of a sensor to reproduce output readings at room temperature, unless otherwise specified, when the same measurand is applied to it consecutively, under the same conditions and in the same direction (Fig. 4). It is expressed as the maximum difference between output readings as determined by two calibration cycles (Fig. 4). It is usually stated as "within $x\%$ FSO".

Resolution. The minimal change of the measurand value necessary to produce a detectable change at the output. When the measurand increment is from zero, it is called the *threshold*.

Selectivity. The ability of a sensor to measure one measurand (e.g., one chemical component) in the presence of others.

Sensitivity. The ratio of the change in sensor output to the change in the value of the measurand. It is the slope of the calibration curve (Fig. 3). For a sensor in which the output y is related to the measurand x by the equation $y = f(x)$, the sensitivy $S(x_a)$, at point x_a, is

$$S(x_a) = \frac{dy}{dx}\bigg|_{x=x_a}.$$

It is desirable to have a high and, if possible, constant sensitivity. For a sensor having $y = kx + b$, where k and b are constants, the sensitivity is $S = k$ for the entire measurand range. For a sensor having $y = kx^2 + b$, the sensitivity is $S = 2kx$ and changes from one point to another over the measurand range. For the calibration curve shown in Fig. 3, the sensor has a constant sensitivity and it can be expressed in units of volt/pascal.

Speed of Response. The time at which the output reaches 63% (i.e., $1/e$) of its final value in response to a step change in the measurand.

Stability. The ability of a sensor to maintain its performance characteristics for a certain period of time. Unless otherwise stated, stability is the ability of a sensor to reproduce output readings, obtained during the original calibration, and constant room conditions, for a specified period of time. It is typically expressed as a percentage of FSO.

1.4 EVOLUTION OF SEMICONDUCTOR SENSORS

According to the basic definition of sensors given in Section 1.1, all semiconductor devices are sensors. In 1874, Braun did the earliest systematic investigation of a metal–semiconductor rectifying system. He noted the dependence of the resistance on the polarity of the applied voltage.[9] Thus, the measurand for this device was electrical—the voltage, and the output was also electrical—the resistance variation.

If we restrict ourselves to nonelectrical measurands, then one of the earliest semiconductor sensors was the point-contact rectifier for detecting radiowaves in 1904.[10] The measurand was a radiant signal, and the output was electrical.

The roots of modern semiconductor sensors date back to the development of transistor technology. We shall use the diaphragm pressure sensor as an example to describe the evolution of sensors,[11] since this pressure sensor has the largest market share of all semiconductor sensors.

1.4.1 Discovery Phase (1947–1960)

After the invention of the bipolar transistor in 1947, extensive efforts were devoted to the study of semiconductor properties and device characteristics. In 1954, Smith discovered the piezoresistive effect in silicon and germanium.[12] This effect is responsible for an appreciable change in the resistivity when a

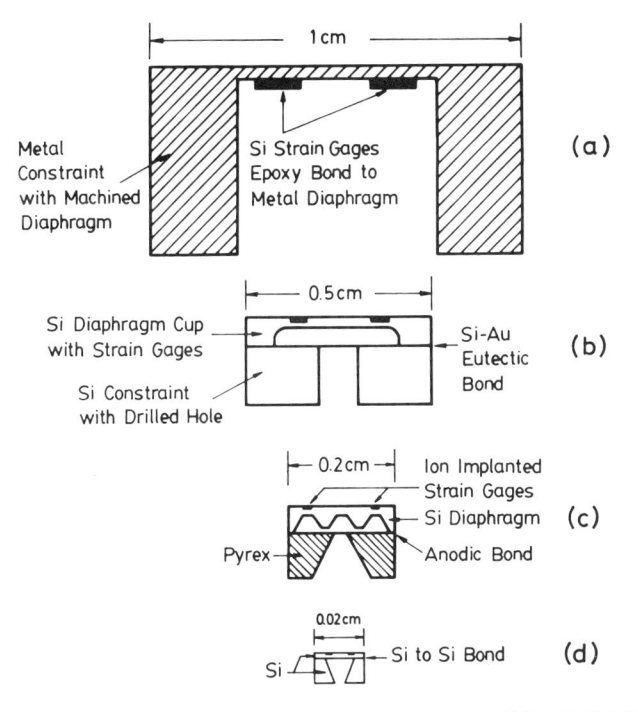

Fig. 5 Evolution of diaphragm pressure sensors. (After Ref. 11)

mechanical force is applied to the semiconductor. The resistance change can be larger by two orders of magnitude than the corresponding resistance change in metals used previously for strain-gage applications.

The first semiconductor pressure sensor was made by cutting a silicon bar out of a wafer to form a resistive strain gage. These bars were then adhesively bonded, manually, onto a metal diaphragm that was in direct contact with the media to be measured, Fig. 5a. The applied stress causes a deflection of the diaphragm, which in turn transmits the stress to the bonded strain gages. The change in resistance is linearly proportional to the applied force. The first design suffered from low yield and poor stability due to thermal mismatch of the silicon-glue-metal interface. The first industrial applications using this approach was introduced in 1958.

1.4.2 Basic Technology Development Phase (1960–1970)

To improve the sensor performance, strain gages were diffused directly into a silicon diaphragm. The entire diaphragm had to be bonded to a constraint that provided package stress isolation. A flat silicon diaphragm was subsequently replaced by a diaphragm cup where a cavity was formed by mechanical milling supplemented by a chemical etching process (Fig. 5b). The metal constraint was also replaced by a silicon constraint that was bonded to the diaphragm cup through a gold–silicon eutectic bond.

This approach improved the sensor performance and reduced the sensor size. The major drawback was the cost. Cavity formation was done in one-at-a-time mode. Also, it was difficult to control accurately the location of the diffused strain gage with respect to the stress region of the diaphragm. This phase was also the "commercialization and market development period" with major effort by several companies to bring the technology to practical applications.

1.4.3 Batch Process Phase (1970–1980)

The diaphragm was improved by using a selective anisotropic chemical-etching process. This was a batch process, that is, the etching was performed on an entire wafer, so that hundreds of sensor diaphragms were made simultaneously. Using steps similar to those in integrated-circuit fabrication, cavity size was precisely defined by lithography steps. Further improvements were made by using ion implantation to form the strain gage. A new technology of electrostatic bonding of the silicon wafer to special glass was also adopted to minimize thermal miscmatch (Fig. 5c).

The batch process substantially reduced the sensor size and the manufacturing cost. In addition, the sensors had improved stability, lowered temperature errors, and more accurate electrical parameters. This phase was also considered as the "cost reduction and application expansion period". Practical applications were feasible, especially in the aerospace and industrial-control areas.

1.4.4 Micromachining Phase (1980–present)

Micromachining is a new approach to constructing sensors, and other structures with dimensions on a micrometer-scale (10^{-6} m), based on process technologies developed for the microelectronic (or integrated circuit) industry.[13] By combining the aforementioned technologies with specially developed processing steps to etch trenches and grooves inside bulk semiconductor or to form free-standing, micron-dimensional beams, bridges, or membranes, we have moved into the micromachining phase of sensor development.

In the present phase, micromachining technology is being extensively used to reduce device size, to lower production cost, to improve sensor performance (e.g., higher sensitivity, higher selectivity, faster response time, etc), and to make monolithic integration with microelectronic circuits and/or micro-mechanical systems to broaden sensor applications. A silicon-to-silicon wafer-lamination process has also been developed.[11] This process allows the creation of a molecular bond between two silicon wafers (Fig. 5d). Note that the dimension of the sensor has been reduced by about two orders of magnitude compared to Fig. 5a. A discussion on the evolution of sensor interface with electronic circuits is presented in Section 10.1.2 of Chapter 10.

In this phase, the technology has started to respond to widespread application of microprocessors and their decreasing cost. We have realized that further penetration of microelectronics into new products and new markets is seriously hampered by the lack of semiconductor sensors that have a comparable performance/cost ratio to that of microelectronic components. The only technology capable of providing the required ratio is micromachining of semiconductor.

Semiconductor sensors have entered a new era of development, an era marked by rapid growth of new applications. While the previous ten years can be called the "Decade of Microprocessors", it is conceivable that the next ten years will be the "Decade of the Sensors".[11]

1.5 ORGANIZATION OF THE BOOK

The book is organized based on the measurands listed in Table 1. In this book, the types of semiconductor sensors we are mainly concerned with are the types whose output is electrical. We will consider sensor devices that can detect or measure any measurands in Table 1 except the electric measurands. This is because when electrical parameters such as current and voltage have to be measured, we only need a "modifier" and a display unit; besides, most such devices have already been covered in standard semiconductor-device text or reference books.[14]

Chapter 2 is concerned with sensor technology. Semiconductor sensors are often fabricated using processes similar to those developed for integrated circuits. However, many specialized fabrication steps are needed. These include anisotropic ethcing to form diaphragms, and the use of sacrificial layers to form free-standing beams or membranes. These processes are referred to as *micromachining*, since the feature dimenions of the sensors are of the order of microns (10^{-6} m), and spacings between sensor parts may be one micron or less. The sensor technology covered in this chapter will be extensively used in the subsequent chapters.

In Chapters 3 through 9, each chapter considers a sensor family related to a special measurand group. Acoustic sensors are considered in Chapter 3, with special emphasis on piezoelectric materials and surface-acoustic-wave (SAW) sensors. Chapter 4 treats the mechanical sensors, which include piezoresistive sensors and capacitive sensors for pressure, acceleration, and flow-rate measurements. The above two sensor families belong to the mechanical-signal type.

Chapter 5 considers magnetic sensors, especially those with low permeabilities such as Hall devices, magnetotransistors, magnetodiodes, and carrier-domain magnetometers. Radiation sensors are discussed in Chapter 6; these sensors are used to detect both electromagnetic waves and nuclear-particle radiations. Important radiation sensors include HgCdTe infrared sensors, visible-light color sensors, high-energy photodiodes and drift-chamber X-ray sensors. Chapter 7

14 CLASSIFICATION AND TERMINOLOGY OF SENSORS

treats thermal sensors including thermal-conductivity sensors, humidity sensors, and micro-calorimeters.

Chapters 8 and 9 consider chemical sensors and biosensors, respectively. Chemical sensors are mainly used to detect gaseous species, with primary emphasis on detecting combustible gases such as CO, H_2, alcohols, propane, and other hydrocarbons. The detection is done primarily with semiconducting metal-oxides (e.g., SnO_2) and field-effect transistors (e.g., ion-sensitive field-effect transistors, ISFET). Biosensors are a special class of chemical sensors which take advantage of the high selectivity and sensitivity of biologically active materials. In Chapter 9 we consider the enzyme electrode, pH-ISFET biosensor, and the packaging considerations for these sensors.

The last chapter, Chapter 10, deals with integrated sensors. Most semiconductor sensors can be monolithically integrated with specially designed electronic circuits on the same semiconductor substrate. Such an approach improves the signal-to-noise ratio by eliminating added noise due to signal tranmission through long leads; it can provide a more linear output than that of the sensor itself by using on-chip feedback systems, and it can minimize sensor parameter drift by employing on-chip, accurate current or voltage sources. In addition, integrated sensors can perform analog-to-digital conversion, impedance transformation, and many other signal-conditioning functions.[3,7] Clearly, integrated sensors offer many important advantages, which will be discussed in Chapter 10.

To keep the notation simple in this book, we have sometimes found it necessary to use a simple symbol more than once, with different meanings. For example, in Chapter 5 α represents the Hooge noise parameter, in Chapter 7

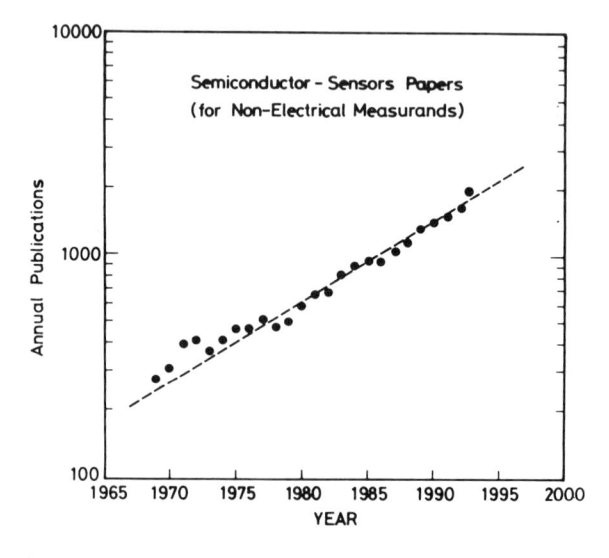

Fig. 6 Annual publication of semiconductor-sensor papers.

it represents the Seebeck coefficient, and in Chapter 9 it represents the enzyme-loading factor. Within each chapter, however, a symbol has only one meaning and is defined the first time it appears. Many symbols do have the same or similar meanings consistently throughout this book; they are summarized in Appendix A. We have also included the international systems of units, physical constants, and properties of Si and GaAs in Appendixes B, C, and D, respectively.

At present, the semiconductor-sensor field is moving at a rapid pace. The number of semiconductor-sensor publications has grown from about 270 papers in 1969 to over 1900 papers in 1993 with a seven-fold increase in 24 years (Fig. 6). The annual growth rate of the semiconductor sensor is almost twice that of the general sensor field. Note that many topics, such as micromachining and integrated sensors, are still under intensive study. Their ultimate capabilities are not yet fully understood. The material presented in this book is intended to serve as a foundation. The references listed at the end of each chapter can supply more information.

REFERENCES

1. R. S. Muller, R. T. Howe, S. D. Senturia, R. L. Smith, and R. M. White, Eds., *Microsensors*, IEEE Press, New York, 1991, p. viii.
2. M. J. Usher, *Sensors and Transducers*, McMillian, Hampshire, 1985.
3. S. Middelhoek and S. A. Audet, *Silicon Sensors*, Academic Press, New York, 1989.
4. J. R. Carstens, *Electrical Sensors and Transducers*, Prentice-Hall, Englewood Cliffs, 1993.
5. R. M. White, "A sensor classification scheme", *IEEE Trans. Ultrason. Ferroelec. Freq. Contr.* **UFFC-34**, 124 (1987).
6. H. M. Norton, *Handbook of Transducers*, Prentice-Hall, Englewood Cliffs, 1989.
7. R. Passal-Areny and J. G. Webster, *Sensors and Signal Conditioning*, Wiley-Interscience, New York, 1991.
8. I. R. Sinclair, *Sensors and Transducers*, 2nd ed., Newnes, Oxford, 1992.
9. F. Braun, "Uber die Stromleitung durch Schwefelmetalle", *Ann. Phys. Chem.*, **153**, 556 (1874). (This paper is translated into English and collected in S. M. Sze, Ed., *Semiconductor Devices: Pioneering Papers,* World Scientific, Singapore, 1991.)
10. J. C. Bose, *U. S. Patent* 775,840 (1904).
11. J. Brysek, K. Petersen, J. R. Mallon, L. Christel, and F. Pourahmadi, *Silicon Sensors and Microstructures*, NovaSensor, Fremont, 1990.
12. C. S. Smith, "Piezoresistive effect in germanium and silicon", *Phys. Rev.* **94**, 1 (1954).
13. For a review on integrated-circuit technology, see, for example, S. M. Sze, Ed., *VLSI Technology*, 2nd ed., McGraw–Hill, New York, 1988.
14. For example, see S. M. Sze, *Physics of Semiconductor Devices*, 2nd ed, Wiley-Interscience, New York, 1981; and *Semiconductor Devices: Physics and Technology*, Wiley, New York, 1985.

2 Semiconductor Sensor Technologies

C. H. MASTRANGELO
Center for Integrated Sensors & Circuits
University of Michigan
Ann Arbor, MI 48109-2122, USA

W. C. TANG
Ford Microelectronics Inc.
Colorado Springs, CO 80921-3698, USA

2.1 INTRODUCTION

Semiconductor sensors are transducers that convert mechanical signals into electrical signals. These devices are widely used for the measurement and control of physical variables. Microphones are used in audio systems. Pressure sensors are used in fluidic, pneumatic, and tactile detection systems. Accelerometers are used in navigational and air-bag deployment. Magnetic sensors are used in positional control. Infrared and visible light sensors are used in cameras and night-vision systems. Temperature and flow sensors are used in air conditioning and automotive systems. Chemical sensors are used in biological diagnostic systems. The list of applications of these devices is enormous, and it is growing on a yearly basis. Currently, there is a large demand for low-cost, accurate, and reliable sensors for industrial and consumer product applications.

In the past twenty years, the application of microelectronic technology to the fabrication of mechanical devices greatly stimulated research in semiconductor sensors. Such microfabricated devices are micromachined sensors. Micromachining technology takes advantage of the benefits of semiconductor technology to address the manufacturing and performance requirements of the sensor industry. The versatility of semiconducting materials and the miniaturization of VLSI patterning techniques promise new sensors with better

---

Semiconductor Sensors, Edited by S. M. Sze.
ISBN 0-471-54609-7 © 1994 John Wiley & Sons, Inc.

capabilities and improved performance-to-cost ratio over those of conventionally machined devices. Figure 1 shows an example of a microelectromechanical sensing system (MEMS) used in the deployment of air-bags which illustrates the integration of electrical and mechanical devices.

A major factor that contributes to the cost of manufactured products is the overhead expense on production facilities. Technology-based products such as precision electronic and mechanical devices require expensive facilities and highly skilled laborers. These costs are largely independent of the number of products produced. Therefore, the per-unit cost of manufactured goods decreases as the production volume increases. Maximizing throughputs without sacrificing product quality is one of the major goals of manufacturers.

An example that illustrates this point occurs in the microelectronics industry. Integrated-circuit technology allows thousands of electronic circuits to be batch-fabricated simultaneously through a single pass of processing sequences. Batch-fabrication of microelectronic circuits was made possible through the invention of planar technology. In the planar manufacturing process, three-dimensional devices are built on a wafer substrate using stacked layers of planar materials with different but coordinated two-dimensional patterns.

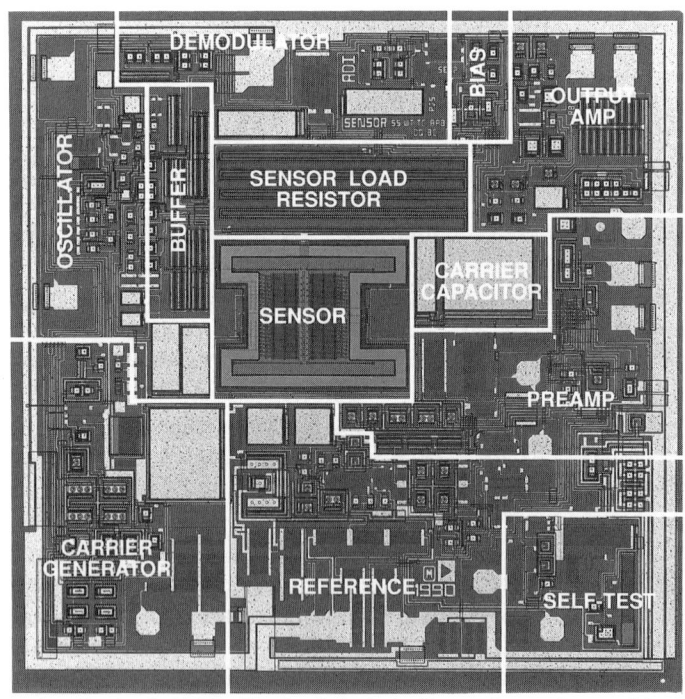

Analog Devices' ADXL-50, the industry's first surface micromachined accelerometer, includes signal conditioning on chip.

Fig. 1 State of the art surface micromachined accelerometer that integrates micromechanical sensors with BICMOS technology. (Courtesy of Analog Devices.)

By optically repeating the patterns on the wafer, many units are fabricated with just one pass of the process (Fig. 2). Micromachined sensors benefit from the same planar manufacturing processes.

Because sensors receptive to different physical variables are structurally different, in general, there is no single technology that allows for the fabrication of a wide variety of sensors. However, there are two major classifications of microsensor technologies. *Bulk-micromachined* sensors are primarily made by the accurate machining of a relatively thick substrate. *Surface-micrimachined* sensors are primarily constructed from stacked thin films. Both technologies use materials and processes borrowed from VLSI technology. The three processes of deposition, lithography, and etching are sufficient to construct a wide variety of mechanical structures required for specific sensors. A fundamental sensor-fabrication problem is the development of a suitable fabrication-process sequence of these basic machining steps that define the desired shape and function of the device. The description of the many micromachining techniques available and their trade-offs is the main topic of this chapter.

Since there is a great deal of literature on VLSI processing, here we only present a cursory review of the basic fabrication techniques. These techniques are discussed in great detail in many excellent textbooks.[1-5] Specific technology issues relevant to sensor design and performance are discussed in detail. Although this chapter is intended as a general reference for microsensor technology, most of its contents are directed toward silicon micromachining.

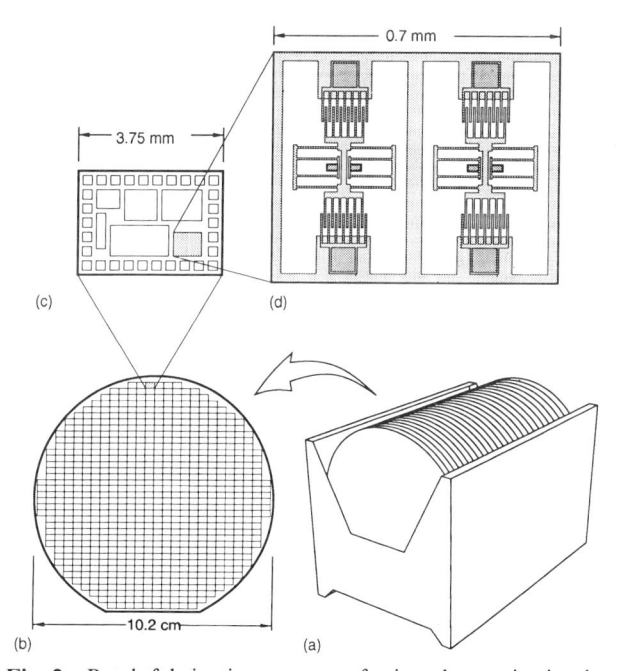

Fig. 2 Batch-fabrication process of microelectronic circuits.

The same techniques are applicable to other materials, and these are briefly cited at the end of the chapter.

2.2 BASIC FABRICATION PROCESSES

2.2.1 Deposition

Thin films are essential building materials in semiconductor microsensors. Surface-micromachined sensors are constructed by the successive deposition and patterning of thin solid films 0.1–5 μm thick while larger bulk-micromachined sensors use thin films for passivation and dielectric functions. These structural materials are direct spin-offs from the semiconductor industry. Thin films may be deposited on a substrate by both physical and chemical means. In this section we describe some of the most common deposition techniques used in microsensor technology.

Spin Casting. In this process, the thin-film material is in solution in a volatile liquid solvent. The dissolved material is poured on the sample and the sample is rotated at high speed. As the liquid spreads, the volatile solvent evaporates leaving a uniform thin layer of solid material behind as shown in Fig. 3. Spin-casting is a commonly used technique for the deposition of organic materials such as photosensitive resists and polyimides as well as inorganic spin-on glasses. The thickness of the solid film depends on the degree of solubility and spin speed, and typically ranges from 0.1 to 50 μm. The flow of the material during the spin-casting process blurs or planarizes the existent topography of the surface, yielding a smoother interface. This property is desirable for structures that are composed of many levels. Because the spin-casting process relies on the solution flow, it does not yield continuous films when the surface has pronounced steps higher than two to three times the film thickness. Spin-cast materials suffer from severe shrinkage when the film coalesces after solvent

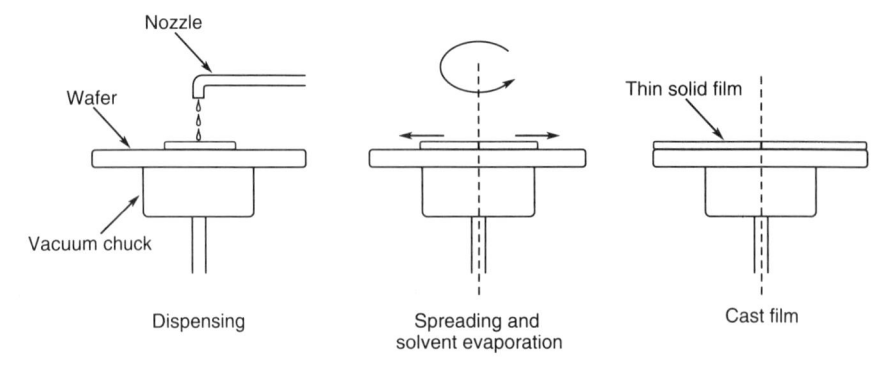

Fig. 3 Spin-casting process.

removal and post-bake; hence spin-cast films intrinsically have a high-stress state. Spin-cast films are less dense and more susceptible to chemical attack than materials deposited by other means.

Evaporation. Thin films can be evaporated from a hot source onto a substrate as shown in Fig. 4. The evaporation system consists of a vacuum chamber, pump, holding frame for the samples, crucible, and a shutter. A sample of the material to be deposited is placed in the crucible, and the chamber is evacuated to 10^{-6}–10^{-7} Torr. The crucible is then heated using a tungsten filament or an electron beam to flash-evaporate the material from the crucible onto the sample. The film thickness is determined by the time that the shutter is opened. The evaporation rate of the material is a function of the vapor pressure of the material. Hence, materials that have a low melting point (such as aluminum) are easily evaporated, but refractory materials (such as tungsten) require much higher temperatures that cause burning of organic films where they land on the sample. Since the evaporated material originates from a point source, evaporated films experience shadowing effects yielding poor step coverage and nonuniform thickness. A second factor affecting the coverage is the surface mobility of the species on the substrate. Surface mobility is high in evaporated gold which shows excellent step coverage. In general, evaporated films are highly disordered and have large residual stresses; thus, only thin layers of the material can be evaporated.

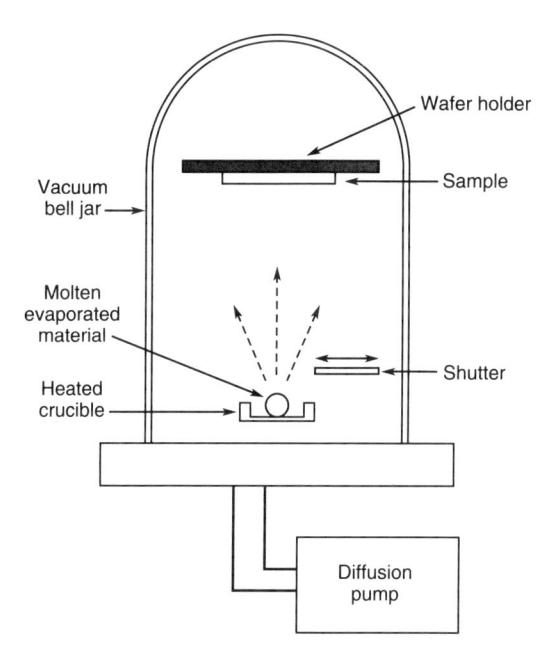

Fig. 4 Schematic of an evaporation system.

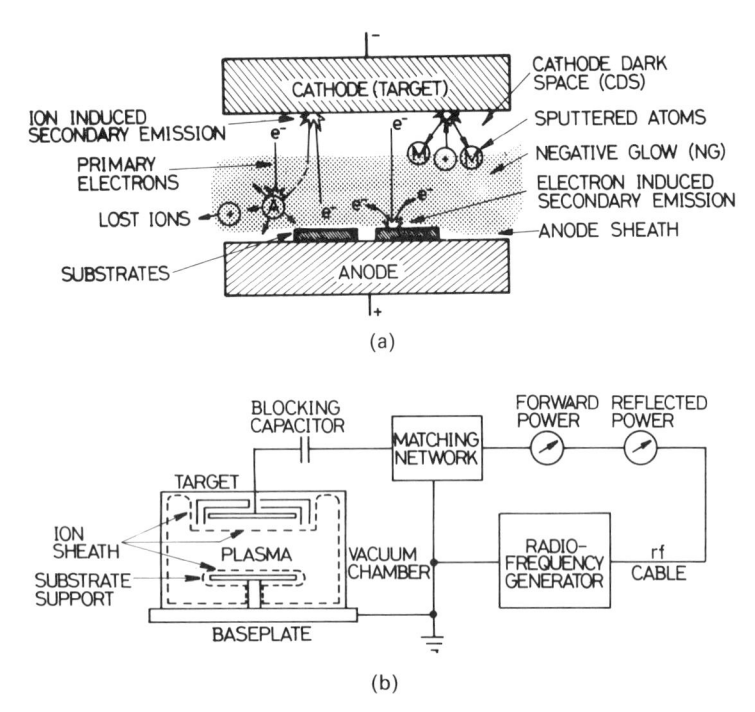

Fig. 5 (a) Sputtering process. (b) Schematic of a typical sputtering system. (After Refs. 1 and 5)

Sputtering. This technique overcomes many of the problems of flash-evaporation. The sputtering system shown in Fig. 5 consists of a vacuum chamber, a sputtering target of the desired film, a sample holder, and a high-voltage DC or RF power supply. After evacuating the chamber to 10^{-6}–10^{-8} Torr, an inert gas such as Ar or He is introduced in the chamber at a few mT of pressure. A plasma of the inert gas is next ignited. The energetic ions of the plasma bombard the target. The kinetic energy of the impinging ions is sufficient to make some of the target atoms escape from the surface. These atoms land on the sample and form a thin film. The sputtering process creates a continuous planar flux of the species landing on the substrate, thus, it is preferred for mass production. Sputtered films have better step coverages and uniformity than evaporated films. The addition of magnetic fields to the plasma improves step coverage. The high-energy plasma overcomes the temperature limitations of evaporation. Most materials from the periodic table can be easily sputtered as well as inorganic and organic compounds. Refractory materials can be sputtered with ease. Materials from more than one target can be sputtered at the same time. Such a process (known as co-sputtering) allows one to control the atomic ratio of the species in the thin film. This is particularly important when the thin-film stoichiometry needs to be well controlled.

2.2 BASIC FABRICATION PROCESSES

The structure of the sputtered films is disorganized, and its stress and mechanical properties are sensitive to specific sputtering conditions. Some of the inert gas can be trapped causing anomalies in its mechanical and structural characteristics.

Reactive Growth. The preceding were physical methods of deposition because the chemistry of the deposted material did not play a substantial role in the deposition process. It is also possible to deposit films from a chemical species that reacts with the sample substrate. In reactive growth, the chemically reactive species combines with the substrate to form a new thin film. The most common example of reactive growth is oxidation. When oxygen is in contact with a silicon surface at high temperatures, it forms silicon dioxide. This film is used in the manufacture of microsensors as a dielectric and etch-stop layer. The basic kinetics of the reaction involve two limiting factors: (a) the reaction rate at the sample interface, and (b) the diffusion of new species through the already-formed films. In general, for thick films the latter will dominate the growth, yielding a nonlinear growth rate. Another instance of reactive growth occurs in the formation of silicides. Reactivity grown films are of excellent quality but, in general, suffer from very-large stresses due to volume changes in the end products. For example, in silicon dioxide the volume increases by about 45% whereas silicides shrink by a comparable amount. These changes induce mechanical warping of the sample and restrict the film thickness to a few micrometers to minimize warpage.

Chemical Vapour Deposition (CVD). In CVD, one or several gaseous species are thermally broken down into their components. When they impinge on the substrate, some of these components nucleate on the substrate surface eventually forming the thin film. The process of thermal breakdown of molecules is known as pyrolysis. CVD deposition is controlled by (a) mass transport, and (b) reaction-limited processes. The latter is preferred as it yields good uniformity. Figure 6 shows a schematic of a CVD furnace. The furnace consists of a heated quartz tube, a sample holder, a pump and a set of gas injectors. The furnace is heated in an inert gas until it reaches the deposition temperature. Next, the gas is evacuated, and the reactive species are introduced through the injectors at the deposition pressure. Many materials including polycrystalline silicon, silicon nitride, silicon dioxide, and refractory metals are deposited by low-pressure thermal CVD (or LPCVD). LPCVD films are the highest-quality films available yielding the most controllable mechanical characteristics. Unlike evaporated or sputtered films, CVD films can be deposited conformally on the sample. This characteristic is highly desirable for refilling and sealing cavities. CVD films have amorphous or polycrystalline structure. The stress of the films can in general be controlled through tight control of deposition conditions and subsequent annealing. A special CVD process known as epitaxial growth allows for the growth of monocrystalline films on crystalline substrates. Since epitaxial films have the same characteristics as bulk crystals, these films are highly

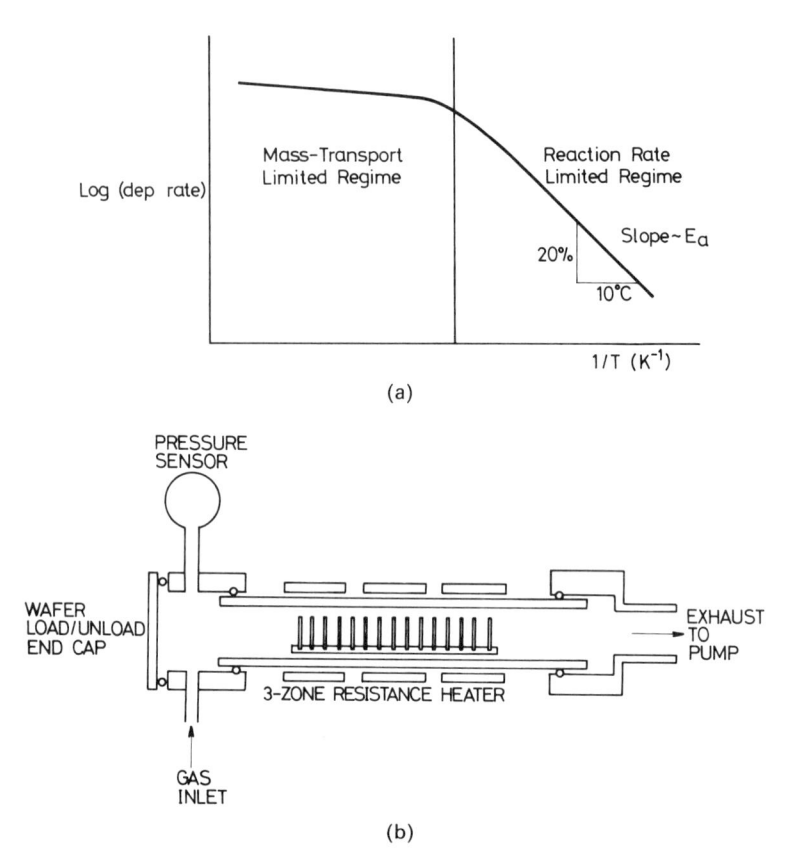

Fig. 6 (a) Regimes of the CVD process. At low temperatures the deposition is controlled by the surface reaction rate. (b) Schematic of a typical CVD deposition system. The system consist of a cylindrical heated tube with inlets and outlets for the reaction and exhausts gases. (After Refs. 1 and 5)

desirable for mechanical sensors. Epitaxial films are widely used in the fabrication of diaphragms for pressure sensors.

Plasma Deposition. Plasma-induced reactions are commonly used for the deposition of microsensor materials. The decomposition of gaseous compounds into reactive species can be induced by the presence of a plasma. Such a process is known as plasma-enhanced CVD or PECVD. Figure 7 shows a schematic of a PECVD system consisting of a vacuum chamber, a pump, two electrodes, a gas source for the chemical species, and an RF source. The system is first evacuated and a gas containing the atomic components of the film is delivered to the chamber at a pressure of a few mT. The plasma is then ignited creating many ionized species; some of these are deposited on the substrate forming a

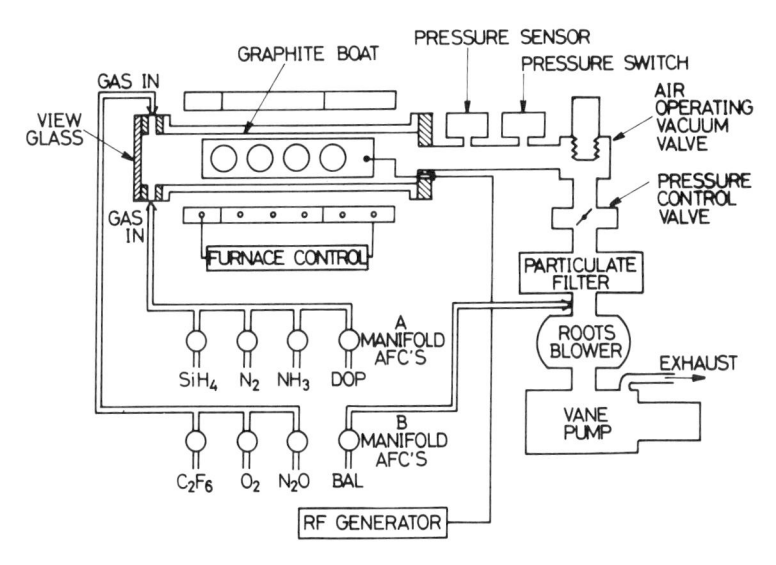

Fig. 7 Schematic of a plasma-enhanced CVD deposition system. (After Ref. 1)

solid film. PECVD films are deposited at a faster rate and require a lower deposition temperature than thermal CVD films, thus permitting the deposition on low-melting point substrates. PECVD films are commonly used for intermetal dielectric layers where a low-temperature film of modest quality is adequate. Plasma CVD films contain cracks and pinholes. Accurate control of stoichiometry is difficult as the deposited film contains trapped reaction byproduct species (especially H_2) that affect its mechanical integrity and residual stress. Because of the directionality of the plasma, PECVD films are in general nonconformal. A variety of organic films can be deposited with the aid of a plasma, which induces polymerization of monomers. This process is known as plasma polymerization. Poly-tetraflurorethylene, methyl-methacrylate, and hexamethyldichlorosilazane films can be deposited by this technique. Some of these films are useful as resists for nonplanar substrates.

2.2.2 Lithography

The patterns on each layer are defined by a processing sequence called photolithography (Fig. 8), which begins with a computer-generated quartz mask with the planar geometry of the corresponding layer defined by transparent and opaque regions. A thin layer of photosensitive organic material (photoresist), typically 1 μm thick, is then deposited onto the wafer, covering the film to be defined. After exposing the photoresist with a light source flashed through the mask onto the sample, the mask image is then transferred onto the photoresist following chemical development. The patterned photoresist, in turn, acts as a

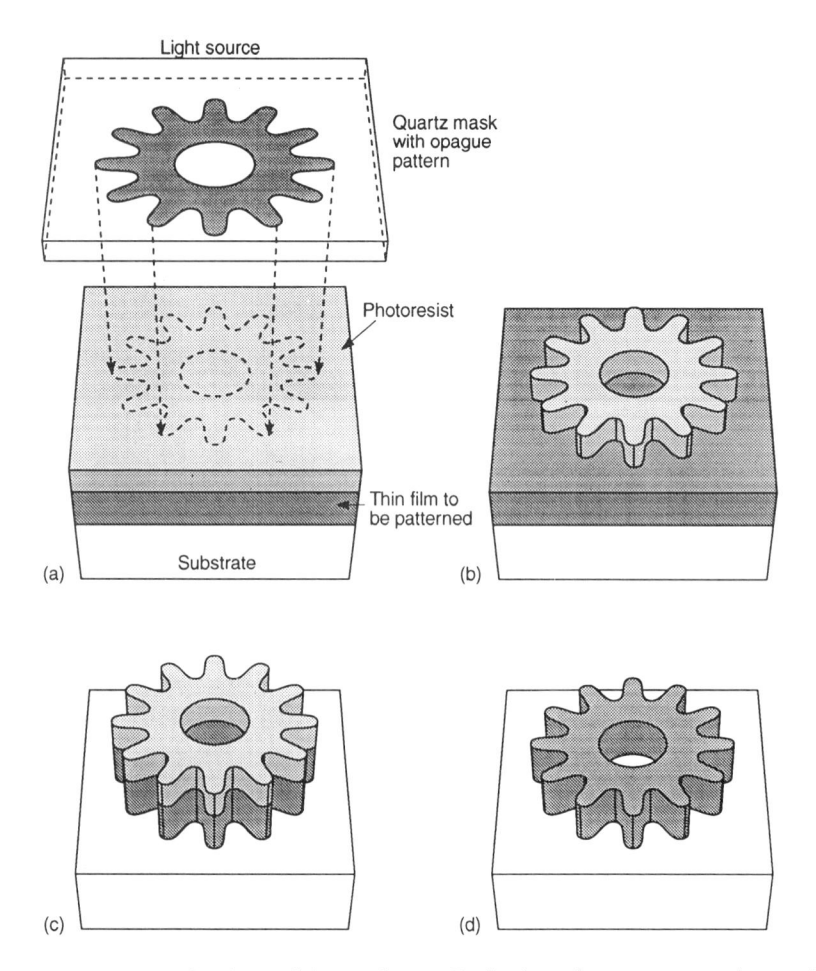

Fig. 8 Process steps in planar lithography. (a) Projection of pattern onto photoresist. (b) Photoresist development. (c) Etching of film beneath photoresist. (d) Photoresist removal leaving patterned film behind.

resistant mask when the underlying material is processed with chemical treatment, ion implantation, or other appropriate processes. Finally, the photoresist is stripped, leaving the layer with transferred patterns. This photolithography step is repeated for subsequent layers with the corresponding masks.

Mask Making. The quartz mask is also generated by a lithographic process from a glass plate with chromium patterns. The information containing the geometrical features corresponding to a particular mask is electronically entered with the aid of a layout editor system. The geometric features of the layout are next broken down into small rectangular regions of fixed dimensions. The

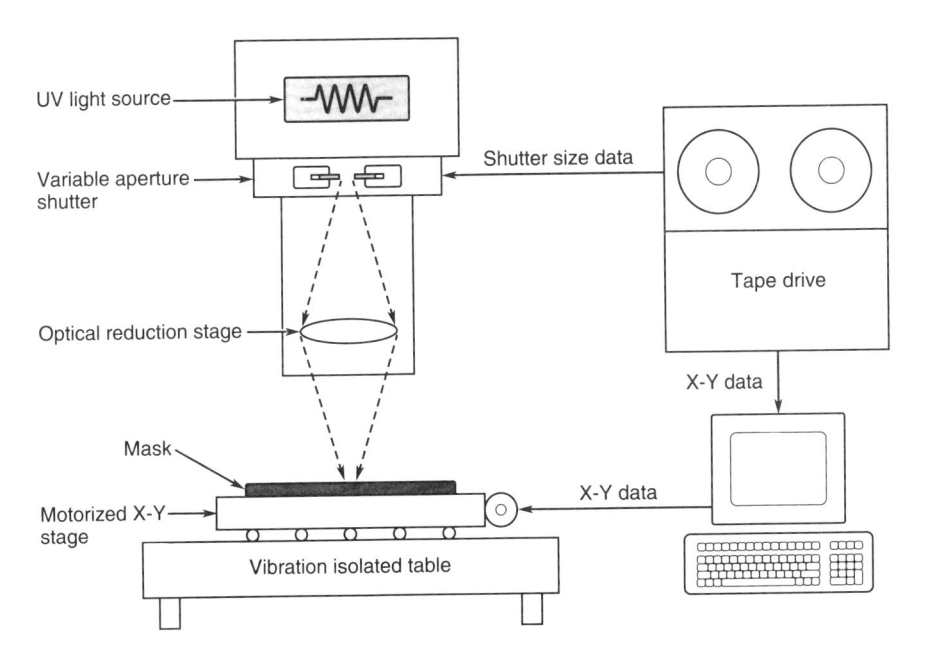

[Figure: Schematic diagram of a photomask-making machine showing UV light source, variable aperture shutter, optical reduction stage, mask on motorized X-Y stage on vibration isolated table, connected to tape drive and computer providing shutter size data and X-Y data.]

Fig. 9 Schematic of a photomask-making machine. The patterns on the mask are exposed one at a time. Typical pattern counts of 100,000 are common.

fractured mask data is stored on a tape, which is transferred to a mask-making machine. Figure 9 shows a schematic of a mask maker showing a precision x–y stage, a lamp, and an electronically controlled shutter. A reticle mask plate consisting of a glass plate coated with a light blocking material (such as chromium) and a photoresist coating is placed on the positioning stage. The tape data is then read by the mask maker and specifies the position of the stage and the aperture of the shutter blades. Each of the fractured boxes is exposed onto the plate one at a time. Depending on the complexity of the design as many as several hundred thousand boxes per die are exposed. An enlarged reticle of the chip is first generated; subsequently the projection is reduced and the image repeated to create the final mask.

Alignment and Exposure. In a multilayered planar process, the mask features must match the substrate features. Thus the mask for the current level is aligned to special patterns etched on the wafer (alignment keys) before exposure. This procedure is normally accomplished in a mechanical aligner. A typical photolithographic aligner consists of a sample and mask holder, a stereoscopic microscope, an ultraviolet light source, and a precision positioning stage. There are several types of aligners. Steppers align and project the image of each chip one at a time. Contact aligners expose the entire wafer at the same time. In

28 SEMICONDUCTOR SENSOR TECHNOLOGIES

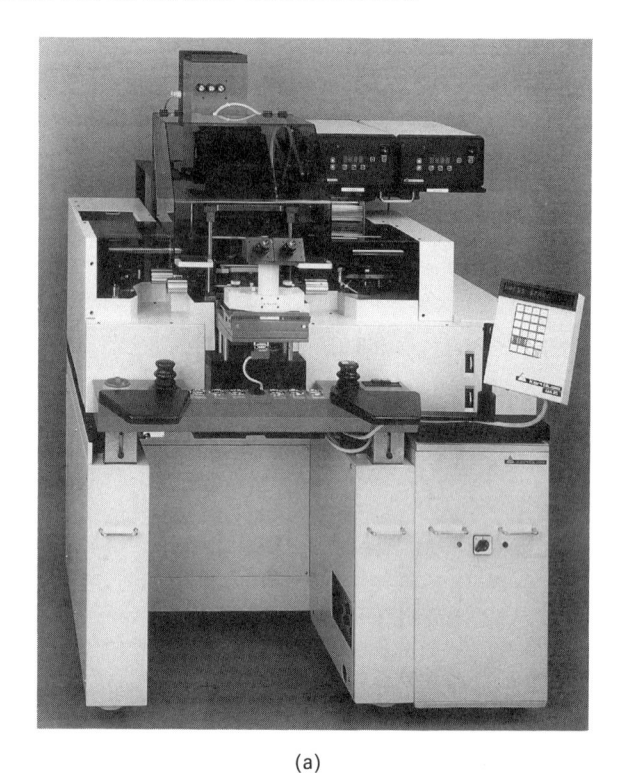

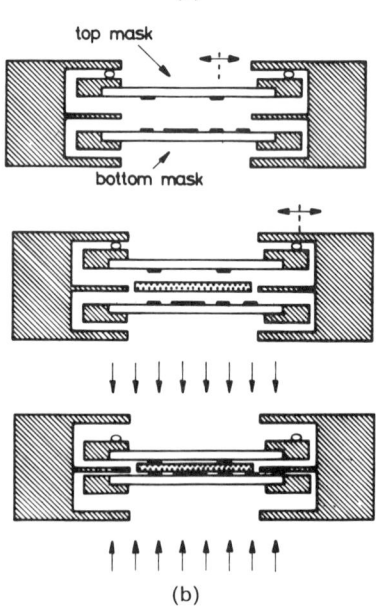

Fig. 10 (a) Photograph of a double-sided mask aligner. Courtesy Karl Suss Co. (b) Top and bottom masks are aligned first. Next the wafer is inserted between the masks and the mask assembly is aligned to the wafer before exposure.

microsensor technology, it is commonly required to perform lithography on both sides of the wafers, requiring two lithographic masks. Figure 10 shows the basic structure of a double-sided aligner. In this system, the two masks are first aligned to each other and the mask assembly is securely fixed. The wafer is inserted between the masks and either the top or bottom mask is used for the mask-to-wafer alignment. The resist is exposed on both sides of the wafer using the ultraviolet light source. A second method for double-sided alignment uses infrared-light microscopes to locate the alignment keys on the back-side of the wafer.

2.2.3 Etching and Other Patterning Techniques

The electrical response of mechanical sensors is very much dependent on their dimensions. A central requirement for patterning structures using lithographic techniques is the faithful reproduction of the defining mask patterns on the structural film since this determines the accuracy of the device characteristic. Patterning of these materials is accomplished by selective chemical etching in exposed areas using photoresist as a mask. In this section we discuss some of the most common patterning techniques for microsensors.

Lift-off. Lift-off is one of the simplest patterning techniques. In this process, a layer of an intermediate film (such as photoresist) is first deposited and patterned on the substrate. The structural film is next evaporated on the sample. This film is patterned by selectively removing the intermediate layer (and the structural material above it) as shown in Fig. 11a. Lift-off techniques are commonly used for patterning low-melting point metal interconnects.

Wet Chemical Etching. Most chemical etchants are liquids. In wet etching, samples with patterned photoresist are immersed in the etching solution for a specified time. The etching solution only removes the film on the exposed areas (Fig. 11b). The profile of the patterned material depends on the anisotropy of the etch. Wet etchants are in general isotropic, that is they etch the film with equal etch rates in all directions. Isotropic etching creates undercuts that make the patterned structure smaller than the resist mask. The profiles and etch rates are also controlled by the diffusion of active species to the exposed areas. These aspects of wet etching are discussed in Section 2.4. Anisotropic etchants for crystalline substrates are described in Section 2.5. The list of wet chemical etchants is very large. The reader is referred to Vossen's book[5] for a comprehensive list.

Dry Etching. In dry etching,[6-8] the chemically active species is in a gaseous state. Very reactive species can be created in an ionized plasma. There is a broad appeal in plasma-etching techniques because of the anisotropic etching behavior displayed by some important etch systems. Plasma etching is carried out inside a chamber constantly fed with gases at moderately low pressure

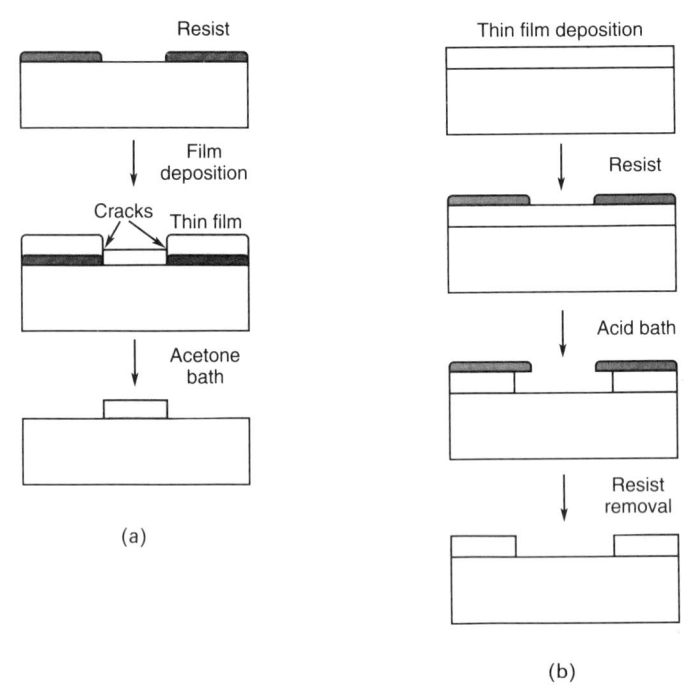

Fig. 11 Common thin-film patterning techniques: (a) lift-off and (b) wet etching. In lift-off, all metal that is above the resist is removed. In wet etching, all metal that is not below the resist is removed.

(10 mTorr to 1 Torr) with the sample placed between two electrodes as shown in Fig. 12. An RF voltage is applied between the electrodes forming a glowing plasma that contains highly reactive free radicals and ionized species. The principal mechanism of plasma etching is a chemical reaction between the vapor-phase free radicals in the plasma and the solid-phase material of the film enhanced by the ionized species. The reaction products are volatile and are readily removed from the low-pressure reaction chamber. The high degree of anisotropy in dry etching is a result of the chemical reaction preferentially enhanced on the surface parallel to the electrodes by ion bombardment. Under the influence of an RF field, the highly energized ions impinge on the surface either to stimulate reaction, or to prevent inhibitor species from coating the surface. Therefore, the vertical sidewalls, being parallel to the direction of ion bombardment, are little affected by the plasma. Some plasma systems are isotropic and rely only on the diffusion of ionized species to etch the substrate films.

Dry-Etching Parameters. Selectivity over the masking film, anisotropy, etch rate, and uniformity are important parameters for a plasma-etch system. These parameters are influenced by a number of factors, some of which are specific

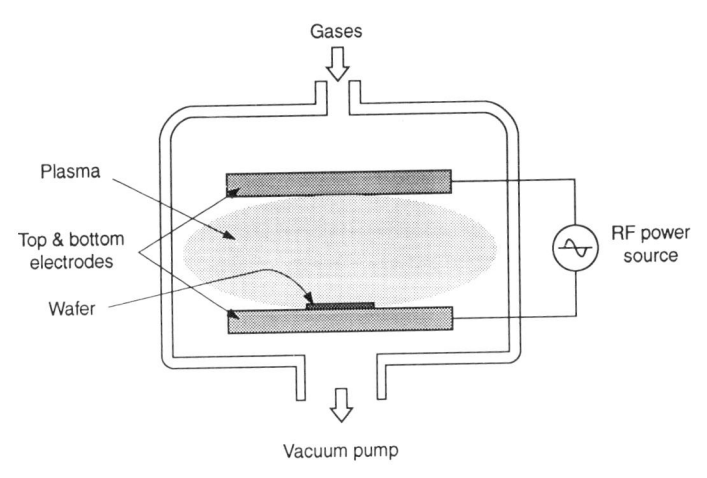

Fig. 12 Schematic of a typical plasma-etching system. The ionized species from the plasma reach the wafer surface and attack it. High directionality is achieved through ion bombardment and sidewall deposition.

to the choices of etch chemistry and equipment set-up. However, pressure, temperature, the frequency of the RF voltage, and the presence of an applied magnetic field have roughly the same effects on most anisotropic etch systems.[7,9] In general, a higher pressure (close to 1 Torr) implies enriched presence of reactive radicals, which promote a first-order, isotropic chemical etch rate. However, the degree of ionization is only a weak function of pressure. A lower pressure increases the ratio of energetic ions to reactive radicals by decreasing the number of reactive radicals. Therefore, at low pressure (10 mTorr or so), ion-enhanced etching becomes dominant, improving anisotropy. Higher temperature enhances the overall reactions of the free radicals. In some cases, the etch rate on the mask becomes too high for practical etching. Selectivity is primarily preferred over etch rate by lowering the temperature in those cases. The RF frequency of the applied voltage is tuned to the relaxation time constant of the etch reaction, which is usually between 100 KHz and 10 MHz. Some etchers have an option to set up a magnetic field through the plasma, typically at around 100 Gauss. The presence of a magnetic field confines the energetic ions during etching, resulting in a higher etch rate and degree of anisotropy.[9] A possible drawback is the premature erosion of the etch mask caused by the accelerated etch rate.

2.3 MICROMECHANICAL PROCESS DESIGN

Semiconductor sensors are accurate analog transduction devices. The characteristics and precision of the devices are influenced by (a) the building materials, and (b) the machining accuracy. In this section, we discuss how these factors affect device performance.

2.3.1 Materials Property Control

The output characteristics of microsensors are largely dependent on the mechanical properties of its constitutive materials. In the past, device designers have overlooked the importance of using good-quality building materials resulting in poor device yields and irreproducible characteristics. A set of reproducible, well-characterized building materials is essential to yield satisfactory device performance. In this section, we discuss some of the most relevant material parameters that affect the performance of microsensors.

Elastic Properties. Many semiconductor sensors rely on the mechanical deformation of a structure to detect physical variables such as pressure, force, and acceleration. The Young's modulus of a material characterizes its mechanical deformation with respect to applied stresses, $E = \sigma/\varepsilon$, where σ is the normal stress and ε is the strain. In crystalline materials, the periodic atomic lattice of crystals yields very repeatable Young's moduli which are, in general, anisotropic. The Young's modulus of thin films is much less understood. Mechanical properties in thin films are dependent on film microstructure (grain size, orientation, density, stochiometry) which is determined by specific deposition conditions. Thin films of a material are often polycrystalline or amorphous depending upon these conditions. The film microstructure changes with heat cycles often resulting in drifting mechanical characteristics. The influence of growth mechanisms on the microstructure and on its utlimate mechanical properties is not understood and is a subject of current research.

Over the last decade, a number of researchers have developed methods for the characterization of elastic properties of micromechanical materials. The Young's modulus of a thin film can be measured using several methods. The most widely used method is the measurement of load-deformation characteristics. Figure 13 shows a schematic of a nanoindentation apparatus.[10] A hard diamond

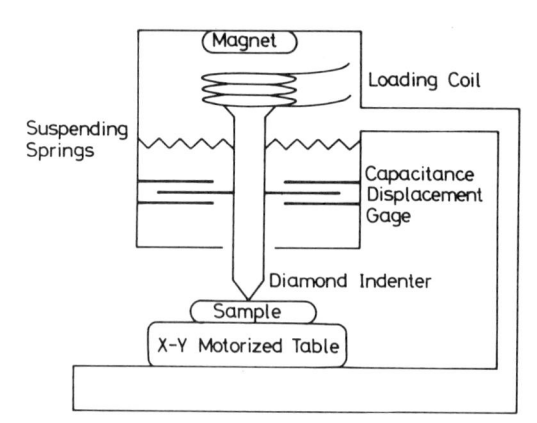

Fig. 13 Schematic of a nanoindentation apparatus. The force of the plunger is controlled by a magnet. (After Ref. 10)

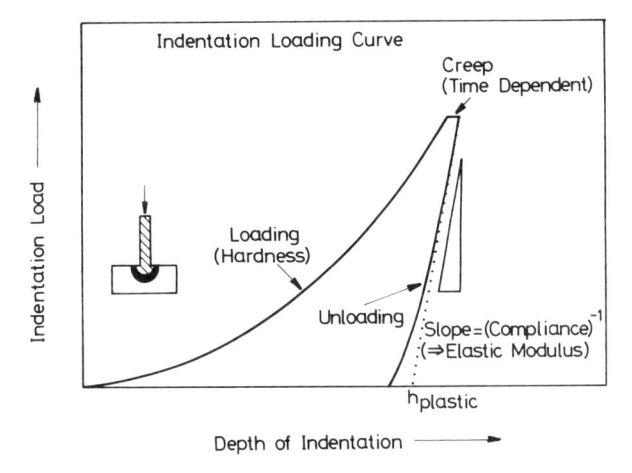

Fig. 14 Example of the load-deflection data extracted from a nanoindenter experiment. The slope of the curve is indicative of the Young's modulus.

indenter is plunged into the thin film, and the film's deformation is recorded. The Young's modulus of the sample in compression is determined from these parameters (see Fig. 14). Nanoindentation methods are commonly used for the measurement of deformation of soft materials such as metals.

For harder materials, it is convenient to machine a flexible part that is used to measure the load deflection. Flexible beams and plates are commonly used for this purpose. In beams, the load-deflection measurement is made with a stylus profiler of known load.[11-13] Direct tensile tests[14] have been performed on beams and whiskers. In plates, the load is provided by the pressure of a gas on one side of the plate while the deflection is measured optically on the other. Several researchers have measured the Young's modulus from mechanical resonance of beams.[14] In general, beam and plate methods have limited accuracy because they require tight control of sample dimensions.

Fracture and cyclic fatigue are important mechanical parameters that determine the device durability and operational limits. Static fracture strains of thin films have been measured by two methods. Doubly supported overhanging structures are patterned to self-rupture at known internal strains.[15-17] Slider-type[18] devices with externally applied strains have also been used. The fatigue and fracture of micromechanical materials have been studied under cyclic loading in both polycrystalline[19,20] and monocrystalline[21,22] thin films. Measurements on monocrystalline materials suggest that brittle fracture can occur under cyclic stresses that are substantially lower than the static fracture stress. Great care in the shape of the design is necessary to avoid material fracture along crystal planes.[23]

Stray Stresses. The behavior and structural integrity of the micromechanical sensor is very much influenced by the presence of stray stresses. Stray stresses are stresses that are present in the film in the absence of external forces. In mechanical sensors, small stray stresses are manifested as noise signals that are detected by the sensor yielding incorrect results while large stresses cause severe structural deformations. There are two sources of stray stresses. Thermal stresses are the product of mismatches in the thermal expansion coefficients of different films. In thin films, thermal stresses develop because these films are grown at higher than ambient temperature. Thermal stresses can cause undesirable bimetallic warping effects (Fig. 15). Typically, thermal strains in the order of 5×10^{-4} are observable in silicon micromachined sensors. Thermal stresses also develop at the microsensor–package interface. Metal-can packages expand at a much higher rate (about 10 times higher) than the microsensor substrate causing deformation and sometimes fracture. Stress isolation techniques that use soft buffer layers[24] and stress relaxation[25,26] structures are essential for the proper operation of microsensors.

A much larger component of stray stress is residual stress. Residual stresses develop because as-deposited thin films are not in the most favorable energetic configuration. Residual stresses can be compressive, which makes the film expand, or tensile, which makes the film shrink. Since these films are firmly

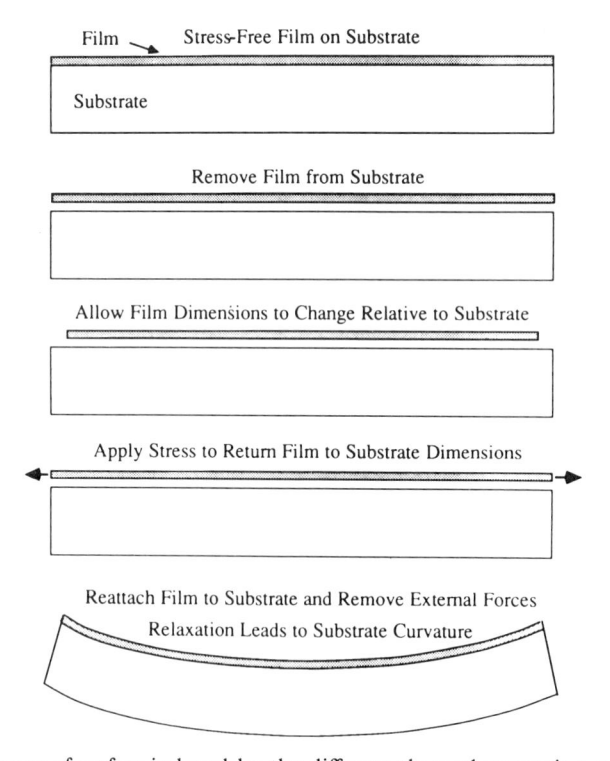

Fig. 15 Warpage of wafers induced by the different thermal expansion coefficients of the substrate and the deposited films.

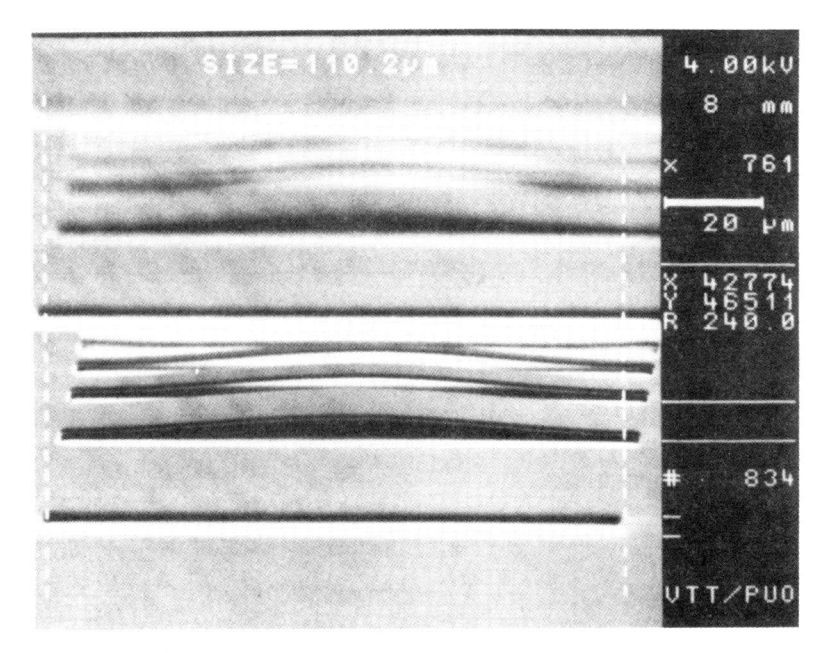

Fig. 16 Example showing sample warpage of doubly clamped beams caused by a compressive residual stress. (After Ref. 143)

attached to the substrate, the internal strain cannot be relaxed, hence causing mechanical deformations. Residual stresses can be relaxed by high-temperature anneals; however, the anneal temperatures are quite high and may not be practical for the production of micromechanical devices. Rapid thermal annealing has proved to be useful in this respect with limited success.[27]

In bulk-micromachined sensors, stray stresses can produce mechanical warping and offsets. In surface-micromachined sensors, the effects of residual stresses can be catastrophic. Residual stresses are responsible for large bending, warping (Fig. 16), and even fracture of thin film materials; therefore, only low-stress, thin films are used in surface-micromachined devices. Most thin films deposited at low temperatures have large residual stresses. Diffused etch-stop layers also have substantial tensile stresses[28-30]. Fortunately, in LPCVD materials the stress can be tightly controlled. Near-zero-stress polysilicon films are readily deposited. Similarly silicon-rich silicon nitride can be deposited in a compressive or tensile stage. The as-deposited stress in LPCVD can be kept under 50 MPa but cannot be completely eliminated. Typically, microsensors are constructed with films that are slightly tensile as-deposited, and, at some stage in the process, the stress is relieved by an anneal step.

A number of experimental techniques are available for the measurement of residual stresses. The earliest technique is based upon curvature measurements. In this method, a thin layer of the film is grown on one side of a flat substrate.

If the film and substrate thicknesses are known, the curvature of the substrate is related to the stray stress by the Stoney equation

$$\sigma_r = \frac{Et_s^2}{6(1-v)t}\left(\frac{1}{R_0} - \frac{1}{R_f}\right) \quad (1)$$

where E, and v are the Young's modulus and Poisson's ratio of the substrate, R_0 and R_f its initial and final curvature, t_s the wafer thickness, and t the film thickness. Stresses as low as 10 MPa can be measured by this technique. For lower stresses the measurement becomes dependent on boundary conditions and gravity.[31] Figure 17 shows an illustration of a commercial stress gage.

In many instances, the deposited film must be patterned and subject to various thermal cycles. *In situ* stress diagnostic structures (Fig. 18) provide an indication of the stress of the films on the device wafer. These structures rely on the warping or mechanical buckling that is easily observed under microscopic examination. Clamped beams are used to determine compressive stresses. If the structural film is under compressive stress, the stress may exceed the buckling stress of the beam. The buckling is easily observed by either interference or SEM inspection. The beam buckles when its compressive residual stress σ_r is

$$\sigma_r = -\frac{\pi^2 h^2 E}{3L^2} \quad (2)$$

where L is the beam length and h its thickness. The negative sign indicates that

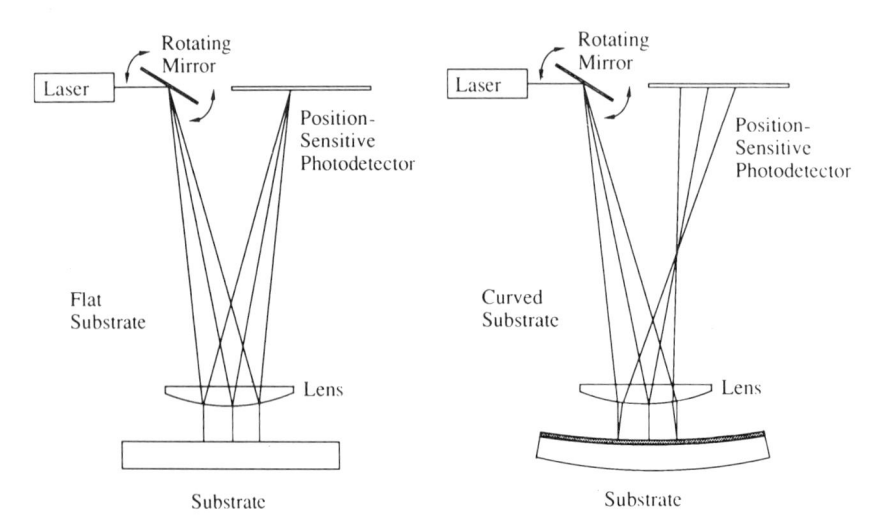

Fig. 17 Schematic of an optical stress gage machine. The sample is scanned with a laser through a rotating mirror. If the sample is warped, the extent of the reflected image is an indication of its radius of curvature.

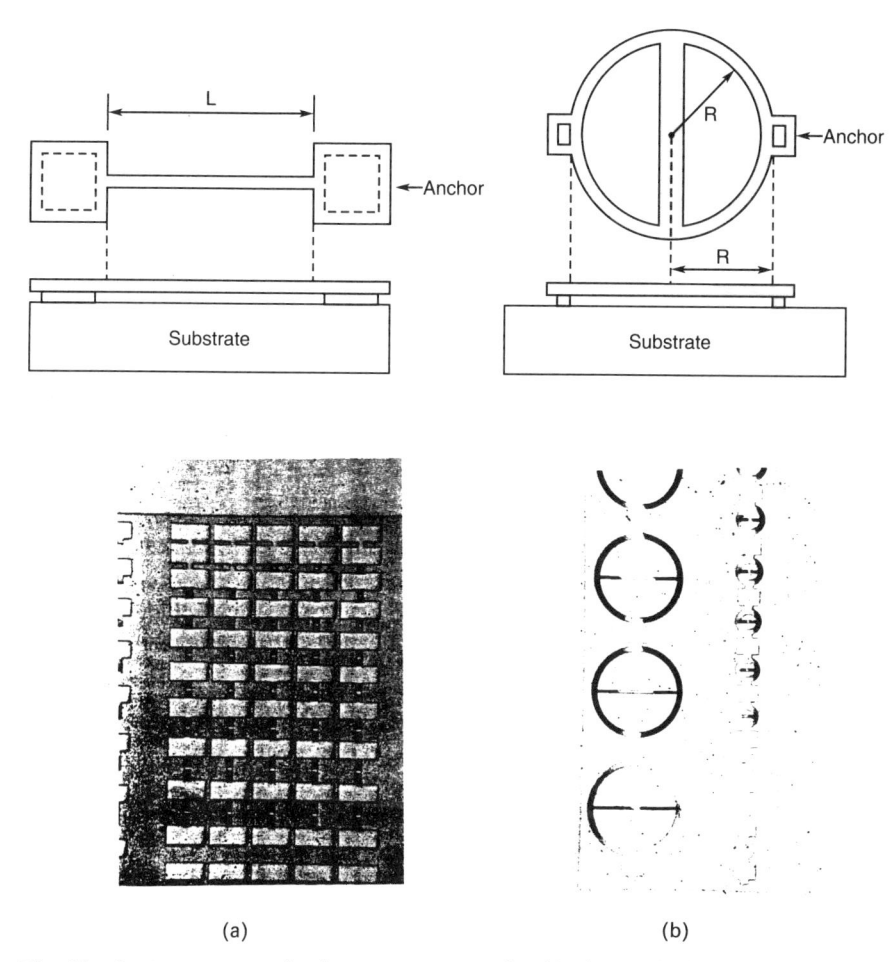

Fig. 18 *In situ* structures for the measurement of residual stress in thin films. (a) Doubly supported beams buckle under compressive stress. Top view: beam structure. Bottom: array of doubly supported beams of increasing length for diagnostics. The third beam on the second row is the first to buckle. (b) Structure of Guckel rings. The central beam of Guckel rings[32] buckle under tensile stress. Bottom: array of Guckel rings of increasing size. The fourth ring from the bottom on the second column is the first to buckle. (After Ref. 218)

the stress is compressive. For long beams, the compressive buckling stresses can be quite small (<1 MPa). Buckling stresses are also used for the measurement of tensile stresses. The most common diagnostic structures are Guckel rings (Fig. 18b). Buckling in these structures occurs when the tensile residual stress exceeds

$$\sigma_r = \frac{\pi^2 h^2 E}{12 g(R) R^2} \tag{3}$$

where R is the radius of the ring and $g(R)$ is a function of the inner and outer ring diameter with $g(R) \leqslant 0.918$.[32] Both beam and ring structures[33-35] are routinely used to check the quality of the films grown. Other less common methods of stress measurement include membrane deflection[36] and X-ray diffraction[37]. The nonuniformity of these stress components is also an important parameter. Residual stress gradients are measured using straight and spiral cantilever beams.[16,38,39]

Thermal Parameters. In thermal sensors, it is desired to raise the temperature locally on certain areas of the sensor. Thermal sensors such as radiation detectors and micro-calorimeters rely on good thermal isolation of the sensor from its substrate. The material parameters that determine the sensitivity and time response of these devices are the thermal conductivity, density, and heat capacity. Tight control over these parameters is important for reproducible thermal characteristics. The thermal conductivity of thin films has been measured from static heating of microbridges[40] and suspended plates.[41] Film heat capacity and thermal diffusivity are measured from their transient characteristics[42] while the film density is easily determined using microbalance measurements. Most micromechanical materials have repeatable densities and heat capacities. However, the thermal condudtivity seems to be dependent on film microstructure. Differences by a factor of four or more have been observed between bulk and thin films of a particular material.

Dealing with Material Uncertainties. Sensor designers use statistical process-control experimental techniques to grow these materials in a reasonably reproducible manner. Typically, the mechanical parameters of thin films can be controlled within 10-20% of the target values. In general, these tolerances are too high to yield sensors that have accurate characteristics. The use of ratiometric techniques, where the unknown variable is scaled to a known reference, help to eliminate material dependencies, but do not eliminate them completely. The most common approach to deal with material uncertainties in sensors is through calibration of each device against a reference. Elaborate procedures involving multiple point and temperature measurements are required in the sensing environment. This operation is usually performed at the individual sensor level; hence the operation is costly and, in many cases, economically prohibitive. In some devices, it is possible to perform the calibration by electrical means (autocalibration). This advanced topic is a current area of research.

2.3.2 Dimensional Control

The accuracy and reproducibility of microsensors is also determined by the precision and uniformity of the machining processes across the sample. Figure 19 shows the various errors that occur both locally and across the substrate. These are errors contributed by lithographic and micromachining processes.

2.3 MICROMECHANICAL PROCESS DESIGN 39

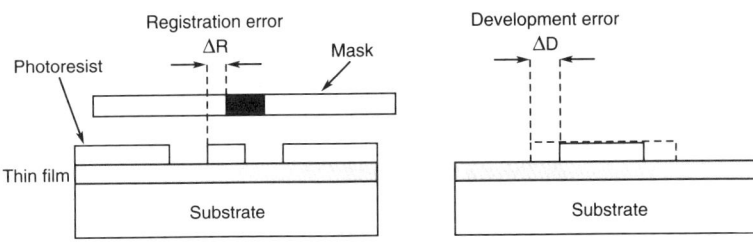

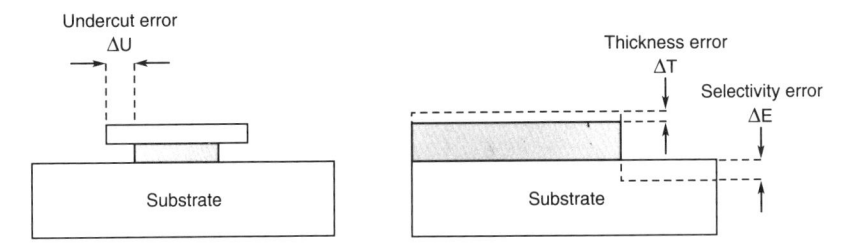

Fig. 19 Dimensional errors in micromachining. (a) Lithographic transfer errors. (b) Micromachining errors.

Lithographic Transfer Errors. During the lithographic pattern transfer, two errors may occur. Registration errors are caused by the misalignment of the lighographic masks to the features on the substrate. Resolution errors are caused by the wavelength limitations and developer selectivity of the photolithographic transfer process. The actual pattern on the resist differs from the projected mask image. For optical UV lithography, the edge of the pattern may be blurred by as much as 0.5 μm.

Micromachining Errors. The pattern of the resist is not transferred to the film perfectly. The accuracy of the transfer depends on the etchant anisotropy. Reactive-ion etching generates the most anisotropic profiles with nearly vertical sidewalls. Wet etches, on the other hand, are isotropic, producing semicircular walls and undercuts.

In the direction normal to the surface of the wafer, dimensional errors are generated by variations in the deposition rates of thin films. When the films are patterned, the vertical dimensions will depend on the accuracy of the etch rate and its selectivity, that is, the ratio of the film etch rate to that of the material below it. Most etches have finite selectivities, hence, requiring tight etch controls and relatively thick etch-stop layers to prevent over-etching.

All the errors mentioned above are local errors at a particular point in the wafer. The process parameters are subject to spatial uniformity variations across the wafer surface. Uniformity variations in deposition and etching are difficult to control, and are a major source of errors in device characteristics. Nonuniformity in deposition and etch rates as large as 20% across 100 mm silicon wafers is not uncommon.

Selectivity and Uniformity Trade-off. In many etch processes, the selectivity of the etch is not high (1:3–1:10). Hence, it is necessary to assure that the thickness of the underlying etch-stop layer is sufficient to withstand over-etch cycles. This problem is self-evident in the presence of thickness and etch-rate nonuniformities. For example, to successfully etch a nonuniform film on the wafer, the etch time must be long enough to clear the thickest areas at the lowest etch rate. The etch-stop film must withstand the etch for a worst-case time equal to the etch time minus the time required to etch the thinnest areas at the fastest etch rate. This assures the integrity of the etch-stop.

Etching Profiles. The shape of the micromachined structure is determined by the characteristics of the etch. The shape of the sidewalls of the patterned film

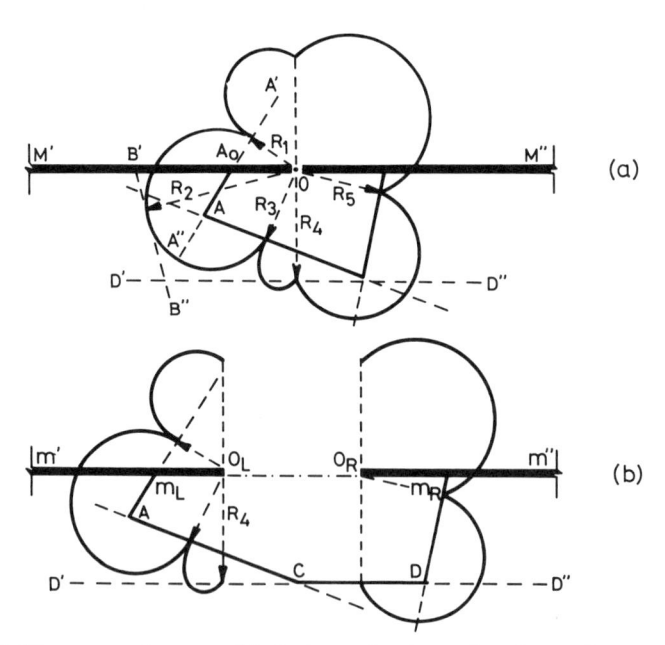

Fig. 20 Wulff construction graphical determination of etch profiles. The curved contours show the etch rates at a particular direction. The shape of the etch front is determined by the intersection of normals to the rate vectors at a particular time. In (a) the profile is determined by the shortest vectors R_1, R_3, and R_5. (b) In a larger slit, R_4 also determines the shape of the etch front. (After Ref. 43)

can be calculated from first principles. Several dissolution models have been studied for both wet and dry etching. In wet etching, the profiles are controlled by the etching reaction at the solid–liquid interface and the availability of reactants at the etch-front surface.

If the etch solution contains a large concentration of reactants in direct contact with the substrate surface, the etch profile is dominated by the reaction rate. Etch-front profiles are calculated by propagating the geometry of the substrate interface in the direction normal to the exposed surfaces using appropriate etch-rate vectors. The etch front is determined[43] by the contour of intersection of lines normal to the rate vectors or an "envelope of normals" (Fig. 20).[44,45] This procedure is known as the Wulff construction. This simple procedure has been modified[46–49] to model both wet and dry etching. Rate vectors can be determined experimentally using a wheel-wagon test pattern technique developed by Jaccodine.[44]

If the replishment of reactive species near the surface is insufficient to keep up with the surface reaction rate, the overall reaction is limited by the diffusion of reactant species from the bulk of the solution. The effects of etchant diffusion on the profile of a simple, two-dimensional slit mask have been studied.[50–52] Enhanced etching occurs at the slit corners (Fig. 21) where the local consumption of diffusing species is smaller than on exposed surfaces. Diffusion effects are

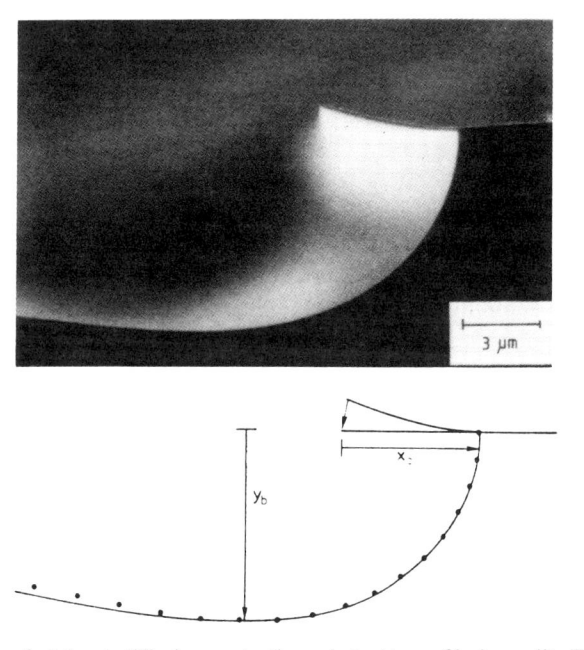

Fig. 21 Effects of etchant diffusion onto the substrate profile in a slit. The etch rates increase near the edge of the mask since the etchant is less depleted in this area. (After Ref. 52)

important in cases where the etching solution has difficulty penetrating the structures (such as sacrificial etching), and can affect the integrity of the mechanical structures[53,54] due to excessively-long etch times.

2.4 BULK MICROMACHINING

In bulk micromachining, the sensors are shaped by etching a large single-crystal substrate. Thin films are patterned on the bulk to perform isolation and transducer functions. Anisotropic etching techniques provide a high-resolution etch and tight dimensional control. Often, bulk-micromachined sensors use two-sided processing, creating a self-isolated structure with one side exposed to the measured variables while the device side is enclosed in a clean package. Two-sided structures are very robust for operation in environments hostile to microelectronic devices. Simple mechanical devices such as diaphragm pressure sensors and cantilever-beam piezoresistive acceleration sensors are fabricated commercially by this technique. Bulk micromachining techniques have been used for the past 20 years and are still the most popular sensor fabrication technology.

More complex sensors such as capacitive devices require more than one rigid piece. Bulk samples can be bonded together at the wafer level to form these devices. Many wafer bonding techniques are currently available. Nevertheless, bonding requires wafer alignment and extreme cleanliness of the bonded surfaces to prevent the formation of bubbles and voids. These are major problems to be resolved before bonding is widely accepted as an economically feasible technology. By far the most developed semiconductor-sensor technology is bulk silicon micromachining, and it is the main focus of this section.

2.4.1 Bulk Materials

In this section we describe some of the most commonly used substrate materials in bulk micromachining. Single-crystal substrates are used for the fabrication of the mechanical parts and dielectric substrates are used for structural supports. The most common substrates are silicon and amorphous glass. Other materials such as gallium arsenide and quartz have been used to a lesser extent.

Single-crystal silicon is the most commonly used material for solid-state micromechanical sensors and actuators. Silicon single-crystal substrates of high purity are readily available at low cost for semiconductor manufacturing. Because of its high purity and crystalline structure, the mechanical properties of silicon are well controlled, yielding sensors with reproducible mechanical characteristics. Mechanically, silicon has a higher strength than steel but it is rather brittle. A good review on the various mechanical uses of bulk silicon is given in Ref. 55. The electrical properties of silicon are sensitive to stress, temperature, magnetic fields, and radiation which permit the design of a variety of sensors. Its piezoresistive property[56,57] is the most widely used mechanism for mechanical sensors. Silicon is machined precisely by anisotropic etchants that selectively attack crystal planes to define the shape of the sensor. Anisotropic etchants for silicon are discussed in the next section.

Amorphous-glass substrates are bonded to other substrates to form sealed devices. Amorphous glass is a mixture of SiO_2 doped with impurities. Sodium-rich substrates are commonly used in bulk micromachining because of the high strength of anodic bonding. Glasses with varying compositions, mechanical properties, and melting points are available. To assure low stray stresses, it is desirable to select glass samples with a thermal expansion coefficient that matches that of the bonded substrate. Anodically bonded 7740 Pyrex glass matches the thermal expansion of silicon, and it is the most commonly used glass substrate in bulk micromachining. Doped glasses have low melting points, thus preventing further high-temperature processing of the sample. Quartz is the crystalline form of glass. Quartz substrates can be anisotropically machined in much the same way as silicon. Quartz is also used as a structural material because of its high piezoelectric coefficient and its superb thermal stability. Micromachined quartz resonators are commonly used. Unlike amorphous glass, quartz is very brittle and can fracture very easily. Extreme care is required to avoid fracture along crystal planes.

Other, less common, bulk materials include single-crystal gallium arsenide substrates for optical devices and aluminum oxide substrates. Recently, gallium arsenide substrates have been used for pressure sensors.

2.4.2 Anisotropic Silicon Etching

The ability to selectively etch into the bulk of crystalline silicon provides a powerful technique in creating precise three-dimensional structures. Most anisotropic etchings of crystalline silicon are based on liquid-phase chemical reactions, with a few techniques developed using vapor-phase and plasma etchings.[58-60] Since the early 1960s, strong alkaline solutions have been used to etch silicon selectively according to crystallographic orientations.[61] Generally, the etch rate is slowest in the $\langle 111 \rangle$ directions, and fastest in the $\langle 100 \rangle$ and $\langle 110 \rangle$ directons. Selectivity is defined as the ratio of the etch rates of the desired direction to those of the undesired one. The higher the selectivity, the better defined the finished geometry. Some silicon etchants display a reduced etch rate in regions that are heavily doped with boron, adding flexibility in defining the finished structures. Furthermore, an externally-applied electrical potential on one side of a p–n junction within the bulk silicon can also be used to influence the etch rate in the direction perpendicular (electrochemical etching) to the junction. Finally most of the silicon etchants have very low etch rates on silicon oxide or silicon nitride, making the latter suitable masking films for defining the desired geometry.

Examples. Figure 22 illustrates how three different structures can be created in a single processing sequence: a square opening through the thickness of the

44 SEMICONDUCTOR SENSOR TECHNOLOGIES

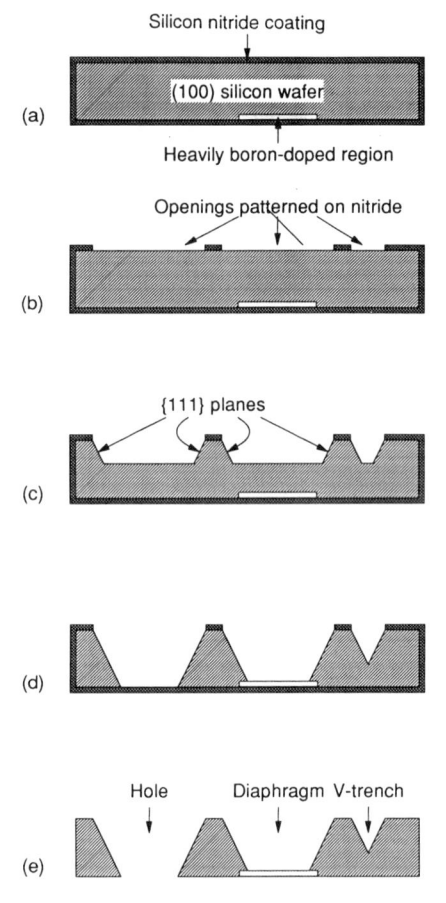

Fig. 22 Processing steps of a typical anisotropic wet chemical etching of bulk silicon: (a) a silicon wafer with a heavily boron-doped region is protected on both sides by a layer of deposited silicon nitride; (b) etch windows of various patterns are opened on one side; (c) the wafer is progressively etched in a heated alkaline path, with sloped sidewalls gradually forming along the etch-resistant {111} planes; (d) etching stops when only {111} planes and the heavily boron-doped region are exposed; (e) the protective nitride layer is removed, resulting in a square hole through the wafer, a V-groove, and a thin boron-doped silicon diaphragm on the bottom

wafer, a V-groove with a rectangular top view, and a square diaphragm on the bottom of the wafer. A (100) silicon wafer is doped heavily with boron to define the region where a diaphragm is to be formed. The wafer is then protected on both sides by depositing a thin layer of silicon nitride (Fig. 22a). Etch windows corresponding to the three desired structures are patterned and opened on the top, with the edges of the patterns aligned along the $\langle 111 \rangle$ directions (Fig. 22b). During etching in a heated alkaline bath, sloped sidewalls are gradually formed along the etch-resistant {111} planes, while the $\langle 100 \rangle$ etch fronts progress downward with time (Fig. 22c). The four {111} planes each form a 54.74° arctan($\sqrt{2}$) slope with the surface of the wafer, the (100) plane. The wafer continues to be etched until only {111} planes and the heavily boron-doped region are exposed (Fig. 22d). Finally, after the nitride layer is removed, the three desired structures are achieved. Because of the constraint of the angular relationship between the {111} and the {100} planes, any holes formed with this procedure will always have a bigger opening on the top than on the bottom, with the difference proportional to the thickness of the wafer. Also, the width of the V-groove is always $\sqrt{2}$ times the depth of the groove.

Design Considerations. To achieve precise dimensional control, one must carefully align the masking patterns to the ⟨111⟩ directions. Figure 23 illustrates that a misaligned square pattern results in an oversized opening, bound by four {111} planes that encompass the original design. Note that special precautions should be taken when attempting the create structures with convex corners such as mesas. Although etching stops very well at concave corners formed at intersections {111} planes, convex corners are inherently unstable. Fast etching planes, in addition to the {100} planes, become exposed easily at convex corners, allowing rounding or faceted etching at those locations. Given prolonged etching, the intended mesa can be completely etched away. Techniques are available to compensate for convex corner rounding by modifying the original design and carefully timed etching.[62-66]. Rectangular patches, beams, or triangular patterns can be placed at the convex corners of the mesa to slow the exposure of fast-etching surfaces. Also, by combining multiple masking and etching steps, one can create sophisticated three-dimensional structures.[67]

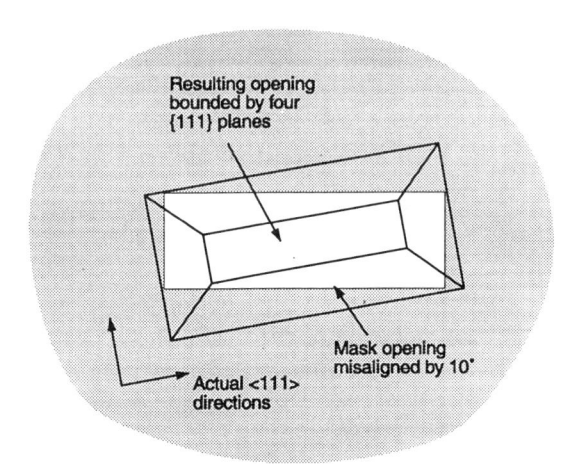

Fig. 23 Effect of misalignment between the patterned mask and the actual ⟨111⟩ direction. After etching is completed, the resulting opening is formed by four bounding {111} planes that encompass the original design.

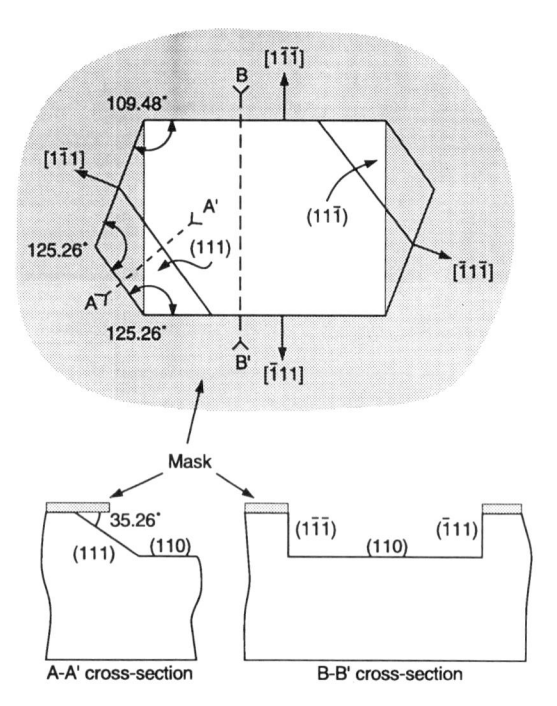

Fig. 24 Top view and cross-sectional views of an etched cavity in a (110) silicon wafer, showing angular relationships between adjacent sidewalls

(110)-Silicon.[68-70] In contrast to a (100)-silicon wafer, where the four sidewalls of an etched cavity always intersect the surface at 54.74°, a (110) wafer yields different geometrical results when anisotropically etched. There are six {111} planes that can be exposed and intersect with the (110) surface, and therefore the top view of an etched cavity is hexagonal with three pairs of parallel sides. Figure 24 illustrates the angular relationship between these planes. Two of the {111} planes, the $(11\bar{1})$ and the $(\bar{1}11)$ planes, each forms a 35.26° (arctan$\sqrt{2}/2$) angle with the (110) surface. The other four {111} planes [$(1\bar{1}1)$, $(\bar{1}11)$, $(1\bar{1}\bar{1})$, and $(\bar{1}1\bar{1})$] all form vertical sidewalls of the cavity.

Parallel trenches with vertical sidewalls can be created from a surface pattern made up of parallel lines aligned along the $(1\bar{1}\bar{1})$ or $(\bar{1}11)$ planes (Fig. 25). On the other hand, if the line patterns are aligned parallel to the $(11\bar{1})$ and $(\bar{1}11)$ planes, shallow V-grooves with sidewalls at an angle of 35.26° with the surface will be created (Fig. 25).

Alkaline Etchants and Etch Masks.[71-76] Different alkaline solutions display different etching characteristics and selectivities, depending on the chemical compositions and temperature. A commonly used alkaline etchant is a mixture

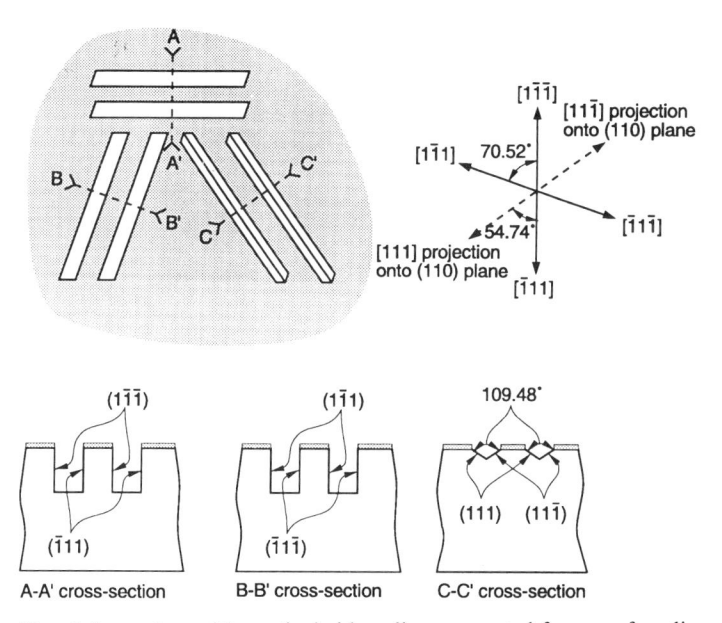

Fig. 25 Parallel trenches with vertical sidewalls are created from surface line patterns parallel to the $(1\bar{1}1)$ or $(\bar{1}11)$ planes. The line patterns parallel to the $(\bar{1}\bar{1}1)$ and $(11\bar{1})$ planes result in shallow V-grooves with sidewalls at angles of 35.26° with the surface.

TABLE 1 Compositions of Two Different EDP Solutions Proposed by Reismann *et al.*

Type	Ethylenediamine (ml)	Pyrocatechol (g)	Water (ml)	Pyrazine (g)	Temperature (°C)	(100) etch rate (μm/h)
F	1000	320	320	6	115	80
S	1000	160	133	6	50–115	5–60

of ethylene diamine $NH_2(CH_2)_2NH_2$, pyrocatechol $C_6H_4(OH)_2$, and water (commonly known as EDP).[77-79] A small amount of pyrazine ($C_4H_4N_2$) may be added to increase both the etch rate and selectivity. The composition can be varied to achieve different desired effects. For example, Reisman *et al.*[78] proposed two specific compositions optimized for fast-etching rate (type-F) and slow or low-temperature etching (type-S) (Table 1). During etching, the chemical composition of the solution is kept constant with a reflux system. In addition, a nitrogen purge may be necessary to prevent premature aging of the EDP solution due to atmospheric oxidation. The type-F EDP etchant should be used only at the prescribed temperature, otherwise, solid residue may form. There is no solid residue with the type-S etchant over the range 50 to 115°C. The etch rate (R) is temperature dependent, obeying the Arrhenius law

$$R = R_0 \exp(-E_a/kT) \qquad (4)$$

TABLE 2 Experimentally Determined Activation Energies (E_a) and Pre-exponential Factors (R_0) for Etch Rate Calculation with the Arrhenius Equation: $R = R_0 \exp(-E_a/kT)$

Etchants	⟨100⟩ Si		⟨110⟩ Si		SiO$_2$	
	E_a(eV)	R_0 (μm h)	E_a(eV)	R_0 (μm h)	E_a(eV)	R_0 (μm h)
Type-S EDP	0.40	9.33×10^6	0.33	1.16×10^6	0.80	1.36×10^8
KOH, 20%	0.57	1.23×10^{10}	0.59	3.17×10^{10}	0.85	3.52×10^{11}
a-KOH, 20%	0.62	4.08×10^{10}	0.58	4.28×10^9	0.90	1.72×10^{12}
KOH, 34%	0.61	3.10×10^{10}	0.60	3.66×10^{10}	0.89	2.34×10^{12}
NaOH, 24%	0.65	1.59×10^{11}	0.68	7.00×10^{11}	0.90	3.20×10^{12}
LiOH, 10%	0.60	3.12×10^{10}	0.62	8.03×10^{10}	0.86	2.34×10^{11}

a-KOH contains isopropyl alcohol at 250 ml/l

where R_0, the pre-exponential factor, and E_a, the activation energy, are experimentally determined, k, Boltzmann's constant, is 86.1×10^{-6} eV/K, and T is the temperature in K. Table 2 shows the values for E_a and R_0 for a number of popular etchants.[75] The etch-rate ratio among the three most important crystallographic directions, ⟨100⟩:⟨110⟩:⟨111⟩, varies with temperature because of the difference in the activation energies. For type-S EDP, it is estimated to be 30:30:1 at 115°C, rising to about 100:150:1 at 50°C.[75] Therefore, one can optimize the selectivity and etch rate by selecting an appropriate etch temperature. Etching at temperatures below 50°C is too slow to be practical.

A distinct advantage of EDP is its high etch selectivities with some materials that can be used conveniently as masking films. For example, type-S EDP etches thermally grown SiO$_2$ extremely slowly, at around 55 Å/h at 115°C, resulting in an etch selectivity of (100) Si over SiO$_2$ in excess of 10,000. Silicon nitride shows no measurable etch rate at all. Some metals including Au, Cr, Ag, and Cu are also resistant to EDP. Therefore, great flexibility exists for choosing a masking film. Finally, etch rates in all crystal planes decrease sharply in regions doped with boron at higher than 3×10^{19} cm^{-3} (Fig. 26),[76,79] making it an effective etch-stop for EDP.

Another popular etchant is an aqueous solution of potassium hydroxide, with concentration ranging from 10 to 50%.[61] Isopropyl alcohol is frequently added to improve selectivity.[80] The merit of KOH etchant is that its selectivity of {110} over {111} crystal planes can be much higher than EDP. It was reported that an aqueous solution of 55% KOH by weight exhibited a selectivity of {110} over {111} planes as high as 500.[70] Therefore, KOH is often used to etch deep trenches in (110) silicon wafers. Unfortunately, the selectivity over SiO$_2$ is less than 500 at various concentrations, which is not quite enough for masking during long etches.[81] Additionally, KOH etching does not stop quite as well in boron-doped regions (Fig. 26). Nevertheless, silicon nitride is still a convenient and effective masking film even for KOH etchants. It should be noted that in timed-etch applications, the bottom surface of an etched (100) pit tends to be

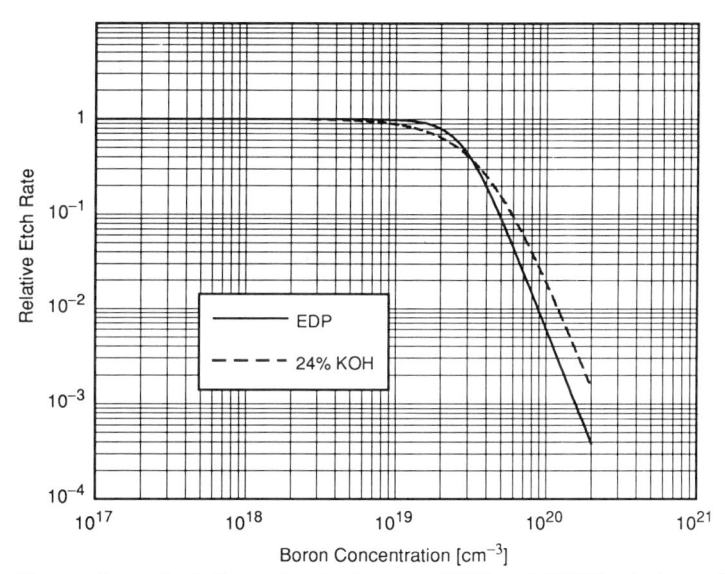

Fig. 26 Comparison of relative etch rates between EDP and KOH solutions of silicon in the $\langle 100 \rangle$ direction as a function of boron-doping level.

rough if the KOH concentration is below 30%. Hillocks as high as 10 μm have been observed.[81,82]

Other strong alkaline solutions, such as cesium hydroxide (CsOH),[83] ammonium hydroxide (NH_4OH),[84,85] sodium hydroxide (NaOH),[61,86] hydrazine ($N_2H_4 \cdot H_2O$),[87] and choline ((CH_3)$_3$N(CH_2OH)OH)[88] have been studied for specific anisotropic etch applications, with variable effectiveness and trade-offs. An important property of the etch is its effect on metals. Most of the etchants mentioned above will attack interconnect layers; hence, they demand special passivation and isolation techniques. In particular, two etchants will not disturb metals. EDP does not attack gold metallizations but it does attach aluminum. Recently a new type of etchant, tetramethylammonium hydroxide (TMAW)[89] has been found to leave aluminum undisturbed at the expense of an increased hillock density.

Acidic Etchants and Etch Masks. A different etch system consisting of hydrofluoric, nitric, and acetic acids (HF, HNO_3, and CH_3COOH) does not show crystal-orientation dependency.[90–92] However, in a certain composition, it etches heavily doped regions (p^+ or n^+) much faster than lightly doped ones. With a 1:3:8 volume ratio (HF:HNO_3:CH_3COOH) at room temperature, the etch rate in the heavily doped region ($>5 \times 10^{18}$ cm^{-3}) is between 50 and 200 μm/h, with a selectivity over lightly doped region ($<10^{17}$ cm^{-3}) being 150.[73] Since its dopant-dependent selectivity is the opposite of that of strong alkaline systems, the acidic system can be used as a complementary etchant to enhance the flexibility of creating chemically etched structures.

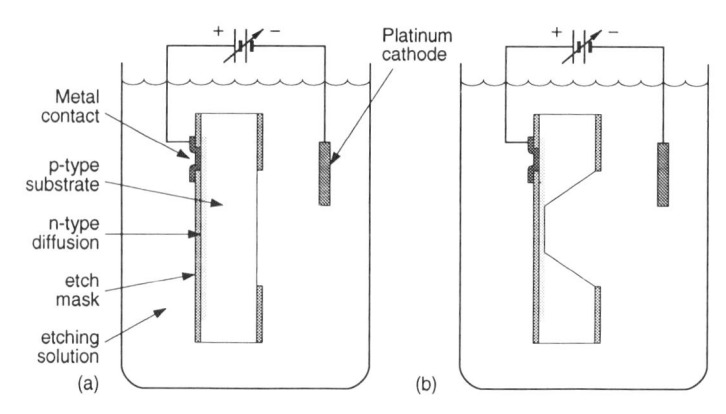

Fig. 27 The electrochemical etch-stop technique used to form a diaphragm.

However, the acidic system etches SiO_2 somewhat slowly (around 2 μm/h); hence a SiO_2 mask can be used only for brief etching times. For longer etch times, either Si_3N_4 or Au should be used. The etching behavior varies considerably with the mixture ratios. For example, a 1:2:1 volume ratio exhibits a much faster etch rate of up to 250 μm/h in all dopant concentrations.

Electrochemical Etch-Stops.[93-100] Electrochemical passivation techniques can be used as an alternative to heavy boron-doping to create diaphragms. In this technique, a small positive voltage of 0.5 to 0.6 V is applied to one side of a p–n junction of a wafer in an etch bath of a strong alkaline such as EDP or KOH (Fig. 27). The cathode, typically a platinum electrode, is immersed in the etchant. Either the p or n side of the junction can be protected from etching by attaching it to the anode, as long as the anodic potential is not high enough to forward bias the p–n junction when the p side is protected. The etching stops when the exposed side of the p–n junction is completely etched away, with an etch-rate drop equivalent to the selectivity over SiO_2. In fact, it has been suggested that the electrochemical passivation phenomenon is due to the formation of an SiO_2 passivation layer by anodic oxidation when the etchant reaches the junction.

The electrochemical etch-stop technique avoids the use of heavy dopants, which may cause mechanical stress problems within the finished diaphragm,[101-103] and precludes the fabrication of electronic devices within it.

2.4.3 Wafer Bonding

Wafer-bonding techniques are commonly used for sealing microsensors and in the construction of composite bulk-micromachined sensors. Anodic bonds, metallic seals, low-temperature glass bonding, and fusion bonding are the most common wafer-bonding techniques. Good reviews of these bonding techniques and other more exotic bonding methods are given in Refs. 24, 104.

Anodic Bonding. The technique of anodic or electrostatic bonding[105] was discovered in 1969. In this technique, the bond is accomplished between a conductive substrate and a sodium-rich glass substrate. Figure 28 shows a schematic of the anodic bond set-up. The conductive and glass substrates are placed in intimate contact. The two-wafer assembly is heated to a temperature in the range 350–450°C. This temperature is sufficient to make the sodium ions of the glass mobile. When a voltage in the order of 400–700 V is applied between the two substrates with the glass substrate as the anode, the sodium ions are

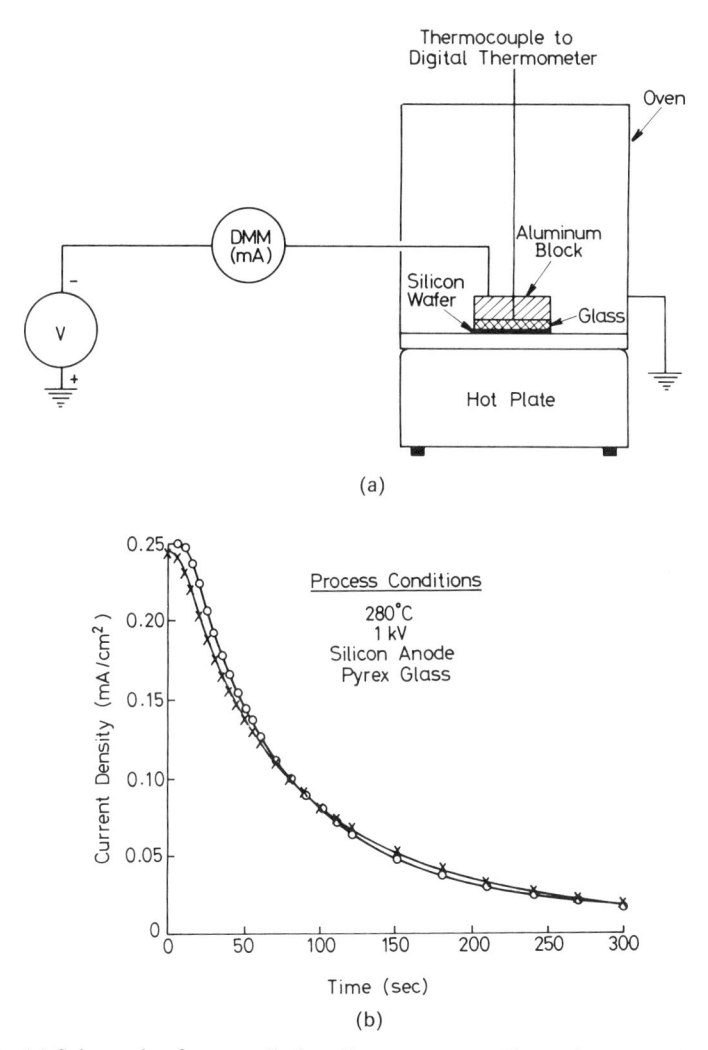

Fig. 28 (a) Schematic of an anodic bonding apparatus. The wafer assembly is pressed against a hot plate when the electric field is applied. (b) Typical current traces. The bond is complete when the current drops to about 10% of its peak value. (After Ref. 217)

depleted from the glass–substrate interface creating a shallow ion-depletion region about 1 μm thick[106–108] with high electric fields in the order of 7×10^6 V/m. The high field induces a large electrostatic pressure of several atmospheres that brings the substrate and glass[109] into intimate contact. A good bond is established in a matter of minutes. The actual mechanism responsible for the bond is still open to speculation. However, there is evidence in the silicon/glass bond suggesting that the formation of a thin layer of SiO_2 is responsible for the strength of the bond. For silicon/glass bonds, the resulting bond is hermetic with bond strength exceeding that of the substrates.[110,111] Because of the strong fields at the interface, anodic bonding of good quality is possible for substrates of poor interface planarity.

Since the bond occurs at temperatures higher than ambient, special attention is needed to avoid bimetallic warping and undesired stray stresses at the interface, which can lead to fracture. Ideally, the glass and substrate should have matched thermal expansion coefficients. For bonding to silicon substrates ($3.2 \times 10^{-6}/°C$), Corning glass 7740 offers the closest match. This glass can be sputtered on other substrate allowing the anodic bonding of various types of substrate. Silicon wafers can be anodically bonded using an intermediate layer of sputtered glass. The procedure yields stable, strong bonds and eliminates most stray stresses. Special care is necessary to assure the composition of the souttered film is the same as that of the sputtering target.

Low-Temperature Glass Bonding. In anodic bonding, the glass–substrate interface experiences high electric fields that can damage active devices. Thermal bonds are desirable when these fields can affect the performance of these devices. In low-temperature glass bonding, the bonding interface is coated with a thin layer of low-temperature glass. The wafers are then brought together under pressure and the whole assembly is heated to establish the bond. Many different types of glass have been used including phosphosilicate[104] and borosilicate[112–115] glasses, sputtered doped[116] glass layers, and spin-on glass slurries[104,117] and frits.[24] The glass for hermetic sealing can be divided into two types: vitreous and devitrifying. Vitreous glasses melt and flow at the bonding temperature when heated. Devitryifying glasses crystallize at the bond temperature, forming a stable film.

Sputtered layers of low-temperature glass, in general, do not make a good bond due to the planarity requirements. Phosphosilicate glass bonds, which occur at temperatures near 1000°C with excellent strength, have been reported.[104] Borosilicate glasses require a much lower bonding temperature (450°C) but they are sensitive to small amounts of contaminants. Sputtered sodium–lead glasses require even lower temperatures (150°C) but provide weaker bonds. In general, these materials require very flat surfaces to perform satisfactory bonds, and can be highly susceptible to surface contamination and roughness.

Glass frits are proprietary combinations of metal oxides in a solvent forming a paste. Under pressure, the frit forms a planarizing film that fills gaps between rough surfaces. The frit bonds the interface when cured by heating. Glass–frit bonding is a very popular bonding method because of the low temperature requirements (≈ 300–$600°C$) and its ability to bond non-flat surfaces. Thermal expansion coefficients of commercially available frits range from 2 to 5 times larger than that of silicon.

Low-temperature glass provides a weaker bond than electrostatic bonding. Careful processing is needed to avoid the formation of bubbles of trapped gas and voids in the interface.

Fusion Bonding. In fusion bonding, wafers are thermally "fused" when in contact at high temperatures in the absence of any intermediate adhesion layers. Fusion bonding of silicon[118] is commonly used for the fabrication of silicon-on-insulator devices and pressure sensors. Figure 29 shows a schematic of fusion-bonding apparatus. The wafers are thoroughly cleaned and placed in physical contact with each other on a quartz holder. If the surfaces are flat, the wafers stick together held by weak Van der Waal forces. The whole assembly

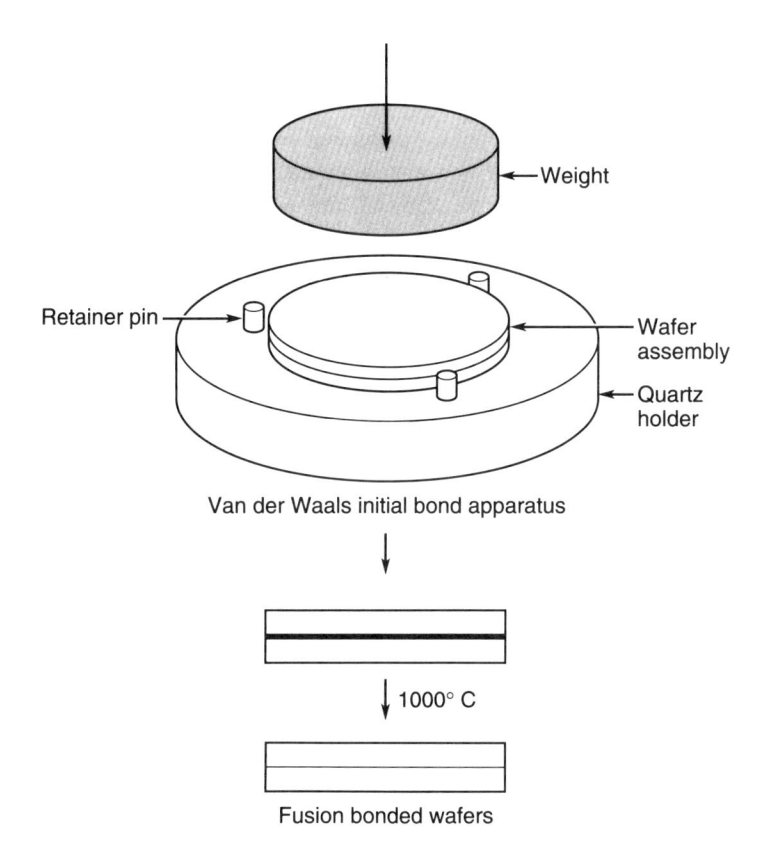

Fig. 29 Schematic of a fusion bonding apparatus. After pressing the wafers together, they are transferred to a high-temperature furnace to establish the bond.

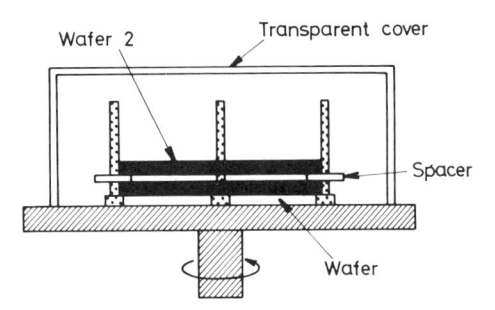

Fig. 30 Hermetic, miniature cleaning chamber for silicon fusion bonding experiments. The wafers are rinsed inside the clean chamber just before they are placed in contact, hence avoiding exposure to particles. (After Ref. 123)

is transferred to a high-temperature furnace to establish the final bond. For silicon wafer bonding, the quality of the bond is excellent with no visible interface[119] even in the presence of native oxide layers. Fusion bonding is an attractive method for applications where intermediate layers cause undesirable stresses. However, this method requires very high bonding temperatures, which do not allow the presence of active devices at the time of bonding. The weak nature of the initial bond makes the quality of the final bond particularly sensitive to the flatness and the cleanliness of the interface. Trapped gas bubbles,[120-122] which lead to partial bonds, can be eliminated by high-temperature annealing and extreme cleanliness. Organic contaminants have been attributed much of blame for the spotty nature of the bond. A number of researchers[123-125] have implemented closed systems (microcleanrooms) where the samples can be cleaned in an ultra-clean chamber before the bonding (Fig. 30).

Reactive Metal Bonding. In reactive metal bonding, a thin layer of metal is sandwiched between layers of silicon. When it is heated, the metal reacts with the silicon surfaces forming an alloy that, when cooled, bonds the two parts together. In eutectic alloys, the melting point is lower than the melting point of the pure metal or the silicon. Gold is a commonly used material for eutectic bonding of silicon. When it is heated, the gold layer diffuses into the silicon forming an eutectic alloy. For the Au/Si system the eutectic composition is 97.1% Au and 2.85% Si, thus the gold is the limiting factor. The temperature required for the formation of the alloy is 363°C. In order to obtain a good-quality bond, it is necessary to eliminate the native oxide from the silicon surface by coating the surfaces with Au. Reactive metal bonding is commonly used for bonding of sensors to packages, but has been used for wafer bonding of pressure sensors.

Organic Bonds. If the parts cannot withstand temperatures high enough for low-temperature glass bonding or anodic bonding, the wafers can be bonded using intermediate glue-like polymer layers. Wafers have been bonded using polyimides and epoxies[126] with moderate strengths. Most organic polymers will not adhere well to silicon substrates and require the use of organosilicon primers such as hexamethyldisilazane (HMDS) and γ-aminopropylsilane (APS). Polymer bonds are less stable than glass or metal bonds due to polymer aging. Furthermore, the plastic nature of the bond induces drifts of device characteristics; nevertheless reasonable-quality bonds are made at temperatures as low as 130°C using this technique.

2.5 SURFACE MICROMACHINING

Surface-micromachined sensors are constructed entirely from thin films. There are several differences and trade-offs between structures made from bulk and thin-film materials. Bulk-machined sensors are in general bigger. The physical dimensions of bulk-micromachined devices are determined by the propagation of patterns along crystal planes. In silicon, the dimensions on the opposite side of the wafer are enlarged by a distance of $\sqrt{2}t$ where t is the thickness of the wafer (Fig. 31). For 100 mm silicon wafers $t = 500\,\mu$m, yielding minimum die sizes 800 μm larger than the surface features. The die enlargement wastes precious silicon real-state and imposes a serious limitation on the device count per wafer.

Surface-micromachined devices do not have enlargement effects and promise order-of-magnitude improvements on device density. The increased throughput reduces the cost of individual sensors and makes the development of merged

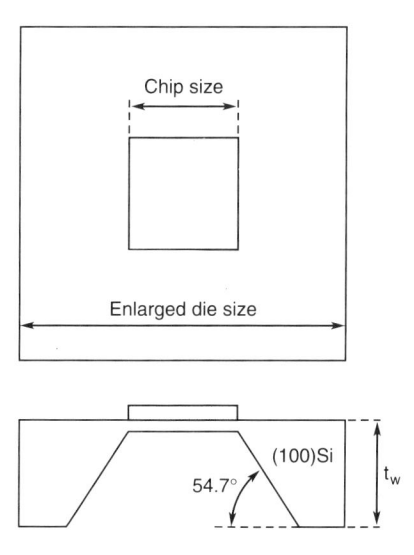

Fig. 31 Die size enlargement in bulk micromachining due to the propagation of the $\langle 111 \rangle$ crystal planes. The opening on the back-side is larger near the front.

integrated sensor/microelectronic systems economically feasible. Whereas typical dimensions for bulk-micromachined sensors are in the millimeter range, surface-micromachined devices are of micrometer dimensions. Surface micromachining permits the fabrication of structurally-complex sensors by stacking and patterning layers or "building blocks" of thin films while multilayered bulk devices are difficult to construct. Free standing and movable parts can be fabricated using sacrificial etching. Although surface micromachining offers increased throughput and complexity, there are important factors that have kept bulk micromachining the current technology.

Many sensors are required to work in environments that are hostile to microstructures. Surface-macromachined features are sensitive to environmental factors such as particulates, humidity, and cleanliness. Special packages are often required to protect the fragile micromechanical elements, which can raise the device cost substantially. Often, designers use merged bulk- and surface-micromachined processes to develop robust inexpensive sensors. Micropackaging techniques are a topic of current research.

Single-crystal materials used in bulk micromachining have well-defined properties in contrast to those of amorphous or polycrystalline thin films, hence yielding sensors with reproducible characteristics. The thin-film microstructure (polycrystalline or amorphous) and its ultimate mechanical properties often depend on specific deposition and post-treatment conditions. A tight control over these parameters is necessary for device reproducibility. Many thin film materials used in surface micromachining often have internal residual stresses that can affect the mechanical integrity of the finished device. Careful material selection is a requirement to avoid device failure. Surface-micromachined sensors made from single-crystal silicon offer the most reproducible characteristics. However, some thin-film devices are starting to be used in commercial applications. The accelerometer chip of Fig. 1 was constructed with surface-micromachining techniques. The ultimate success of surface-micromachined devices will be determined by the availability of thin films of high quality and good stability. In the following sections we discuss some of the common specific materials and processes used in surface micromachining of silicon.

2.5.1 Thin-Film Materials

A variety of thin-film materials is available for the construction of mechanical sensors. High-quality insulators such as silicon dioxide or silicon nitride, conductors such as aluminum, and semiconductors such as silicon are a few examples. In general, CVD films have the lowest stray stress and best reproducibility, hence they are the natural choice when available. Other materials used in microsensors include metals, piezoelectrics, and pyroelectrics.[127-129]

Polycrystalline and Amorphous Silicon. Thin films of silicon can be grown on top of insulators. The structure of the films ranges from randomly oriented crystallites (or polycrystalline) to completely amorphous, depending on the kinetics of the deposition process. Polycrystalline silicon is widely used in the semiconductor industry for the fabrication of self-aligned MOS transistors.[130] Amorphous silicon is used in the fabrication of thin-film transistors in LCD displays. Electrically, polycrystalline silicon has properties similar to bulk silicon for sensing applications. Its piezoresistive coefficient is high;[131] hence, it is particularly attractive for stress-measurement elements that are isolated[132] from the substrate. Dielectrically-isolated polysilicon films are used for high-temperature sensing applications.

Polysilicon films can be deposited by a variety of means. Physical methods of deposition such as evaporation[133] and sputtering yield films with poor mechanical properties that are, in general, amorphous. Polycrystalline films can be formed by subsequent annealing. Chemical methods such as atmospheric CVD[134] and LPCVD yield polycrystalline samples at high temperatures and amorphous samples at low temperatures. The critical parameters that determine the microstructure are the mass transport and nucleation rates.[1] Polycrystalline silicon can be deposited by LPCVD using the silane pyrolysis reaction

$$SiH_4(g) \xrightarrow{heat} Si(s) + 2H_2(g) \qquad (5)$$

at pressures of 300–500 mT and temperatures from 500–700°C. Typical deposition rates range between 3–15 nm/min. These films can be deposited undoped or doped *in situ* with *p*- or *n*-type impurities by introducing PH_3 and B_2H_6 into the furnace. Thin films of silicon can be deposited by plasma CVD at a lower temperature (300°C). In general the plasma films[135] are amorphous[136] and can be changed to polycrystalline by annealing.

The residual stress of LPCVD polycrystalline silicon can be controlled tightly.[137–139] Figure 32 shows the stress of undoped polysilicon measured as a function of deposition pressure and temperature.[27] Two transitions are observed—from compressive to tensile and back to compressive. VLSI polycrystalline silicon deposited at high temperatures (650°C) is in a compressive state[140] and is unsuitable for micromechanical sensors. Polysilicon grown at 605°C has nearly zero stress.[137] The mechanism of low-stress growth is believed to derive from an amorphous state that recrystallizes in a very reproducible way[1] after annealing. In practice, low-stress tensile films are grown between 600–610°C with a successive relaxation anneal. Impurity dopants affect the microstructure of polysilicon and reduce the stress of the film.[141,142] Residual strains less than 30 ppm have been reported in doped films after annealing.[143]

Despite the electrical versatility of this material, its mechanical properties are not fully understood. Several researchers have reported Young's moduli ranging from 140 to 210 GPa depending on the polycrystalline microstructure and orientation.[32] These factors are highly dependent on deposition conditions and thermal history. Despite this drawback, many sensors have been implemented with polysilicon including pressure[32,144] and acceleration

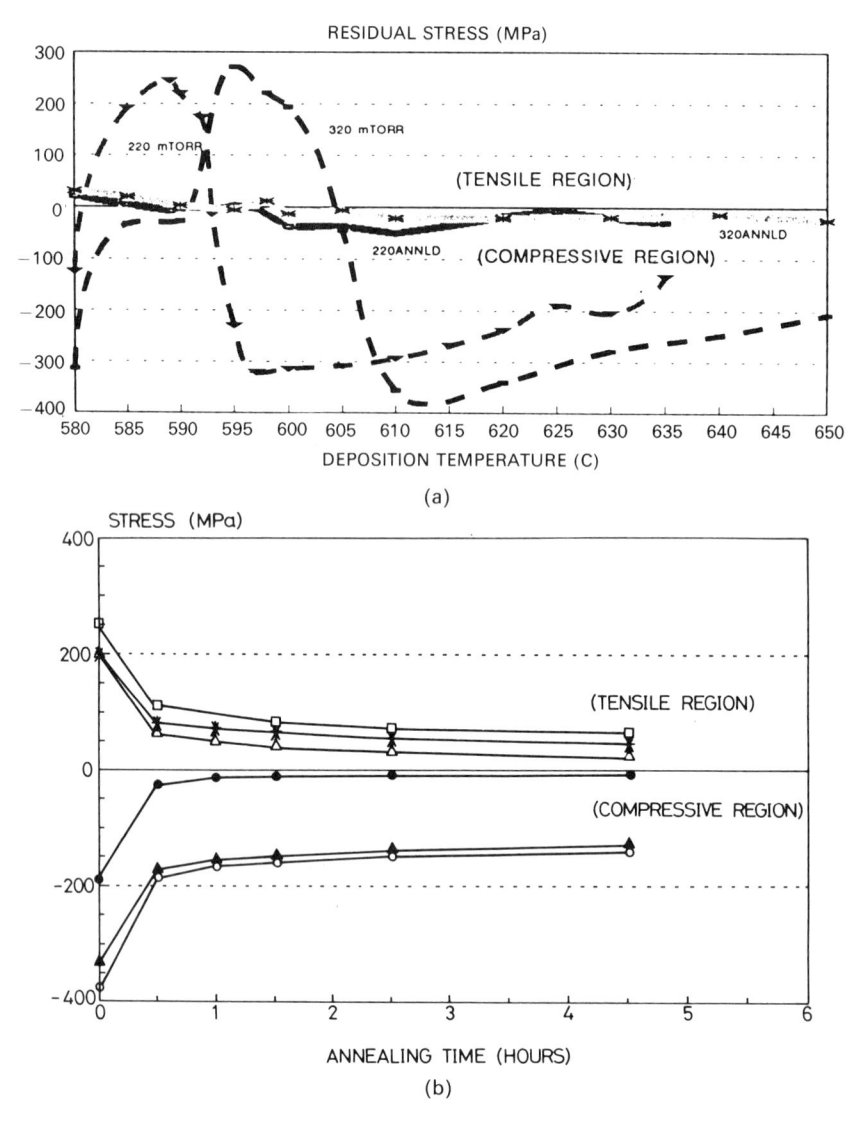

Fig. 32 (a) Residual stress in LPCVD polycrystalline silicon as grown at 220 and 320 mT. (b) After annealing at 900°C. (After Ref. 27)

sensors.[145] PECVD silicon films are attractive because of their low deposition temperature that allows direct integration with CMOS wafers.[146] PECVD films can be grown with both compressive or tensile stresses depending on the deposition temperature.[136]

Single-Crystal Silicon. Single-crystal silicon has a well-defined structure, hence its mechanical properties are very reproducible. For microsensors, high-quality microstructures are fabricated from single-crystal thin films. Unlike polycrystalline and amorphous material, single-crystal films cannot be chemically grown on non-crystalline substrates. There are a few techniques available to form these films on any substrates.

The earliest techniques[147-149] involved wafer bonding and back etching (Fig. 33). In this procedure, the thickness of the film is defined by a p^+ area diffused into the substrate. Next, the wafer is patterned and anodically bonded to another substrate. The wafer containing the diffused layer is etched from the back leaving the heavily doped layer intact. Layers of crystalline films as thick as 25 μm can be fabricated using this process. Contacts to the diffused areas are made by a slight overlap in the metallization. Although this technique is simple and effective, there are difficulties associated with the diffusion. The high level of impurities is known to affect the mechanical stress[14] in the material as well as its mechanical properties. Undoped and n-type films cannot be made using this technique.

Single-crystal films with low doping can be formed using silicon-on-insulator (SOI) techniques[150-154] and epitaxial growth.[155] In back-etched SOI (BESOI) (Fig. 34), a highly doped buried layer is formed on a wafer by ion implantation. The wafer is next bonded to a substrate wafer coated with an insulator such as silicon dioxide. The top wafer is back etched, stopping at the highly implanted region, which is subsequently removed by a selective etch. Thin layers of silicon 0.1 μm thick are formed by this process. The layer thickness is increased by subsequent epitaxial growth. Thin layers of single-crystal silicon can also be formed by bonding and grinding techniques.[156-158] Other SOI techniques such as high-energy oxide implantation are feasible but prohibitively expensive. Other less common techniques include merged epitaxial lateral overgrowth (MELO),[159,160] which requires the presence of a seed surface connected to the underlying substrate. Compared with the deposited thin-film approaches, the bonding techniques are susceptible to problems due to surface cleanliness and flatness as well as proper alignment. The ultimate factor that will determine the future of this technology is the availability of low-cost, high-quality SOI substrates.

Silicon Nitride. Thin-film silicon nitride is an insulating material commonly used in the sensor industry as a mask and a high-temperature protective film. Silicon nitride can withstand strong etching solutions such as concentrated HF and KOH, and it is an excellent diffusion mask to prevent impurity diffusion and ionic contamination. Its high mechanical strength ($E \approx 73$ GPa) makes this film suitable for friction and dust barriers. The nitride films used in semiconductor sensors are amorphous but crystalline forms do exist. Stoichiometric (Si_3N_4) nitride has an index of refraction of 2.1 and a resistivity greater than 10^{15} Ω-cm. Silicon nitride is also a good thermal insulator compared to polysilicon.

Silicon nitride films can be sputtered, or deposited by plasma and thermal CVD. LPCVD silicon nitride films are the highest quality films available. The

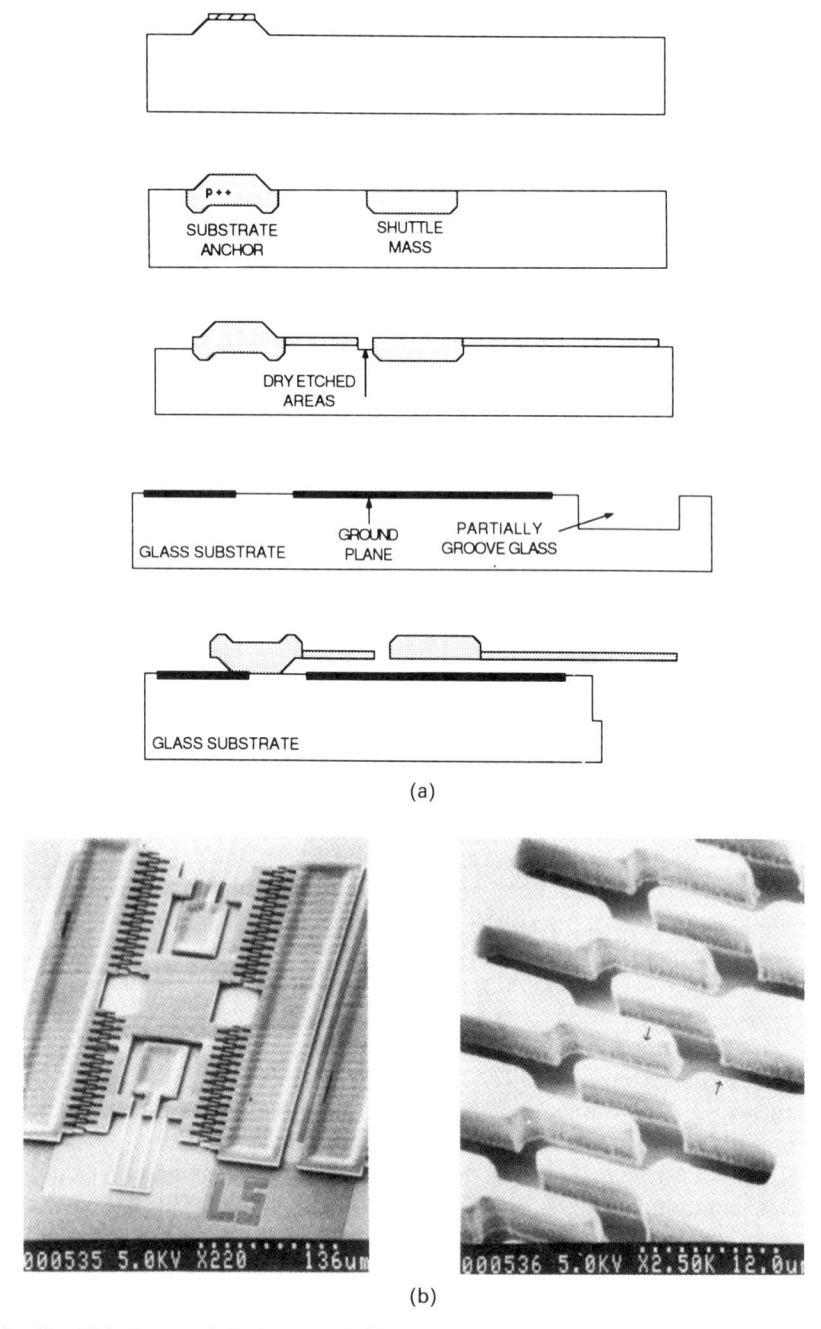

Fig. 33 Thin layers of single-crystal silicon made by bonding and back etching. (a) The process involves the selective doping of silicon with high doses of boron, anodic bonding and release etch. (b) The photographs show micromechanical actuators fabricated with this process. (After Ref. 149)

2.5 SURFACE MICROMACHINING 61

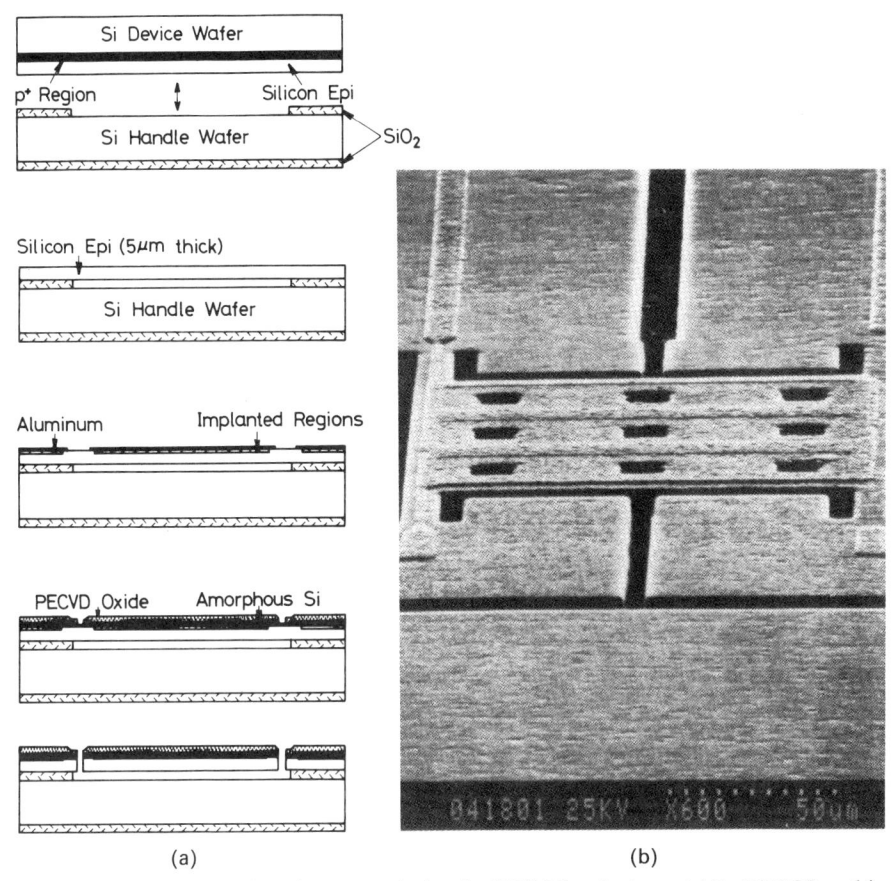

Fig. 34 Crystalline silicon layers made by the BESOI technique. (a) In BESOI, a thin layer of epi is grown on top of an etch-stop layer. After bonding to another substrate the wafer is back-etched until the etch-stop is removed. The etch stop is next removed leaving a low-doped layer of single-crystal silicon behind. (b) The photograph shows an example shear force sensor fabricated using this process. (After Ref. 155)

deposition takes place in a furnace with dichlorosilane ($SiCl_2H_2$) and ammonia (NH_3), which react as follows

$$3SiH_2Cl_2(g) + 4NH_3(g) \xrightarrow{heat} Si_3N_4(s) + 6HCl(g) + 6H_2(g). \tag{6}$$

The reaction takes place at a few mT (300–500 mT) and at temperatures between 700–900°C. Stochiometric nitride films require a gas ratio of $SiH_2Cl_2:NH_3$ of 1:3 to 1:4. These films have a large tensile stress (1–2 GPa); hence, only a few hundred nanometers can be grown without excessive warpage and cracking. Silicon-rich[161] nitride films are commonly used in semiconductor sensors. The stress of these films can be tightly controlled by adjusting the ratio of

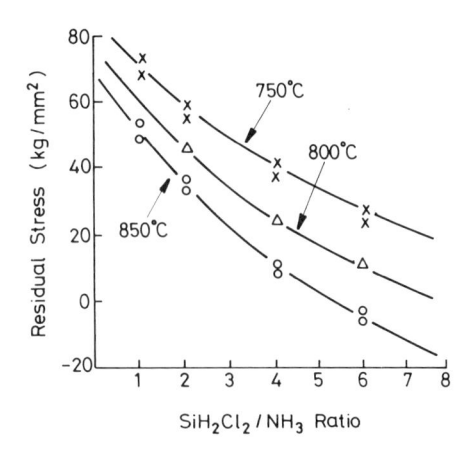

Fig. 35 Residual stress of silicon-rich silicon nitride for various gas ratios. (After Ref. 161)

dichlorisilane to ammonia. Nearly-zero-stress films are grown at 835°C and ratios of 4:1, allowing for the deposition of μm-thick films. If the ratio is further increased, compressive films are grown. Figure 35 shows the stress of the film as a function of the gas ratio and temperature. The stochiometry of the low-stress film is approximately $Si_{1.0}N_{1.1}$. Silicon-rich films exhibit a higher index of refraction (typically 2.4–2.5) and a much higher etch resistance in HF than stoichiometric films. However, the film resistivity is affected adversely by the excess silicon.[162]

Plasma-deposited films are formed at low temperatures (200–350°C) by reacting silane and ammonia or silane and nitrogen.[1] PECVD films tend to be compressive with residual stresses of 100–800 MPa. The film stress is compressive or tensile, controlled by the deposition frequency of the plasma RF source[163] (Fig. 36). PECVD films are structurally weaker than LPCVD films and cannot be used as masks for strong etchants due to the presence of pinholes and point defects. PECVD films have trapped hydrogen with varying $Si_xN_yH_z$ compositions. Silicon nitride films can be combined with oxygen during deposition to form oxynitride film $Si_xO_yN_z$. The tensile stress of the nitride is compensated by the compressive stress of the oxide, resulting in a low-stress film. Oxynitride films are attacked by both silicon dioxide and nitride etchants.

Silicon Oxide. Silicon combines chemically with oxygen to form silicon dioxide (SiO_2). Thin-film silicon dioxide is amorphous and has a lower density than silicon. Silicon dioxide is an excellent electrical and thermal insulator. Its resistivity is high (10^{12} Ω-cm), and its thermal conductivity is low (1.4×10^{-2} W/cm°C); hence, this material is useful for thermal detectors (bolometers), gas, and flow sensors.

Silicon dioxide can be grown or deposited by a number of means. Exposure of silicon surfaces at high temperatures in an oxygen environment grows

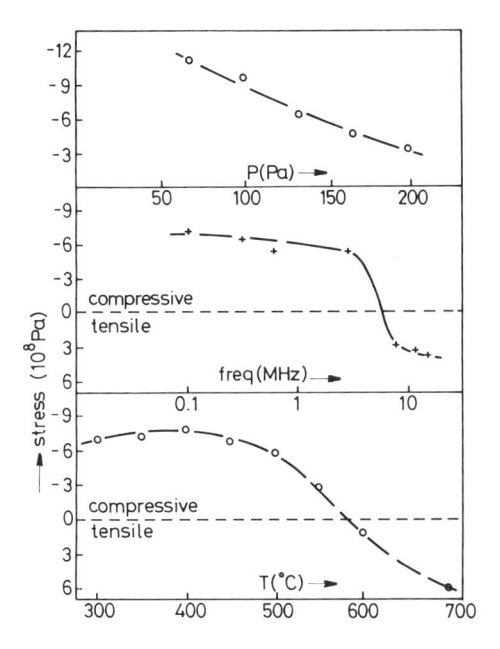

Fig. 36 Residual stress of plasma-enhanced CVD silicon nitride under various growth conditions. (After Ref. 163)

high-quality silicon dioxide films. This reactive growth process is known as oxidation.[3] During reactive growth, the oxide film expands in volume by about 45%, thus it creates a compressive stress in the silicon underneath. Silicon dioxide films can be deposited by LPCVD using the pyrolitic reaction

$$\mathrm{SiH_4(g) + O_2(g) \xrightarrow{heat} SiO_2(s) + 2H_2(g)} \tag{7}$$

at a pressure of 300–500 mT, and a temperature of 450°C. Typical deposition rates are 20–40 nm/min. These films can be *in situ* doped with phosphorus or boron impurities producing phosphosilicate and borosilicate glasses (PSG and BSG). Oxide films can be deposited by a plasma from the reaction

$$\mathrm{SiH_4(g) + 2N_2O(g) \xrightarrow{plasma} SiO_2(s) + 2N_2(g) + 2H_2(g)} \tag{8}$$

Silicon oxide films can also be sputtered and spin-cast (spin-on glass). The highest-quality films are reactively grown and the worst films are spin-cast. In general, the quality of the film increases with growth temperature.

The stress of oxide films is compressive (≈ 1 GPa) and cannot be annealed out; hence, silicon dioxide is not an adequate structural film for mechanical sensors. However, because this film is removed easily from the substrate, it is commonly used as a sacrificial layer (see Section 2.5.3). The mechanical properties of oxides can be affected by the introduction of dopants. Adding high doses of boron or phosphorus makes the oxide flow at relatively low temperatures. This feature is useful for planarization purposes.

Organic Films. A variety of soft organic films are used in semiconductor sensors. The most common organic films are polyimides. Polyimide films are spin-cast and evaporated onto substrates.[164-166] This polymer (as with most plastics) swells when it is exposed to a high-humidity environment. A number of polyimide-based devices have been used for humidity sensors. The mechanical properties of polyimides have been measured.[17] This material is normally in a tensile state due to shrinkage during curing. Polyimides are also used as sacrificial layers, which are removed by oxygen plasma.

Parylenes (polymerized *p*-xylylene) are used for passivation and contamination barriers for sensors that are exposed to harsh biological environments. Parylene deposits conformally from the vapor phase at low temperatures. Other polymers used for passivation are hexamethyldisilazane,[167] polystyrene, tetrafluoroethylene, and latex films. Many of these polymers are deposited by plasma polymerization, spin-casting, and evaporation. A good survey of the plastic films is given in Refs. 168 and 169. Active-sensing polymers such as pizeoelectric polyvinyldifluoride in PVDF[128] are used in pyroelectric and piezoelectric sensors. A common problem of polymer materials is their poor adhesion to the substrates. Adhesion promoters such as APS are used on the sample prior to the polymer deposition to improve its adhesion.

2.5.2 Thin-Film Etching for Surface Micromachining

Etching Silicon Thin Films. Polysilicon thin films can be etched isotropically with the common wet-chemistry etchants and masking techniques used for bulk micromachining, or with dry plasma etching in a fluorine-based chemistry. In a fluorine-based etcher, CF_4 or SF_6 gases are commonly used as a source for F atoms in the plasma. The F atoms are so reactive that they react with silicon without ion assistance, and, thus, the etching is isotropic.[8] Oxygen is often added as a 10 to 20% level to suppress the formation of polymers, which contaminate the sample and deplete F atoms. SiO_2 and Si_3N_4 films are commonly used as etch masks, with etch selectivities higher than 15. The etch rate is dependent on many parameters including etch temperature, pressure, flow rates of the gases, RF power, and electrode separation, and is, typically, between 50 and 300 nm/min.

An important consideration for F-based etching is called the loading effect,[6,8,170] which manifests itself as inconsistent etch rates at different locations on the same wafer, as well as between wafers with different patterns. F atoms are depleted faster at locations with larger exposed areas of etchable materials. Since chemical reaction is limited to the first order by the availability of reactive atoms, those locations exhibit a lower etch rate. In a typical plasma etcher with parallel-plate electrodes, the center of the wafer often exhibits a 5 to 25% lower etch rate than the edge, resulting in a bull's-eye pattern.[6] Also, in a multiple-wafer etcher, the etch rate decreases as more wafers are etched simultaneously.[8] The loading effect implies that one must allow for substantial over-etching in order to clear the whole wafer, which necessarily results in excessive undercutting near the perimeter of the wafer. Lowering the gas-flow rates and raising the pressure help to alleviate the intra-wafer nonuniformity, in addition to increasing the etch rate.[6] Certain single-wafer plasma etchers are designed to counter radial nonuniformity by feeding gases through a shower-head-like top electrode, with more openings at and around the center.

In applications where anisotropic etching of polysilicon is required, chlorine-based plasma etching is commonly employed. Cl atoms are not as reactive as F atoms, and, under most etching conditions, reaction with silicon must be initiated by ion bombardment. Therefore, a high degree of anisotropy can be obtained. In addition, the concentration of reactive Cl atoms is determined primarily by the rate that Cl atoms recombine to form nonreactive molecules and not as much on depletion due to reaction. As a result, compared to F-based etching, the loading effect is less prominent. Popular masking materials include SiO_2 and/or photoresist,[171] as well as refractory metals such as nickel,[58] chromium,[172] or a Ni–Cr combination.[173] The presence of photoresist on top of a SiO_2 mask tends to enhance protective carbon-based polymer formation on the sidewalls of the etched structure, and thus minimizes undercutting. The selectivities of Cl-based plasma on silicon over the refractory metals are usually quite good, ranging from 10 to 15. Hence thin metal layers patterned by a lift-off technique are excellent masks for Cl plasma etching.

A common source for Cl atoms is CCl_4, with oxygen optionally added at up to a 10% level to suppress polymer formation and enhance the etch rate.[8] However, excess O_2 may slow the reaction or even stop etching completely, mainly due to plasma oxidation of polysilicon to form an etch-resistant silicon oxide film.[174] If photoresist is used as an etch mask, the presence of O_2 accelerates mask erosion. Another common additive is He, which has a high thermal conductivity and assists heat dissipation to stabilize the reaction. He can be added at up to a 50% level, Pure Cl_2 gas is also used for etching, mixed with $SiCl_4$[175] or BCl_3[173] to optimize sidewall profiles and to etch deep trenches into single-crystal silicon.

In addition to the influence of etching parameters such as pressure, temperature, and RF power, the etch rate in Cl-based plasma is affected by the doping level in silicon. A high doping concentration of phosphorus (10^{20} cm^{-3}) can raise the etch rate by as much as 10 times compared to undoped silicon.[8] Since Cl atoms etch aluminum as well, the etch chamber for Cl-based plasma must be anodized. Note that native oxide on the polysilicon film may hinder the onset of etching reaction, which can be circumvented by briefly dipping the wafer in a diluted HF solution prior to plasma etching or by adding a brief plasma-etch step with CF_4 at the beginning.[176]

The sidewall of the etched polysilicon structure tends to exhibit a sloped profile, mainly due to the gradual, isotropic erosion of the etch mask during the etch period. This phenomenon is especially prominent with photoresist as the masking material. It is possible to optimize the etching by mixing Cl and F plasma to obtain a desired sidewall profile with controlled lateral undercutting.[177] Control over of the sidewall angle has been reported[58] with a mixture of SF_6 and CCl_2F_2.

Bromine-based plasma etching is ion-assisted and exhibits etching characteristics similar to Cl-based plasma etching.[9,176] Br is generally the least reactive of the three halogens, implying that ion bombardment is almost certainly needed to initiate plasma etching. The degree of anisotropy achievable with Br-based etching is potentially superior to chlorine plasma. The source for reactive Br atoms is usually HBr gas, with or without magnetic-field confinement. Besides being more anisotropic than chlorine plasma, bromine plasma has much higher selectivity on silicon over oxide or photoresist than chlorine plasma, making it attractive for relatively-deep silicon etching. Achievable selectivities for silicon over photoresist and SiO_2 have been reported to be 60 and 100, respectively. As a result, there is almost no measurable linewidth degradation even after a 200% over-etch.[176] However, because of the toxic etch products from bromine compounds, the operation and maintenance of Br-based plasma requires substantial safety and health precautions.

Etching Silicon Oxide. Isotropic etching of SiO_2 is usually done in wet chemistry with diluted HF or buffered HF (BHF). BHF consists of an aqueous solution of HF with the addition of NH_4F to control the pH and to slow the depletion of F^- ions. Therefore, it is superior to HF in etch uniformity and etch-rate consistency. Since photoresist is commonly used as the masking material for an oxide etch, it is important that both the photoresist and the resist-oxide interface remain intact in the etchant.[4] There is a tendency for unbuffered HF to degrade the integrity of the interface, resulting in photoresist peeling in some extreme cases. On the other hand, if a quick or blanket etch is needed, concentrated HF, with an etch rate of up to 50 $\mu m/min$, may be used. Like polysilicon, the etch rate of SiO_2 is also dependent on the dopant concentration and film quality. Phosphosilicate glass (low-temperature deposited SiO_2 with heavy phosphorus content) that contains 6 mole% of P_2O_5 etches more than 10 times faster than thermally grown oxide,[178] while oxide densified at high temperature shows a decreased etch rate. HF-based etchants are highly selective over silicon, Si_3N_4, and photoresist, showing almost no sign of attack, However, when photoresist is used for masking, good adhesion to the underlying oxide layer must be ensured to prevent photoresist peeling in the etch solution. Since SiO_2 is hydrophilic, a dehydration step followed by coating with a surfactant layer should be performed before spinning-on the photoresist. If exposed aluminum is on the wafer, dihydroxyalcohol or glycerol may be added to slow the attack on the metal.[178]

Ansiotropic etching of SiO_2 is accomplished by ion-assisted plasma etching. With a 1:1 combination of C_2F_6 and CHF_3, the resulting sidewalls are acceptably vertical. Etching selectivity of oxide over silicon is also quite good, but selectivity over nitride is very poor. Thus, if the material underlying the oxide film is silicon nitride, a practical procedure is to stop plasma etching before the nitride layer is exposed, and switch to wet chemical etching with either HF or BHF to complete the etch.

Etching Silicon Nitride. A common wet etchant for silicon nitride is heated H_3PO_4 at 140–200°C.[178] Like silicon oxide, the etch rate for silicon nitride is a function of the film quality and composition, ranging between 50–200 Å/min.[178] For example, the presence of oxygen in the nitride film (silicon oxynitride) slows the etch to a variable degree.[180] A good etch mask for a H_3PO_4 etch is CVD SiO_2, with selectivity of 10:1 or better. The etch also stops at exposed silicon. Stochiometric Si_3N_4 may also be etched with concentrated HF. However, oxide cannot be used as mask, and the etch rate drops drastically for silicon-rich nitride (Si_xN_y).

Anisotropic etching of Si_3N_4 can be accomplished with the same plasma formulation as SiO_2 with good selectivity over silicon but without selectivity over oxide. It is possible to achieve selectivity over SiO_2 using wet chemistry and sacrificing anisotropy. By using isotropic F-based plasma etching, such as those with SF_6 as a gas source, some selectivity may be achieved.[8]

2.5.3 Sacrificial Etching

In suspended sensor elements, some areas of their thin-film structural layers are floating. Nevertheless, the deposition processes require that the deposited film be anchored to a solid surface everywhere it grows. Suspended structures are made by forming the structural material on a solid sacrificial layer, which serves as a spacer and a temporary anchor surface for the deposition of the sensor structural material. The sacrificial layer is completely removed after all the structural depositions are completed, as shown in Fig. 37, freeing the element.[73,181,182]

Figure 38 illustrates how sacrificial etching techniques can be used to create an electrostatic micromotor with well-defined, sub-micron tolerance between the rotor and the center hub.[183–185] Surface-micromachined pressure and acceleration sensors are fabricated by this technique. A wet release is a sacrificial etch that takes place in a liquid solution. Conversely, in dry releases the sacrificial layer is removed using a reactive gas.

Sacrificial Materials. An adequate sacrificial material must satisfy stringent requirements. The film thickness must be grown within acceptable tolerances. Nonuniform depositions lead to suspended structures with undesirable curvature or roughness. This parameter is especially important when the spacer gap is small ($\leqslant 0.5$ μm). In the release, the sacrificial layer is removed entirely. The selectivity of the sacrificial etch and its etch rate must be very high so that the rest of the structure is not significantly attacked.[186] In sacrificial etching,

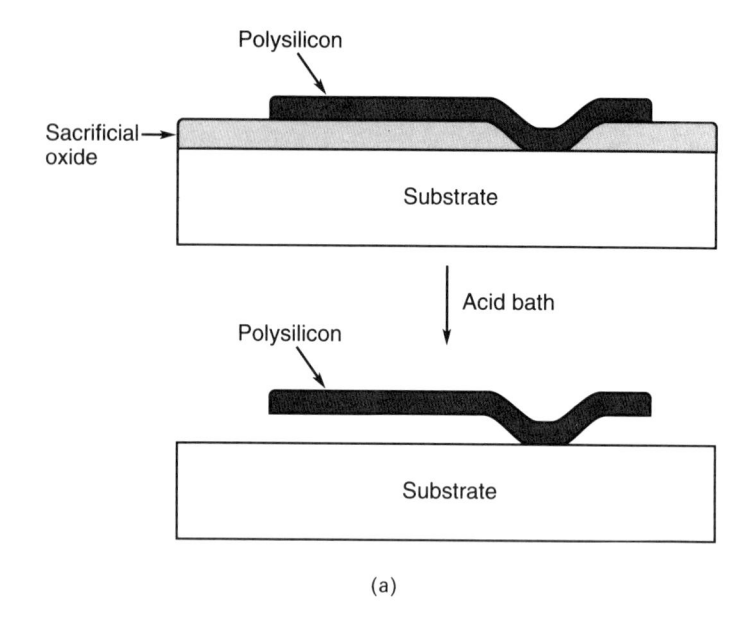

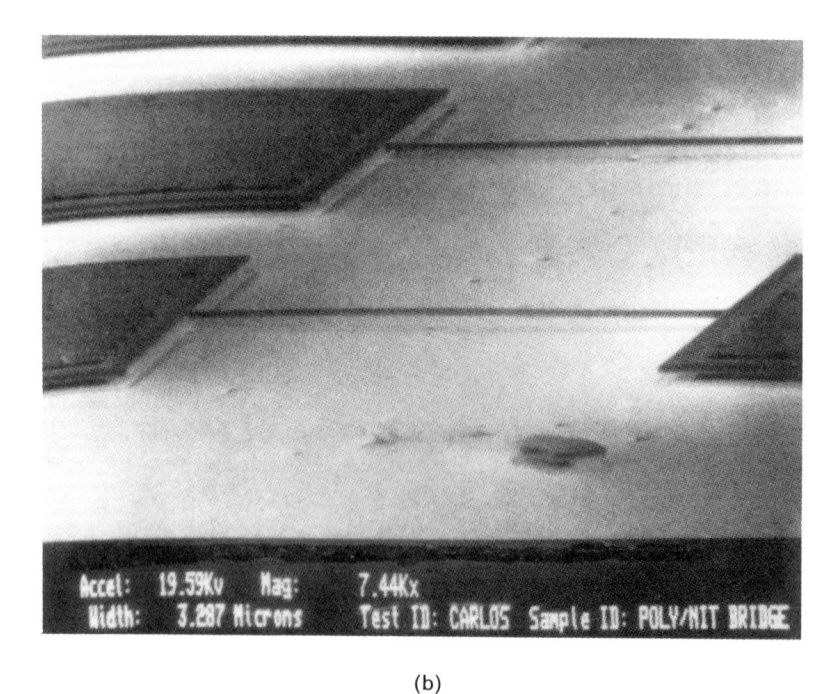

Fig. 37 (a) Example of a sacrificial etch-release for a cantilever beam. (b) Photograph of a released microbridge $200 \times 3 \times 3$ μm^3. (After Ref. 40)

2.5 SURFACE MICROMACHINING   69

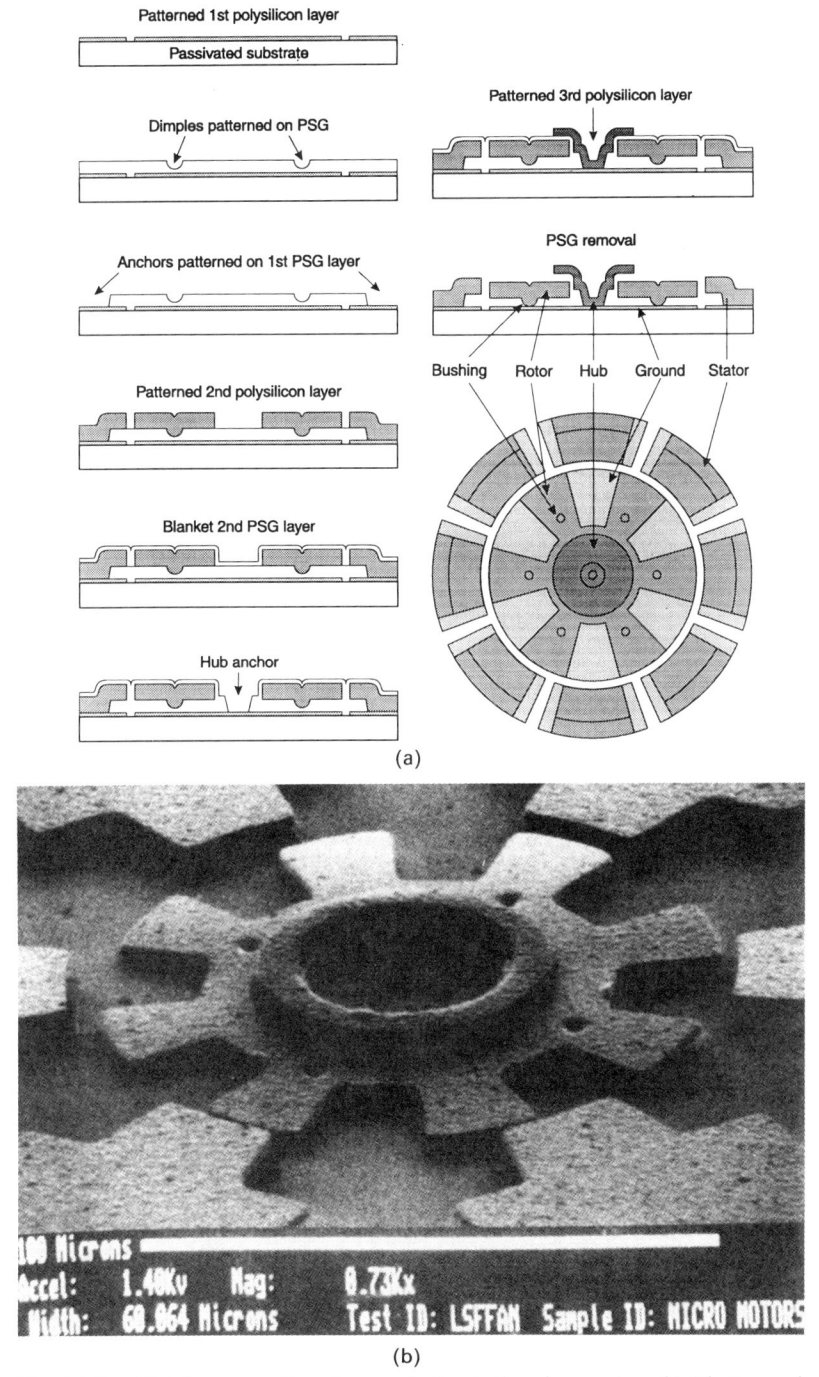

Fig. 38 (a) Sacrificial process flow for an electrostatic micromotor. (b) Photograph of micromotor. (After Ref. 183)

it is not uncommon to etch through narrow channels several hundred microns long, requiring long etch times.

There are few etching systems and materials that can meet these requirements. In silicon sensors, a good candidate is deposited low-temperature oxide (LTO) or PSG, which can be conformally deposited over a moderately complex surface topography. LTO or PSG can be readily removed by concentrated 49% HF without affecting the structural polysilicon.[183,187] The selectivity of the etch for the LTO/polysilicon system is approximately $S \approx 10^5$. The selectivity for the LTO/nitride system is much lower ($S \approx 10^2$). Silicon-rich nitride films are attacked at a much lower rate ($S \approx 10^3$) hence, they are preferred for HF-released microstructures.

Organic films such as polyimide and photoresist are also used as sacrificial materials because they are easily removed by oxygen plasmas. These films limit the deposition temperature of materials deposited subsequently, severely restricting the choice of structural material.

Design Considerations. The sacrificial layer is usually removed by opening a hole or channel to the surface in some nonessential sensor area. Both vertical and horizontal channels are used (Fig. 39). The etch solution penetrates through the openings slowly removing the sacrificial layer. The opening should be large enough to allow the etch to proceed without requiring an unduly-long etch process. In long and narrow channels, diffusion effects will limit the etch rate considerably.[53] Multiple openings cut the etch time dramatically and should be used whenever possible.

If a wet etch is selected to remove the sacrificial material, or if any wet processes are required subsequent to releasing the structure, the effects of capillary forces must be considered. Liquid surface tension can pull a flexible structure towards an adjacent surface (the substrate) while the liquid is being removed. After contact, the microstructure can adhere to the surface (pinning) even after the liquid is completely driven off.[188,189] The exact mechanism for the pinning is not yet understood.

Pinning may be eliminated by several methods. One method uses a dry-etching release avoiding any subsequent wet processes.[190] This technique is restricted to isotropic dry etches with high selectivities and fast diffusing species, because the plasma does not penetrate very far into undercuts of the sacrificial layer. Ozone-based organic etches are excellent for this purpose. If wet etching cannot be avoided, the capillary force can be eliminated by freeze-drying and supercritical drying techniques. In freeze-drying, the rinse liquid is frozen in place, and sublimated in a vacuum chamber.[172,191] In critical-point drying,[192] the liquid is removed at its critical-point pressure and temperature (where its surface tension is zero).

Pinning is also avoided by minimizing the contact surface. A simple release method uses an intermediate sacrificial material which holds surfaces apart during the wet-etching step of the original sacrificial layer. The intermediate material is then plasma etched to release the structures.[193] Tiny dimples, which

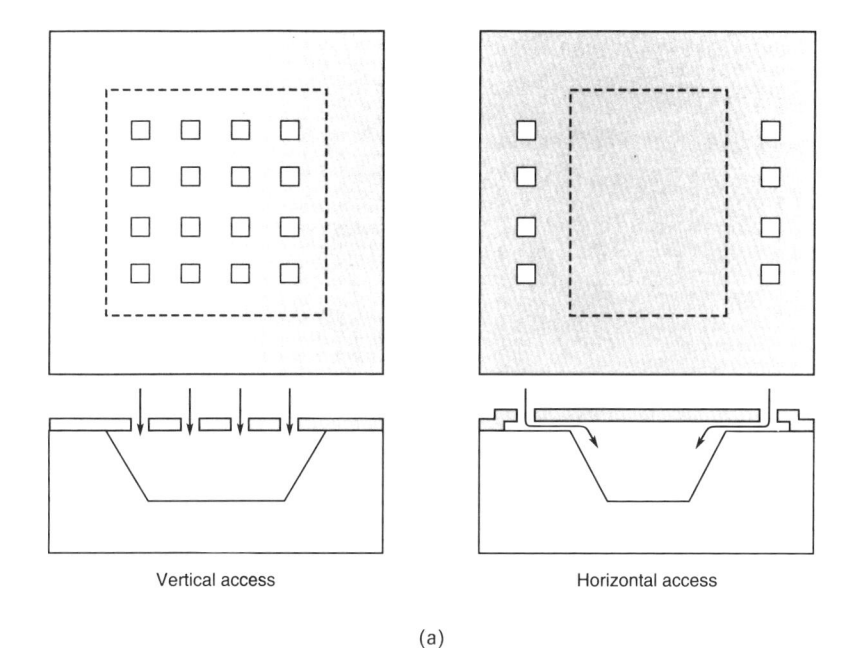

Fig. 39 (a) Vertical and horizontal access channels used in sacrificial etching of cavities. (b) The horizontal channel can be very shallow (0.1 μm). (After Ref. 196)

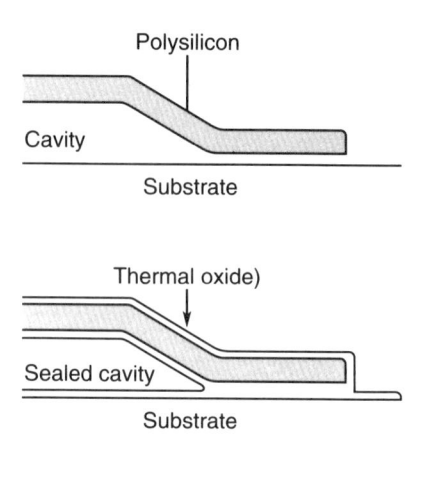

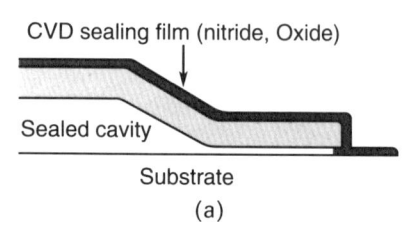

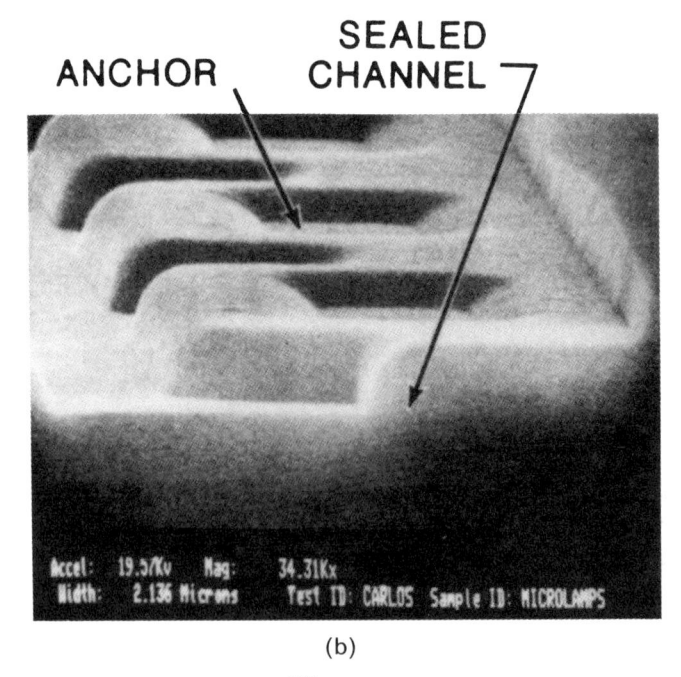

Fig. 40 (a) Examples of reactive seals[194] and shadow plugs. (b) Lateral channel sealed with silicon nitride. (c) Basic method of shadow plugging. (After Ref. 195)

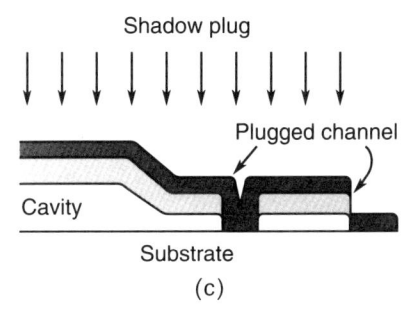

```
                    Shadow plug
            ↓ ↓ ↓ ↓ ↓ ↓ ↓ ↓ ↓ ↓
                           Plugged channel
    ┌─────────┐         ┌──┐
    │ Cavity  └─────────┘  └──────
    ├────────────────────────────
                Substrate
                  (c)
```

Fig. 40 *continued*

act as stand-off bumps to minimize the contact area when surfaces are pulled together,[187] are often constructed.

2.5.4 Sealing

Sacrificial-etching techniques form cavities that are open to the surface through a small opening. The cavity can be hermetically sealed by plugging the opening. Sealed cavities are used in surface-micromachined pressure sensors, resonators, and infrared emitters. There are a few sealing techniques available.[194] The earliest sealing technique[32] used the thermal reactive growth of silicon oxide on the interior of a polysilicon cavity. The enlarged volume of the oxide on the cavity walls eventually seals the opening. The remaining oxygen in the cavity is consumed by the reaction creating a vacuum. In order to establish a satisfactory seal, the permeability of air through the seal must be low. For thin layers of glass, this parameter is relatively high; hence, the seal is coated with a diffusion barrier such as silicon nitride to prevent air leakage. High-quality vacuum seals with internal pressures as low as 10^{-6} Torr have been reported. Figure 40a illustrates this method.

The cavity can also be sealed by plugging the hole with a deposited material. In this method, a thin film is conformally coated on the samples until the opening is sealed. Figure 40b shows the cross-section of a deposited LPCVD Si_xN_y seal.[195] In both reactive and deposited sealing techniques, the interior of the cavity is coated with the plug material. This is a problem when the mechanical characteristics of the cavity walls must be tightly controlled (as in pressure sensors). Shallow lateral channels introduce the least amount of sealing material into the cavity.

A sealing technique that does not coat the cavity interior is shadow plugging. In this method, the thin film is deposited from a source that deposits films with poor or line-of-sight step coverage. Sputtered, evaporated, and PECVD films[196] are adequate for this purpose. The plugging occurs when enough material is deposited to block the entrance of the opening. Since these materials have pinholes and cracks, it is necessary to coat the sealed areas with a better quality film on top of the deposited films (such as SiN). Figure 40c shows the basic idea behind this method. Little or no material penetrates the cavity after the initial seal.

Sealing of channels with spin-cast organic polyimides has been accomplished. The channels used in these seals are sufficiently narrow to prevent the liquid polyimide from penetrating the chamber.[197,198]

2.5.5 Lithographic Constraints and Planarization

Microsensors are three-dimensional structures fabricated using planar processing. In many cases, the surface topography determined by the thickness of the films needed can exceed the maximum that a lithographic process tolerates. Regular optical lithography with thin spin-on resists fails if the topography of the wafer exceeds 5 μm. Larger step heights yield very nonuniform resist with thick layers near concave corners and thin layers on convex corners as in Fig. 41. If thicker layers and taller features are desired, special resist-casting processes, including the use of multiple coatings, spray, and plasma polymerization, permit step heights exceeding 100 μm[199,200] at the expense of an out-of-focus projected image, resist overexposure, and overdevelopment. The steepness of the sidewalls is limited by the developer's ability to remove thicker segments of exposed resist selectively.

Many of these problems are solved if the sample is planarized regularly. Planarization[2] is a process that leaves the wafer surface flat or substantially reduces its topography. Planarization processes are used regularly for multilevel metallizations in semiconductor devices. A common planarization procedure involves the deposition of a thick spacer layer and subsequent mechanical and chemical polishing.

Planarization is also accomplished by reactive-ion etching. In this process, a layer of spin-on or PECVD silicon oxide is deposited on the sample. Next, a thick layer of photoresist is spin-cast. The thick resist leaves the wafer surface relatively flat. The photoresist and oxide are then etched by a plasma that attacks both materials at the same rate.[2] Excellent planarization is obtained by this means. Other methods make use of the anisotropy in the etch to produce flatter surfaces. Materials that "flow" such as spin-on glasses provide the best planarization characteristics.

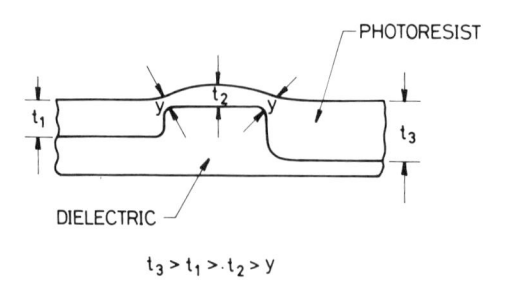

$t_3 > t_1 > t_2 > y$

Fig. 41 Nonuniform thickness of spin-cast resist near the corners. (After Ref. 1)

2.6 OTHER MICROMACHINING TECHNIQUES

2.6.1 LIGA

LIGA[201-206] (a German acronym for Lithographie, Galvanoformung, Abformung) consists of three basic processing steps: lithography, electroplating, and molding. It begins with coating a thick photoresist ranging from 300 μm to more than 500 μm in thickness on a substrate with an electrically conductive surface. In order to penetrate the thick resist with well-defined sidewalls, lithographic patterning is done with extended exposure from highly collimated X-ray radiation from a synchrotron through an X-ray mask (Fig. 42a). The desired structures are formed from the thick photoresist after developer treatment (Fig. 42b). Metal is then electroplated on the exposed conductive surface of the substrate, filling the space and covering the top surface of the resist (Fig. 42c). After the photoresist is removed, the metal structure is formed (Fig. 42d), and can be used repeatedly as a mold insert for injection molding to form multiple plastic replicas of the original plating base (Fig. 42e). The

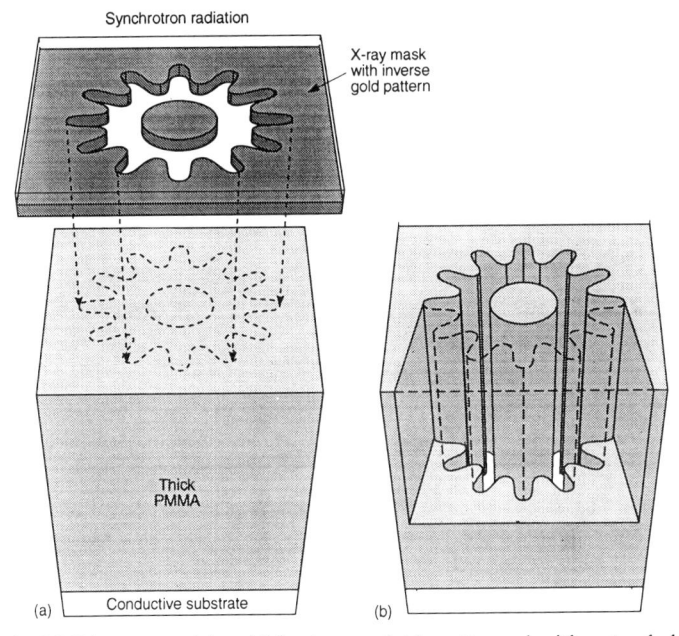

Fig. 42 The LIGA process: (a) a thick photoresist is patterned with extended exposure to X-ray radiation; (b) developing results in the desired structures formed from the thick photoresist; (c) metal is electroplated onto the exposed conductive surface of the substrate, filling in the space and covering the top surface of the resist; (d) the photoresist is removed, forming a complementary metal structure; (e) the metal structure is then used as a mold insert for injection molding to form multiple plating bases; (f) the plating base replica is, in turn, used to electroplate additional metal parts; (g) after removing the plastic part, the final product is formed.

76 SEMICONDUCTOR SENSOR TECHNOLOGIES

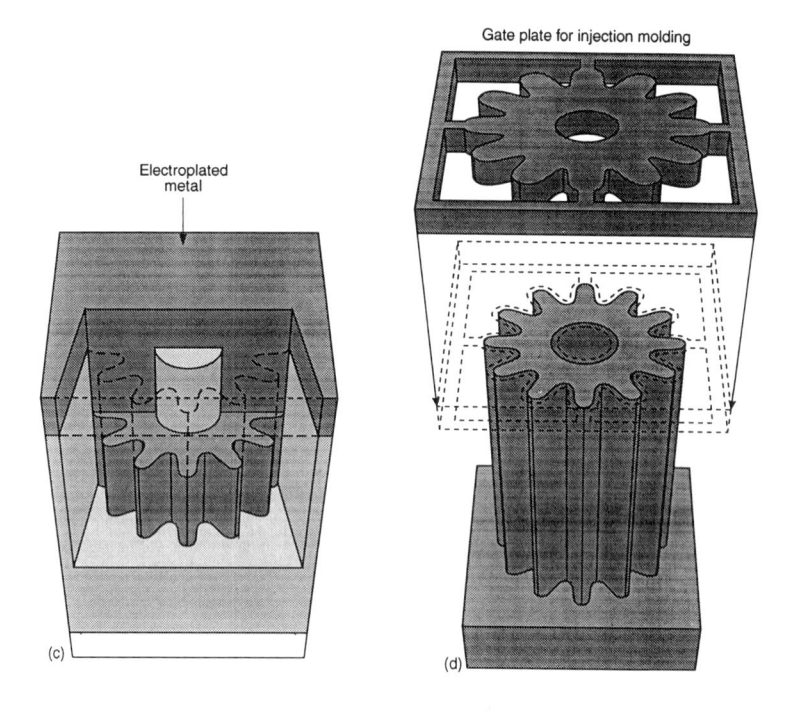

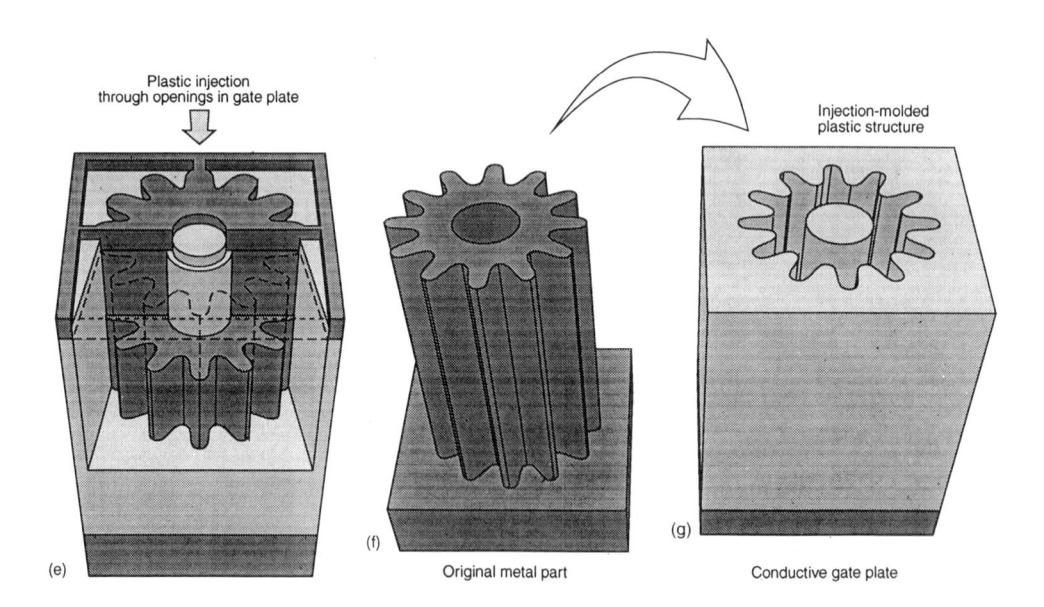

Fig. 42 *continued*

plating-base replicas, in turn, are then used to electroplate many metal structures as the final products (Fig. 42f and g).

Sacrificial techniques are combined with the basic LIGA process to create partially freed, flexure-suspended structures or completely freed devices.[207] The sacrificial layer may be patterned titanium film or polyimide removed by selective wet etching following the electroplating and photoresist removal steps.

Process Requirements. Beyond conventional photolithography, there are several unique requirements for the LIGA process. The first one is the need for X-ray radiation. The synchrotron should be capable of delivering at least 1 GeV energy, with a wavelength on the order of 7 Å or shorter. The exposure must be sustained for several hours or longer. Due to the high-energy, long exposure time, the X-ray mask must be made of materials with high atomic numbers, to form the opaque region for effective absorption of X-ray, and low-atomic-number materials in transparent regions. A supported, 1 to 2 μm-thick, low-stress diaphragm made of silicon nitride or titanium foil, with a 3 to 5 μm-thick, electroplated-gold pattern, has been used successfully. The thick photoresist, typically poly-methyl methacrylate (PMMA), cannot be spun on to the substrate to achieve the 500 μm thickness. Instead, *in situ* polymerization and casting with well-controlled thickness and minimal stress is used. The developing solution must retain a high selectivity to the unexposed PMMA during extended development periods. An excellent developer has the composition: 60 vol.% of 2-(2-butoxyethoxy)ethanol, 20 vol.% of tetrahydro-1, 4-oxazin (morpholine), 5 vol.% of 2-aminoethanol, and 15 vol.% water.

The electroplating step requires precise controls on current density, temperature, concentration, and composition of the plating solution to avoid formation of hydrogen bubbles, which may result in fatal defects in the plated structures. The built-in stress of the finished structures is a function of these plating conditions. Finally, injection molding is done in vacuum to completely fill the voids in the mold insert and special agents are added during the process to facilitate subsequent mold separation.

The distinct advantage of the LIGA process is the ability to create three-dimensional structures as thick as bulk-micromachined devices, while retaining the same degree of design freedom as surface micromachining. Microstructures with feature sizes of several microns have been made with a thickness in excess of 300 μm with the LIGA process. Although the initial synchrotron radiation is a costly step, using the original metal structure to create multiple replicas may be a viable manufacturing process, provided that the injection molding and mold-separation steps can be repeated at production volume without serious degradation to the original mold insert.

LIGA-Like Process.[208] For applications that require a lower definition quality than LIGA, thick layers can be patterned using conventional lithographic equipment. This adaptation aims at replacing the costly and scantly available synchrotron with a UV light source and suitable photosensitive polyimides to replace PMMA. Thick polyimides are commercially available, and can be spin-coated to a thickness of 60 μm or more. A G-line ($\lambda = 436$ nm) mask aligner with an exposure energy of 350 mJ/cm and an ordinary optical mask has been used successfully to expose a 40 μm-thick film, resulting in structures with a minimum feature size of 7 μm. Although the resolution and precision in the finished structures are inferior to LIGA, this approach offers a very cost-effective alternative for creating moderately thick planar structures. Multiple coatings of thick polyimides allow some degree of design freedom in the vertical dimension.

2.6.2 Quartz Processing

Quartz is an attractive material for microsensors because of its piezoelectric properties. Quartz has a crystalline structure belonging to the trigonal trapezohedral class of the rhombohedral subsystem. Figure 43 shows a representation of the crystal structure of quartz.

Quartz substrates are widely used in the production of resonant and ultrasonic devices. Like silicon, quartz can be micromachined by selectively etching some of the crystal planes. Common anisotropic etchants of quartz are HF and NH_4F. These etchants (and others) have been studied by Danel[209] and other researchers.[210,211] Figure 44 shows some of the rates for Z-cut quartz etching. Other mechanical sensors have been constructed using quartz, including gyroscopic devices[212] and acceleration sensors.[209,213]

2.6.3 GaAs Processing

This material has mechanical properties similar to silicon. The fact that GaAs is a direct bandgap semiconductor makes it very important for the fabrication of optical devices. A large variety of optical sensors has been developed. This is a large topic by itself, and is the main subject of several major textbooks.

GaAs can be etched anisotropically like silicon using H_2O_2 solutions. Figure 45 shows some anisotropically etched GaAs samples. Many of these etchants are catalogued in William's book.[214] Recently, the effects of piezoresistance of GaAs have been studied. A number of microstructures have been fabricated for microactuator and optomechanical purposes.[215,216]

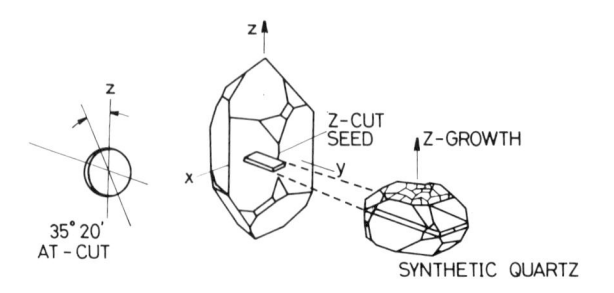

Fig. 43 Crystaline structure of Z-cut quartz. (After Ref. 209)

2.6 OTHER MICROMACHINING TECHNIQUES   79

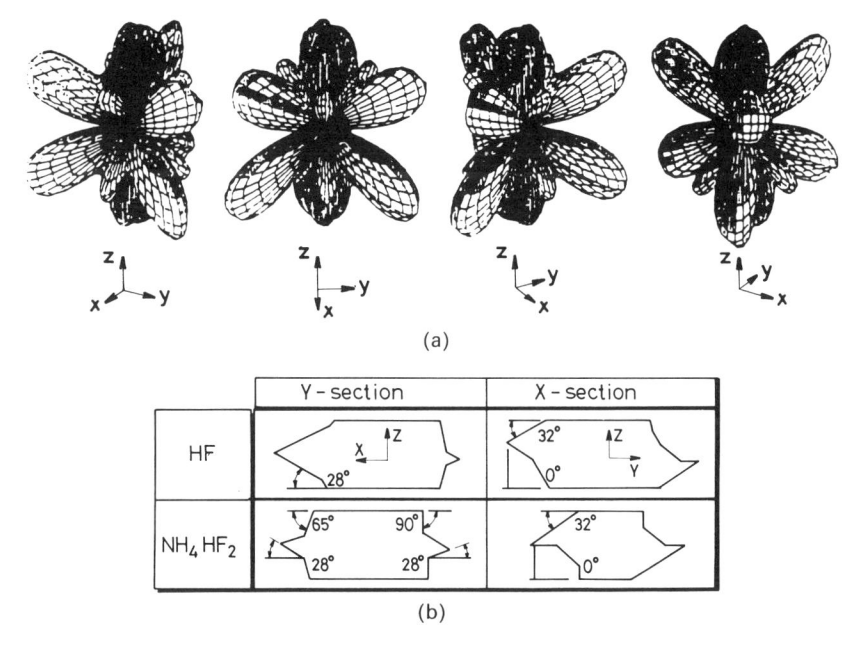

Fig. 44 (a) Etch-rate vectors for Z-cut quartz using ammonium bifluoride NH_4HF_2. (b) Sample profiles for two etchants. (After Ref. 209)

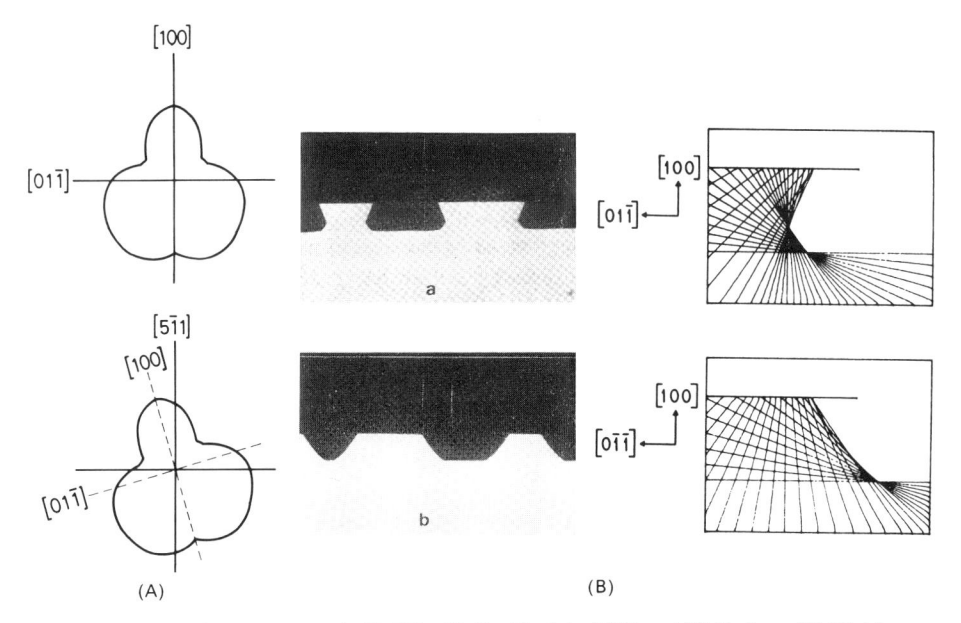

Fig. 45 (A) Etch-rate vectors in $H_2SO_4:H_2O_2:H_2O$ in (100) and (511) slices. (B) Etching profiles in GaAs samples. (After Ref. 43)

2.7 SUMMARY AND FUTURE TRENDS

This chapter outlines the numerous adaptations of conventional planar technology for the fabrication of micromechanical structures suitable for semiconductor-sensor applications. It provides a broad overview of IC-process materials and their relevant mechanical properties, which are crucial information for the design strategy. The choice of materials is tightly coupled with the fabrication procedures. The fabrication approach and technology chosen for a particular sensor affect not only the various performance parameters, but also the manufacturability, yield, and ultimately the cost.

The two most prominent techniques currently in use are bulk and surface micromachining. Bulk micromachining takes advantage of the extremely high quality of the bulk materials, such as single-crystal silicon, to form a mechanically reliable sensor. Subtractive processes such as anisotropic chemical etching coupled with additive processes such as wafer-to-glass bonding enable the creation of three-dimensional mechanical parts with a high degree of precision. The resulting structures are usually robust even with the small dimensions achieved with bulk micromachining. On the other hand, surface micromachining involves forming and etching various thin films on the substrate surface to create structures much smaller than those fabricated by bulk micromachining. Sacrificial etching is commonly employed in surface micromachining to form suspended or constrained structures that can move with certain degrees of freedom. Film thicknesses are usually on the order of a micron to no more than 10 microns, and are limited by practical consideration of the time required for depositing the films, the technical limitation on how well the films can be etched, and the planar resolution required. Sealing techniques are available for both bulk and surface micromachining to encapsulate part or all of the sensor for protection against a harsh environment.

Surface micromachining of thin films is especially advantageous when extremely small sensors are desired. In addition, on-chip integration with signal-processing circuits may be easier to accomplish when the substrate is left intact. However, the mechanical properties of thin films are difficult to control; therefore, sensor designs based on surface micromachining must be tolerant of the variations in processing conditions. LIGA and LIGA-like techniques aim at creating high-aspect-ratio microstructures not achievable with either bulk or surface micromachining. Other materials used in sensor technology include quartz and GaAs. The high quality factor of quartz and its piezoelectric property are attractive for resonant sensors. The piezoelectricity of quartz allows simple detection of resonance and therefore simplifies signal processing. GaAs microstructures, because of their optoelectric and piezoelectric properties, promise applications in micromechanical photonic-sensing devices.

2.8 PROBLEMS

1. Suppose that you are bulk micromachining a square silicon diaphragm of width w for a pressure sensor. The relationship between the deflection d and the applied pressure P is

$$d = \frac{\alpha w^4 P}{D} \qquad D = \frac{Et^3}{12(1-v^2)} \qquad (9)$$

where t is the diaphragm thickness, D the flexural rigidity, E the Young's modulus, v its Poisson ratio, and $\alpha = 0.00582$. The diaphragm is defined by photolithographically defining an opening on a (100) silicon wafer of thickness t_w. The etch stops at a diffused layer of thickness t_i and depth d_i below the surface. What is the maximum allowable variation of wafer thickness if the diaphragm deflection must not change by more than 2%? Use $t_w = 500$ μm, $t_i = 0.5$ μm, $d_i = 9.75$ μm, $w = 500$ μm, $t = 10$ μm, $E = 180$ GPa, and $v = 0.18$. The (100) plane etch rate is 40 μm/h.

2. All realistic process flows have nonuniformities. Suppose that t_w has a variance of 1 μm across the substrate, the etch rate varies by 15%, the implant depth varies by 0.2 μm, and the selectivity of the etch solution with respect to the etch stop is 500:1. (a) What is the maximum variance of the diaphragm width w and thickness t across the substrate? (b) Describe the variance of the deflection characteristics? (c) Can you keep the deflection within 10% of its nominal value?

3. Suppose that you are etching layer L_1 of thickness t_1 and stopping at layer L_2 of thickness t_2. Layer L_1 has a thickness variance of σ_{t_1} and layer L_2 a thickness variance of σ_{t_2}. If the etchant has selectivity S_{12}, etch rate R_1 and variance σ_{R_1}: (a) calculate the minimum thickness of L_2 such that it will not be completely removed anywhere in the substrate, and (b) what is the nominal value and variance of the etched step across the substrate?

4. The feature depicted in Fig. P1 is patterned onto a (100) wafer surface with an ideal mask. The wafer is then etched in a solution with infinite selectivities

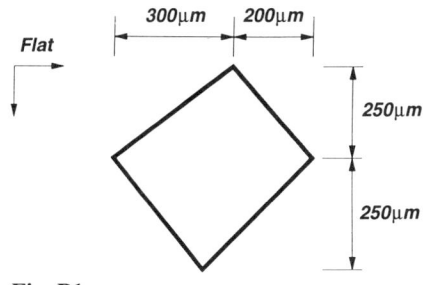

Fig. P1

for $\langle 100 \rangle$ over $\langle 111 \rangle$, and $\langle 110 \rangle$ over $\langle 111 \rangle$ for a long period of time. (a) Draw the top view of the final etched pattern, indicating the angles between all sides as well as the length of the sides. (b) Now assume a finite selectivity of 20 for $\langle 100 \rangle$ over $\langle 111 \rangle$.

5. a) Repeat part (a) of the previous problem for the case of a (110) wafer.
   b) Assume a finite selectivity for $\langle 110 \rangle$ over $\langle 111 \rangle$, qualitatively repeat part (b) of the previous problem for a (110) wafer.

6. Using the data in Table 2 and the etch-rate ratio for type-S EDP provided in the text, i.e., 30:30:1 at 115°C and 100:150:1 at 50°C for $\langle 100 \rangle:\langle 110 \rangle:\langle 111 \rangle$, calculate the activation energy, E_a, and the pre-exponential factor, R_0, for $\langle 111 \rangle$ etching with type-S EDP.

7. A 1 μm-thick photoresist mask is used to pattern a 5 μm-wide line onto a 4 μm-thick polysilicon film to be etched anisotropically in a Cl-based plasma. Assume an etch selectivity of 10 for polysilicon over the photoresist, and that the etch on polysilicon is perfectly anisotropic, i.e., etching occurs only in the vertical direction, while the photoresist is attacked with perfect isotropy. Also, assume that the initial sidewall profile of the photoresist is perfectly vertical. Draw the cross-section across the 5 μm-wide line, showing the etch profiles of the photoresist and the polysilicon film after a 3 μm-deep etch of the polysilicon film is completed. Indicate the dimensions and angles of the resulting profiles.

8. Repeat the previous problem, but assume a finite anisotropy for the polysilcon etch, such that the horizontal direction is etched at 20% of the rate of the vertical direction.

9. In the structure of Fig. P2 the sacrificial layer is removed by undercutting through the access hole. The top structural layer is uniform and 2 μm thick. The length of the sacrificial layer is 100 μm. During the release, the structural layers are exposed to the sacrificial etch. Assume that the sacrificial-layer etch proceeds at a uniform rate $R = 1$ μm/min, and the etch has a selectivity of 500:1. (a) Calculate the time required to remove the sacrificial layer.

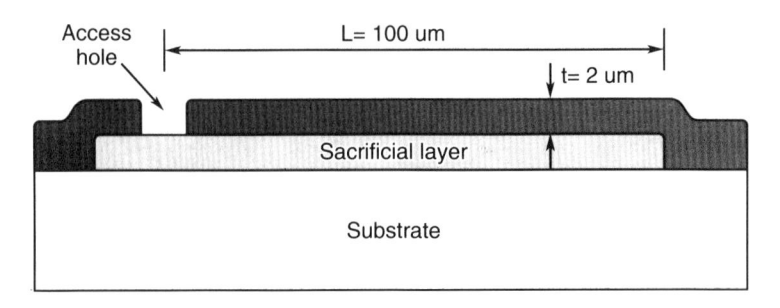

Fig. P2

(b) Calculate the profile of the walls inside the channel. (c) What is the final shape of the channel? (d) What is the minimum and maximum thickness?

10. In Problem 9, (a) calculate the minimum selectivity required such that the structural layer is not removed. (b) Calculate the selectivity such that the wall thickness does not change by more than 15%.

11. In Problem 9, assume that the etch rate and thickness of the channel walls have variances of 10% across the wafer. Calculate the minimum selectivity needed to assure that the wall thickness does not change more than 20%.

12. For very long channels, the process is limited by diffusion effects. The etch rate at the etch front is dependent on the etchant concentration $N(x)$.

$$R = R_0 \left(\frac{N(x)}{N_0} \right) \qquad (10)$$

where N_0 is the concentration of the etchant in the bulk of the liquid. The flux of etchant consumed at the etch front is

$$J_f = R(x) N_s \qquad (11)$$

where N_s is the atomic concentration of the sacrificial material. The flux of fresh etchant coming into the channel is

$$J_0 = -D \frac{\partial N}{\partial x} \bigg|_0 \qquad (12)$$

where D is the diffusion coefficient. (a) What is the differential equation that determines the position of the etch front? (b) Solve the equation for the case where $R \to \infty$. (*Hint:* Use flux balance.)

13. Similar depletion effects occur in the sealing process. In the sealing process, the reaction takes place at all exposed surfaces. Assume that the channel cavity is infinitely long, and after the cavity is etched, we wish to seal it. If the deposition rate $R_d(x)$ depends on the gas concentration $N(x)$ as

$$R_d = R_{d0} \left(\frac{N(x)}{N_0} \right) \qquad (13)$$

the flux of gas being depleted by the deposition at the walls as the gas penetrates the cavity is

$$J_w = R(x) N_s \qquad (14)$$

where N_s is the density of the solid. (a) Calculate the rate $R_d(x)$ and steady-state concentration $N(x)$ assuming a concentration at the opening of N_0. Ignoring two-dimensional diffusion and necking effects, what is the profile of the deposited material on the inside of the channel after it is sealed? (*Hint:* Assume that the deposition stops after the channel is blocked.)

14. The permeability K_p of a layer of thickness t is defined such that the flux of gas permeating across it is

$$J = -\frac{K_p(P_1 - P_0)}{t} \tag{15}$$

where P_1 and P_0 are the pressures at both sides of the layer. (a) Assuming that we are using a silicon oxide plug to seal a channel that is 0.2 μm thick, 10 μm wide, and 10 μm long, P_1 is at room pressure and P_0 at a high vacuum, and $K_p = 2.4 \times 10^{-3}$ m/N-s, what is the leakage rate in sccm into the cavity? If the cavity has a volume of 10^3 μm^3, what is the leakage rate in Pa/s. How long will it take for P_0 to reach 0.2 atmospheres? Can you say anything about the effectiveness of the oxide seal?

REFERENCES

1. S. Wolf and R. N. Tauber, *Silicon Processing for the VLSI Era, Volume 1—Process Technology*, Lattice Press, Sunset Beach, CA, 1986.
2. S. Wolf, *Silicon Processing for the VLSI Era, Volume 2—Process Integration*, Lattice Press, Sunset Beach, CA, 1989.
3. S. M. Sze, *VLSI Technology*, McGraw–Hill, New York, 1988.
4. L. I. Maissel and R. Glang, *Handbook of Thin Film Technology*, McGraw–Hill, New York, 1970.
5. J. L. Vossen, *Thin Film Processes*, Academic Press, New York, 1978.
6. A. S. Kao and H. G. Stenger, Jr., "Analysis of nonuniformtities in the plasma etching of silicon with CF_4/O_2," *J. Electrochem. Soc.* **137**, 954 (1990).
7. D. L. Flamm and G. K. Herb, "Plasma etching technology—an overview," in *Plasma Etching: An Introduction*, D. M. Manos and D. L. Flamm, Eds., ch. 1, p. 1.
8. D. L. Flamm, "Introduction to plasma chemistry," in *Plasma Etching: An Introduction*, D. M. Manos and D. L. Flamm, Eds., ch. 2, p. 91.
9. D. R. Sparks, "Plasma etching of Si, SiO_2, Si_3N_4, and resist with fluorine, chlorine, and bromine compounds," *J. Electrochem. Soc.* **139**, 1736 (1992).
10. W. D. Nix, "Mechanical properties of thin films," *Metall. Trans.* **20A**, 2217 (1989).
11. Y. C. Tai, *IC-Processed Polysilicon Micromechanics: Technology, Materials and Devices*, PhD Thesis, University of California, Berkeley (1989).
12. Y. C. Tai and R. S. Muller, "Integrated stylus force gauge," *Sens. Actuators* **A21–A23**, 410 (1990).

13. Y. C. Tai and R. S. Muller, "Measurements of Young's modulus on microfabricated structures using a surface profiler," in *1990 Int. Workshop on Micro Electromechanical Systems (MEMS 90)*, p. 147
14. X. Ding, W. H. Ko, and J. M. Mansour, "Residual stress and mechanical properties of boron-doped $p+$ silicon films," *Sens. Actuators* **A21–A23**, 866 (1990).
15. L.-S. Fan, R. T. Howe, and R. S. Muller, "Microstructures for fracture toughness characterization of brittle thin films," in *1989 Int. Workshop on Micro Electromechanical Systems (MEMS 89)*, p. 40.
16. L.-S. Fan, *Integrated Micromachinery—Moving Structures in Silicon Chips*, PhD Thesis, University of California, Berkeley (1990).
17. R. T. Howe, S. D. Senturia, and M. Mehregany, "Novel microstructure for the in-situ measurement of mechanical properties of thin films," *J. Appl. Phys.* **62**, 3579 (1987).
18. Y. C. Tai and R. S. Muller, "Fracture strain of LPCVD polysilicon," in *Int. Workshop on Solid-State Sensors and Actuators (Hilton Head '88)*, p. 88.
19. A. P. Lee, L. Lin, and A. P. Pisano, "Normal and tangential impact in micro electromechanical structures," in *1991 Int. Workshop on Micro Electromechanical Systems (MEMS 91)*, p. 21.
20. A. P. Lee, A. Pisano, and M. Lim, "Impact, friction, and wear testing of microsamples of polycrystalline silicon," in *1992 Smart Materials Fabrication and Materials for Microelectromechanical Systems Conference, Pittsburg, PA*, p. 67.
21. J. A. Connally and S. B. Brown, "Micromechanical fatigue testing," in *1991 Int. Conf. Solid-State Sensors Actuators (Transducers '91)*, p. 953.
22. J. A. Connaly and S. B. Brown, "Slow crack growth in single-crystal silicon," *Science* **256**, 1537 (1992).
23. F. Pourahmadi, D. Gee, and K. Petersen, "The effect of corner radius of curvature on the mechanical strength of micromachined single-crystal silicon structures," in *1991 Int. Conf. Solid-State Sensors Actuators (Transducers '91)*, p. 197.
24. T. A. Kneckt, "Bonding techniques for solid-state pressure sensors," in *1991 Int. Conf. Solid-State Sensors Actuators (Transducers '87)*, p. 95.
25. H. L. Offereins, H. Sandmaier, B., Folkmer, U. Steger, and W. Lang, "Stress free assembly technique for a silicon based pressure sensor," in *1991 Int. Conf. Solid-State Sensors Actuators (Transducers '91)*, p. 986.
26. V. L. Spiering, S. Bouwstra, R. M. F. J. Spiering, and M. Elwenspoek, "On-chip decoupling zone for package-stress reduction," in *1991 Int. Conf. Solid-State Sensors Actuators (Transducers '91)*, p. 982.
27. D. G. Oei and S. McCarthy, "The effect of temperature and pressure on residual stress in LPCVD polysilicon films," in *1992 Materials Research Socety Conf.*, p. 397.
28. X. Ding, W. H. Ko, and W. He, "A study on silicon-diaphragm buckling," in *Int. Workshop on Solid-State Sensors and Actuators (Hilton Head '90)*, p. 128.
29. X. Ding and W. H. Ko, "Buckling behavior of boron-doped $p+$ silicon diaphragms," *1991 Int. Conf. Solid-State Sensors Actuators (Transducers '91)*, p. 201.
30. L. B. Wilner, "Strain and strain relief in highly doped silicon," in *1992 Int. Workshop on Solid-State Sensors and Actuators (Hilton Head '92)*, p. 76.
31. F. J. von Pressig, "Applicability of the classical curve-stress relation for thin films on plate substrates." *J. Appl. Phys.* **66**, 4262 (1989).

32. D. W. Burns, *Micromechanics of Integrated Sensors and the Planar Processed Pressure Transducer*, PhD Thesis, University of Winconsin, Madison (1988).
33. K. Najafi and K. Suzuki, "A novel technique and structure for the measurement of intrinsic stress and Young's modulus of thin films." in *1989 Int. Workshop on Micro Electromechanical Systems (MEMS 89)*, p. 96.
34. J. F. J. Goosens, B. P. van Drieenhuizen, P. J. French, and R. F. Wolfenbuttel, "Stress measurement structures for micromachined sensors," in *1993 Int. Conf. Solid-State Sensors Actuators (Transducers '93)*, p. 783.
35. L. Lin, R. T. Howe, and A. P. Pisano, "A passive in-situ strain gauge," in *Int. Workshop on Micro Electromechanical Systems (MEMS 93)*, p. 201.
36. O. Tabata, K. Kawahata, S. Sugiyama, and I. Igarashi, "Mechanical property measurements of thin films using load-deflection of composite rectangular membranes," *Sens. Actuators* **20**, 135 (1989).
37. I. C. Noyan, *Residual Stress: Measurement by Diffraction and Interpretation*, Springer-Verlag, New York, 1987.
38. L.-S. Fan, R. S. Muller, W. Yun, R. T. Howe, and J. Huang, "Spiral microstructures for the measurement of average strain gradients in thin films," in *1990 Int. Workshop on Micro Electromechanical Systems (MEMS 90)*, p. 177.
39. T. A. Lober, J. Huang, M. A., Schmidt, and S. D. Senturia, "Characterization of the mechanisms producing bending moments in polysilicon micro-cantilever beams by interferometric deflection measurements," in *1988 Int. Workshop on Solid-State Sensors and Actuators (Hilton Head '88)*, p. 92.
40. C. H. Mastrangelo, Y.-C. Tai, and R. S. Muller, "Thermophysical properties of low-residual stress, silicon-rich, LPCVD silicon nitride films," *Sens. Actuators* **A21–A23**, 856 (1990).
41. F. Volklein and H. Baltes, "A microstructure for measurement of thermal conductivity of polysilicon thin films," *J. Microelectromech. Syst.* **1**, 193 (1992).
42. C. H. Mastrangelo and R. S. Muller, "Fabrication and performance of a fully integrated μ-pirani pressure gauge with digital readout," in *1991 Int. Conf. Solid-State Sensors Actuators (Transducers '91)*, p. 245.
43. D. W. Shaw, "Morphology analysis in localized crystal growth and dissolution," *J. Cryst. Growth* **47**, 509 (1979).
44. R. J. Jaccodine, "Use of modified free energy theorems to predict equilibrium growing and etching shapes," *J. Appl. Phys.* **33**, 2643 (1962).
45. C. Herring, "Some theorems on the free energies of crystal surfaces," *Phys. Rev.* **82**, 87 (1951).
46. J. L. Reynolds, A. R. Neureuther, and W. G. Oldham, "Simulation of dry-etched line-edge profiles," *J. Vac. Sci. Technol.* **16**, 1772 (1979).
47. A. R. Neureuther, C. Y. Liu, and C. H. Ting, "Modelling ion milling," *J. Vac. Sci. Technol.* **16**, 1767 (1979).
48. C. J. Mogab and W. R. Harshberger, "Plasma processes set to etch finer lines with less undercutting," *Electronics* **51**, 117 (1982).
49. G. C. Schwartz, L. B. Rothman, and T. J. Schopen, "Competitive mechanisms in reactive ion etching in a CF_4 plasma," *J. Electrochem. Soc.* **126**, 464 (1979).
50. H. K. Kuiken, "Etching through a slit," *Proc. R. Soc. London* **A396**, 95 (1984).
51. H. K. Kuiken, "Etching: a two-dimensional mathematical approach," *Proc. R. Soc. London* **A392**, 199 (1084).

52. H. K. Kuiken, J. J. Kelly, and P. H. L. Notten, "Etching profiles at resist edges," *J. Electrochem. Soc.* **133**, 1217 (1986).
53. J. Liu, Y. C. Tai, J. Lee, K.-C. Pong, Y. Zohar, and C.-M. Ho, "In-situ monitoring and universal modelling of sacrificial PSG etching using hydrofluroic acid," in *1993 Int. Workshop on Micro Electromechanical Systems (MEMS 93)*, p. 71.
54. D. J. Monk, D. S. Soane, and R. T. Howe, "A diffusion/chemical reaction model for HF etching of LPCVD phosphosilicate glass sacrificial layers," in *Int. Workshop on Solid-State Sensors and Actuators (Hilton Head '92)*, p. 46.
55. K. E. Petersen, "Silicon as a mechanical material," *IEEE Proc.* **70**, 420 (1982).
56. C. S. Smith, "Piezoresistance effect in germanium and silicon," *Phys. Rev.* **94**, 42 (1954).
57. O. N. Tufte and E. L. Stelzer, "Piezoresistive properties of silicon diffused layers," *J. Appl. Phys.* **34**, 313 (1963).
58. Y. Gianchandani and K. Najafi, "Micron-sized, high aspect ratio bulk silicon micromechanical devices," in *1992 Int. Workshop on Micro Electromechanical Systems (MEMS 92)*, p. 208.
59. M. Takinami, K. Minami, and M. Esashi, "High-speed directional low-temperature dry etching for bulk silicon micromachining," *Tech. Dig. 11th Sensor Symp.*, 1992, p. 15.
60. C. Lihnder, T. Schan, and N. F. de Rooij, "Deep dry etching techniques as a new IC compatible tool for silicon micromachining," in *1992 Int. Conf. Solid-State Sensors Actuators (Transducers '91)*, p. 524.
61. P. J. Holmes, "Practical applications of chemical etching," in *The Electrochemistry of Semiconductors* P. J. Holmes, Ed., p. 329.
62. M. M. Abu-Zeid, "Corner undercutting in anisotropically etched isolation contours," *J. Electrochem. Soc.* **131**, 2138 (1984).
63. X. Wu and W. Ko, "A study on compensating undercutting in anisotropic etching of silicon," in *1987 Int. Conf. Solid-State Sensors Actuators (Transducers '87)*, p. 126.
64. S.-C. Chang and D. B. Hicks, "Mesa structure formation using potassium hydroxide and ethylene diamine based etchants," in *1988 Int. Workshop on Solid-State Sensors and Actuators (Hilton Head '88)*, p. 102.
65. B. Puers and W. Sansen, "Compensation structures for convex corner micromachining in silicon," *Sens. Actuators* **A21–A23**, 1036 (1990).
66. H. Sandmaier, H. L., Offereins, K. Kuhl, and W. Lang, "Corner compensation techniques in anisotropic etching of (100)-silicon using aqueous KOH," in *Int. Conf. Solid-State Sensors Actuators (Transducers '91)*, p. 456.
67. A. Koide, K. Sato, S. Suzuki, and M. Miki, "A multistep anisotropic etching process for producing 3-D silicon accelerometers," *Tech. Dig. 11th Sensor Symp.*, 1992, p. 23.
68. A. I. Stoller, "The etching of deep vertical-walled patterns in silicon," *RCA Rev.* **31**, 271 (1970).
69. D. L. Kendall, "On etching very narrow grooves in silicon," *Appl. Phys. Lett.* **26**, 195 (1975).
70. D. L. Kendall, "Vertical etching of silicon at very high aspect ratios," *Ann. Rev. Mater. Sci.* **9** (1979).
71. K. E. Bean, "Anisotropic etching of silicon," *IEEE Trans. Electron, Devices,* **ED-25**, 1185 (1978).
72. E. Bassous, "Fabrication of novel three-dimensional microstructures by the

anisotropic etching of (100) and (110) silicon," *IEEE Trans. Electron Devices* **ED-25**, 1178 (1978).
73. K. E. Petersen, "Silicon as a mechanical material," *Proc. IEEE* **70**, 420 (1982).
74. Y. Lindén, L. Tenerz, and B. Hök, "Fabrication of three-dimensional silicon structures by means of doping-selective etching (DSE)," *Sens. Actuators* **16**, 67 (1989).
75. H. Seidel, L. Csepregi, A. Heuberger, and H. Baumgärtel, "Anisotropic etching of crystalline silicon in alkaline solutions: I. orientation dependence and behavior of passivation layers," *J. Electrochem. Soc.* **137**, 3612 (1990).
76. H. Seidel, L. Csepregi, A. Heuberger, and H. Baumgärtel, "Anisotropic etching of crystalline silicon in alkaline solutions: II. influence of dopants," *J. Electrochem. Soc.* **137**, 3626 (1990).
77. R. M. Finne and D. L. Klein, "A water-amine-complexing agent system for etching silicon," *J. Electrochem. Soc.* **114**, 965 (1967).
78. A. Reisman, M. Berkenblit, S. A., Chan, F. B. Kaufman, and D. C. Green, "The controlled etching of silicon in catalyzed ethylenediamine-pyrocatechol-water solutions," *J. Electrochem. Soc.* **126**, 1406 (1979).
79. N. F. Raley, Y. Sugiyama, and T. van Duzer, "(100) silicon etch-rate dependence on boron concentration in ethylenediamine-pyrocatechol-water solutions," *J. Electrochem. Soc.* **131**, 161 (1984).
80. J. B. Price, "Anisotropic etching of silicon with potassium hydroxide-water-isopropyl alcohol," in *Semiconductor Silicon 1973*, H. R. Huff and R. R. Burgess, Ed., 339.
81. L. D. Clark, Jr. and D. J. Edell, "KOH:H_2O etching of (110) Si, (111) Si, SiO_2, and Ta: an experimental study," *Proc. IEEE Micro Robots and Teleoperators Workshop, Hyannis, MA, Nov. 1987*
82. G. Findler, J. Muchow, M. Koch, and H. Münzel, "Temporal evolution of silicon surface roughness during anisotropic etching processes," in *1992 Int. Workshop on Micro Electromechanical Systems (MEMS 92)*, p. 62.
83. L. D. Clark, Jr., J. L. Lund, and D. J. Edell, "Cesium hydroxide (CsOH): a useful etchant for micromachining silicon," *Tech. Dig. IEEE Solid-State Sensor and Actuator Workshop, Hilton Head, SC*, 1988, p. 5
84. W. Kern, "Chemical etching of silicon, germanium, gallium arsenide, and gallium phosphate," *RCA Rev.* **39**, 278 (1978).
85. U. Schnakenberg, W. Benecke, and B. Löchel, "NH_4OH-based etchants for silicon micromachining", *Sens. Actuators* **A21–A23**, 1031 (1990).
86. I. J. Pugacz-Muraszkiewicz, "Detection of discontinuities in passivating layers on silicon by NaOH anisotropic etch," *IBM J. Res. Dev.* **16**, 523 (1972).
87. M. Mehregany and S. D. Senturia, "Anisotropic etching of silicon in hydrazine," *Sens. Actuators* **13**, 375 (1988).
88. M. Asano, T. Cho, and H. Muraoka, "Application of choline in semiconductor technology," *Electrochem. Soc. Ext. Abs.*, no. 354, 911 (1976).
89. U. Schnakenberg, W. Benecke, and P. Lange, "TMAHW etchants for silicon micromachining," in *1991 Int. Conf. Solid-State Sensors Actuators (Transducers '91)*, p. 815.
90. H. Robbins and B. Schwartz, "Chemical etching of silicon: I. The system HF, HNO_3, and H_2O," *J. Electrochem. Soc.* **106**, 505 (1959).

REFERENCES

91. H. Huraoka, T. Ohhashi, and Y. Sumitomo, "Controlled preferential etching technology," in *Semiconductor Silicon 1973*, H. R. Huff and R. R. Burgess, Eds., p. 327.
92. B. Schwartz and H. Robbins, "Chemical etching of silicon: IV, Etching technology," *J. Electrochem. Soc.* **123**, 1903 (1976).
93. H. A. Waggener, "Electrochemically controlled thinning of silicon," *Bell Syst. Tech. J.* **50**, 473 (1970).
94. W. K. Zwicker and S. K. Kurtz, "Anisotropic etching of silicon using electrochemical displacement reactions," in *Semiconductor Silicon 1973*, H. R. Huff and R. R. Burgess, Eds., p. 315.
95. T. N. Jackson, M. A. Tischler, and K. D. Wise, "An electrochemical p-n junction etch stop for the formation of silicon microstructures," *IEEE Electron. Device Lett.* **EDLM-2**, 44 (1981).
96. O. J. Glembocki, R. E. Stahlbush, and M. Tomkiewicz, "Bias-dependent etching of silicon in aqueous KOH," *J. Electrochem. Soc.* **132**, 145 (1985).
97. B. Kloech and N. F. de Rooij, "A novel four electrode electrochemical etch-stop method for silicon membrane formation," in *1987 Int. Conf. Solid-State Sensors Actuators (Transducers '87)*, p. 116.
98. B. Kloeck, S. D. Collins, N. F. de Rooij, and R. L. Smith, "Study of electrochemical etch-stop for high-precision thickness control of silicon membranes," *IEEE Trans. Electron Devices*, **ED-36**, 663 (1989).
99. Y. P. Xu and R. S. Huang, "Anodic dissolution and passivation of silicon in hydrazine," *J. Electrochem. Soc.* **137**, 948 (1990).
100. V. M. McNeil, S. S. Wang, K.-Y. Ng, and M. A. Schmidt, "An investigation of the electrochemical etching of (100) silicon in CsOH and KOH", *Tech. Dig. IEEE Solid-State Sensor and Actuator Workshop, Hilton, Head, SC,* 1990, p. 92.
101. H. Seidel and L. Csepregi, "Studies on the anisotropy and selectivity of etchants used for the fabrication of stress-free structures," *Electrochem. Soc. Ext. Abs. Montreal, Canada,* 1982, p. 194.
102. X. Ding, W. H. Ko, Y. Niu, and W. He, "A study on silicon-diaphragm buckling," *Tech. Dig. IEEE Solid-State Sensor and Actuator Workshop, Hilton, Head, SC,* 1990, p. 128.
103. S. T. Cho, K. Najafi, and K. D. Wise, "Scaling and dielectric stress compensation of ultrasensitive boron-doped silicon microstructures," in *1990 Int. Workshop on Micro Electromechanical Systems (MEMS 90)*, p. 50.
104. W. H. Ko, J. T. Suminto, and G. J. Yeh, "Bonding techniques for microsensors," in *Micromachining and Micropackaging of Transducers,* C. D. Fung, P. W. Cheung, W. H. Ko, and D. G. Fleming, Eds., p. 41.
105. G. Wallis and D. L. Pomerantz, "Field assisted glass-metal sealing," *J. Appl. Phys.* **40**, 3946 (1969).
106. P. M. Sutton, "Space charge and electrode polarization in glass, I," *J. Am. Ceram. Soc.* **47**, 188 (1964).
107. P. M. Sutton, "Space charge and electrode polarization in glass, II," *J. Am. Ceram. Soc.* **47**, 219 (1964).
108. E. H. Snow and M. E. Dumesnil, "Space-charge polarization in glass films," *J. Appl. Phys.* **37**, 2123 (1966).

109. G. Wallis, "Direct-current polarization during field-assisted glass-metal sealing," *J. Am. Ceram. Soc.* **53**, 563 (1970).
110. S. Johansson, K. Gustafsson, and J.-A. Schweitz, "Strength evaluation of field assisted bond seals between silicon and Pyrex glass," *Sens. Mater.* **3**, 143 (1988).
111. S. Johansson, K. Gustafsson, and J.-A. Schweitz, "Influence of bonded area ratio on the strength of FAB seals between silicon microstructures and glass," *Sens. Mater.* **4**, 209 (1988).
112. L. A. Field and R. S. Muller, "Fusing silicon wafers with low melting temperature glass," *Sens. Actuators* **A21–A23**, 935 (1990).
113. A. D. Brooks, R. P. Donovan, and C. A. Hardesty, "Low-temperature electrostatic silicon-to-silicon seals using sputtered borosilicate glass," *J. Electrochem. Soc.* **119**, 545 (1972).
114. R. Legtenberg, S. Bouwstra, and M. Elwenspoek, "Low-temperature glass bonding for sensor applications using boron oxide thin films," *J. Micromech. Microeng.* **1**, 157 (1991).
115. A. Hanneborg, M. Nese, and P. Ohickers, "Silicon-to-silicon anodic bonding with a borosilicate glass layer," *J. Micromech. Microeng.* **1**, 139 (1991).
116. M. Esahi, A. Nakano, S. Shoji, and H. Hebiguchi, "Low-temperature silicon-to-silicon bonding with intermediate low melting temperature glass," *Sens. Actuators* **A21–A23**, 931 (1990).
117. H. J. Quenzer, W. Benecke, and C. Dell, "Low temperature wafer bonding for micromechanical applications," in *1992 Int. Workshop on Micro Electromechanical Systems (MEMS 92)*, p. 49.
118. M. Shimbo, F. Kurukawa, F. Fukuda, and K. Tanzawa, "Silicon-to-silicon direct bonding method," *J. Appl. Phys.* **60**, 2987 (1986).
119. K. Y. Ahn, R. Stengl, T.-Y. Tan, and U. Gosele, "Growth, shrinkage, and stability of interfacial oxide layers between directly bonded silicon wafers," *Appl. Phys.* **50(A)**, 85 (1990).
120. K. Mitani, V. Lehmann, and U. Gosele, "Bubble formation during silicon wafer bonding: causes and remedies," in *1990 Int. Workshop on Solid-State Sensors and Actuators (Hilton Head '90)*, p. 74.
121. K. M. V. Lehmann, R. Stengl, and D. Feijoo, "Causes and prevention of temperature-dependent bubbles in silicon wafer bonding," *Jpn. J. Appl. Phys.* **30**, 615 (1991).
122. K. Mitani and U. Gosele, "Formation of interface bubbles in bonded silicon wafers—a thermodynamic model," *Appl. Phys.* **54**, 543 (1992).
123. R. Stengl, T. Tan, and U. Gosele, "A model for the silicon wafer bonding process," *Jpn. J. Appl. Phys.* **28**, 1735 (1990).
124. V. L. K. Mitani, R. Stengl, and T. Mii, "Bubble-free wafer bonding of GaAs and InP on silicon in a microcleanroom," *Jpn. J. Appl. Phys.* **28**, L2141 (1989).
125. V. Lehman, O. Iwk, and G. U. Stengl, "Semiconductor wafer bonding," *Adv. Mater.* **2**, 372 (1990).
126. C. den Besten, J. M. R. E. G. van Hal, and P. Bergveld, "Polymer bonding of micromachined silicon structures," in *1992 Int. Workshop on Micro Electromechanical Systems (MEMS 92)*, p. 104.
127. R. H. Brown, "Piezo film: form and function," *Sens. Actuators* **A21–A23**, 729 (1990).

128. G. Mader and H. Meixner, "Pyroelectric infrared sensor arrays based on the polymer PVDF," *Sens. Actuators* **A21–A23**, 503 (1990).
129. R. Takayama, Y. Tomita, J. Asayama, K. Nomura, and H. Ogawa, "Pyroelectric infrared array sensors made of c-axis oriented la-modified $PbTiO_3$ thin films," *Sens. Actuators* **A21–A23**, 508 (1990).
130. T. Kamins, *Polycrystalline Silicon for Integrated Circuit Applications*, Kluwer Academic, Boston, MA, 1988.
131. D. Schubert, W. Jenschke, T. Uhlig, and F. M. Schmidt, "Piezoresistive properties of polycrystalline and crystalline silicon films," *Sens. Actuators* **11**, 145 (1987).
132. Y. Onuma, K. Kamimura, and Y. Homma, "Piezoresistive elements of polycrystalline semiconductor thin films," *Sens. Actuators* **13**, 71 (1988).
133. G. W. Racette and R. T. Frost, "Deposition of polycrystalline silicon films under ultrahigh vacuum," *J. Cryst. Growth* **47**, 384 (1979).
134. J. Adamczewska and T. Budzynski, "Stress in chemically vapour-deposited silicon films," *Thin Solid Films* **113**, 271 (1984).
135. Z. Iqbal and A. P. W. abd S. Veprek, "Polycrystalline silicon films deposited in a glow discharge at temperatures below 250°C," *Appl. Phys. Lett.* **36**, 163 (1980).
136. S. Chang, W. Eaton, C. Gonzalez, B. Underwood, J. Wong, and R. L. Smith, "Micromechanical structures in amorphous silicon," in *1991 Int. Conf. Solid-State Sensors Actuators (Transducers '91)*, p. 751.
137. L.-S. Fan and R. S. Muller, "As-deposited low-strain LPCVD polysilicon," in *1988 Int. Workshop on Solid-State Sensors and Actuators (Hilton Head '88)*, p. 55.
138. H. Guckel, D. W. Burns, H. A. C. Tilmans, C. C. G. Visse, D. W. DeRoo, T. R. Christenson, P. J. Klomberg, J. J. Sniegowski, and D. H. Jones, "Processing conditions for polysilicon films with tensile strain for large aspect ratio microstructures," in *1988 Int. Workshop on Solid-State Sensors and Actuators (Hilton Head '88)*, p. 51.
139. P. Krulevitch, R. T. Howe, G. Johnson, and J. Huang, "Stress in undoped LPCVD polycrystalline silicon," in *1991 Int. Conf. Solid-State Sensors Actuators (Transducers '91)*, p. 949.
140. I. T. Toncheva and I. S. Vassilev, "Stress polycrystalline silicon films," *Thin Solid Films* **60**, 353 (1979).
141. S. P. Muraka and T. F. Retajczyk, "Effect of phosphorpous doping on stress in silicon and polycrystalline silicon," *J. Appl. Phys.* **54**, 2069 (1983).
142. M. S. Choi and E. W. Hearn, "Stress effects in boron-implanted polysilicon films," *J. Electrochem. Soc.* **131**, 2443 (1984).
143. M. Orpana and A. O. Korhonen, "Control of residual stress of polysilicon thin films by heavy doping in surface micromachining," in *1991 Int. Conf. Solid-State Sensors Actuators (Transducers '91)*, p. 957.
144. K. Chau, C. Fung, P. R. Harris, and G. Dahrooge, "A versatile polysilicon diaphragm pressure sensor chip," in *1991 Int. Electron Devices Meeting (IEDM 91)*, p. 695.
145. R. S. Payne and K. Al Dinsmore, "Surface micromachined accelerometer: a technology update," *SAE Paper 910496* (1991).
146. D. L. Flowers, L. Ristic, and H. G. Hughes, "Mechanical and structural characterization of in-situ phosphorous doped enhanced alpha silicon films," in *1991 Int. Conf. Solid-State Sensors Actuators (Transducers '91)*, p. 961.

147. H.-L. Chau and K. D. Wise, "An ultraminiature solid-state pressure sensor for a cardiovascular catheter," *IEEE Trans. Electron Devices* **ED-35**, 2355 (1988).
148. K. Suzuki, "Single-crystal silicon microactuators," in *1991 Int. Electron Devices Meeting (IEDM 90)*, p. 625.
149. Y. B. Gianchandani and K. Najafi, "A bulk silicon dissolved wafer process for microelectromechanical devices," *J. Microelectromech. Syst.* **1**, 77 (1992).
150. J. B. Lasky, "Wafer bonding for silicon-on-insulator technologies," *Appl. Phys. Lett.* **48**, 78 (1986).
151. W. P. Maszara, G. Goetz, and J. M. McKitterick, "Bonding of silicon wafers for silicon-on-insulator," *J. Appl. Phys.* **64**, 4943 (1988).
152. J. Haisma, G. A. C. M. Spierings, U. K. P. Biermann, and J. A. Pals, "Silicon-on-insulator wafer bonding-wafer thinning technological evaluations," *Jpn. J. Appl. Phys.* **28**, 1426 (1989).
153. W. P. Maszara, "Silicon-on-insulator by wafer bonding: a review," *J. Electrochem. Soc.* **138**, 341 (1991).
154. J.-P. Colinge, *Silicon-on-Insulator Technology: Materials to VLSI*, Kluwer Academic, Boston, MA, 1991.
155. J. Shajii, K.-Y. Ng, and M. A. Schmidt, "A microfabricated floating element shear stress sensor using wafer bonding technology," *J. Microelectromech. Syst.* **1**, 89 (1992).
156. K. Petersen, P. Barth, J. Poydock, J. Brown, J. Mallon, J. Bryzek, "Silicon fusion bonding for pressure sensors," in *1988 Int. Workshop on Solid-State Sensors and Actuators (Hilton Head '88)*, p. 144.
157. K. Petersen, F. Pourahmadi, J. Brown, P. Parsons, M. Skinner, and J. Tudor, "Resonant beam pressure sensor fabricated with silicon fusion bonding," in *Int. Conf. Solid-State Sensors Actuators (Transducers '91)*, p. 664.
158. K. Petersen, D. Gee, F. Pourahmadi, R. Craddock, J. Brown, and L. Christel, "Surface micromachined structures fabricated with silicon fusion bonding," in *Int. Conf. Solid-State Sensors Actuators (Transducers '91)*, p. 397.
159. G. W. Neudeck, P. J. Schubert, J. L. Glenn, J. A. Friedrich, W. A. Klaasen, R. P. Zingg, and J. P. Denton, "Three dimensional devices fabricated by silicon epitaxial lateral overgrowth," *J. Electronic Mater.* **19**, 1111 (1990).
160. J. J. Pak, A. E. Kabir, G. W. Neudeck, J. H. Logsdon, D. R. DeRoo, and S. E. Saller, "A micromachining technique for a thin silicon membrane using merged epitaxial lateral overgrowth of silicon and SiO_2 for an etch stop," in *1991 Int. Conf. Solid-State Sensors Actuators (Transducers '91)*, p. 1028.
161. M. Sekimoto, H. Yoshihara, and T. Ohkubo, "Silicon nitride single-layer x-ray mask." *J. Vac. Sci. Technol.* **21**, 1017 (1982).
162. E. S. Kim, *Integrated Microphone with CMOS Circuits on a Single Chip*, PhD Thesis, University of California, Berkeley (1990).
163. W. G. Valkenburg, M. F. C. Williamsen, and W. A. P. Clasen, "Influence of deposition temperature, gas pressure, gas phase composition, and RF frequency on composition and mechanical stress of plasma silicon nitride layers," *J. Electrochem. Soc.* **132**, 893 (1985).
164. L. B. Rothman, "Properties of thin polyimide films," *J. Electrochem. Soc.* **127**, 2216 (1980).

165. Y. Takahashi, M. Iijima, K. Inagawa, and A. Itoh, "Synthesis of aromatic polyimide film by vacuum deposition polymerization," *J. Vac. Sci. Technol.* **A5**, 2253 (1986).

166. J. R. Salem, F. O. Sequeda, J. Duran, W. Y. Lee, and R. M. Yang, "Solventless polyimide films by vapor deposition," *J. Vac. Sci. Technol.* **A4**, 369 (1986).

167. S. K. Ray, C. L. Maiti, and N. B. Charkrabarti, "Low-temperature deposition of dielectric films by microwave plasma enhanced decomposition of hexamethyldisilazane," *J. Electronic Mater.* **20**, 907 (1991).

168. J. J. Licari, *Plastic Coatings for Electronics,* Krieger, New York, 1981.

169. H. Yasuda, *Plasma Polymerization,* Academic Press, New York, 1985.

170. M. Dalvie and K. F. Jensen, "Combined experimental and modeling study of spatial effects in plasma etching: CF_4/O_2 etching of silicon," *J. Electrochem. Soc.* **137**, 1062 (1990).

171. L. Y. Tsou, "Effect of photoresist on plasma etching," *J. Electrochem. Soc.* **136**, 2354 (1989).

172. H. Guckel, J. J. Sniegowski, and T. R. Christenson, "Advances in processing techniques for silicon micromechanical devices with smooth surfaces," in *1989 Int. Workshop on Micro Electromechanical Systems (MEMS 89),* p. 71.

173. A. Kassam, C. Meadowcroft, C. A. T. Salama, and P. Ratnam, "Characterization of BCl_3-Cl_2 silicon trench etching," *J. Electrochem. Soc.* **137**, 1613 (1990).

174. G. C. H. Zau and H. H. Sawin, "Effects of O_2 feed gas impurity on Cl_2 based plasma etching of polysilicon," *J. Electrochem. Soc.* **139**, 250 (1992).

175. M. Sato and T. Arita, "Etched shape control of single-crystal silicon in reactive ion etching using chloride," *J. Electrochem. Soc.* **134**, 2856 (1987).

176. L. Y. Tsou, "Highly selective reactive ion etching of polysilicon with hydrogen bromide," *J. Electrochem. Soc.* **136** 3003 (1989).

177. J. P. McVittie and C. Gonzales, "Anisotropic etching of Si using SF_6 with C_2ClF_5 and other mixed halocarbons," in *Proc. 5th Symp. on Plasma Processing,* G. S. Mathad, G. C. Schwartz, and G. Smolinsky, Eds., Vol. 85-1, 552.

178. W. Kern and C. A. Deckert, "Chemical etching," in *Thin film processes,* J. L. Vossen and W. Kern, Eds., ch. V-1, 401.

179. W. van Gelder and V. E. Hauser, "The etching of silicon nitride in phosphoric acid with silicon dioxide as a mask," *J. Electrochem. Soc.* **114**, 869 (1967).

180. J. T. Milek, *Silicon Nitride for Microelectronic Applications, part 1—Preparation and Properties,* IFI/Plenum, New York, 1971.

181. H. C. Nathanson, W. E. Newell, R. A. Wickstrom, and J. R. Davis, Jr., "The resonant gate transistor," *IEEE Trans. Electron Devices* **ED-14**, 117 (1967).

182. R. T. Howe and R. S. Muller, "Polycrystalline and amorphous silicon micromechanical beams: annealing and mechanical properties," *Sens. Actuators* **4**, 447 (1983).

183. L. S. Fan, Y. C. Tai, and R. S. Muller, "IC-processed electrostatic micromotors," in *1988 Int. Electron Devices Meeting (IEDM 91),* p. 666.

184. M. Mehregany, S. D. Senturia, and J. H. Lang, "Friction and wear in microfabricated harmonic side-drive motors," *Tech. Dig. IEEE Solid-State Sensor and Actuator Workshop, Hilton Head, SC,* 1990, p. 17.

185. V. R. Dhuler, M. Mehregany, and S. M. Phillips, "Micromotor operation in a liquid

environment," *Tech. Dig. IEEE Solid-State Sensor and Actuator Workshop, Hilton Head, SC,* 1992, p. 10.
186. M. A. Schmidt, R. T. Howe, S. D. Senturia, and J. H. Haritonidis, "Surface micromachining of polyimide/metal composites for a shear-stress sensor," *Proc. IEEE Micro Robots and Teleoperators Workshop, Hyannis, MA, Nov. 1987*
187. W. C. Tang, T.-C. H. Nguyen, and R. T. Howe, "Laterally driven polysilicon resonant microstructures," in *1989 Int. Workshop on Micro Electromechanical Systems (MEMS 89),* p. 53.
188. R. L. Alley, G. J. Cuan, R. T. Howe, and K. Komvopoulos, "The effect of release-etch processing on surface microstructure striction," *Tech. Dig. IEEE Solid-State Sensor and Actuator Workshop, Hilton Head, SC,* 1992, p. 202.
189. C. H. Mastrangelo and C. H. Hsu, "A simple experimental technique for the measurement of the work of adhesion of microstructures," *Tech. Dig. IEEE Solid-State Sensor and Actuator Workshop, Hilton Head, SC,* 1992, p. 208.
190. T. Hirano, T. Furuhata, and H. Fujita, "Dry releasing of electroplated rotational and overhanging structures," in *1993 Int. Workshop on Micro Electromechanical Systems (MEMS 93),* p. 278.
191. N. Takeshimo, K. J. Gabriel, M. Ozaki, J. Takashashi, H. Horiguichi, and H. Fujita, "Electrostatic parallelogram actuators," in *1991 Int. Conf. Solid-State Sensors Actuators (Transducers '91),* p. 63.
192. G. T. Mulhern, S. Soane, and R. T. Howe, "Supercritical carbon dioxide drying for microstructures," in *1993 Int. Conf. Solid-State Sensors Actuators (Transducers '93),* p. 296.
193. C. Mastrangelo and G. Saloka, "A dry-release method based on polymer columns for microstructure fabrication," in *1990 Int. Workshop on Micro Electromechanical Systems (MEMS 90),* p. 77.
194. R. T. Howe, "Surface micromachining for microsensors and actuators," *J. Vac. Sci. Technol.* **B6,** 1809 (1988).
195. C. H. Mastrangelo and R. S. Muller, "Vacuum-sealed silicon micromachined incandescent light source," in *1989 Int. Electron Devices Meeting (IEDM 89),* p. 503.
196. S. Sugiyama, T. Suzuki, K. Kawahata, K. Shimaoka, and M. Takigawa, "Micro-diaphragm pressure sensor," in *1986 Int. Electron Devices Meeting (IEDM 86),* p. 184.
197. L. A. Field, *Fluid-Actuated Micromachined Rotors and Gears,* PhD Thesis, University of California, Berkeley (1991).
198. K. J. Gabriel, O. Tabata, K. Simaoka, S. Sugiyama, and H. Fujita, "Surface-normal electrostatic/pneumatic microactuator," in *1992 Int. Workshop on Micro Electromechanical Systems (MEMS 92),* p. 128.
199. H. Guckel, J. Uglow, M. Lin, D. Denton, J. Tobin, K. Euch, and M. Juda, "Plasma polymerization of methyl methacrylate: a photoresist for 3D applications," in *1988 Int. Workshop on Solid-State Sensors and Actuators (Hilton Head '88),* p. 9.
200. S. Kawahito, Y. Saski, M. Ashiki, and T. Nakamura, "Micromachined solenoids for highly sensitive magnetic sensors," in *1991 Int. Conf. Solid-State Sensors Actuators (Transducers '91),* p. 1077.
201. W. Ehrfeld, P. Bley, F. Götz, P. Hagmann, A. Maner, J. Mohr, H. O. Moser, D. Münchmeyer, W. Schelb, D. Schmidt, and E. W. Becker, "Fabrication of microstructures using the LIGA process," *Proc. IEEE Micro Robots and*

Teleoperators Workshop, Hyannis, MA, Nov. 1987

202. H. Guckel, T. R. Christenson, K. Skrobis, D. Denton, B. Choi, E. G. Lovell, J. W. Lee, and T. W. Chapman, "Deep X-ray lithography for micromechanics," *Tech. Dig. IEEE Solid-State Sensor and Actuator Workshop, Hilton Head, SC,* 1990, p. 118.

203. W. Menz, W. Bacher, M. Harmening, and A. Michel, "The LIGA technique—a novel concept for microstructures and the combination with Si-technologies by injection molding," in *1990 Int. Workshop on Micro Electromechanical Systems (MEMS 91),* p. 69.

204. H. Guckel, K. J. Skrobis, T. R. Christenson, J. Klein, S. Han, B. Choi, and E. G. Lovell, "Fabrication of assembled micromechanical components via deep X-ray lithography," in *1991 Int. Workshop on Micro Electromechanical Systems (MEMS 91),* p. 74.

205. M. Harmening, W. Bacher, P. Bley, A. El-Kholi, H. Kalb, B. Kowanz, W. Menz, A. Michel, and J. Mohr, "Molding of three-dimensional microstructures by the LIGA process," in *1992 Int. Workshop on Micro Electromechanical Systems (MEMS 91),* p. 202.

206. T. R. Christenson, H. Guckel, K. J. Skrobis, and T. S. Jung, "Preliminary results for a planar microdynamometer," *Tech. Dig. IEEE Solid-State Sensor and Actuator Workshop, Hilton Head, SC,* 1992, p. 6.

207. H. Guckel, J. Klein, T. Christenson, K. Skrobis, M. Landon, and E. G. Lovell, "Thermo-magnetic metal flexure actuators," *Tech. Dig. IEEE Solid-State Sensor and Actuator Workshop, Hilton Head, SC,* 1992, p. 73.

208. A. B. Frazier and M. G. Allen, "High aspect ratio electroplated microstructures using a photosensitive polyimide process," in *1992 Int. Workshop on Micro Electromechanical Systems (MEMS 92),* p. 87.

209. J. S. Danel, F. Michel, and G. Delapierre, "Micromachining of quartz and its applications to an acceleration sensor," *Sens. Actuators* **A21–A23**, 971 (1990).

210. C. R. Tellier, "Some results on chemical etching of AT-cut quartz wafers in ammonium bifluoride solutions," *J. Mater. Sci.* **17**, 1348 (1982).

211. J. K. Vondeling, "Fluoride-based etchants of quartz," *J. Mater. Sci.* **18**, 304 (1983).

212. L. D. Clayton, E. P. Eernisse, R. W. Ward, and R. B. Wiggins, "Miniature crystalline quartz electromechanical structures," *Sens. Actuators* **A21–A23**, 171 (1990).

213. J. Söderkvist, "Design of a solid-state gyroscopic sensor made of quartz," *Sens. Actuators,* **A21–A23**, 293 (1990).

214. R. E. Williams, *Modern GaAs Processing Methods*, Artech House, Boston, 1990.

215. K. Hjort, J.-A. Scheitz, S. Andersson, O. Kordina, and E. Janzen, "Epitaxial regrowth in surface micromaching of gaas," in *1992 Int. Workshop on Micro Electromechanical Systems (MEMS 92),* p. 83.

216. Z. L. Zhang, G. A. Porkolab, and N. C. MacDonald, "Submicron, movable gallium arsenide mechanical structures and actuators," in *1992 Int. Workshop on Micro Electromechanical Systems (MEMS 92),* p. 72.

217. K. B. Albaugh, P. E. Cade, and D. H. Rasmussen, "Mechanisms of anodic bonding of silicon to pyrex glass," in *Int. Workshop on Solid-State Sensors and Actuators (Hilton Head '88),* p. 109.

218. H. Guckel, "Surface micromachined pressure transducers," *Sen. Actuators,* **A28**, 133 (1991).

3 Acoustic Sensors

M. E. MOTAMEDI
Rockwell Science Center
Thousand Oaks, CA, USA

R. M. WHITE
Department of Electrical Engineering and Computer Sciences, and
The Berkeley Sensor & Actuator Center
University of California, Berkeley, CA, USA

3.1. INTRODUCTION

Acoustic sensors are devices that employ elastic waves at frequencies in the megahertz to low gigahertz range to measure physical, chemical, or biological quantities. Their high sensitivity makes these devices particularly attractive for chemical vapor and gas sensing. In many cases, the output of these sensors is a frequency, which can be measured simply and accurately with an electronic counter. With proper design, these sensors can be quite stable, permitting a large dynamic range to be realized.

The first ultrasonic sensors employed vibrating piezoelectric crystal plates[1,2] fabricated for use as the frequency-determining elements of electronic oscillators (Fig. 1a). This device, referred to in the literature as the quartz crystal microbalance (QCM), is more appropriately denoted, after its particle motion, as a thickness shear-mode (TSM) sensor. The addition of a special coating to absorb vapor or gas molecules forms a sensor whose resonant frequency depends on the number of molecules absorbed. More recently, the advantages of other ultrasonic sensor configurations have become evident, particularly ones that use surface acoustic waves (SAWs) traveling on the surface of a solid (Fig. 1b), elastic flexural plate waves (FPWs) in a very thin membrane (Fig. 1c), or the so-called acoustic plate mode (APM) arrangement in which waves bounce at

---

Semiconductor Sensors, Edited by S. M. Sze.
ISBN 0-471-54609-7 © 1994 John Wiley & Sons, Inc.

98 ACOUSTIC SENSORS

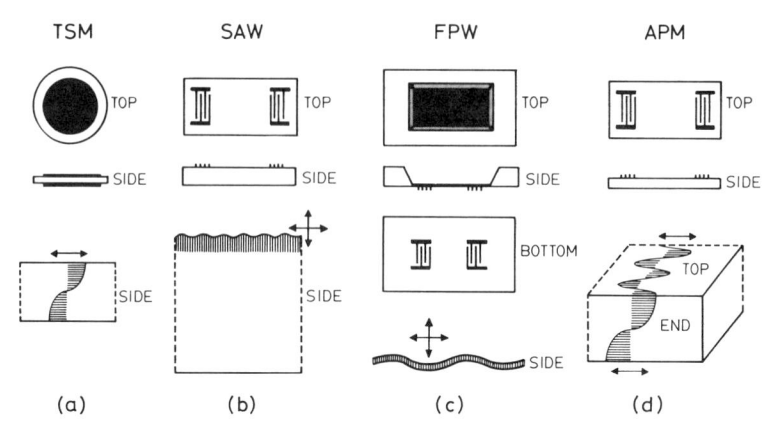

Fig. 1 Chief forms of acoustic wave sensors (from Ref. 1). Upper sketches show device configuration in top view and cross-section. Bottom sketches indicate particle motion and the mode distribution in each device. TSM = thickness shear mode; SAW = surface acoustic wave; FPW = flexural plate wave; APM = acoustic plate mode. (After Ref. 1)

an acute angle between bounding planes of a plate (Fig. 1d). With the advent of micromachining, ultrasonic acceleration sensors have been fabricated from etched cantilever and other structures, as we shall see.

This chapter presents the principles of the basic acoustic sensors most commonly fabricated by the planar processing methods used to make integrated circuits, as well as some specialized techniques such as the deposition of thin piezoelectric films.

All the sensors that we discuss are based on elastic motions in solid members of the sensor. In many of these devices, as illustrated in Fig. 1, propagating elastic waves are used. Elastic waves in a solid are produced when atoms of the solid are forced into vibratory motion about their equilibrium positions. The neighboring atoms then produce a restoring force tending to bring the displaced atoms back to their original positions. (The different types of elastic waves will be discussed in Section 3.2.) These sensors are designed so that the propagation characteristics of these waves—their phase velocity and/or their attenuation coefficient—are affected by the measurands of interest. Thus, in some of these devices, mechanical forces induced by an applied pressure or an acceleration of the sensor increase the wave velocity. In others, gravimetric effects such as the sorption of molecules or the attachment of bacteria cause a reduction of wave velocity. Finally, when a viscous liquid contacts the active region of an elastic wave sensor, the wave is attenuated.

After discussing elastic wave motion briefly, in Section 3.3 we will consider acoustic materials. Some bulk piezoelectric materials that are used as the sensor substrate will be discussed. The emphasis here is mostly on surface-acoustic-wave devices such as SAW delay lines and SAW resonators/filters. Piezoelectric thin films of materials such as ZnO and AlN are considered crucial for a variety of low-cost acoustic sensors where VLSI process compatibility is an essential requirement. In the case of sensors and actuators requiring very high piezoelectric coupling, lead-zirconate-titanate (PZT) materials are recommended. We will discuss a range of PZT materials processed by sputtering and sol-gel methods.

Then, in Section 3.4, we will briefly discuss acoustic sensing with emphasis on sensor fabrication and piezoelectric materials processing. We will also discuss methods of thin-film deposition and sputtering, reactive-ion etching (RIE), PZT materials preparation, substrate selection, monolithic processing, and electrochemical and orientation-dependent etching.

A major section of this chapter is allocated to piezoelectric semiconductors. We will first introduce the theory of operation of monolithic sensors, and then we will discuss several very attractive sensors based on ZnO-on-Si and monolithic micromachining technology.

Since surface-acoustic-wave devices play an important role in the performance of high sensitivity acoustic sensors, in Section 3.5 we will cover SAW propagation, transducer design, and SAW resonators and oscillators, including several sensor examples.

Pressure sensors and accelerometers are currently the major applications of microsensor manufacturing. Although many sensor companies have manufactured pressure sensors during the last decade, we will not discuss pressure sensors here because of the limited space available, but in Section 3.6 we will concentrate instead on some newer elastic-wave sensors. We briefly mention the measurement of vapors and liquids with acoustic sensors. A summary and a discussion of future trends are presented in Section 3.7.

3.2 ACOUSTIC WAVES

3.2.1 Acoustic Waves in Solids

Acoustic waves propagating in solids have been used in a variety of devices such as electrical filters, acoustic sensors, actuators, medical and biological instruments, acoustic bulk oscillators, and delay lines. During the 1960s, rapid growth of research and development of a sophisticated acoustic device called the "interdigital transducer" (IDT) introduced a new horizon into the domain of electrical engineers. The acoustic transduction of IDTs was based on SAW propagation and its reality was demonstrated first by White and Voltmer[3] during 1965 using a planar process compatible with VLSI technology. SAW technology reached maturity during the 1970s, when research and development were applied to actual systems and many practical and commercial system components of radar and spread-spectrum communication were reported.

The most useful property of acoustic waves is their low velocity compared with that of electromagnetic waves. Typical velocities in solids range from

1.5×10^5 cm/s to 12×10^5 cm/s, with SAW velocities on the order of 3.8×10^5 cm/s to 4.2×10^5 cm/s. Therefore, acoustic velocities are five orders of magnitude smaller than those of electromagnetic waves, which makes acoustic-wave devices inherently miniature as a rule. Acoustic devices based on SAW technology can be fabricated with fundamental frequencies as high as 5 GHz. These devices can have areas as small as a few square millimeters, and can be fabricated monolithically with all required electronic circuitry.

3.2.2 Elastic Wave Motions and Phase Velocity

The elastic waves that propagate in solids are illustrated schematically in Fig. 2. In an unbounded solid, only bulk longitudinal and transverse waves can propagate (Figs. 2a and b). As a bulk longitudinal wave progresses, the particles

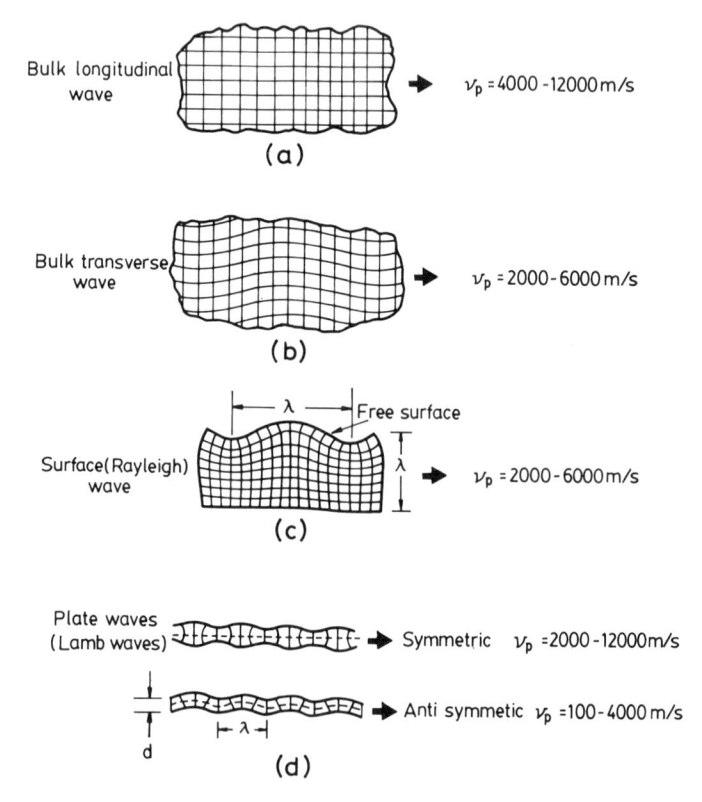

Fig. 2 Schematic illustration of motions of groups of atoms shown in cross-sectional views of solids as plane elastic waves propagate to the right. Vertical and horizontal displacements are exaggerated for clarity; typical wave velocities are shown at right: (a) bulk longitudinal wave in unbounded solid; (b) bulk transverse wave in unbounded solid; (c) surface acoustic wave (SAW) in semi-infinite solid, disturbance extends below surface to a depth of about one wavelength, λ; (d) waves in thin solid plates (Lamb waves).

of the solid move parallel to the propagation direction, indicated by the heavy arrow. This wave is similar to the pressure wave in a fluid. One obtains the phase velocity of each bulk wave by assuming small wave amplitudes, so that Hooke's law applies where one can write an expression of Newton's force law, and then seek a phase velocity that satisfies the resulting equations.

A SAW can propagate along the surface of a semi-infinite solid, which has a single boundary (Fig. 2c). At all practical frequencies, the velocity of the SAW on a homogeneous substrate is independent of frequency, as is the case for bulk waves as well. As the sketch of Fig. 2c suggests, the particle motion of a SAW is largest near the surface and decreases to nearly zero at a depth of about one wavelength.

There is no stress acting on the free boundary of the solid. Again, by assuming small wave amplitudes so that Hooke's law applies, and using Newton's force law, one arrives at a set of equations whose solution yields the phase velocity for the SAW as well as the relative amplitudes, phases, and distributions with depth of the particle displacement components.

The typical self-consistent solution—which usually must be found with a computer—is sketched in Fig. 3. The SAW velocity in any solid is slightly smaller than that of transverse bulk waves propagating in the same direction in an unbounded sample of that material.

When a solid is bounded by two surfaces, forming a plate, the stresses on both boundaries must vanish. Analysis then yields a doubly-infinite set of modes, known as Lamb waves,[5] with velocities that depend on the plate's material properties and on the ratio of plate thickness to wavelength. These plate waves may be divided into symmetric and antisymmetric sets (Fig. 2d) according to the symmetry of the component of motion parallel to the plate's surface about the center plane of the plate. Imposing boundary conditions at the two plate surfaces yields expressions for the wave velocities. One finds a trigonometric, rather than exponential, dependence of amplitude of motion on the distance

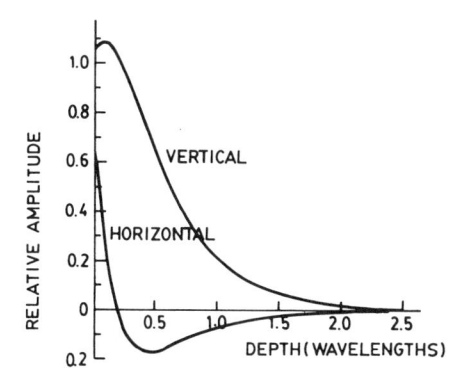

Fig. 3 Typical SAW particle displacements in horizontal and vertical directions for waves propagating horizontally in an isotropic medium.

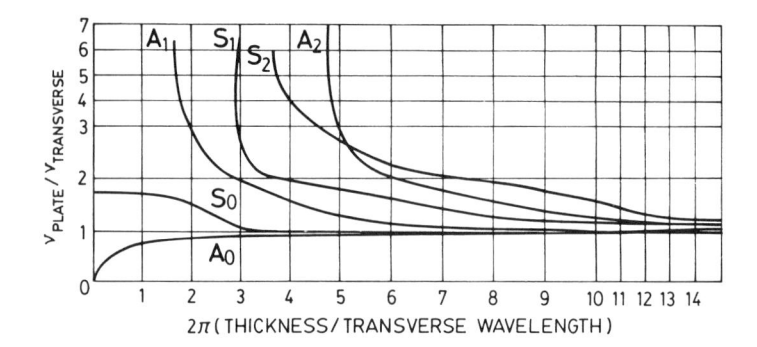

Fig. 4 Plate-wave velocities for symmetric (S) and antisymmetric (A) modes in an isotropic solid plotted as a function of plate thickness.[5] Assumed Poisson's ratio is 0.34.

from the midplane of the plate. The velocity, plotted for several of the modes in Fig. 4, is a strong function of the plate thickness, as measured in wavelengths. We shall show that for vapor sensing with these plate waves it is advantageous to use the lowest-order antisymmetric mode (A_0) in a very thin plate, to achieve high sensitivity.

Piezoelectric Coupling to Elastic Waves. The most convenient way to couple electric circuits to elastic waves is by using the piezoelectric effect. This effect results in the production of electric polarization charge when a piezoelectric crystal is stressed. Conversely, an electric field causes a piezoelectric substance to deform (undergo strain). The earliest bulk-wave sensors were plates of crystalline quartz (SiO_2), a naturally piezoelectric substance. For surface waves the choice is greater: crystalline quartz, the artificial ferroelectric single-crystal lithium niobate ($LiNbO_3$), deposited thin films of ZnO, and other piezoelectrics.

In Section 3.3 the topic of piezoelectric materials is dealt with at length. Here, we summarize these features of piezoelectric transduction that are of most importance in acoustic sensors.

The proper piezoelectric material and its orientation depend on the type of motion desired, the allowable variation with temperature, and fabrication considerations. Popular bulk-mode crystals are AT and BT cuts of crystalline quartz. Electrically excited AT- and BT-cut quartz plates vibrate parallel to their plane faces in a shearing motion, and have zero linear temperature coefficients of velocity at 25°C. Similarly, the ST cut of crystalline quartz is popular for SAW devices because its linear temperature coefficient of velocity is zero at room temperature.

The strength of the piezoelectric effect in elastic-wave devices also depends on the configuration of the transducing electrodes used. To excite a bulk mode electrically, an alternating voltage is applied between broad-area electrodes on the two opposing faces of a crystal plate. One can show that piezoelectric stresses are generated at the interface of the plate with the conducting electrodes, which are usually one-half wavelength apart at the frequency of operation. Hence, the waves launched at each electrode add in phase on traversing the thickness of the crystal, and resonant vibrations build up. For SAW and plate-mode sensors, transducing electrodes are either interdigitated or unipolar comb-like electrodes whose spatial period, p, equals the wavelength or some integral multiple thereof (see Figs. 1b and c). The strength of coupling of a given piezoelectric material in this situation depends on several factors:

1. the piezoelectric and elastic properties of the material itself;
2. the orientation of the applied electric field relative to the crystalline axes;
3. the relative locations of the comb-like electrode and the ground plane, when a piezoelectric film is employed on a non-piezoelectric substrate; and
4. for SAW and plate modes, the ratio of spatial period, p, to operating wavelength, λ.

Crystalline quartz has relatively low piezoelectric coupling but good uniformity, and the material can be cut in zero-temperature-coefficient orientations. Lithium niobate has strong coupling directions, but a relatively high temperature coefficient of velocity. Sputter-deposited zinc oxide has moderate coupling and can be applied in thin-film form. With additional non-piezoelectric films, it is possible to compensate lithium niobate and zinc oxide transducers for temperature-induced velocity changes over a narrow wavelength range on certain substrate materials. PZT films have strong piezoelectric coupling and a high dielectric constant that is often advantageous for electrical design reasons.

Phase Velocities and the Effects of Measurands. The phase velocity, v_p, of an elastic wave in a solid usually has the functional form $v_p = $ (elastic stiffness/ material density)$^{1/2}$. From this functional form we see that a measurand change that decreases stiffness, such as a rise of temperature, lowers the phase velocity. Increasing temperature also causes thermal expansion that increases the spacing between transducing electrodes in all these sensors, which is equivalent to increasing the wavelength of operation. A change that increases density also reduces the phase velocity. In piezoelectric solids, a wave that is piezoelectrically active (one that couples to the piezoelectric fields) is stiffened; hence, piezoelectric coupling raises the phase velocity. In sensors where the electric fields produced by the piezoelectric effect and a propagating wave can interact with the medium adjacent to the crystal, an increase in conductivity produces a shielding effect, reducing the stored electric energy, and so reducing the piezoelectric stiffening and lowering the phase velocity. Table 1 lists some factors that can affect wave velocities and be measured with acoustic sensors. To determine sensitivity to measurands, one must examine these dependencies quantitatively.

104 ACOUSTIC SENSORS

TABLE 1 Factors that Affect Ultrasonic Wave Velocity

Influence	Mode	Examples
Change elastic stiffness	B, S, P	Temperature change; sorption
Change density	B, S, P	Temperature change; polymer curing; sorption
Change piezoelectric stiffness	B, S, P	Dielectric loading; illumination of semiconductor
Change thickness	B, P	Etching; deposition; sorption
Change length	B, S, P	Temperature change; change of transducer position
Change tension	P	Pressure; acceleration force
Reactive surface	B, S, P	Sorption
Dissipative surface loading	B, S, P	Viscous fluid loading

B, S, and P denote bulk, surface, and flexural plate modes respectively.

3.3 PIEZOELECTRIC MATERIALS

3.3.1 Piezoelectricity

Piezoelectric properties of a medium can be explained by the concept of piezoelectricity, which determines the distribution of the electric polarization and demonstrates how a piezoelectric field reacts to an electrical stress by emitting depolarization waves. The polarization field is linear with respect to mechanical strain in crystals belonging to certain symmetry classes. Examples of piezoelectric crystals and their related classes are: quartz, class 32; $LiNbO_3$, class 3m; $LiTaO_3$, class 3m; $Li_2B_4O_7$, class 4mm; and GaAs, 43m.[5] In equilibrium, crystal strain force is balanced by the internal polarization force. When this equilibrium is disturbed either by application of an external electric field or external mechanical stress, the emitting depolarization field will create a rebalance force to maintain the initial equilibrium. If the external force is from an electric field (applied voltage), a displacement will occur, but if the external force is from a mechanical displacement (vibration), an electric field will be produced. This phenomenon is called piezoelectricity.[7,8] The definition and methods of measurement of piezoelectric crystal units, in detail, are reported in the literature.[9]

Although, piezoelectricity has been known for more than a century, the first understanding of piezoelectric crystals was in 1910 after Voigt's publications on the physics of crystals.[10] He showed which classes of crystals (32 excited classes) are piezoelectric and he determined the nonzero coefficients of those classes. Later, in France, Langevin[11] conceived the idea of electrically exciting the quartz plate vibrators, and originated the development of modern ultrasonic devices.

Acoustic sensors can be designed based on piezoelectric crystals combined with hybrid signal-processing technology using available electrical components.

Examples of these devices are high-stability acoustic resonators and accelerometers.[12] The important design constants of piezo materials for SAW applications is given in Ref. 13. For low-cost acoustic sensors, a monolithic technology is required, where the sensor elements and electronics are processed on the same substrate. The monolithic devices have the advantage of high sensitivity. To design and fabricate monolithic acoustic sensors compatible with standard VLSI technology, a piezoelectric film is required. Processing a single-crystal piezoelectric film on the semiconductor substrate is very difficult and in some cases is impractical. Fortunately, in most cases, an oriented (z-axis) sputtered piezoelectric film is adequate for processing practical sensors.

3.3.2 Piezoelectric Thin Films

For the design of semiconductor sensors, performance characteristics, reliability, and acoustic parameters of the piezo-films are considered essential. During the past decade, many researchers reported useful characteristics of piezo-films for different methods of growth and different applications.[13,14] There are many piezoelectric films reported. The three films most popular with the sensor industry are ZnO, AlN and PZT. The major properties considered in design trade-offs for monolithic sensors are as follows:

1. value of electromechanical coupling;
2. good adhesion to substrate;
3. resistance to environmental effects (e.g., humidity, temperature);
4. VLSI process compatible (e.g., deposition methods and etching);
5. temperature and acceleration sensitivity;
6. cost effectiveness.

These key films are considered essential for producing high-performance acoustic sensors. Table 2 shows a summary of possible applications and potential

TABLE 2 Application Summary of Three Major Piezo-Films ZnO, AlN, and PZT

Applications	ZnO	AlN	PZT	Others
Pressure sensors	√			√
Gas sensors				√
Bulk acoustic resonators	√	√		
Plate mode sensors	√			√
Accelerometers	√			√
TV VIF filters	√		√	
SAW devices	√	√	√	
Actuator/translator	√		√	

uses of these films. In the following we will highlight the properties and methods of processing of these three piezoelectric films. A variety of methods have been used. Depending on what material and what substrate are used, vacuum deposition, sputtering, and sol-gel are the major methods currently used for preparing piezo films. Sol-gel is used exclusively for piezoceramic films (PZT, PLZT). Vacuum deposition of piezoelectric films can be achieved by direct-evaporation,[15] indirect evaporation,[16] or more commonly, by a reactive evaporation method.[17] For our suggested key piezoelectric films, reactive evaporation is used for processing ZnO and AlN. Sputtering is the second approach to vacuum deposition of thin piezo-films and several variations of this process have been reported. Examples of some variations are: reactive diode sputtering;[18] triode sputtering;[19] RF sputtering,[20] and magnetron and reactive magnetron sputtering.[21,22] Most of the sputtering methods are used for ZnO and AlN. In the following sections we will discuss processing, properties, process compatibility, and fundamental design rules for each individual piezoelectric film (ZnO, AlN and PZT). As part of this discussion, we will explain some details of sputtering in the section on ZnO and AlN, and sol-gel in the section on PZT.

3.3.3 Zinc Oxide (ZnO) Thin Films

ZnO was the first piezoelectric material to be used for commercial applications (e.g., TV SAW filters[23]). Because of its high piezoelectric coupling (compared with non-ceramic materials), great stability of its hexagonal phase, and its pyroelectric property, it plays an important role in research and development of acoustic sensors. Another advantage of ZnO is the ease of chemical etching, which makes it ideal for many sensor applications.

Bulk acoustic or SAW measurements are usually used to study ZnO films. In the case of bulk acoustic waves, the tuned round-trip insertion loss and, in the case of SAW delay lines and SAW convolvers and correlator, the output insertion losses and phase characteristics are used to measure the piezoelectric coupling of sputtered ZnO films.[22,24]

The first sputtered piezoelectric ZnO film, reported[18] in 1965, employed diode sputtering of a pure zinc target that was reactively sputtered in an oxygen/argon mixture at 50 microns pressure. Improved deposits were obtained with triode sputtering[25-27] with submicron pressures. Various improvements and inclusion of a post-annealing step have produced high-quality ZnO films having a c-axis orientation normal to the substrate to within 2° and sputtering rates up to 3 microns/hour.

Higher sputtering rates and excellent film quality have resulted from the use of DC[25] and RF[28] magnetron sputtering. ZnO films have also been made[29] by laser-assisted evaporation employing a CO_2 laser and a ZnO powder source in a vacuum chamber. In most sensor applications, ZnO film deposition employs a planar magnetron and so-called S-gun systems, in which the ZnO source is the planar cathode, surrounded by a dark space shield and backed by permanent magnets (see Ch. II-4 of Ref. 30). The advantage of planar magnetron sputtering is that electron bombardment of the substrate, such as a silicon wafer, is greatly reduced, so that the wafer temperature can be better controlled with an external heater coupled to the wafer. Good-quality sputtered ZnO films have been deposited on substrates such as gold, aluminum, platinum, crystalline quartz and oxidized silicon.

The sputter S-gun is a circular magnetron in which an intense plasma is formed near the cathode due to ionization resulting from bombardment because of the $\mathscr{E} \times B$ configuration.[27] This source is a high-current, low-voltage source operating at relatively low pressure, as are other magnetron systems, while conventional diode sources tend to be high-voltage, low-current sources operating at higher pressures.

Laser-assisted evaporation of thin-film ZnO has some unique advantages.[31] First, the process is clean as the source acts as its own crucible. Multiple and large-area sources can be evaporated by scanning the laser beam with external mirrors. The process can be controlled well by controlling power density and pulse duration. Compounds or mixtures can be evaporated concurrently and *in situ* laser annealing is possible. Lithium-doped ZnO films having excellent orientation and a smooth surface have been deposited on quartz and oxidized silicon at temperatures as low as 150°C.

3.3.4 Aluminum Nitride (AlN) Thin Films

Aluminum nitride is another promising thin-film piezoelectric material due to its high acoustic velocity and its endurance in humidity and high temperature. The piezoelectric coupling factor of AlN is relatively large; it is lower than ZnO but higher than most other thin-film piezoelectric materials, excluding thin-film piezoelectric ceramics. The high acoustic velocity of AlN makes it very attractive at frequencies in the GHz range where the device feature sizes are limited by VLSI critical dimensions (0.2–0.5 μm). The AlN endurance to high temperature makes it compatible with silicon and GaAs monolithic processes required for diffusion, annealing and ion implantation. The AlN resistance to humidity makes it suitable for commercial non-hermetic packaging. The initial investigations for preparing AlN films were by chemical vapor deposition (CVD).[32,33]

During the early years of the past decade, high-frequency SAW devices were reported[34,35] using metal-organic CVD (MOCVD) methods for preparing AlN. In these reports, material constants of AlN and SAW properties of AlN/Al$_2$O$_3$ and AlN/Si devices are presented. Most of these material constants were unknown at that time. Reference 35 provided information for designing SAW devices and acoustic bulk resonators based on AlN piezo films. A SAW delay line with zero temperature coefficient was reported[36] that was an introduction to the design and fabrication of high-frequency, high-stability resonators/filters using thin-film piezo-AlN. Another method processing of AlN is reactive molecular beam epitaxy (MBE).[37]

The drawback of AlN thin films, produced by CVD or MBE methods, is the high temperature (1000–1300°C) requirement for substrate heating. This requirement constrains the substrate choices and restricts standard VLSI metallization for interface electroding of the acoustic sensor devices.

In the early 1980s, RF planar-magnetron sputtering technology, which was well known for ZnO, was applied to AlN. Researchers were reporting low-temperature deposition of c-axis AlN films on glass substrates by RF magnetron sputtering.[38,39] Later, it was demonstrated that both c-axis and single-crystalline AlN films can be grown at substrate temperatures as low as 50 to 500°C.[40] A summary of magnetron sputtering conditions for low temperature processing of AlN is shown below:[40]

Atmospheric gas	$Ar + N_2$ (1:1) or N_2
Gas pressure	$10^{-2} - 3 \times 10^{-3}$ Torr
Substrate temperature	50–500°C
Target material	99.6–99.99% pure Al
Target size	diameter 100 mm, thickness 6 (mm)
Target-substrate spacing	40 mm
Input RF power	100–200 W
Film-thickness range	1–7 μm
Sputtering rate	0.2–0.8 (μm/h)

Figure 5 shows an X-ray diffraction pattern and rocking curve of a 0.8 μm-thick AlN film sputtered on a basal plane of Al_2O_3. The minimum value of s, the standard deviation of c-axis orientation, calculated from Fig. 5 is 0.72°.

During the past decade many researchers have worked on the development of piezoelectric AlN films for applications in SAW devices and acoustic sensors. The AlN films are most attractive for high-frequency, bulk acoustic resonantors

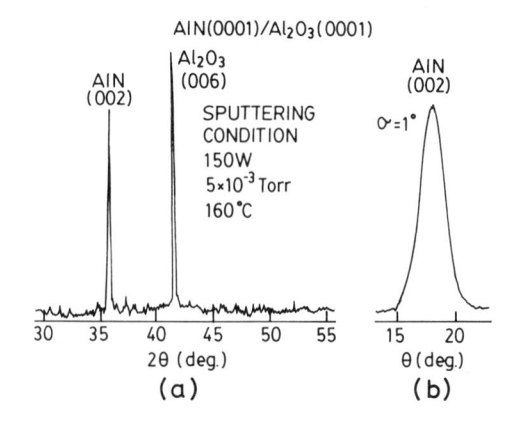

Fig. 5 Sputtering deposition of AlN films on basal plane of Al_2O_3. (a) X-ray diffraction pattern and rocking curve of a 0.8 μm thick AlN film; (b) related rocking curve. (After Ref. 40)

and SAW sensors. For more information about AlN films refer to the references cited in this section.

3.3.5 Pb(Zr, Ti)O$_3$ (PZT) Thin Films

High-coupling piezoelectric materials for fabrication of semiconductor integrated sensors are in great demand. The thin-film form of the solid solution system Pb(Zr, Ti)O$_3$, called PZT, has high piezo-coupling, very high dielectric constant[41] and its semiconductor application is suitable for the development of sensors. The piezoelectric coupling of PZT, in some cases, is more than an order of magnitude larger than either of ZnO or AlN. In addition to high piezoelectric coupling, PZT has a large pyroelectric response and a large spontaneous polarization, which makes it a good candidate material for IR detectors and non-volatile memory devices. Since we have limited space to cover the vast range of sensor applications in this chapter, we will only emphasize a few applications related to acoustic sensors such as SAW sensors and actuators.

Many potential applications of PZT have been suggested and some have already been investigated. Some examples are: SAW delay lines,[42,43] pyroelectric sensors[44] and memory devices. Growing interest in PZT applications during the past five years suggests a need for refining industrial PZT processing.

For more than a decade, the fabrication of PZT thin films has been investigated by many researchers. Many methods and techniques have been reported such as electron-beam evaporation,[45] RF sputtering,[46] ion-beam deposition,[47] epitaxial growth by RF sputtering,[48] magnetron sputtering,[49] MOCVD,[50] laser ablation,[51] and sol-gel.[52] Among these methods, more work has been done in RF sputtering and chemical sol-gel techniques.

RF sputtering requires the geometry between the source material and the substrate to be under very tight control. In addition, to process a good-quality thin-film PZT, deposition parameters such as substrate temperature, gas combination and pressure inside the chamber, RF power, and deposition intervals should be controlled precisely during the deposition. These tight requirements for sputtering make it unattractive for industrial processing. The advantage of the sputtering method is the high-quality surface topology with fine grain boundaries.

The sol-gel method is based on spinning of a chemical solution, allowing easier composition control and film homogeneity. The method has the potential of low cost and can be used to process large-area substrates. In the sol-gel method, the substrate material, additional buffer layers, interface metallization film for electrodes and thermal mismatch coefficients of all the layers of the deposition film should be carefully studied. For this reason it is very difficult to process high-quality PZT thin films that are thicker than 1 μm. If the thermal mismatch is not carefully considered, localized microscopic cracks will form during the firing cycles and thermal annealing, which will be propagated through the multi-layers. PZT with localized cracks has a low piezoelectric coupling and results in very poor device performance. For a more detailed understanding of PZT thin-film processing, refer to the references cited in this section.

3.4 ACOUSTIC SENSING

Several important issues apply to all of the types of acoustic sensors considered in this chapter. These are the methods used to obtain information about measurands, the relative sensitivity of gravimetric acoustic sensors, and the fabrication techniques used to make acoustic sensors.

3.4.1 Measurement Methods

Let us consider how the phase velocity or attenuation changes of an acoustic sensor might be detected. Figure 6 summarizes the main options. We may use either a resonator or a delay line; for example, Fig. 1a shows a bulk-wave resonator, and Fig. 1b shows a SAW delay line. With either a resonator or a delay line, we may make measurements on the device itself or incorporate it in a circuit involving electrical feedback to form an oscillator. We will refer to these two options as passive or active circuit approaches, respectively.

For passive bulk-wave resonators, we can measure resonant frequency, f_{res}, to infer the wavelength, λ_{res}, and, hence the velocity from the relation

$$v_p = f_{res}\lambda_{res} = f_{res}d/2, \tag{1}$$

where d is the resonator thickness. For a passive delay line, the phase shift between input and output transducers, which are separated by a known distance, yields the velocity. For resonator or delay-line oscillators, the oscillator frequency is usually desired; it can be measured easily and precisely with a digital counter.

For investigating wave attenuation due to dissipative processes, such as that

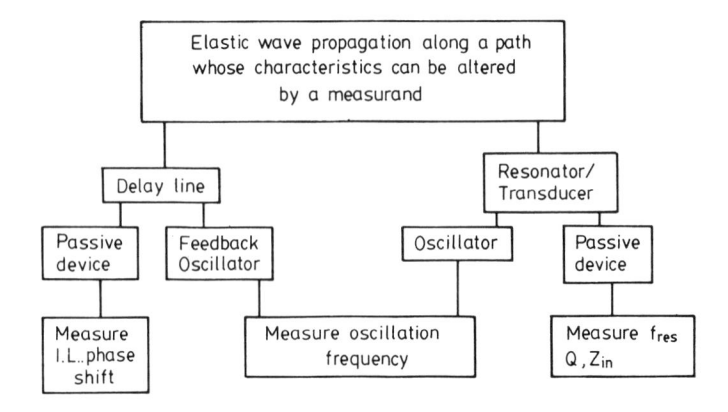

Fig. 6 Measurement options for ultrasonic resonators and delay lines.

caused by a viscous liquid in contact with one of these sensors, we measure the quality factor, Q (or, equivalently, the width of the passband of the resonator), the insertion loss of the delay line, or the gain of the active circuit used for the oscillator. It is often useful to measure, simultaneously, the changes in both the phase velocity and the attenuation.

3.4.2 Sensitivity of Gravimetric Acoustic Sensors

The use of acoustic sensors to measure amounts of mass deposited on their surfaces was first analyzed by Sauerbrey who showed that the fractional frequency shift, $\Delta f/f_{res}$, of a quartz bulk-wave resonator crystal caused by addition of a mass per unit area Δm could be written in terms of a gravimetric sensitivity factor, S_m, as

$$\Delta f/f_{res} = S_m \Delta m, \qquad (2)$$

where $\Delta f = f_{loaded} - f_{res}$ is the frequency shift from the unloaded resonant frequency value, f_{res}, produced by the loading. The factor of proportionality S_m, is a negative quantity for pure mass loading (i.e., loading that does not stiffen the sensor). The numerical value of S_m depends upon the design, material, and operating frequency or wavelength of the acoustic sensor. As an example, it was found that the frequency of a 6-MHz, AT-cut quartz crystal resonator decreases 1 Hz upon sorption of just 12 nanograms (10^{-9} g) of gas molecules per square centimeter onto one of its faces. Response to sub-monolayer coatings can readily be observed with suitably designed acoustic sensors.

The gravimetric sensitivity factors for the main types of acoustic sensors (shown in Fig. 1) are summarized in Table 3, along with experimentally determined values of S_m. The physical principle underlying these results is that the response of gravimetric acoustic sensors to surface loading depends on the relative amounts of mass added to the surface and the mass of the portion of the sensor in which the wave energy propagates. In bulk-wave sensors the energy travels in a half-wavelength-thick region, and in a SAW sensor the energy is confined to a depth of about one wavelength. By contrast, with microfabrication, membranes for flexural-plate-wave (FPW) sensors that are only a few microns thick can be made. Hence, it is reasonable that FPW sensors should have a high gravimetric sensitivity, and that the gravimetric sensitivity of SAW sensors can be increased by designing them for high-frequency, small-wavelength operation. The advantage of SAW sensors over bulk-wave gravimetric sensors lies in the ability to fabricate SAW sensors that operate at much higher frequencies than conventional bulk-wave resonator sensors.

3.4.3 Fabrication Techniques

In Chapter 2 the topic of fabrication is covered in detail under semiconductor technology. Here, we summarize those topics related to acoustic sensors, SAW

TABLE 3 Mass Sensitivities S_m of TSM, SAW, APM and FPW Sensors (From Ref. 80)

Sensor	Theoretical Sensitivity S_m	Experimental Devices – Device Description	Operating Frequency (MHz)	Experimental S_m Value (cm²/g)	Calculated S_m Value (cm²/g)	Reference
Bulk-wave resonator	$-2/\rho\lambda$ or $n/\rho d$	AT-cut quartz TSM	6	-14	-14	54
Surface-wave device (assumed elastically isotropic)	$-K(\sigma)/\rho\lambda$	ST-cut SAW delay line	112	-91^a	$-151^b(-155)^c$	55
Acoustic plate-mode device	$-1/\rho d$	ST-cut quartz APM	97	-9.5 (zeroth mode); -19.4 (higher modes)	-9.3 (zeroth mode); -18.6 (higher modes)	Sec. 3.3.2 of this Chapter
Flexural plate device	$-1/2\rho d$	ZnO-on-silicon nitride FPW delay line	4.7 2.6	-442^d -990^d	-450^e -951^e	80

ρ, σ and d are the density, Poisson's ratio and thickness of the device material respectively, λ is the wavelength and n is an integer corresponding to operation at the nth harmonic. The factor $K(\sigma)$ ranges from 0.8 (for $\sigma = 0.5$) to 2.2 (for $\sigma = 0$).
[a] Experimental value determined by deposition of Langmuir–Blodgett films.
[b] Experimental value for mass loading alone.
[c] Mass-loading value calculated for SAW in isotropic solid using $S_m = K(\sigma)/\rho d$ assuming Poisson's ratio $\sigma = 0.35$.
[d] Liquid loading experiments and verified functional dependences gave value of M, membrane mass per unit area; experimental value of S_m is then $S_m = -1/2M$.
[e] Value of M is based on composite membrane thickness and densities. S_m is then given by $S_m = -1/2M$.

sensors, based on micromachining of silicon. Micromachining of silicon has been used to fabricate miniature three-dimensional acoustic sensors.[53,54] Most of these devices use thin membranes or cantilever beams integrated with piezoresistive or piezoelectric sensing elements. For such integrated sensors, it is more economical to make the micromechanical structures completely by planar processes on one surface of the wafer to assure processing compatibility with conventional integrated circuits. But for applications which require much thicker structures (e.g., for high-pressure sensors or for very high g-level accelerometers), the thickness of thermal oxide diaphragms or beams may be inadequate, and it becomes necessary to fabricate the device by a two-sided etching technique using silicon as part of the device structure.

In this section we consider fabrication techniques for silicon acoustic sensors, using controlled etching for two-sided processing. Micromachining can be performed by dry etching using SF_6 gas, chemical anisotropic etching (e.g., using EDP, KOH, NaOH), or isotropic etching using acid-based solutions. Anisotropic etching, covered in Chapter 2, is a key process for micromachining silicon acoustic sensors.

Two-sided processes include a back-side etch in which etching takes place nearly through the entire wafer thickness. Because wafer thicknesses vary somewhat, this etch process cannot be controlled accurately by timing alone. Two control methods are used. In one, a layer of heavily doped p^+ silicon serves as an etch-stop for EDP. In the other, an electrochemical etching process is used that employs current monitoring to reach a desired limiting thickness.

Back-side Etch-Stop for EDP Using p^+ Doping. Heavily boron-doped (p^+) silicon effectively resists EDP etching. A wafer with such an etch-stop at a preselected depth can be fabricated using an epitaxial layer of silicon (with accurately controlled thickness) over a p^+ surface. (Patterning of the p^+ layer can be used to make elaborate 3D structures.) When such a wafer is etched from the back-side, the p^+ layer acts as an etch-stop, leaving the epitaxial layer intact. This fabrication technique has been used for obtaining thick diaphragms and beams in acoustic sensors.

Electrochemical Control (ECC) Etch-Stop. In this technique,[56] two epitaxial layers of silicon are used: the one closer to the substrate that has very high resistivity is required for accurate control of the etch depth, and the epi layer, with moderate resistivity, on the top of the first layer, is used for VLSI device processing. The front of the wafer is orientationally etched through the epitaxial layers as a compatible process step during the monolithic IC processing of the acoustic sensor. ECC etching is done after the back of the sensor is patterned. All process steps are designed to maximize the degree of compatibility with the current integrated-circuit processing technology.

ECC Etch Example. Figure 7 shows a simplified process flowchart for the fabrication of a cantilever beam to illustrate the ECC etch-stop process. The starting material is a 3-inch silicon wafer, p-type, with 0.01 Ω-cm resistivity and (100) orientation. Two layers of n-type epi are deposited on the front of the wafer: the epi layer closer to the substrate is (30 Ω-cm) and 20 microns thick; the epi layer on the top has resistivity of (10 Ω-cm) and it is about 10 microns thick. The total thickness of the epi layer is about 30 microns. The front-side anisotropic etching process is performed by EDP solution inside a controlled reflux system at 95°C for approximately 1.5 h. The etch rate at this temperature is about 25 μm/h assuring the passing for 1.5 h through the 30 micron epi layers.

Figure 7 is representative of the device processing steps and, for simplicity, only the piezoelectric capacitor element is shown in the cross-sections. Figure 7e and f are representative of the back-side isotropic etching process. This etching step is performed by the electrochemical control etching technique. At this stage the front of the wafer is completely processed. A layer of 3000 Å gold is deposited on the back of the wafer and both sides of the wafer are coated with a thick resist. The resist is patterned aligned to the front-side beam and the front-side resist is completely exposed to be protected in the developing process. After the photoresist is patterned, the gold is etched (the device at this stage is shown in Figure 7e) and is ready to be mounted for ECC etching. Figure 7f shows the end view of the beam after the ECC etching process is successfully performed.

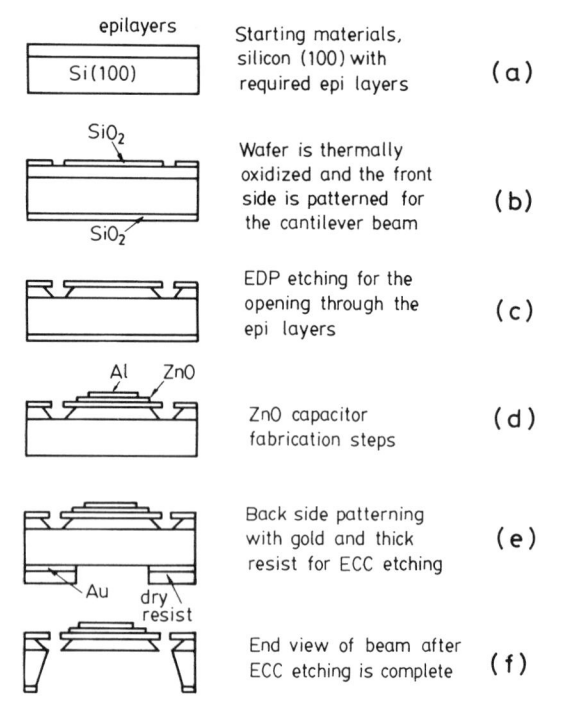

Fig. 7 Simplified process flow for processing silicon cantilever beam. (See text for description.)

ECC Etch Apparatus. The experimental arrangement used for the electrochemical control etching is shown in Fig. 8. A 3-inch wafer, which is successfully processed to the stage shown in Fig. 7e, is mounted face down with wax. A platinum wire makes contact to the gold on the back of the wafer causing the low-resistivity silicon substrate to act as an anode when the wafer holder is immersed in the solution. The cathode electrode is a platinum screen connected through a platinum wire to the negative terminal of the power supply. A digital meter monitors the current, which it records on the strip chart recorded during the etching process. The etchant solution is a mixture of $H_2SO_4:HF:H_2O$, 10:10:80. A special acid-resistant pump with speed control is used to circulate the solution and pass it through the screen for better etch rate and etch uniformity. Initially, when the wafer is immersed in the solution, the current is steady and proportional to the exposed silicon. The initial current path is closed through the low-resistivity substrate, but when the etch surface reaches the high-resistivity epi, the resistance will restrict the current and cause a drop of 50–90% from the initial current level. The results of this method indicate that the beam thickness can be controlled to about a fraction of a μm, and is dependent on the RMS roughness of the back-side of the wafer. The initial current density varies, ranging from 0.2 to 0.7 A/cm^2. As the current is increased, the etch rate will increase. We were able to demonstrate etch rates from 6 mil/hr to 20 mil/hr (1 mil = 2.54×10^{-3} cm). The etch rate is also a function of the pump speed.

This method has been used to fabricate pressure sensors and accelerometers with different g-range performances. The accelerometer devices that are processed by this method can be designed for either high-g or low-g acceleration

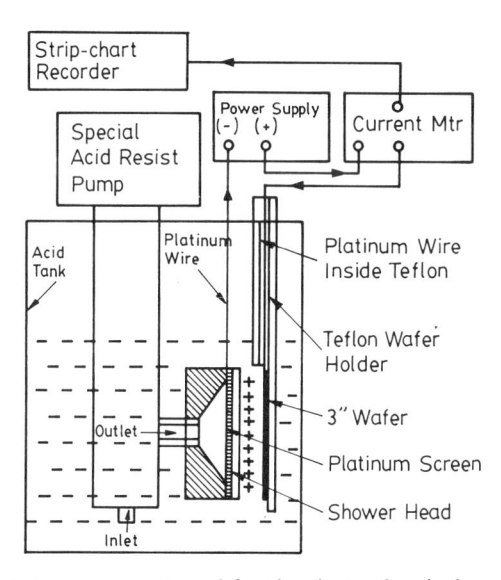

Fig. 8 Experimental arrangement used for the electrochemical control etching (ECC) technique.

performances. Beam dimensions are 1×2 mm horizontally, and the thickness varies from 5 to 30 μm depending on the g acceleration to be measured.

Accelerometer Process. In processing a silicon cantilever beam, etching steps are important phases of the processing. If the beam material is SiO_2 or Si a few microns thick, it is possible to fabricate the beam by a single-side etch process.[55] This etch step is best done after the device is completed, and usually uses an anisotropic etchant like EDP, NaOH or KOH.

If a thicker silicon cantilever beam is required (high g-level accelerometers) a two-sided etching technique is essential. EDP solution is used for two-sided etching, or electrochemical etching[56] can be used to etch the back. A simplified process chart for the two-sided etching technique is shown in Fig. 7. For a more detailed understanding of the sensor process refer to the previous section (Section 3.3). A complete processed device on the front-side is shown in Fig. 9.

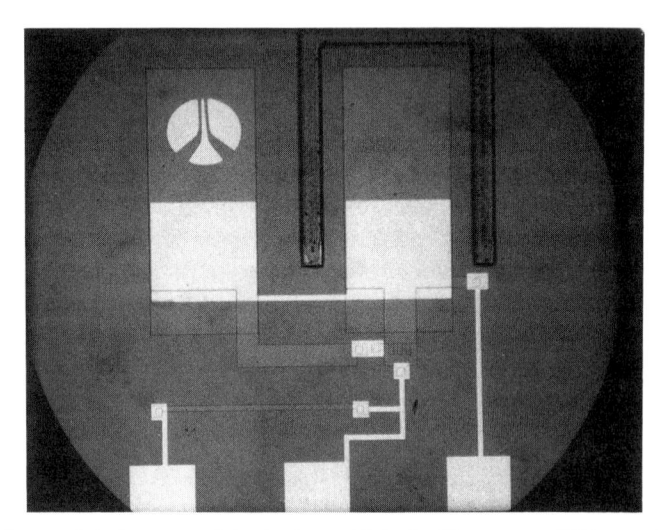

Fig. 9 A completed processed device (one chip) of a silicon monolithic accelerometer.

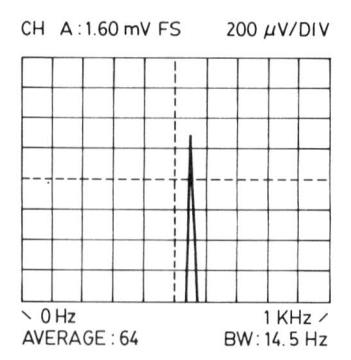

Fig. 10 Output response of the accelerometer device for an input of 50-g sinusoidal signal.

Figure 10 shows the output response of this device for an input of a 50-g sinusoidal signal. The dynamic response indicates a high signal-to-noise ratio and very good sensitivity.

3.5 SAW SENSORS

3.5.1 Monolithic SAW Technology

The recent development of monolithic SAW devices represents a breakthrough in the applications of acoustic sensors in navigation and communications signal processing. A piezoelectric thin-film material, such as ZnO, AlN or PZT, permits both acousto-electric SAW and semiconductor electronic device components to be fabricated on the same monolithic substrate.[57,58] The development of such monolithic signal processors is expected to result in very large time-bandwidth devices monolithically integrated on a single substrate with temperature compensation[59] and compatibility with low-cost batch-fabrication techniques.

The basic operating principle for SAW devices on Si substrates does not differ greatly from those on piezoelectric-type substrates. For monolithic SAW transductions, the substrate is generally a semiconductor such as Si or GaAs. All SAW devices are based on some form of piezoelectric effect, which is necessary to transform an electrical signal into a mechanical or acoustic signal. Traditionally, the substrate for a SAW device has been a large single-crystal substrate using a piezoelectric insulating material. The single-crystal material has a dual function. It generates the acoustic waves and also provides a propagation medium for the generated waves. Analysis of the physics of SAW on such a substrate is straightforward. Since the substrate is much thicker than the SAW wavelength and the acoustic energy is confined to within a few wavelengths of the surface, the substrate can be approximated as a semi-infinite half-space consisting of the single crystal.

In contrast, the analysis of the SAW propagating on non-piezoelectric substrates such as Si or GaAs is complex. Although the wave solutions on the bare substrate are simple, the bare substrate cannot be used alone. It requires some form of thin-film piezoelectric layer to generate the acoustic waves. By using a thin-film piezoelectric material such as ZnO on Si or AlN on GaAs, a double-layer structure consisting of two materials of different properties is formed. Because of these conditions, both the SAW velocity and piezoelectric coupling will be dispersive. Since the piezoelectric film is applied by methods such as chemical vapor deposition or sputtering, the structure obtained is generally polycrystalline with a relatively rough surface. In spite of this, the fact that SAW sensor elements and co-processors can be integrated on the same chip makes the device extremely attractive.

Interdigital transducers (IDT) formed by photolithography on the surface of the piezoelectric thin-film overlay generate the surface waves. In their simplest form, IDTs consist of many parallel fingers alternately connected to opposite electrodes to which the signal is applied. The sputtered ZnO thin films on Si are used for generating SAWs at practical frequency ranges from 100 to 500 MHz. We will discuss a general IDT design that has been fabricated and tested to optimize parameters associated with the IDTs, such as insertion loss and bandwidth. The chip in this design covers an area of 2.5×2.5 mm^2. Each chip contains transducers with 10 and 20 finger pairs and three different acoustic wavelengths, λ_a. One of the transducers (20 finger pairs with $\lambda_a = 12$ μm) is selected for discussion and it is shown in Fig. 11. Figure 11 shows the three process layers superimposed.

In a typical ZnO-on-Si device, the substrate is (111) Si which is covered by thermal oxide (SiO$_2$) 0.6–1.0 μm in thickness. Cr–Au pads, 600 Å thick are deposited under the IDT areas. This Au film not only establishes a vertical electric field within the IDT structure to increase mechanical coupling, but at the same time improves the c-axis orientation of ZnO which, in turn, also increases piezoelectric coupling.

The Cr–Au films are carefully inspected for surface texture and extremely low defect density. After a careful cleaning step, the ZnO and Al films for the IDT electrodes are deposited in vacuum on the wafer. For SAW transducer applications, the ZnO thickness is in the range of 3000 to 5000 Å. This thickness

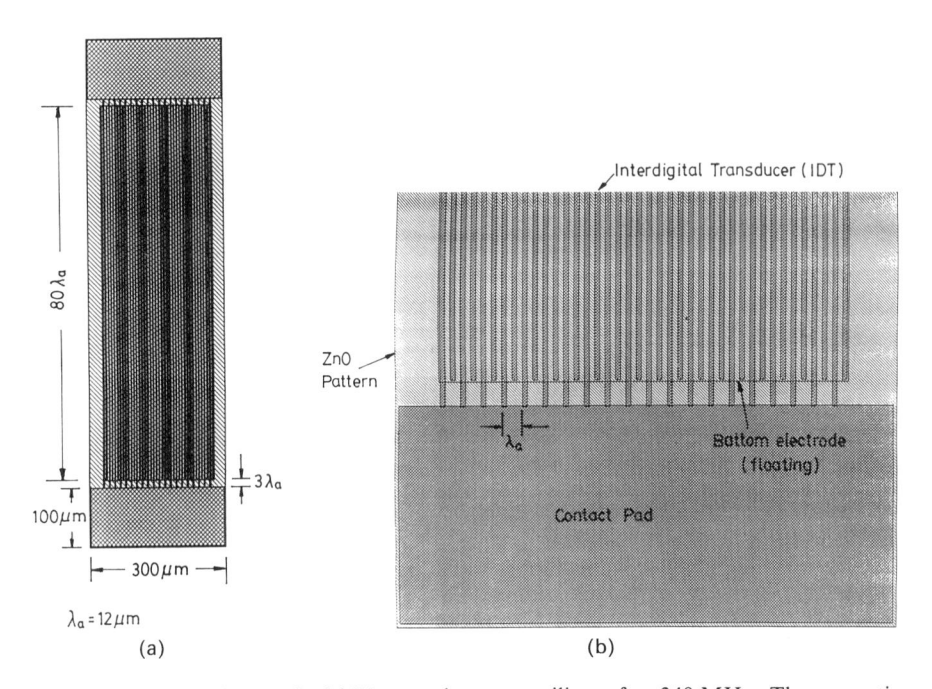

Fig. 11 Design of a ZnO SAW transducer on silicon for 340 MHz. The acoustic wavelength, $\lambda_a = 12$ μm. The interdigital transducer (IDT) dimensions are shown in λ_a. (a) Transducer dimensions; (b) a close up of electrodes.

has been determined as the optimum range for tuning the $\Delta v/v$ curve to the first mechanical coupling peak for a ZnO–SiO$_2$–Si structure. After the films are deposited, the IDT is patterned using conventional photolithographic techniques. This is followed by a second photolithographic patterning that defines the ZnO pad areas. The transducers have an acoustic wavelength (λ_a) of 12 μm and an electrode overlap of $W = 80\ \lambda_a$.

Two different thicknesses of the ZnO layer have been used successfully in the transducer structures. The thicker film, typically 0.5–1.0 μm in thickness, is used to generate the second spatial harmonic or Sezawa mode of the layered structure. The thinner film is typically five percent of a wavelength in thickness and is used to generate the fundamental mode of a layered wave.[60] However, the higher thickness requirements for the Sezawa mode are not compatible with standard semiconductor processing. Therefore, the fundamental mode is normally used. Figure 12 shows the amplitude of transmission of the fundamental mode through a 1.4-inch delay-line device that has a center frequency at 340 MHz. In this device, insertion loss is better than −40 dB (not tuned) at a crosstalk level of −90 dB, resulting in a dynamic range of better than 50 dB. The delay response of this SAW device is shown in Fig. 13. In this figure, the signal information pulse is produced after a 8.4 μs delay.

3.5.2 SAW Characterization

SAW devices such as delay lines, filters, convolvers, resonators, and oscillators are key components for acoustic sensors. All these devices are in some way related to elasto-electric conversion phenomena. The elasto-electric conversion in SAW devices is produced by interdigital transducers. IDTs are the SAW key components and their characterization and testing is essential to study the design and performance of SAW devices.

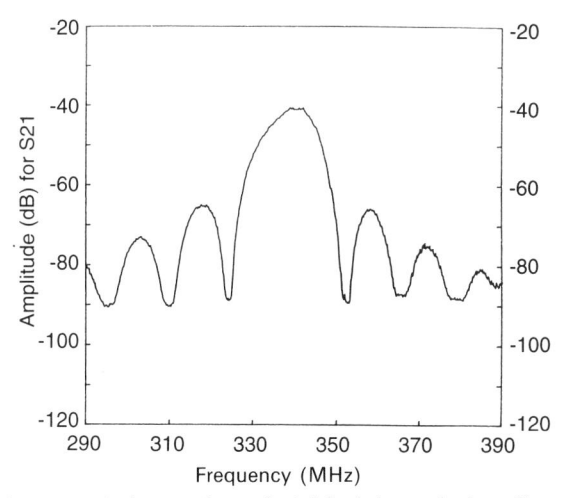

Fig. 12 Acoustic transmission a through 1.4-inch-long device. Center frequency is 340 MHz. (Not tuned for circuit stray elements.)

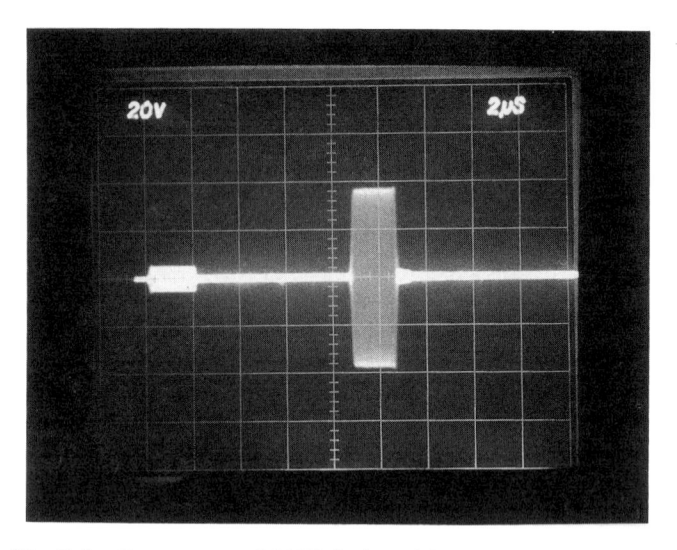

Fig. 13   Delay line output of SAW device with 8.4 μm delay at 340 MHz.

Traditional methods of evaluating acoustic devices in terms of material parameters require a measurement of the transducer electrical parameters at resonance. For example, the electromechanical coupling coefficient for a simple SAW transducer is expressed in terms of the measured parallel transducer resistance at resonance[61]

$$Rp = (8k^2 F_0 C_s N^2)^{-1}, \tag{3}$$

where N is the number of finger-pair electrodes, C_s is the capacitance per finger-pair, k is the piezoelectric coupling and F_0 is the resonant frequency.

In principle, only one measurement at F_0 is required, however, when parasitic reactances are not small and/or the frequency is high, significant measurement errors can result in anomalous values for k^2. Here, an approach is described that utilizes automatic-network analysis techniques. Recent advances in computational interface hardware and associated measurement hardware have made it possible to design software that measures and models IDT behavior over a wide frequency range. The network-analysis method measures complex return loss spectra, commonly referred to as scattering parameters, S_{ij}. These parameters can then in turn be transformed through software to represent any emittance characteristic desired.

We will discuss a general numerical method for determining equivalent circuit parameters from measurement data and the results of measurements on ZnO overlay transducers.

The method described here requires a network measurement system and a network analyzer. The important elements of the total characterization system include a frequency synthesizer, for phase-locked resonator measurements, as

3.5 SAW SENSORS   121

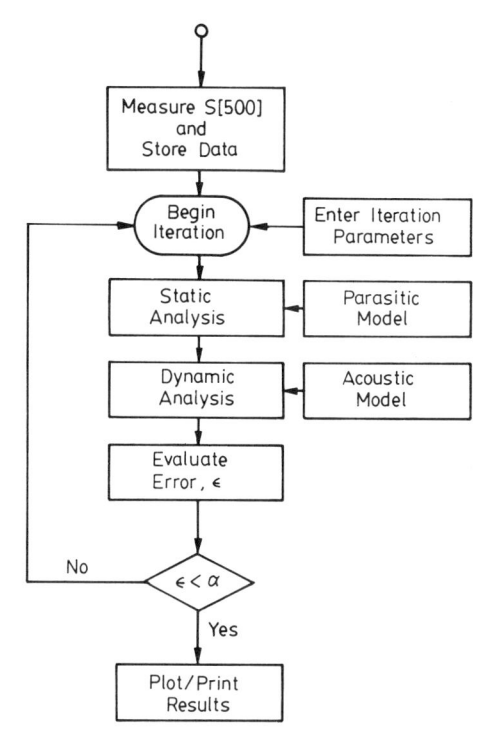

Fig. 14 Block diagram of numerical method used in software to determine transducer equivalent circuit values from stored numerical values of return loss.

well as a normalizer for averaging and digital data storage. The entire system can be interconnected via a data bus and can be controlled by a computer.

A flowchart of the numerical method is shown in Fig. 14. The device to be evaluated (transducer) is connected to the network analyzer and the measurement is controlled by software. Scattering parameters are measured, stored, and transmitted over the data bus to the computer where analysis and modeling is performed. Prior to actual circuit modeling, the data are corrected for analyzer reference errors, such as might be encountered by devices at the end of long cables, using stored scattering parameter data taken on the analyzer system without the device under test being connected. Equivalent-circuit analysis consists of an iteration loop wherein parasitic-model circuit elements and acoustic-model parameters are evaluated. After the equivalent-circuit elements have been iteratively evaluated, the difference between circuit-model predictions and experimental data is summed at all of the data points being tested. The sum is considered as a cumulative error that is then used to determine whether the fit between modeling and experiment is satisfactory. If not, the gradient with respect to each model parameter being fitted is evaluated and used to generate corrected elemental values. The process is repeated until the cumulative error is smaller than a desired value α or until a maximum number of iterations have been performed. The output is then displayed.

In general, multi-parameter, gradient-search routines are needed to fit the data to equivalent-circuit models. However, the use of selected "frequency points" can greatly simplify the search routines by working in an area of the spectrum where the electrical characteristics are independent of all but a few equivalent-circuit parameters. For example, shunt capacitance is iteratively determined at a frequency that is the lowest analysis frequency. Similarly, series inductance is iteratively determined at the maximum frequency. Any of several equivalent-circuit models can be selected. Parasitics are normally modeled by a series inductance and resistance shunted by parallel capacitance and resistance. The parasitic elements are referred to as static elements because they do not change rapidly with frequency.

After the iterations on the static elements of the equivalent circuit are completed, a similar search is used to determine the acoustic parameters such as electromechanical coupling, k^2. These parameters lead to circuit elements that are referred to as dynamic in the sense that they cause the electrical characteristics to change rapidly with frequency in the neighborhood of the acoustic models, which can be either in-line or cross-field.[61] In the case of SAW resonators, simple tuned-circuit elements are used to fit experimental data in polar form.

To demonstrate the model, several results are presented using the parallel form of the crossed-field model for SAW transducers. The ability to determine coupling coefficients in the presence of loss is an important advantage of this method. As the frequency of operation is raised into the UHF range, parasitic reactances become increasingly difficult to avoid and large errors can result from unintentionally matching the transducer parameters with lead inductance.

Figure 15 shows the experimental and modeled return loss of a ZnO overlay transducer on fused quartz operating at 247 MHz. Considerable loss is indicated by a background return loss magnitude of -0.85 dB. Nevertheless, the analysis method correctly determined that $k^2 = 0.006$ for this transducer. The method also found parasitics to be caused by 11.93 nH in series with a 2.49 Ω resistance and a shunt resistance of 5300 Ω.

An important class of SAW transducers is ZnO overlays on semiconductor substrates such as Si and GaAs. Because the substrate is a semiconductor, broad-band parasitic conduction losses are inevitable. Furthermore, dispersion is known to occur and parasitic reactances can be a problem at high frequencies. Figure 16 is an analysis of a $ZnO/SiO_2/Si$ overlay transducer. The scattering parameter losses were difficult to remove experimentally in view of the high frequency, 320 MHz, nevertheless the coupling was found to be $k^2 = 0.007$ with a series inductance of 14 nH, series resistance of 2.7 Ω, and shunt resistance of 2737 Ω. The inclusion of wire inductance was crucial to the determination of coupling because the wire inductance partially matched the transducer at this frequency.

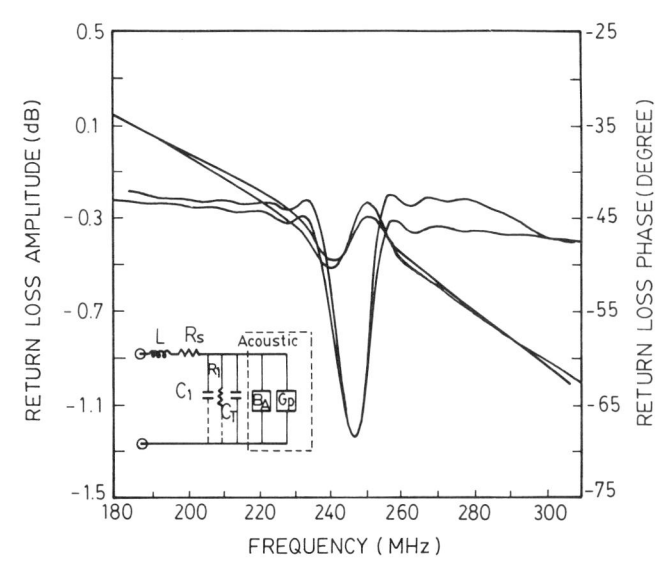

Fig. 15 Experimental and theoretical plots of return loss for a 20 finger-pair ZnO overlay transducer on fused quartz. Equivalent-circuit parameters are determined to characterize the piezoelectric coupling. Determined parameters are: $k^2 = 0.006$, $L = 12$ nH, $R_s = 2.5$ Ω, $R_1 = 5300$ Ω.

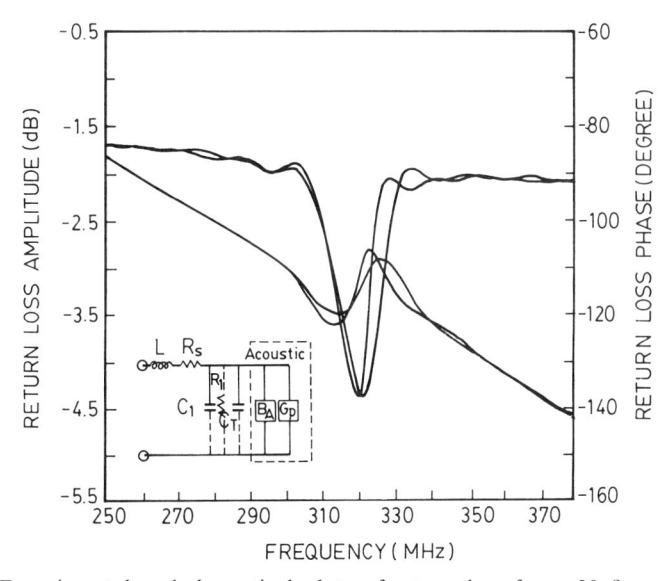

Fig. 16 Experimental and theoretical plots of return loss for a 20 finger-pair ZnO overlay transducer on thermally oxidized silicon. Equivalent-circuit parameters are determined to characterize the piezoelectric coupling. Determined parameters are: $k^2 = 0.007$, $L = 14$ nH, $R_s = 2.7$ Ω, $R_1 = 2737$ Ω.

3.5.3 SAW Gas Sensors

Devices that can accurately monitor concentrations of various gases and vapors are required for many applications in science and engineering. Many of the sensing devices in use today are rather large and bulky, and often lack the precision, speed, or reliability needed for advanced applications. Consequently, there has been increasing interest in the last 10–15 years in developing miniature solid-state devices for use as gas sensors. Three types of device are being intensively studied. These are: (1) semiconductor sensors fabricated on silicon substrates; (2) optical sensors using integrated optics or optical fibers; and (3) acoustic-wave sensors.

The earliest acoustic-wave gas sensors were those utilizing bulk acoustic waves. The bulk-wave piezoelectric gas (BWPG) detector was first introduced by King.[62] The BWPG detector is a simple piezoelectric device consisting of a bulk-wave resonator coated with selectively sorbent film. Detection occurs when mass loading, due to the selective sorption of a particular gas, produces a shift in the resonant frequency of the device. The BWPG detector has been used to detect microconcentrations of gases such as hydrogen sulfide, ammonia, hydrogen chloride, and sulfur dioxide.[63] The basic principle of the bulk-wave gas detector can be extended to surface acoustic waves. The first report of a chemical vapor sensor based on SAW technology appeared in 1979.[64] Since that time, a number of workers have investigated SAW sensors for detecting a variety of gases and vapors.

The basic structure of a SAW gas sensor is shown in Fig. 17. It consists of a SAW delay line connected in the feedback path of a suitable amplifier resulting

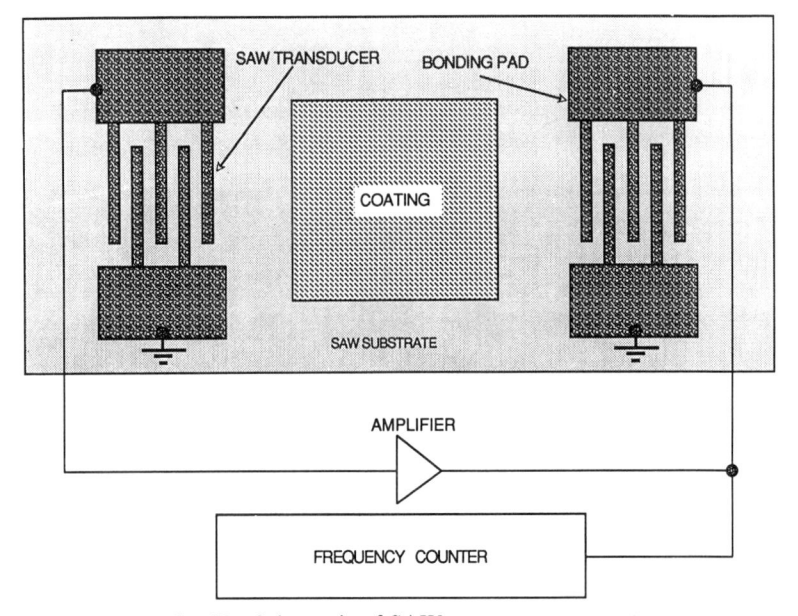

Fig. 17 Schematic of SAW gas sensor concept.

in a delay line stabilized SAW oscillator. For use as a gas sensor, the propagation path of the delay line is coated with a thin film of a suitable material (the chemical interface or membrane), which can selectively sorb the gas of interest. Sorption of gas results in a change in the mechanical and/or electrical properties of the interface. This results in a change in the time delay of the delay line and thereby in the frequency of the oscillator. Thus, the sensor is essentially an oscillator whose frequency is modulated by the gas concentration. Attractive features of the SAW chemical sensor include high sensitivity, wide dynamic range, direct digital (frequency) output, and potentially low cost due to fabrication compatibility with silicon microelectronic circuits.

The SAW vapor sensor is similar to the bulk-wave piezoelectric quartz-crystal vapor detector originally developed by King[62] and later extensively investigated by others.[63] Both these approaches are mass sensitive and require a selective coating; both use a shift in resonant frequency as the indicating signal. However, there are several significant differences which deserve mention. First, the SAW device is capable of operating at frequencies at least two orders of magnitude higher than the bulk-wave device and is, therefore, capable of much greater sensitivity. Second, the planar geometry of the SAW device allows one side to be rigidly mounted, making it even more rugged than bulk-wave devices. Third, the device is intrinsically a microsensor with its ability to occupy a volume as small as a few nanoliters. Finally, the SAW device allows multiple sensors to be fabricated close together on the same substrate. Thus, temperature drift compensation using a reference device can be achieved with great precision.

Considerable work has been reported in the last 10–15 years on SAW sensors for detecting a variety of gases and vapors. A summary of the various sensors reported to date is given in Table 4. The table lists the measurand (the gas or

TABLE 4 Summary of Various SAW Chemical Sensors

Measurand	Chemical Interface	SAW Substrate	Reference
Organic vapor	Polymer film	Quartz	81–83
SO_2	TEA[a]	Lithium niobate	84
H_2	Pd	Lithium niobate, silicon	85, 86
NH_3	Pt	Quartz	87
H_2S	WO_3	Lithium niobate	88
Water vapor	Hygroscopic	Lithium niobate	89, 90
NO_2	PC[b]	Lithium niobate, quartz	91–93
$NO_2, NH_3, CO, SO_2, CH_4$	PC[b]	Lithium niobate	94
Vapors of explosives, drugs	Polymer films	Quartz	95
CO_2, Methane	C[c]	Lithium niobate	96

[a] TEA = Triethanolamine.
[b] PC = Phthalocyanine.
[c] C = No chemical interface used. Detection based on changes in thermal conductivity produced by the gas.

vapor to be detected), the chemical interface used to detect the gas (vapor), and the substrate used for the SAW device. In most cases the substrate has been quartz or lithium niobate. In a few cases the SAW device has been fabricated on a silicon substrate. In this case, a thin film of piezoelectric zinc oxide is deposited on the substrate to enable generation and detection of SAWs by the IDT. Details about the performance (sensitivity, selectivity, response time, etc.) of the various sensors can be found in the references listed in the table.

An interesting feature associated with SAWs in relation to the detection of chemical species is the fact that for the absorption–desorption processes they do not require the presence of either a charge or heat variation; rather they require changes in the mass and/or elastic constants of the membrane absorbing the measurand. This particular feature gives SAW sensors the potential for wider use and makes them complementary to other sensors based on different working principles.

The main problem in SAW chemical sensors, a problem shared by other types of chemical sensors, is related to the difficulty of fabricating organic or inorganic coatings capable of selectively absorbing a given gas. It is a formidable task to identify or synthesize coatings that are selective to one and only one gas, and show no response (cross-sensitivity) to other interfering gas. This problem can be overcome by using an array of multiple sensors and coating each device with a different sensing material. The output of each sensor is then used as the input to a suitable pattern-recognition algorithm. This method can overcome not only the selectivity problem, but it can also allow the determination of the concentrations of individual components in a given gas mixture.[65]

Another interesting approach to the detection of various components in a multicomponent gas or vapor mixture has been described, in which both the velocity and the attenuation of the surface acoustic wave are measured as a function of gas concentration.[66] Information about both the velocity and attenuation allows species identification and quantification using a single SAW sensor.

3.6 SENSOR APPLICATIONS

3.6.1 Acceleration Sensors

The need for less costly, light-weight, high-performance acoustic sensors to read acceleration data of a moving vehicle has grown rapidly during the past decade. Some of the demanding applications are in rotation sensors (gyroscope), automated manufacture, automotive and other vehicles, vibration and seismic monitoring, scientific measurement, and in military and space systems. In some of these applications, discrete or hybrid structures are acceptable. In others, where low weight, small size, and superior performance are essential, monolithic integration is desirable. It is possible to achieve self-calibrated accelerometers by including custom VLSI gate arrays with the accelerometer elements.

One of the applications of acoustic sensors is an integrated microbeam accelerometer that can be used on a moving vehicle to measure acceleration. The acceleration data with on-board information of initial values makes the accelerometer suitable for use as a position and velocity estimator such as a global positioning system (GPS) or car air-bag actuators. In general, the accelerometer is a miniature cantilevel beam fabricated as an integrated circuit. The sensing element is a piezoelectric thin-film ZnO capacitor, which is built into the beam and generates signal proportional to acceleration loading. This signal, after processing through on-chip circuitry, is the output of the accelerometer.

Acceleration loading of the cantilever beam causes strain in the piezoelectric capacitor, which results in an output voltage signal proportional to input acceleration. This signal is conditioned by a temperature-compensation capacitor and an isolation amplifier, both of which are integrated with the acceleration sensor on a silicon chip. The buffered signal is then converted to digital format by a voltage-to-frequency converter, which may be electronically integrated on-chip or off-chip (hybrid). Several process techniques for fabrication of such accelerometers have been reported.[55,56,67,68] When the required sensitivity is low (approximately microvolts per gram), the cantilever beam can be fabricated with a single-side etch.

For many applications it is possible to make an oxide beam by a completely planar process on one surface of the wafer, assuring processing compatibility with conventional integrated circuits, but for applications that require very high acceleration levels, the thickness of an oxide beam is inadequate, and it is necessary to fabricate the beam by a two-sided etching technique using silicon as part of the beam structure. Because several layers of dissimilar films are used for the construction of the cantilever beam, it is necessary to consider the effects of varying the physical dimensions of the beam and varying the thickness of those layers that have the most influence on the mechanical properties of the cantilever beam.

A cantilever-beam accelerometer is supported (or "clamped") at one end and has its other end free from all except gravitational forces. We will discuss two types of cantilever-beam accelerometers: piezoelectric, and SAW-oscillator devices.

In the first type of device, a piezoelectric film deposited on the beam produces a voltage when the beam deflects as a result of acceleration.[55,68] In this device, ICs on the device condition signals for acceptable output. In the second type, a SAW delay line or a resonator on a cantilever beam with a piezoelectric substrate (typically quartz, $LiNbO_3$) is connected to a feedback amplifier so that oscillation results.[69,70] The oscillation frequency is altered by the change of path length caused by acceleration-induced strain.

The piezoelectric cantilever-beam devices provide larger outputs than the resonant devices, but require deposition of a piezoelectric film. High-yield production techniques for fabricating these devices have, however, been developed during the past decade. The SAW accelerometer is characterized by high sensitivy and high dynamic range. We shall discuss just the piezoelectric and SAW cantilever devices.

Piezoelectric Cantilever-Beam Accelerometer. We will discuss a miniature monolithic accelerometer designed and fabricated to be compatible with standard VLSI technology. Figure 18 shows a schematic of the device. The accelerometer IC (integrated circuit) contains a small cantilever beam on which is a thin-film piezoelectric capacitor. The beam is produced by etching the Si "chip". Acceleration forces normal to the surface of the chip cause the beam to bend and increase strain in the piezoelectric capacitor. The piezoelectric effect converts this strain into an electrical charge that is proportional to the acceleration. The on-chip IC also includes a temperature-compensating capacitor, which is electrically back-to-back and physically identical to the sensor capacitor, and a p^+ load resistor to linearize the input amplifier. The input amplifier is a depletion-mode PMOS device, which is completely passivated by Si glass to eliminate any possible surface-charge leakage.

The novel process steps in making an accelerometer chip are the formation of the cantilever beam by a micro-electro-mechanical (MEM) process. The process involves micromachining, which is the three-dimensional sculpting of the silicon to form a mechanical structure. In the following, we will discuss two major techniques for processing cantilever-beam structures.

In the first technique, the beam formation is by top-surface etching, commonly by a chemical process like a solution of EDP to anisotropically etch the Si from underneath the layered structure.[54] The cantilever beam is a composite structure of metal, ZnO, SiO$_2$, and Si, with thicknesses ranging from 2 to 5 μm. When the beam thickness is required to be in the 2 μm range, the Si portion of the

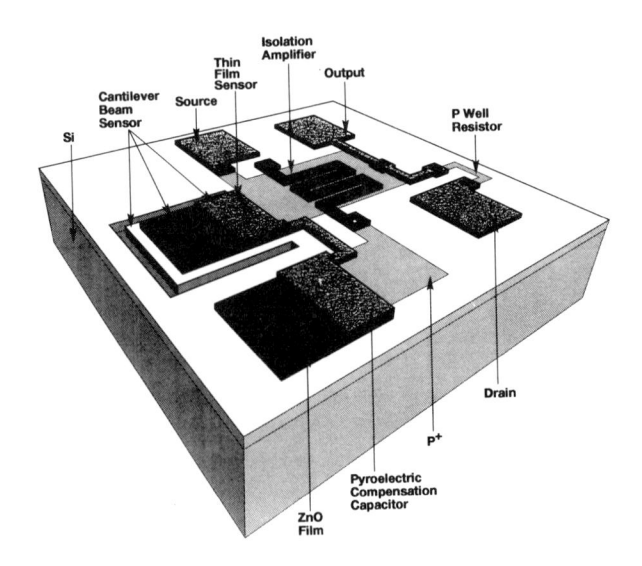

Fig. 18 Schematic of a silicon monolithic cantilever beam.

3.6 SENSOR APPLICATIONS   129

structure may be eliminated. Figure 19 shows a picture of several SiO_2 cantilever beams with different beam dimensions and a thickness of 1 μm. The larger beam has a dimension of 12×38 mil^2. The beams are fabricated using an EDP etching process. The pattern is designed for studying anisotropic etching.

In the second technique of cantilever-beam formation, two etching steps, one on the front and the other on the back, are required. This process is called the two-sided etching technique. This method is commonly used for a beam thickness larger than 5 μm.

For high-g Si accelerometers, a two-sided etching process has been developed based on an electrochemical control method.[56] Figure 20 shows cross-sections of a Si cantilever beam designed for the two-sided etching process. For fabrication of such a structure, an etch depth of practically the entire wafer thickness is required, which makes process control of uniformity, sidewalls and etch-stop extremely difficult. To fabricate a cantilever beam such as shown in Fig. 20, a low-resistivity Si substrate is coated with two layers of epitaxial Si films. The top-layer epitaxial Si is for the MOS circuitry and signal conditioning of the sensor. The buried-layer epitaxial Si has very high resistivity for

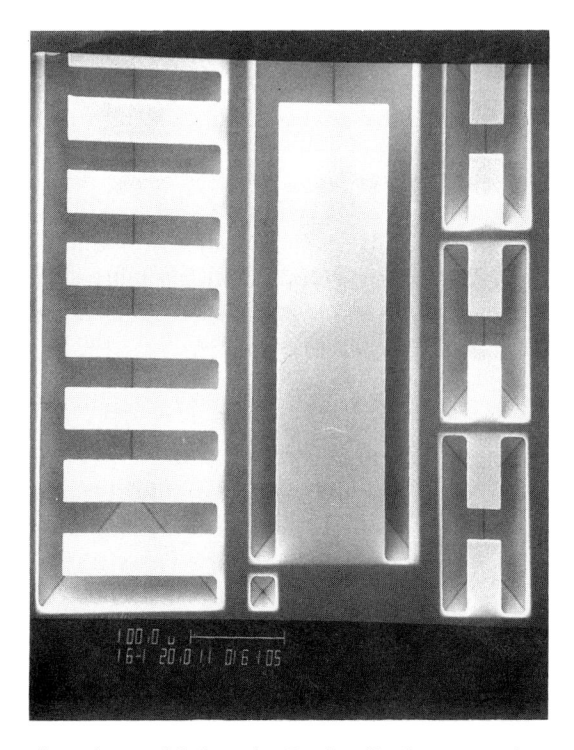

Fig. 19 SiO_2 cantilever beams fabricated at Rockwell using top-surface micromachining. Patterns are designed for studying EDP anisotropic etching. The larger beam has dimensions 12×38 mil^2.

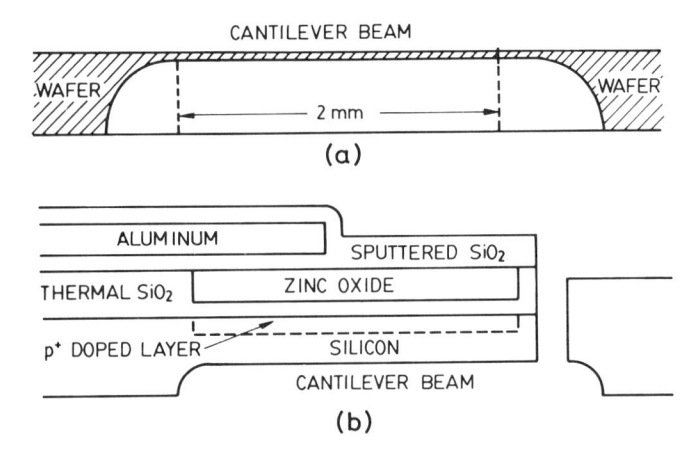

Fig. 20 Cross-section of Si cantilever-beam accelerometer designed for two-sided etching. (a) Drawn to scale. (b) Vertical scale increased to show detail of composite structure.

electrochemical etching control. The back-side etching is achieved by nitric-acid-based solutions. Results of the etch profile demonstrate a perfect isotropic process. The detail of the process was discussed earlier in Section 3.4.3. Figure 21 shows a scanning electron micrograph of a cantilever beam processed by the two-sided etching technique. The beam is 25 microns thick, and it can survive a 20,000 g acceleration.

The piezoelectric cantilever-beam accelerometer has the attributes of small size and low weight, and it requires low power. It uses the low-cost photolithographic fabrication methods developed for the IC industry and is well suited to high acceleration levels. Initial feasibility studies of the device concentrated on performance analysis and fabrication processing. The principal problem in earlier studies was the incompatibility of the processing methods for forming the beam and those forming the rest of the IC. Advanced MEM technology demonstrated the possibility of using a standard VLSI process and producing micromechanic structures like accelerometers with high yield and excellent performance.

A VLSI-compatible micromachining dry process can be used for two-sided cantilever beam processing. In this technique, both front and back etching is done by a dry controlled-etching process. Figure 22 shows a complete 3-inch processed accelerometer wafer using a two-sided dry etching process at the Rockwell VLSI lab.

A design was optimized to performance requirements for high-g applications by computer-aided design studies. These studies use parametric models of the critical performance criteria as functions of physical dimensions of the beam and its composite elements. In this design, the beam dimension is 40×20 mils. Beam thickness is 5 mils, and the survival g-level requirement is more than 10^5 g's. Figure 23 shows a computer-generated plot of all of the front

Fig. 21 SEM micrograph of Si cantilever beam processed by two-sided etching technique. Beam thickness is 25 μm. The beam can survive 20,000 g acceleration level.

design layers superimposed. Some new features of this design are low-impedance output level ($<100\,\Omega$) for increasing signal transmission efficiency by using an on-chip buffer amplifier, and cross-axis compensation by designing the sensor to be composed of two identical accelerometers with a 180° relative rotation. The first feature is extremely important for high-g packaging, while the second feature is important to minimize the complexity of the signal-conditioning circuitry.

Two typical output responses of cantilever-beam piezoelectric accelerometers obtained in tests on a dynamic shaker, are shown in Figs. 24 and 25. Figure 24 shows the output variation over a 40-dB dynamic range with inputs of 1-, 10-, and 100-g sinusoidal acceleration. Figure 25 shows a g^2 error term about 50 dB below the linear term. The linearity is limited by the ability of PMOS design more than piezoelectric sensor response. The frequency of the lowest resonant mode of the typical cantilever beam was found to be 8.4 kHz.

132 ACOUSTIC SENSORS

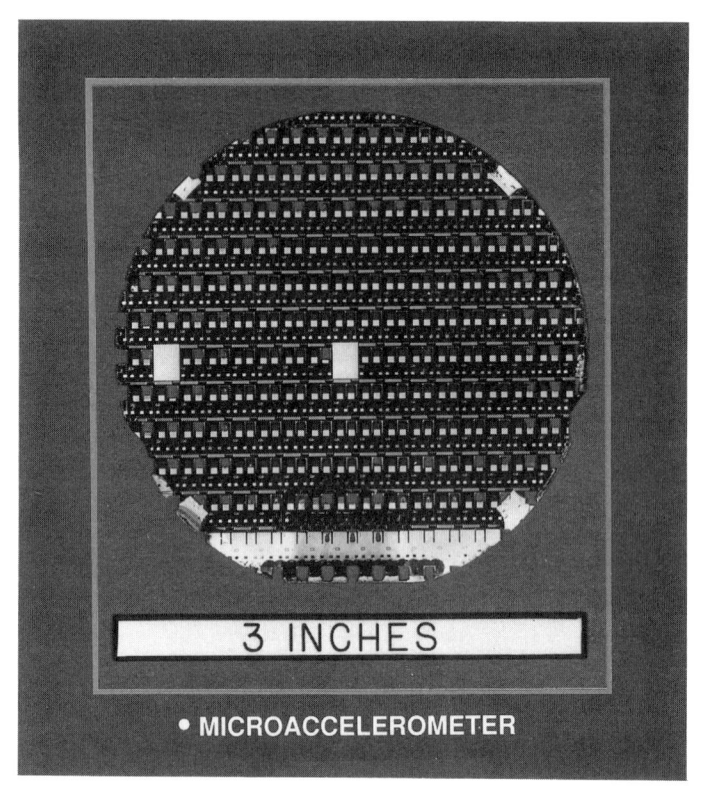

Fig. 22 Complete 3-in processed wafer of monolithic accelerometers with acceptable yield (wafer is processed in the VLSI Lab, Rockwell).

Theoretical Development. Analysis, which we will not include here for the sake of brevity, yields the sensitivity of the output voltage of the accelerometer as a function of device dimensions and material parameters. Of particular interest are the resonant frequencies of the accelerometer beams. One may want to adjust these frequencies to obtain additional gain at selected frequencies, or to eliminate resonant effects within the pass band of interest. The resonant frequencies for the beam are best described as the resonant frequencies of a flat plate which is clamped at one end. These frequencies are relatively insensitive to beam width. For homogeneous, isotropic beams, the lowest frequency is given by the following formula:

$$f_{min} = (2\pi)^{-1}(3.491)\sqrt{\frac{\varepsilon(1-\sigma^2)h^2}{12\rho l^4}}$$

$$= (0.1604)\sqrt{\frac{\varepsilon(1-\sigma^2)}{\rho}}\left(\frac{h}{l^2}\right) \quad (4)$$

3.6 SENSOR APPLICATIONS   **133**

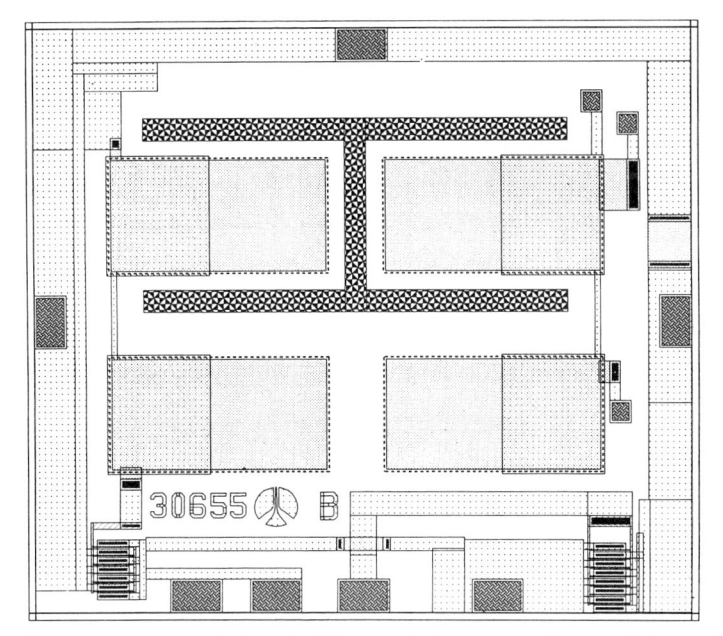

Fig. 23 Computer-generated plot of all front-side layers superimposed. Device includes temperature compensation, buffer amplifier, and cross-axis compensation. Chip size is 120×140 mil^2.

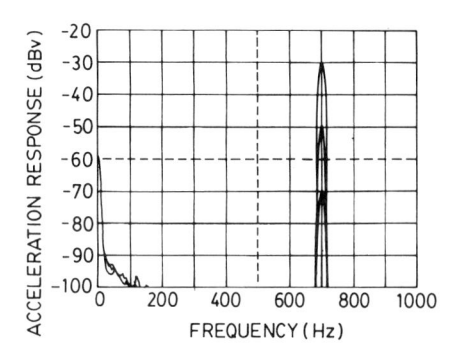

Fig. 24 Output response of typical monolithic accelerometer for three input levels; 1, 10, and 100 g. Results demonstrate excellent linearity over 40 dB dynamic range.

where ε is the elastic modulus, ρ is mass density and σ is Poisson's ratio of the beam material; h and l are the height and length of the beam, respectively. For most of the devices that have been fabricated and tested, the minimum resonant frequency has been in the range from 20 to 40 kHz.

Design Issues. Design trade-offs generally require the derivation of mathematical models that relate performance characteristics (such as scale factor, nonlinearity,

134 ACOUSTIC SENSORS

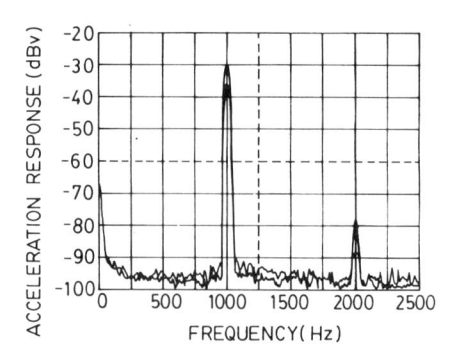

Fig. 25 Output response of a typical monolithic accelerometer, g^2 error term is 50 dB below linear term.

etc.) to design parameters (such as physical dimensions, material properties, etc.). These models were then used in evaluating how the various performance characteristics depend on the design parameters, and how the design parameters must be constrained (or optimized) to make the relevant performance characteristics. These studies showed the theoretical feasibility of meeting the performance requirements, and how a performance characteristic that surpassed its requirements could be traded off to satisfy the requirements of another performance characteristic.

The trade-offs also considered the effects of varying the other physical dimensions of the beam, such as length, width and the relative length (as % of beam length) of the piezoelectric capacitor. The trade-offs were simplified somewhat by reducing the number of parameters to be varied simultaneously. It can be shown that the important performance characteristics are essentially insensitive to the width of the beam. Therefore, it was assumed that the beam width would always equal half the beam length. The length of the piezoelectric capacitor has essentially no influence on the mechanical properties of the beam. The output signal from the piezoelectric capacitor had been shown to be proportional to the average strain in the piezoelectric material. For a given input acceleration, the strain is concentrated more toward the root of the beam and vanishes at the free end. Therefore, the average strain (and the scale factor) is reduced by extending the capacitor farther out on the beam.

Figure 26 shows a perspective of a chip layout that is designed for high g-level acceleration. The silicon beam has dimensions of 1×2 mm^2 with a thickness around 30 microns. The device has a temperature compensated capacitor and an on-chip amplifier, which are monolithically integrated with other circuit elements compatible with standard VLSI processing. The device can be fabricated by the two-sided etching technique using electrochemical-controlled etch-stop.[56] A typical pattern of nine super-imposed (topside) processing masks for the device shown in Fig. 26 is shown in Fig. 27.

After the wafer is processed, the chips are scribed and hybrid-integrated with a V/F (voltage/frequency) converter for the read-out. The chip size is

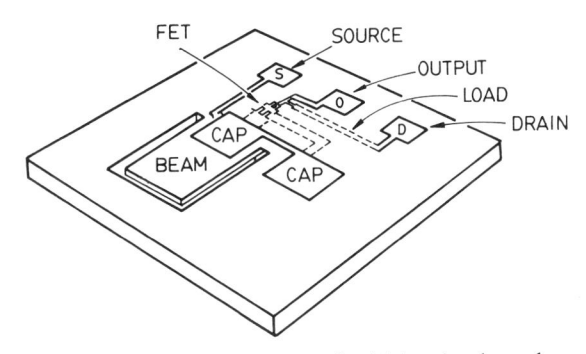

Fig. 26 Schematic of a chip layout for high g-level accelerometers.

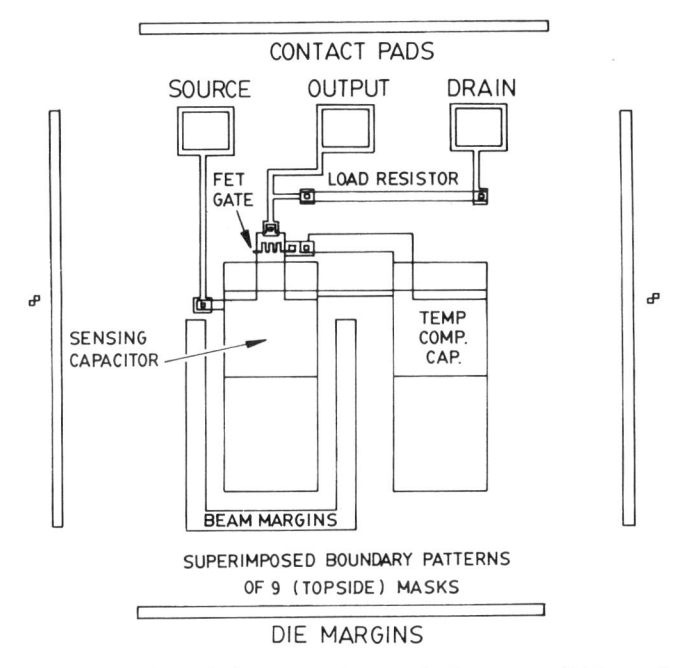

Fig. 27 Superimposition of nine processing masks for a monolithic accelerometer.

240×240 mil^2. Even with hybrid integration of the V/F converter, the accelerometer is very small with a potential for high sensitivity due to monolithic integration of a high-impedance input amplifier. Since the process is designed to be compatible with the VLSI manufacturing process, hybrid components of the acoustic sensor with all the required signal processing/conditioning can be easily integrated on the same chip.

SAW Cantilever-Beam Accelerometer. Production SAW devices are now available for use in many high-performance and high-precision sensors. A cantilever-beam SAW accelerometer contains a SAW delay line or a resonator oscillator acting as a sensing element.[70]

To construct the SAW accelerometer, a hybrid integration technique should be used. Since the sensor element is a SAW oscillator, a piezoelectric quartz substrate is the best choice considering its temperature stability and high-Q performance. When high stability, on the order of 10^{-3} ppm, in frequency is required, a SAW resonator oscillator should be considered. In many applications when frequency stability is not critical, a delay-line oscillator can be well-suited instead of a SAW resonator oscillator.

A SAW accelerometer sensitivity of 10 μg has been achieved using a quartz cantilever-beam accelerometer with a proof mass included in the beam. A unique technology has been developed at Rockwell for forming a cantilever quartz beam shaped to a two-dimensional proof mass structure. For this process, quartz wafers, that were polished on both sides, were coated on both sides with chrome–gold films. The chrome–gold films were patterned into special geometries to form structures of a proof mass self-aligned to the cantilever beam. The patterning is done using standard photolithography and a double-sided alignment technique. Chemical etching was used to etch the quartz from both sides with the gold acting as an etch mask. After successful etching, the beams are still connected together by small unetched bridge sections. These bridge sections can be lightly forced to separate the beams.

This process is mostly reproducible, but there are some critical points. One critical point is the adherence of the Cr–Au film to the quartz to assure a good mask for chemical etching. Another critical point is the quality of the surface finish of the polished quartz to minimize the anisotropic etching process. An alternative to this process uses a special positive resist such as Shipley AZ-4620 instead of a gold mask. This photoresist has excellent selectivity for an HF-based solution and can be spun on with a thickness in excess of several microns.

When lower sensitivity, on the order of 10^{-3} g, is required the cantilever beam of the SAW accelerometer can be a simple rectangular quartz substrate. In this case, the SAW structure, resonator, or delay line can be batch-processed on a 3-inch quartz wafer and then diced to the required beam size. In the case of the SAW accelerometer with a beam shape designed to include a proof mass, it is sometimes cost-effective to produce individual substrates shaped to the beam structure explained earlier, and process the SAW resonator on each substrate. This procedure is sometimes not cost-effective for production, but it is the price that one pays for high-sensitivity (10^{-6} g) and high-stability (10^{-3} ppm in frequency) accelerometers.

SAW technology offers inherently digital acceleration sensors with simple two-to-three-layer processing. The technology uses established planar photolithography for low-cost fabrication. The associated electronics contain only one active amplifier. This simplicity makes it an attractive candidate for digital acoustic-sensor applications.

Examples of SAW accelerometers have the following typical design values: a full-scale range of ± 10 g; sensitivity of 10 μg; scale factor of 5000 Hz/g; and stability of 10^{-3} ppm in frequency. Several prototypes of this design have been fabricated, assembled, and tested. Major problems are: facing device phase noise, which makes it very difficult to achieve stability requirements; and fabricating high-Q SAW resonators (Q in excess of 20,000).

To construct the sensor element of the cantilever beam accelerometer, a two-stage silicon-on-sapphire (SOS) hybrid amplifier has been assembled for use with the high-Q SAW resonator as a feedback element to provide a UHF SAW oscillator with low phase noise. The SAW resonator, fabricated with ST-cut quartz, is mounted in a cantilever beam configuration.

To optimize accelerometer phase-noise performance, a fundamental-frequency oscillator or an oscillator of either one-half or one-fourth the fundamental frequency should be used. Phase-noise contributions in the oscillator circuit will arise not only from the resonator or SAW delay line but also from noise arising from the other circuit components. Of primary concern are the noise figures for the active devices in the oscillator, which may be either bipolar transistors or GaAs devices of the MESFET variety. Passive circuit components also contribute noise. Careful selection must be made, for example, of the types of resistors and capacitors used in the amplifying stage of the oscillator circuit.

3.6.2 Flexural-Plate-Wave Sensors

FPW Sensor Operation. The flexural-plate-wave (FPW) sensor is shown in a schematic, phantom view in Fig. 28. The device has a non-piezoelectric membrane, typically made of silicon-rich, low-stress silicon nitride, deposited

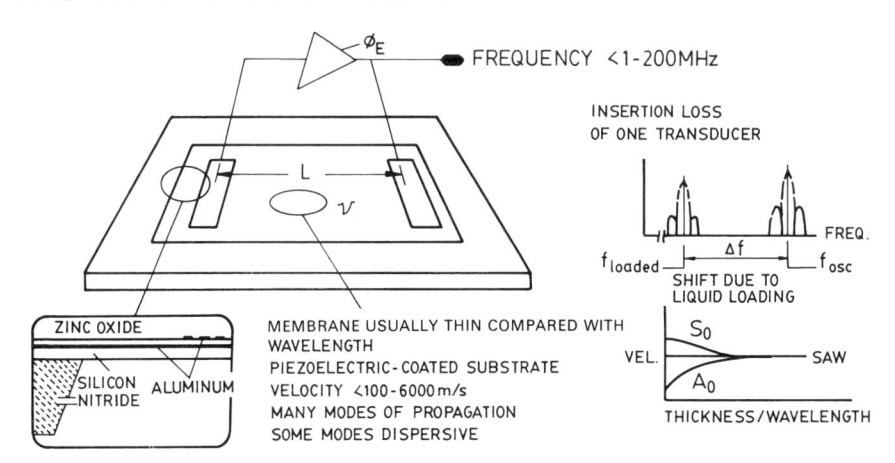

Fig. 28 FPW sensor. Insets show cutaway view of membrane, velocities of A_0 and S_0 modes, and the relatively large decrease of oscillation frequency produced by mass loading.

by the LPCVD process onto a (100)-oriented silicon wafer. The well on the bottom of the wafer is formed with an orientation-dependent etchant, such as KOH. Then, a thin piezoelectric layer is laid down. This layer is typically RF-magnetron-sputtered ZnO. Comb-like electrodes are formed on the membrane, which may have a ground plane between the piezoelectric layer and the supporting membrane to prevent electric fields from reaching liquids in contact with the other side of the membrane. (Because of the aspect ratios involved, it is the electric fields that are *perpendicular* to the membrane that play the primary role in FPW transduction.) If the membrane thickness is much smaller than the wavelength, the lowest-order flexural mode (denoted as the A_0 mode) will propagate in the membrane with a phase velocity of only few hundreds of meters per second. This velocity is less than the speed of sound in most liquids; for example, the speed of sound in water at room temperature is 1480 m/s. Therefore, when the membrane contacts or is immersed in an ideal invisicid fluid, the flexural wave propagating in the membrane produces only an evanescent disturbance in the fluid and no energy radiates from the membrane into the fluid. This is in sharp contrast to an immersed SAW sensor. Because SAW velocities substantially exceed the speed of sound in most fluids, energy will radiate from the SAW into the fluid. The calculated rate of loss for the immersed SAW is about 1 dB per wavelength.[5] Such a high loss would be particularly serious for a biosensor, which usually requires contact with a liquid.

The inset in the upper right of Fig. 28 shows that the bandpass characteristic of the FPW transducer is like that of the SAW interdigital transducer. In the FPW case, however, since the transducers are located on the membrane, any influence that changes the wave velocity also shifts the entire operating band of the transducers, permitting large fractional-frequency shifts to occur. Typically, frequency decreases as large as 30–40% of the initial, air-loaded oscillation frequency are observed upon contacting one side of the membrane of an FPW device with water.

As noted earlier, the A_0 mode is particularly interesting because, as the membrane is made arbitrarily thin, the phase velocity ideally becomes vanishingly small. (Actually, residual stress in the membrane establishes a small but finite minimum phase velocity for vanishingly thin devices.) Thus, the FPW delay line acts as a "slow-wave" transmission line when in contact with a fluid. The evanescent disturbance in the adjacent fluid is confined to the vicinity of the membrane, much as the optical energy in a fiber-optic guide is confined to the central core. When $v_{membrane} \ll v_{fluid}$, the evanescent decay distance, δ, is equal approximately[71] to $\delta/2\pi$. For example, if the wavelength is 100 μm, the evanescent disturbance in a fluid will extend out to a distance of only 16 μm from the membrane.

One can show[71] that the phase velocity, v_p, of FPWs in a very thin plate ($d \ll 1$) can be written as

$$v_p \approx \sqrt{\frac{T + \beta^2 D}{M}} \tag{5}$$

where T is the component of tension in the propagation direction, per unit width of plate perpendicular to this direction, $\beta = 2\pi/\lambda$, D is the flexural rigidity of the plate, and $M =$ mass per unit area of the plate. The term D is equal to $Ed^3/12(1-\sigma^2)$, where $E =$ Young's modulus of the plate, $d =$ plate thickness and $\sigma =$ Poisson's ratio of the plate.

Gravimetric Sensing. If the mass per unit area of the membrane increases by an amount Δm due to vapor sorption, the phase velocity will decrease by a proportionate amount:

$$v_p \approx \sqrt{\frac{T+\beta^2 D}{M+\Delta m}} \approx v_{p0}[1-\Delta m/2M], \tag{6}$$

where v_{p0} is the phase velocity of the unloaded plate. From Eq. 2 we identify the gravimetric sensitivity S_m as

$$S_m = (\Delta f/f)/\Delta m = -1/2M. \tag{7}$$

Figure 29 shows the results of toluene vapor sensing with an 4.3 MHz FPW sensor coated with a 1.5-micron-thick layer of PDMS (poly-dimethylsulfoxane), a rubbery polymer. With an ethyl cellulose coating on a thinner, 2.8 MHz PFW device, the measured sensitivity and noise level of approximately 0.4 Hz indicate that this device should be able to detect concentrations of toluene vapor as low as 0.07 ppm.

For further information about gravimetric vapor sensing with the FPW device see Ref. 72.

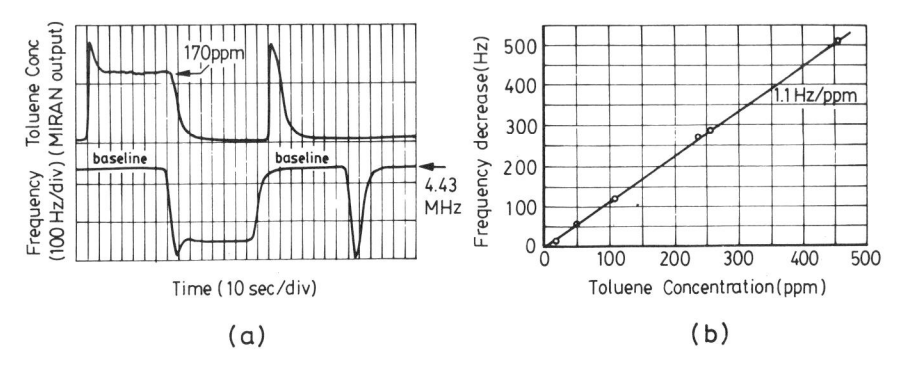

Fig. 29 Response of FPW sensor to toluene vapor.[73] (a) Concentration of toluene in nitrogen carrier gas as measured with a MIRAN infrared spectrophotometer (top) and frequency decrease of FPW sensor (bottom). The delay in response is due to a length of tubing between spectrophotometer and sensor. (b) Plot showing linear dependence of FPW frequency decrease upon toluene concentration.

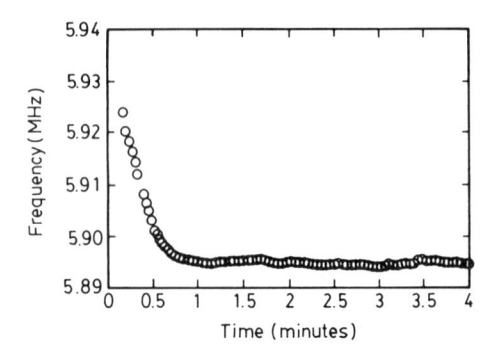

Fig. 30 FPW oscillator frequency versus time after introduction of buffered saline containing 3 mg/ml concentration of bovine serum albumin.[74]

Gravimetric Sensing in the Presence of Fluid Loading. When a fluid contacts the membrane, its mass loads the FPW sensor. It can be shown that the added mass per unit area due to this fluid loading $(\Delta m)_{\text{fluid}}$ is equal to[73]

$$(\Delta m)_{\text{fluid}} = \rho_{\text{fluid}} \delta, \tag{8}$$

where ρ_{fluid} is the fluid density and δ is the evanescent decay distance defined earlier. The gravimetric sensitivity in this case, $(S_m)_{\text{fluid}}$ is approximately equal to

$$(S_m)_{\text{fluid}} = -1/2(M + \rho_{\text{fluid}} \delta). \tag{9}$$

Figure 30 shows the response of an FPW device whose well was filled with buffered saline solution that contained bovine serum albumin (BSA) molecules at 3 mg/ml concentration.

Sensing Viscosity. The inherent attenuation of the FPW device is increased when a viscous liquid is put into the well. The increased insertion loss in decibels is proportional to the square root of the shear viscosity. This effect is shown in Fig. 31 where the insertion loss at 5 MHz was measured for a series of water–glycerol solutions, which have well-known viscosities. Figure 32 shows measurements of the temperature dependence of the viscosity of a six-microliter sample of DMSO (dimethyl sulfoxide). This was a proof-of-concept test to verify that the device would measure viscosity even though the sample volume was no greater than six microliters. For other features of this viscosity sensing the reader should consult the recent literature.[76]

Kinetic Effects—Mixing and Microtransport. Because the FPW membrane can be made quite thin, it takes only a small amount of ultrasonic power propagating in it to produce a wave having a relatively large amplitude. For example, the

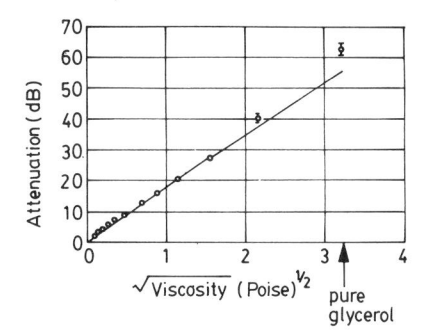

Fig. 31 Response of insertion loss of FPW device to presence of a series of water–glycerol solutions having different viscosities.[75]

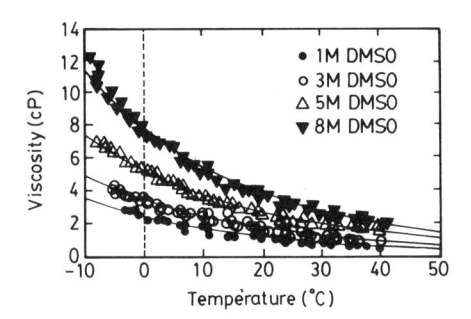

Fig. 32 Temperature dependence of viscosities of water (dashed) and different concentrations of DMSO as measured with FPW device.[76]

peak-to-peak wave amplitude as measured by laser diffraction on a 5 MHz, 3 mm-wide FPW device was approximately 100 nm, even though the power in the wave was only about one milliwatt. Because of this feature, the FPW device can exhibit kinetic effects, such as mixing and the transport of fluids that contact the membrane.

The mixing (stirring) of a liquid by the FPW device was demonstrated in an electrochemical experiment. The electrochemical cell contained an ordinary immobile platinum electrode and a second "activatable" electrode consisting of an FPW device whose membrane was coated with sputtered platinum. A mixture of potassium ferri- and ferro-cyanide was placed in the cell, a DC voltage difference was established between the electrodes, and the current through the cell was monitored as the FPW device was actuated (Fig. 33). With the plate-wave device inactive, the steady current was 4 mA; the current rose when the FPW device was activated. The increase of current observed in this redox reaction is interpreted as indicating that the ultrasonic wave stirs the liquid

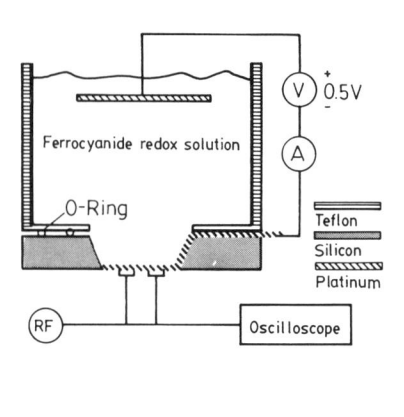

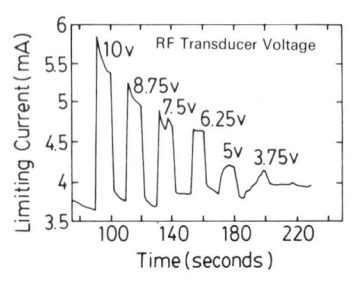

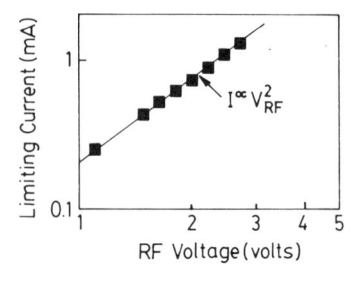

Fig. 33 Electrochemical mixing experiment.[77] Experimental apparatus (top) includes an "active" electrode consisting of a sputtered platinum coating on an FPW device. When the FPW device is energized, stirring near its platinum surface increases rate of contact with ions in solution and so increases the steady current as shown (center). Increase in current (bottom) is proportional to the square of applied transducer voltage, and hence to the square of plate-wave amplitude.

near the bottom electrode, increasing the rate at which ions reach the bottom electrode over that caused by diffusion alone. The increase in current was proportional to the square of the transducer voltage and, hence, to the square of the wave amplitude. This functional dependence substantiates the inter-

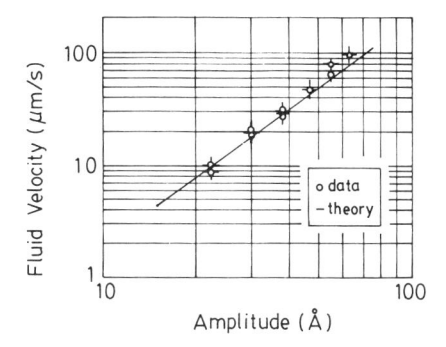

Fig. 34 Experimental and theoretical pumping speeds for water containing polystyrene marker spheres as a function of amplitude of flexural plate wave.[78]

pretation that the mixing and pumping effects are the result of a nonlinear acoustic process occurring in the fluid near the membrane surface.[77] This mixing effect, in which the ultrasonic device acts as a compact stirrer, could be used to increase the speed and signal-to-noise ratio of response of almost any chemical or biochemical sensor whose operation requires mass transport.

In addition to the mixing or stirring, transport of fluids in the direction of wave propagation has been produced by FPW devices.[78] The mechanism responsible appears to be the phenomenon of acoustic streaming, whereby, for sufficiently large intensities of a harmonic wave propagating in a fluid, a steady force in the direction of wave propagation is produced. The amplitude of this steady force is proportional to the square of the amplitude of the harmonic wave.

Figure 34 shows the measured speed of pumping water containing 2.5-micron-diameter polystryene marker spheres as a function of the amplitude of the wave motion in the membrane. The theoretical values derived from acoustic streaming theory are shown.[78]

In summary, we have shown that FPWs can be used in a variety of microsensors, micromixers, and micropumps. These components, possibly along with other elements based on different technologies, may form the basis for microflow systems in which fluids are manipulated for particular purposes.

We have shown that liquids can be pumped, and the transport of granular solids in air has been demonstrated, together with mixing and the sensing of mass per unit area, liquid density and fluid viscosity in micromachined devices. These chips may also contain thin-film resistive heaters that can be used to initiate chemical reactions. Microflow systems employing the ultrasonic elements discussed might perform functions such as biochemical analysis, the reconstitution of a lyophilized drug having a long storage life, or even pumping and mixing to facilitate the amplification of DNA by the polymerase chain reaction in a microfabricated reactor.[79]

3.7 SUMMARY AND FUTURE TRENDS

Several indicators point to the growing use of microfabricated micromechanical devices, which should bode well for the future of all sorts of microfabricated sensors. First micromachined pressure sensors and then accelerometers made their way into automobiles for control and deployment of air-bags. Silicon-based accelerometers and rate gyros may next find use in vehicle navigation systems. In these markets for sensors of purely physical variables there is a growing need for vapor sensors (to monitor pollutant emissions and even air quality in the passenger compartments) that might be met in part by micromachined acoustic sensors. Additional large-volume applications for some of these devices are envisioned in the medical device field. Penetration into these markets should be aided by the growing availability of micromechanical foundries.

The acoustic devices face an initial disadvantage because they employ the piezoelectric effect, with which few electrical or mechanical engineers have experience. Balancing this is the very vigorous activity in developing SAW communications devices, such as filters for use in portable personal communication systems. Some of these devices also employ integrated circuitry on the same chip with the filter. Thin piezoelectric films of zinc oxide have been used in many tens of millions of filters for televisions sets, demonstrating the commercial viability of this piezoelectric technology. The profitable growth of communications SAW technology and commercial use of piezoelectric thin films should stimulate their use in sensing and help familiarize engineers with piezoelectric acoustic techniques.

PROBLEMS

1. **Minimum detectable added mass/area.** Characteristics of hypothetical SAW and FPW gravimetric sensors are given in the table below. Find the minimum detectable added mass per unit area (in g/cm^2) for each sensor.

Sensor Type	Operating Frequency (MHz)	Gravimetric Sensitivity S_m (cm^2/g)	Short-Term Frequency Fluctuation (Hz)
SAW	150	100	10
FPW	2	1000	0.4

2. **Molecular Coverage.** If, for a particular acoustic sensor, the minimum detectable added mass is $1 \text{ ng}/cm^2$, to what fraction of a monomolecular layer of (a) water and (b) an antibody having a molecular weight of 50,000 and a density of $1.35/\text{g-cm}^3$ does this minimum mass correspond?

3. **Viscosity Sensing.** Suppose that when water (whose viscosity is 1 centipoise) is placed in contact with a FPW device the attenuation of the device increases by 1.5 dB. By how much would the attenuation increase if instead glycerol (viscosity 1200 centipoise) contacted the sensor?

4. Assume the minimum longitudinal resonance frequency of a cantilever accelerometer beam has relation to its material properties as,

$$f_{(min)/long} = 0.25 \sqrt{\frac{\varepsilon}{\rho l^2}}.$$

Prove that the f_{min} defined for the transverse resonant frequency given by Eq. (4) is always less than $f_{(min)/long}$ given by the above relation if the cantilever beam thickness and its length satisfy this inequality:

$$h < \frac{1.49}{\sqrt{1-\sigma^2}} l.$$

As an example, for an SiO_2 beam show that as long as that beam thickness is smaller than 1.58 times the beam length, the minimum beam resonant frequency is defined by the transverse resonant mode.

5. Find a formula for acceleration limit as a function of the ratio (h/l^2) where h and l are the height and length of a cantilever beam accelerometer. Plot a beam performance characteristic (design trade-offs) for both acceleration limit and minimum resonant frequency as a function of (h/l^2) where (h/l^2) varies from 10^{-2} to 10. Make a family of plots for SiO_2, Si_3N_4 and silicon. (*Hint*: use formula

$$S_{lim} = \frac{3\rho\alpha_{lim}}{\varepsilon}\left(\frac{h}{l^2}\right)^{-1},$$

where S_{lim} is the strain limit of material, ρ and ε are mass density and elastic modulus of the material respectively.) For material constants use a handbook of materials.

6. Accelerometer cross-axis coupling error reduces the dynamic range and usefulness of a device. This error is proportional to the beam aspect ratio

and it can be formulated as

$$C_{ix} = \frac{6\rho l^3}{5\varepsilon h^2},$$

where ρ, l, ε and h are defined in text.

a) Plot a family of C_{ix} for Si_3N_4 as a function of accelerometer length for beam height h of 0.5, 1, 2, 5, 10 where l, the beam length, varies from 100 to 10,000 μm.

b) Find the limiting values of (h/l) for SiO_2, Si_3N_4 and Si where only 5% cross-axis error per beam unit length is acceptable.

7. Use the design rules of Fig. 11 and sketch a layout for electrode metallization of a 840 MHz SAW transducer. The transducer structure is $ZnO/SiO_2/Si$. Assume a phase velocity of 3.1×10^5 cm/s for the structure and let $Kh_{SiO_2} = 1$ and $Kh_{ZnO} = 0.3$ to achieve temperature stability. Find the thicknesses of ZnO and SiO_2. $K = 2\pi/\lambda_a$ is the propagation vector, where λ_a is the acoustic wavelength, and h_{SiO_2} and h_{ZnO} are the SiO_2 and ZnO thicknesses, respectively.

8. Repeat Problem 7 for an $AlN/SiO_2/Si$ structure. Use the following information related to AlN. v_{ph} (phase velocity) $= 4.4 \times 10^5$ cm/s, $Kh_{SiO_2} = 1$, $Kh_{AlN} = 0.5$.

REFERENCES

1. J. W. Grate, S. J. Martin, and R. M. White, "Acoustic wave microsensors," *Anal. Chem.* **65**, 940A–58A (1993); **65**, 987A–96A (1993).
2. C.-S. Lu and O. Lewis, "Investigation of film-thickness determination by oscillating quartz resonators with large mass load," *J. Appl. Phys.* **43**, 4385 (1972).
3. R. M. White and F. W. Voltmer, "Direct piezoelectric coupling to surface elastic waves," *Appl. Phys. Lett.* **7**, 314 (1965).
4. R. M. White, "Surface elastic waves," *Proc. IEEE* **58**, 1238–1276 (1970).
5. I. A. Viktorov, *Rayleigh and Lamb Waves*, Plenum, New York, 1967.
6. V. M. Ristic, *Principles of Acoustic Devices*, Wiley-Interscience, New York, 1983.
7. B. A. Auld, *Acoustic Fields and Waves in Solids*, Wiley-Interscience, New York, 1973.
8. W. G. Cady, *Piezoelectricity*, McGraw-Hill, New York, printed in revised form, 1964.
9. E. Hafner, "The piezoelectric crystal unit—definitions and methods of measurement," *Proc. of IEEE* **57**, no. 2 (1969).
10. W. Voigt, "General theory of the piezo- and -pyroelectric properties of crystals," *Abh. Gott.* **36**, 1–99 (1890).
11. A. Langevin, "Absolute value of the principal piezoelectric modules of quartz," *C.R.* **209**, 627–30 (1939).

REFERENCES

12. M. E. Motamedi, "Acoustic accelerometers," *IEEE Trans. Ultrason. Ferroelec. Freq. Cont.* **UFFC-34**, 237 (1987).
13. J. G. Gualtieri, J. A. Kosinski, and A. Ballato, "Piezoelectric materials for SAW applications," *Proc. IEEE 1992 Ultrasonics Symp.*
14. T. Shiosaki, "Growth and applications of piezoelectric and ferroelectric thin films." *Proc. IEEE 1990 Ultrasonics Symp.* pp. 537–46.
15. N. F. Foster, *Proc. Joint IEEE-IEE Conf. Application of Thin Film in Electronics Engineering, London, England, July, 1966.*
16. J. de Klerk and E. F. Kelly, *Rev. Sci.* **36**, 506 (1965).
17. R. M. Malbon, D. J. Walsh, and D. K. Winslow, *Appl. Phys. Lett.* **10**, no. 1 (1967).
18. S. Wanuga, et al., "Zinc oxide film transducers," presented at *Ultrasonic Symp., Boston, Dec. 1–4, 1965.*
19. N. F. Foster, et al., *IEEE Trans. Sonic Ultrason.* **SV-15**, 28–41 (1968).
20. Hickernell, *J. Appl. Phys.* **44**, 1061–71 (1973).
21. T. Shiosaki, "High speed fabrication of high-quality sputtered ZnO thin-films for bulk and surface wave applications," *Proc. IEEE 1987 Ultrasonics Symp.* pp. 110–110.
22. B. T. Khuri-Yakub and J. G. Smits, "Reactive magnetron sputtering of ZnO", *J. Appl. Phys.* **52**, no. 7 (1981).
23. H. Fujishima, et al., "Surface acoustic wave VIF filters for TV using ZnO sputtered films," *Proc. 30th Ann. Symp. on Frequency Control,* 1976, pp. 119–22
24. M. E. Motamedi, E. J. Staples, and J. Wise, "Characterization of ZnO/Si SAW transducers and resonators using complex return loss measurements," *Proc. 1981 Sonics and Ultrasonics Symp.*, p. 368.
25. S. Furukawa, et al., *Proc. IEEE Ultrasonic Symp.*, 1979, pp. 940–44.
26. P. J. Clarke, "Processing assembly and testing," *Semicond. Int.* June (1979).
27. P. J. Clarke, *U.S. Patent No. 3,616,450* (1971).
28. T. Yamamoto, T. Shiosaki, and A. Kawabata, "Characterization of ZnO piezoelectric films prepared by RF planar-magnetron sputtering," *J. Appl. Phys.* **51**, 3113–20 (1980).
29. M. Sayer, *Proc. IEEE Symp. on Application of Ferroelectrics, Bethlehem, PA, June 8–11, 1986,* pp. 560–8.
30. J. L. Vossen and K. Werner, *Thin Film Processes,* Academic Press, New York, 1978.
31. H. Sankur and M. E. Motamedi, "Properties of ZnO film deposited by laser assisted evaporation," *Proc. 1983 Sonics and Ultrasonics Symp.* p. 316.
32. J. K. Liu, K. M. Lakin, and K. L. Wang, "Growth morphology and SAW measurement of AlN film on Al_2O_3," *J. Appl. Phys.* **46**, 3703–6 (1975).
33. R. S. Kagiwada, H. H. Yen, and K. F. Lau, "High frequency SAW devices on AlN/Al_2O_3," *Proc. IEEE 1978 Ultrasonics Symp.* pp. 598–601.
34. K. Tsubouchi, K. Sugai, and N. Mikoshiba, "High frequency and low-dispersion SAW devices on AlN/Al_2O_3 and AlN/Si for signal processing," *Proc. IEEE 1980 Ultrasonics Symp.* pp. 446–50.
35. K. Tsubouchi, K. Sugai, and N. Mikoshiba, "AlN material constants evaluations and SAW properties on AlN/Al_2O_3 and AlN/Si," *Proc. IEEE 1981 Ultrasonics Symp.* pp. 375–80.

36. K. Tsubouchi and N. Mikoshiba, "Zero temperature coefficient SAW delay line on AlN epitaxial films," *Proc. IEEE 1983 Ultrasonics Symp.* pp. 299–310.
37. S. Yoshida, et al., "Reactive MBE of AlN," *J. Vac. Sci. Technol.* **16**, 990–3 (1979).
38. F. Takeda and T. Hata, "Low temperature deposition of oriented c-axis AlN films on glass substrates by reactive magnetron sputtering," *Jpn. J. Appl. Phys.* **19**, 1001–2 (1980).
39. T. Shiosaki, et al., "Low temperature growth of piezoelectric AlN film by RF reactive planar magnetron sputtering," *Appl. Phys. Lett.* **36**, 643–5 (1980).
40. T. Shiosaki, et al., "Low temperature growth of piezoelectric AlN film for surface and bulk wave devices by RF reactive planar magnetron sputtering," *Proc. IEEE 1980 Ultrasonics Symp.* pp. 451–4.
41. B. Jaffe, W. R. Cook, and H. Jaffe, *Piezoelectric Ceramics,* Academic Press, New York, 1971.
42. M. Sreenivas, et al., "Surface acoustic wave propagation on PZT thin film," *Appl. Phys. Lett.* **52**, 709–11 (1988).
43. G. Yi, et al., "Ultrasonic experiments with PZT thin films fabricated by sol-gel processing", *Electron. Lett.* **25** (1989).
44. R. Takayama, et al., "Preparation and characteristics of pyroelectric infrared sensor made of c-axis oriented La-modified $PbTiO_3$ thin films," *J. Appl. Phys.* **61**, 411 (1987).
45. A, Oikawa and K. Toda, "Preparation of $Pb(Zr, Ti)O_3$ thin films by an electron beam evaporation technique," *Appl. Phys. Lett.* **29**, 491 (1976).
46. A. Okada, "Some electrical and optical properties of ferroelectric lead-zirconate-lead-titanate thin films," *J. Appl. Phys.* **48**, 2905 (1977).
47. R. N. Castellano and L. G. Feinstein, "Ion-beam deposition of thin films of ferroelectric lead-zirconate-titanate (PZT)," *J. Appl. Phys.* **50**, 4406 (1979).
48. H. Adachi, et al., "Ferroelectric $(Pb, La)(Zr, Ti)O_3$ epitaxial thin films on sapphire grown by RF-planar magnetron sputtering," *J. Appl. Phys.* **60**, 736 (1986).
49. T. Ogawa, S. Senda, and T. Kasanami, "Preparation of ferroelectric thin films by RF sputtering," *Jpn. J. Appl. Phys.* **28**, 11–14 (1989).
50. M. Odada, et al., "Preparation of c-axis oriented $PbTiO_3$ thin film by MOCVD under reduced pressure," *Jpn. J. Appl. Phys.* **28**, 1030 (1989).
51. D. Roy, S. B. Krupanidhi, and J. Dougherty, "Excimer laser ablated lead zirconate titanate thin films," *J. Appl. Phys.* **69**, 1 (1991).
52. G. Yi, Z. Wu, and M. Sayer, "Preparation of PZT thin film by sol-gel processing: electrical, optical, and electro-optic properties," *J. Appl. Phys.* **64**, no. 5 (1989).
53. K. E. Peterson, "Dynamic micromechanics in Si: techniques and devices," *IEEE Trans. Electron. Devices* **ED-25**, 2141 (1978).
54. R. D. Jolly and R. S. Muller, "Miniature cantilever beams fabricated by anisotropic etching of silicon," *J. Electron. Chem. Soc.* **127**, pp. 2750–4.
55. P. Chen, et al., "Integrated silicon microbeam PI-FET accelerometer," *IEEE Trans. Electron. Devices* **ED-29**, no. 1 (1982).
56. M. E. Motamedi, et al., 'Application of electrochemical etch-stop in processing silicon accelerometer," *Proc. 1982 Electrochemical Society Meeting, May 9–14,* p. 188.
57. M. E. Motamedi, "Large scale monolithic silicon convolvers and correlations," *IEEE Trans. Sonic and Ultrason.* **SU-32**, p. 663 (1985).

58. M. J. Vellekoop, *et al.*, "Compatibility of zinc oxide with silicon IC processing," *Sensors & Actuators,* 1989.
59. Ono, S., K. Wasa, and S. Hayakawa, "Surface acoustic wave properties in ZnO–SiO2–Si layered structures," *Wave Electron.* **3**, 35–49 (1977).
60. G. S. Kino and R. S. Wagers, "Theory of interdigital couplers on nonpiezoelectric substrates," *J. Appl. Phys.* **44**, pp. 1440–88 (1973).
61. W. R. Smith, *et al.*, "Analysis of interdigital surface wave transducers by use of an equivalent circuit model," *IEEE Trans. Microwave Theory Techniques* **MTT-17**, pp. 856–64 (1969).
62. W. H. King, "Piezoelectric sorption detector," *Anal. Chem.* **36**, 1735–9 (1964).
63. J. Hlavay and G. G. Guilbault, "Applications of the piezoelectric crystal detector in analytical chemistry," *Anal. Chem.* **49**, 1890–8 (1977).
64. H. Wohltjen and R. Dessy, "Surface acoustic wave probe for chemical analysis," *Anal. Chem.* **51**, 1458–75 (1979).
65. S. L. Rose-Phersson, J. W. Grate, D. S. Ballantine, and P. C. Jurs, "Detection of hazardous vapors including mixtures using pattern recognition analysis of responses from surface acoustic wave devices," *Anal. Chem.* **60**, 2801–11 (1988).
66. G. C. Frye, S. J. Martin, R. W. Cernosek, K. B. Pfeifer, and J. S. Anderson, "Portable acoustic wave sensor systems," *IEEE Ultrasonics. Symp. Proc., Orlando, FL, USA, 1991,* pp. 311–16.
67. P. Chen, *et al.*, "A planar process PI-FET accelerometer," *Proc. 1980 IEDM Conf.,* p. 848.
68. M. E. Motamedi, A. P. Andrews, and E. Brower, "Accelerometer sensors using piezoelectric ZnO thin films," *Proc. 1982 Sonics and Ultrasonics Symp.,* p. 303.
69. P. Harteman and P. L. Meunier, "Surface acoustic wave accelerometers," *Proc. IEEE 1981 Ultrasonics Symp.,* pp. 152–4.
70. D. T. Erikson, "SAW accelerometers, integration of thick and thin film technology," *Proc. IEEE 1985 Ultrasonics Symp.*
71. S. W. Wenzel and R. M. White, "A multisensor employing an ultrasonic Lamb-wave oscillator," *IEEE Trans. Electron Devices (Special Issue on Microsensors and Microactuators),* **ED-35**, 735–43 (1988).
72. B. A. Martin, S. W. Wenzel, and R. M. White, "Viscosity and density sensing with ultrasonic plate waves," *Sens. Actuators,* **A21–A23**, 704–8 (1990).
73. J. W. Grate, S. W. Wenzel, and R. M. White, "Frequency-independent and frequency-dependent polymer transitions observed on flexural plate wave devices and their effects on vapor sensor response mechanisms," *Anal. Chem.* **64**, 413–22 (1992); "Flexural plate wave devices for chemical analysis," *Anal. Chem.* **63**, 1222–32 (1990).
74. B. J. Costello, B. A. Martin, and R. M. White, "Ultrasonic plate waves for biochemical measurements," *Proc. IEEE Ultrasonics Symp., Montreal, Quebec, October 3–6, 1989,* pp. 977–81.
75. B. J. Costello, S. W. Wenzel, and R. M. White, "Ultrasonic flexural-plate waves for liquid-phase sensing," *7th Int. Conf. on Solid-State Sensors and Acuators, Yokohama, Japan, June 7–10, 1993,* pp. 704–7.
76. T. K. Eto, B. Rubinsky, B. J. Costello, S. W. Wenzel, and R. M. White, "Lamb-wave

microsensor measurement of viscosity as a function of temperature of dimethylsulfoxide solution," *1992 Ninth National Heat Transfer Conf. San Diego, CA, August 9–12, 1992.*

77. T. R. Tsao, R. M. Moroney, B. A. Martin, and R. M. White, "Electrochemical detection of localized mixing produced by ultrasonic flexural waves," *Proc. IEEE Ultrasonic Symp., Lake Buena Vista, FL, December 8–11, 1991,* pp. 937–40.

78. R. M. Moroney, R. M. White, and R. T. Howe, "Microtransport induced by ultrasonic Lamb waves," *Appl. Phys. Lett.* **59**, 774–6 (1991).

79. M. A. Northrup, M. T. Ching, R. M. White, and R. T. Watson, "DNA amplification with a microfabricated reaction chamber," *Trans. 7th Int. Conf. on Solid-State Sensors and Acuators, Yokohama, Japan, 1993,* pp. 924–6.

80. S. W. Wenzel and R. M. White, "Analytic comparison of the sensitivities of bulk-, surface-, and flexural plate-mode ultrasonic gravimetric sensors," *Appl. Phys. Lett.* **54**, 1976–8 (1989).

81. C. T. Chuang and R. M. White, "Sensors utilizing thin-membrane SAW oscillators," *IEEE Ultra. Symp. Proc., Chicago, IL, 1981,* pp. 159–62.

82. H. Wohltjen, et al., "Trace chemical vapor detection using SAW delay line oscillators," *IEEE Trans. Ultrason. Ferroelec. Freq. Contr.* **UFFC-34**, 172–8 (1987).

83. S. J. Martin, A. J. Ricco, D. S. Ginley, and T. E. Zipperian, "Isothermal measurements and thermal desorption of organic vapors using SAW devices," *IEEE Trans. Ultrason. Ferroelec. Freq. Contr.* **UFFC-34**, 142–7 (1987).

84. A. Bryant, D. L. Lee, and J. F. Vetelino, "A surface acoustic wave gas detector," *IEEE Ultrason. Symp Proc., Chicago, IL, 1981,* pp. 171–4.

85. A. D'Amico, A. Palma, and E. Verona, "Surface acoustic wave hydrogen sensor," *Sens. Actuators,* **3**, 31–5 (1982).

86. C. Caliendo, A. D'Amico, P. Verardi, and E. Verona, "Surface acoustic wave H_2 sensor on silicon substrate," *IEEE Ultrason. Symp. Proc., Chicago, IL, 1988,* 569–74.

87. A. D'Amico, A. Petri, P. Verardi, and E. Verona, NH_1 surface acoustic wave gas detector," *IEEE Ultra. Symp. Proc., Denver, CO, 1987,* 633–6.

88. J. F. Vetelino, R. K. Lade, and R. S. Falconer, "Hydrogen sulfide surface acoustic wave gas detector," *IEEE Trans. Ultrason. Ferroelec. Freq. Contr.* **UFFC-34**, 156–60 (1987).

89. S. G. Joshi and J. G. Brace, "Measurement of humidity using surface acoustic waves," *IEEE Ultrason. Symp. Proc., San Francisco, CA 1985,* pp. 600–3.

90. J. G. Brace, T. S. Sanfelippo, and S. G. Joshi, "A study of polymer/water interactions using surface acoustic waves," *Sens. Actuators,* **14**, 47–68 (1988).

91. A, J. Ricco, S. J. Martin, and T. E. Zipperian, "Surface acoustic wave gas sensor based on film conductivity changes," *Sens. Actuators* **8**, 319–33 (1985).

92. A. Venema, E. Nieuwkoop, M. J. Vellekoop, W. J. Ghijsen, A. W. Barendsz, and M. S. Nieuwenhuizen, "NO_2 gas-concentration measurement using a SAW-chemosensor," *IEEE Trans. Ultrason. Ferroelec. Freq. Contr.* **UFFC-34**, 148–55 (1987).

93. D. Rebiere, M. Hoummady, D. Hauden, J. Aucouturier, J. Pistre, P. Cunin, and R. Planade, "Surface acoustic wave (SAW) NO_2 sensors: theoretical studies, measurements, and design," *Proc. Transducers 91, San Francisco, CA, USA, 1991,* pp. 351–4.

94. M. S. Nieuwenhuizen, A. W. Barendsz, E. Nieuwkoop, M. J. Vellekoop, and A. Venema, "Transduction mechanisms in SAW gas sensors," *Electron. Lett.* **22**, 184–5 (1986).
95. G. Watson, W. Horton, and E. J. Staples, "Gas chromatography utilizing SAW sensors," *IEEE Ultra. Symp. Proc., Orlando, FL, USA, 1991*, pp. 305–9.
96. G. Watson, R. Ketchpel, and E. J. Staples, "Thermal conductivity gas sensor based on $LiNbO_3$ SAW resonators," *IEEE Ultrason. Symp. Proc., Tucson, AZ, USA, 1992*, pp. 253–6.

4 Mechanical Sensors

B. KLOECK and N. F. DE ROOIJ
University of Neuchatel
Neuchatel, Switzerland

4.1 INTRODUCTION

4.1.1 Mechanical Semiconductor Sensors

Silicon is used for mechanical sensors, because it combines well-established electronic properties with excellent mechanical properties.[1] Other advantages of silicon include drastically reduced dimensions and mass, batch fabrication and easy interfacing or even integration with electronic circuits and microprocessors. Interest in the mechanical properties of silicon and its use for sensors started with the discovery of its piezoresistivity.[2] The first mechanical sensor was the piezoresistive pressure sensor, but since the development of this sensor, a very wide variety of sensors has been conceived and produced.

The strength of silicon is as high as steel but it cannot be deformed plastically. Therefore, phenomena such as creep and hysteresis do not exist in silicon. A well-designed sensor submitted to excessively high overload will either fail or continue to function with the same specifications as before.

This chapter presents the two major classes of mechanical sensors: piezoresistive and capacitive sensors. Three of the most common applications are discussed: pressure sensors, accelerometers and flow sensors. First, a general overview describes fabrication methods specifically designed or adapted for mechanical sensors, and the read-out techniques that are available to transform the mechanical signal into an electric signal.

4.1.2 Fabrication Methods

Most processes used for the fabrication of mechanical sensors are borrowed from integrated circuit technology, including patterning by photolithography,

Semiconductor Sensors, Edited by S. M. Sze.
ISBN 0-471-54609-7 © 1994 John Wiley & Sons, Inc.

oxidation and diffusion, thin-film deposition, metallization, and various other deposition and etching techniques[3]. Some processes have been taken directly, others have developed to respond to the specific needs of sensors. Three-dimensional shaping of silicon to make micromechanical parts is certainly one of them. In general, the goal is to keep these special processes compatible with regular integrated circuit (IC) fabrication technology. A common strategy is to run most of the process cycle in a standard IC facility (lab or silicon foundry), and to finish the device in a dedicated postprocessing lab.

Even though the materials themselves are generally the same as those used in microelectronics and fabricated the same way, they have completely different functions. For instance, silicon oxide may be used as a masking material for deep etching or as a sacrificial layer in surface micromachining; silicon nitride is also an excellent masking material, but it is also used for mechanical suspension. Therefore, in micromachining processes for mechanical sensors, special attention is given to mechanical properties of the materials and thin films, which often means optimizing the fabrication parameters to reduce stress. In microelectronics, stress reduction is important for the long-term stability and reliability of devices, but in mechanical sensors it will directly influence the functionality and operation of the device. This section gives a short overview of relevant silicon micromachining processes for mechanical sensors.

Bulk Micromachining of Silicon. Many methods have been developed to fabricate silicon membranes, cantilevers and other structures that are essential parts of mechanical sensors. A possible classification of bulk silicon micromachining is the following:

Mechanical:	drilling	
Electromechanical:	spark erosion	
Wet chemical:	isotropic	$HF-HNO_3-CH_3COOH$
	anisotropic:	KOH (and other hydroxides)
		EDP
		hydrazine
Electrochemical:	isotropic:	HF
Dry etching:	mechanical:	argon plasma
	reactive:	RIE

Mechanical drilling of silicon can be done on-wafer, but it requires the handling of one hole at a time. Therefore, it is not ideally suited for mass production.

For spark erosion, the metal electrode of the spark machine can be shaped to drill several holes at a time. However, the surface of the resulting structures is relatively rough. This method has been combined with isotropic etching to smooth the surface.

Among the wet chemical-etching methods, isotropic etchants were developed first. A typical isotropic etch solution is the system HF/HNO_3, with H_2O or CH_3COOH as diluents.[4,5] An electrochemical etch system[6] is the solution

HF/H$_2$O. In this case, silicon is oxidized anodically, and the oxide is continuously etched away by the dilute HF solution. A minimal current density is required, hence only conductive silicon, that is, highly doped, can be etched. Evidently, the membranes that are obtained are circular. The lateral under-etch of the etch mask is as large as the depth of the hole. Therefore, isotropic etchants have been combined with the previously mentioned mechanical methods to improve the ratio of lateral dimensions to depth.

By anisotropic etching of silicon, a much wider variety of structures can be made with very precise dimensional control. Anisotropic etches show different etch rates for the different crystal orientations of silicon. For some crystal planes it is easier to break the Si–Si bonds than for others. The differences are determined by the atom density in the plane and by the number of bonds that are exposed to the etch solution or "hidden" behind the silicon atoms. Common anisotropic etchants are solutions of potassium hydroxide (KOH), ethylene diamine/pyrocatechol (EDP), and hydrazine.

Silicon has a diamond cubic crystal structure. The Miller indices of the main crystallographic planes of silicon are (100), (110) and (111). The KOH etch rate in the ⟨100⟩ direction is slightly higher than in the ⟨110⟩ direction and both are much higher than in the ⟨111⟩ direction. If holes, membranes, grooves, pits, etc., have to be etched in silicon, then wafers are taken with surface orientation (100) and (110), depending on the required shape of the etched structures. Figure 1 shows some etch shapes in these two wafers.[1] Parts of the wafers are covered with an etch mask, and the exposed areas are etched. For wafers with ⟨100⟩ surface orientation, etching proceeds along (111) planes which intersect the surface along the mask edges. If the mask opening is a square with a side that is small compared to the etching depth, then a pyramidal pit is formed defined by four (111) planes and etching will continue at the extremely slow etch rate of (111) planes, that is, it will virtually stop. The angle between the ⟨100⟩ and

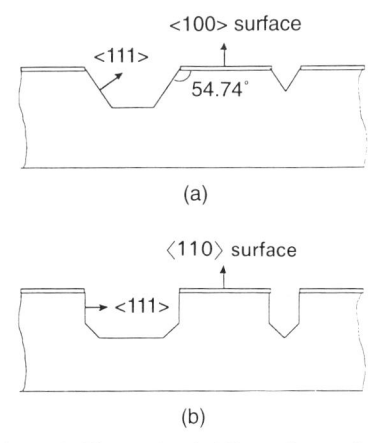

Fig. 1 Anisotropic etching of silicon. (a) ⟨100⟩ wafer surface orientation, (b) ⟨110⟩ wafer surface orientation. (After Ref. 1)

$\langle 111 \rangle$ directions is 54.74 degrees. In $\langle 110 \rangle$ wafers, the (111) planes are vertical, and deep trenches with high aspect ratio can be etched. The materials used to protect the wafer areas that should not be etched are silicon dioxide or silicon nitride.

In microelectronic processes, dry etching or plasma etching is used to pattern thin films (metal, oxide, nitride, polysilicon).[3] This process has been adapted to etch bulk silicon for the fabrication of mechanical sensors. A plasma is an ionized gas composed of ions, electrons and neutrons. It is produced when an electric field of sufficient magnitude is applied to a gaseous atmosphere, causing the gas to break down and become ionized. The plasma is initiated by free electrons that are released by some means such as field emission from a negatively-biased electrode. The free electrons gain kinetic energy from the electric field. In the course of their travel through the gas, the electrons collide with gas molecules and release their energy. The energy transferred in the collisions causes the gas molecules to be ionized (i.e., more electrons are freed). The freed electrons, in turn, gain kinetic energy from the field and the process continues. Therefore, when the applied voltage is larger than the breakdown potential, a sustained plasma is formed throughout the reaction chamber.

It is possible to etch thin films by mere physical impact, namely ion bombardment with relative high-energy noble gas ions, for example by argon (Ar^+) sputter etching. However, for etching silicon, molecular gases are chosen that react chemically with the material which is to be removed or patterned (reactive-ion etching, RIE). Fluorine gases such as SF_6 and CHF_3 and chlorine gases such as $SiCl_4$ and Cl_2 are commonly used. There is a continuous reactant gas supply and a vacuum exit to maintain a low pressure (0.1 to 10 Torr). The ionized reactants impinge on the surface, followed by a chemical reaction, to form volatile compounds. Assisted by ion bombardment, these compounds are desorbed from the surface and pumped out by the vacuum system.

The reproducibility of the operation specifications of mechanical sensors depends strongly on the control of the dimensions during fabrication. Lateral dimensions can be controlled accurately by photolithography masks, but etching depth is independent of the mask design. To control etching depth, a number of automatic etch-stop methods have been developed. Three methods are briefly compared here:

Chemical:	boron etch-stop	
Electrochemical:	isotropic	n^+/n etch-stop
	anisotropic	p/n junction etch-stop

Again, the isotropic etch-stop was developed first.[7] A low-doped, n-type epitaxial layer is grown on a high-doped silicon substrate. The substrate is etched electrochemically as mentioned before. However, the conductivity of the epitaxial layer is too low to reach the required current density and etching stops at the interface. To improve the depth-to-width ratio, this method has been combined with spark erosion, which is facilitated by the relatively high conductivity of the n^+ substrate.[8]

The boron etch-stop method is based on the observation that highly doped silicon (particularly with boron) is etched very slowly by the previously mentioned anisotropic etchants.[9,10] Thus, thin membranes, bridges, or beams can be fabricated by diffusing or implanting boron into a thin layer on one surface of a wafer and by etching the silicon away from the other side through a mask window. As soon as all the silicon is removed, etching is stopped at the diffused layer, which will stay as a free microstructure connected to the thick silicon rim. This is a very versatile technique to make micromechanical devices. However, the high doping concentration precludes them from integrating electronic components.

The most appropriate method to make low-doped, monocrystalline silicon membranes of well-controlled dimensions and thickness is the electrochemical etch-stop on a diffusion or an epitaxial layer. This method is particularly useful for the fabrication of piezoresistive pressure sensors, and, therefore, it will be discussed in that section.

Surface Micromachining of Silicon. The introduction of polysilicon as a mechanical material, has opened new possibilities for micromachining. Polysilicon membranes can be fabricated as follows. An etch-resistant film, for example silicon dioxide or silicon nitride, is grown or deposited on a silicon substrate. Polysilicon is then deposited on top of this film. One fabrication method is to etch away the silicon from the back-side of the wafer until etching is stopped at the inert layer. Smaller sensors can be obtained if the hole is not designed to delineate the membrane, but just to give access to the etch-resistant layer, which is then removed by, for example a HF solution in the case of silicon oxide. The result[11] is shown schematically in Fig. 2.

A second method is surface-micromachining technology.[12] Again, a polysilicon film is deposited and patterned to form a membrane or any other structure. In this approach, the bulk silicon is not etched away, but only the sacrificial layer between the silicon substrate and the polysilicon structure is removed by lateral under-etching. For piezoresistive and capacitive pressure sensors, "via" holes are required on the surface of the wafer to remove the sacrificial layer, and these via holes are closed by oxidation in order to obtain a sealed cavity.[13] Note that only absolute pressure sensors can be made this way. Accelerometers, flow sensors, and a wide variety of other geometries can

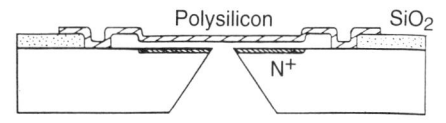

Fig. 2 The combination of polysilicon and a sacrificial layer allows the fabrication of smaller chips than with anisotropic etching of monocrystalline silicon membranes, since less surface is lost by the {111} side planes of the hole. (After Ref. 11)

158 MECHANICAL SENSORS

be made by surface micromachining. The dimensions of polysilicon microstructures can be controlled very accurately, however, the mechanical stability of polysilicon is somewhat inferior to that of monocrystalline silicon.

4.1.3 Read-Out Techniques

All mechanical sensors rely on some physical principle to transform the mechanical signal into an electric signal for display or further electronic treatment. In this section, an overview is given of the read-out techniques. Some of the more common methods will be treated in more detail later. The read-out methods can be classified as follows:

Static methods:
　Stress detection:　　　　piezoresistive strain gages (normal stress)
　　　　　　　　　　　　　transverse voltage gage (shear stress)[14-16]
　　　　　　　　　　　　　piezojunction effect[17,18]
　Deformation detection:　capacitance[19-22]
　　　　　　　　　　　　　optical interferometry[23,24]
Resonant methods:　　　piezoresistive strain gages[25]
　　　　　　　　　　　　　optical[26]

In the class of static methods, the first set of elements are sensitive to the stresses that are induced by deformation in the microstructure. Devices that sense normal stress will be discussed in Section 4.2. The second type of sensors, also called piezo-Hall sensors or X-ducers, is based on the observation that an electric field is developed perpendicular to the current flow if such devices are subjected to a shear stress. The maximum voltage occurs at an angle of 45° to the $\langle 110 \rangle$ directions of a (100) silicon substrate. A Wheatstone bridge is not required to eliminate the temperature coefficient of the resistors, so that only one element is needed, which can yield smaller sensors. The temperature coefficient of the pressure sensitivity, however, is the same as for normal piezoresistive sensors. The piezojunction effect is based on the fact that the current gain and the base–emitter voltage depend on the applied stress. The sensor can be designed so that the base–emitter voltage is a linear function of the stress.

At high temperatures, diffused resistors show p–n junction-leakage problems. To circumvent these problems, polysilicon resistors can be deposited on an isolated membrane or cantilever beam of monocrystalline silicon or other material.[27] Both Wheatstone bridge and transverse voltage sensors have been reported. These devices can be used at temperatures up to 250°C. However, the piezoresistance coefficients of polysilicon are about half as large as those for monocrystalline silicon. By laser recrystallization, they can be increased to two-thirds of the values of monocrystalline silicon.[28]

A second static read-out method is deformation or displacement detection. The microstructure can be used as one electrode of a capacitor with a fixed second electrode. The second electrode is typically a thin metal film deposited on a glass or silicon substrate that is bonded to the rim of the microstructure. In other devices, one polysilicon microstructure can move with respect to another, fixed, polysilicon structure, placed adjacent to the first. In general, the complete microstructure surface contributes to the output signal. The resolution and temperature behavior of capacitive sensors is generally better than that of piezoresistive sensors, however, capacitance is more complicated to measure than voltage, and on-chip circuitry is recommended to avoid stray capacitances. As for the piezoresistive sensors, the on-chip electronics complicates its use at higher temperatures.

Recent developments in optical interference read-out are very promising for high-temperature applications. Interference can be achieved between a semi-transparent optical flat and a thin silicon microstructure. The fringes are measured with a photodetector. A laser beam is applied to this system either directly or via optical fibers. The laser and the read-out equipment are at room temperature and only the fiber-end is close to the sensor, so that high-temperature applications ($>200°C$) are possible.

Finally the sensor can be designed as a resonant structure. For pressure sensors, for instance, the resonance frequency of a silicon membrane depends on the pressure that is applied. The membrane can be excited thermally or electrically by depositing a piezoelectric material such as zinc oxide. Common read-outs are piezoresistive and optical. A frequency signal is interesting from the point of view of data acquisition. Membranes often have high damping characteristics, but there are other mechanical structures that have higher Q-factors,[28] and thus better measurement resolutions.

Each of these systems has its typical fabrication technology and application fields, depending on parameters such as measurement range, resolution, stability, mass production and price, environment (especially temperature), required equipment, and others. In this chapter, emphasis will be placed on non-resonant piezoresistive and capacitive sensors made by bulk micromachining, since these technologies are well established, and the sensors are widely available on the market.

4.1.4 Bonding Techniques

Encapsulation of semiconductor mechanical sensors is a special problem that deserves attention. The way the silicon chip is fixed in its housing is subject to two major conditions: the encapsulation should introduce minimal mechanical stress and, more important, if mechanical stress cannot be avoided, it has to be stable in time to limit drift of the sensor output signal. For some applications, the temperature behavior of the bond must also be considered. These conditions preclude the use of epoxies and other organic adhesives. Considering the thermal expansion coefficients, silicon-to-glass and silicon-to-silicon bonds are the best combinations. A classification of the non-organic bond techniques for these materials is given here:

Silicon to glass: anodic bonding
eutectic bonds
Silicon to silicon: sputtered glass
spin-on glass
glass-frit seals
eutectic bonds
silicon direct bonding

A brief overview of these methods (except for the last one), including the required temperatures and typical applications, is given in Ref. 30. Currently, the most widely used is anodic bonding of silicon to Pyrex 7740 borosilicate glass.[31] The process involves typical temperatures of 400°C and voltages between 600 and 1000 V. The process can be monitored by measuring the current. Depending on the conditions, the bond can be completed in a few seconds to several minutes.

Other techniques have been developed to bond silicon wafers together. In one method, used to fabricate silicon-on-insulator structures, the wafers were bonded (and also isolated) by a silicon oxide layer.[32-34] A first technique[32] used a pair of oxidized wafers and bonded them anodically with 30–50 V applied, at 850–950°C for one hour. A second possibility[33] is to use one oxidized and one bare wafer and to bond them by heat treatment at 1100°C for four hours. Another method[34] works with pairs of oxidized/bare and oxidized/oxidized wafers, and bonding is carried out at 1000°C for 30 minutes.

Bare silicon wafers can be bonded together without any intermediate layer by the so-called silicon fusion bonding (SFB).[35,36] The quality of the interface is such that nearly perfect diodes are obtained with silicon wafers of opposite type. The process involves cleaning of the mirror-polished wafers, a hydrophilic surface treatment, and heating to 1000–1100°C for two hours.[35] This method has been used to fabricate, for instance, small piezoresistive pressure sensors, where the {111} slopes of the etched hole go inwards, instead of outwards as for the classical anisotropically etched sensors.[36] Thus, less chip surface is required for the same membrane dimensions.

4.2 PIEZORESISTIVITY

4.2.1 Introduction

Piezoresistivity is a material property where the bulk resistivity is influenced by the mechanical stresses applied to the material. Many materials exhibit stress dependence through the mobility or the number of charge carriers as a function of the volume of the material.[37] Volume changes effect the energy gap between the valence and the conduction bands. Hence, the number of carriers and, thus, the resistivity changes. The piezoresistance effect, however, is observed to be much larger than predicted by these mechanisms. This has been explained by means of the many-valley model. All materials probably have a piezoresistance effect to some extent, but it is particularly important in some semiconductors. Monocrystalline silicon has a high piezoresistivity, combined with excellent mechanical properties, which makes it particularly suited for the conversion of mechanical deformation to an electrical signal. Needless to say, silicon characteristics are also well known from the electronics industry. Therefore, silicon is widely used as basic material for piezoresistive sensors for mechanical signals such as pressure, flow, force, and acceleration. As a matter of fact, the history of silicon-based sensors started with the discovery of the piezoresistance effect in silicon and germanium[2] in 1954.

After this discovery, extensive studies were made about the solid-state properties of the piezoresistance effect and its potential applications.[38-44] The use of diffusion techniques for the fabrication of piezoresistive sensors for stress, strain, and pressure was proposed by Pfann and Thurston[14] in 1961, and used for the first time on thin, single-crystal silicon membranes by Tufte et al.[45] The idea was soon adopted by others because of the better performance of silicon resistors compared with classical strain gages. Some of the practical advantages that have been recognized from the beginning are listed below:

1. The gage factor of semiconductors is more than an order of magnitude higher than that of metals.
2. Silicon is a very robust material.
3. The integration of gage and membrane eliminates the need to bond the two components together, which eliminates hysteresis and creep.
4. The strain is transmitted perfectly from the membrane to the gage.
5. The resistors are limited to the surface of the element in bending or torsion where the stresses are maximal.
6. Good matching of the resistors can be achieved, which is particularly useful if Wheatstone bridges are used.
7. The technique is very suitable for miniaturization of the sensors.
8. Mass fabrication can profit from the available technology of integrated circuits.
9. It is possible to integrate electronic circuitry directly on the sensor chip, for signal amplification and temperature compensation.

The very first semiconductor strain gages used a homogeneously-doped silicon strip attached to a membrane of other material. Here, the only advantage was the higher gage factor of silicon. Later a whole wafer was used as a membrane under bending, and resistors were diffused to measure the maximum stress at the surface. Further refinement was introduced by etching away part of the silicon under the resistors until a thin membrane was left, so that higher stresses were created and the sensitivity increased.

When the technology for piezoresistive pressure sensors was well established, other applications were investigated. At present, the piezoresistive accelerometer has also found its way to industrial mass production.

In this chapter we will first review the mathematical model that describes the piezoresistance effect; we will give the results of the measurements that were carried out in 1954 to define the piezoresistive coefficients, and then we will briefly discuss the physical explanation of the mechanism. The dependence on doping concentration and on temperature is discussed, since it is very important for sensor design.

These properties will first be described for bulk material with homogeneous doping concentration. The situation becomes more complex when diffused resistors are involved, since the piezoresistance coefficients and their temperature behavior depend on the doping concentration. Hence, we will discuss how effective values can be obtained by integrating over the doping profile. Finally we will show how to apply this knowledge to the design and fabrication of sensors.

4.2.2 Mathematical Description

This section is a formal mathematical description of the piezoresistivity, without considering its physical nature. It starts with the general three-dimensional relation between current and electric field. A method is introduced to describe the influence of stress on this relation. The symmetry of the crystal lattice will help to simplify the rather complex mathematical model.

For a three-dimensional anisotropic crystal, the electric field vector ($\mathscr{E}$) is related to the current vector (i) by a three-by-three resistivity tensor. Experimentally, the nine coefficients are always found to reduce to six and the tensor is symmetrical:

$$\begin{bmatrix} \mathscr{E}_1 \\ \mathscr{E}_2 \\ \mathscr{E}_3 \end{bmatrix} = \begin{bmatrix} \rho_1 & \rho_6 & \rho_5 \\ \rho_6 & \rho_2 & \rho_4 \\ \rho_5 & \rho_4 & \rho_3 \end{bmatrix} \cdot \begin{bmatrix} i_1 \\ i_2 \\ i_3 \end{bmatrix}. \tag{1}$$

Piezoresistance in a Coordinate System Aligned to the Crystal Axes. Both silicon and germanium have a cubic crystal structure. If the Cartesian axes are aligned to the $\langle 100 \rangle$ axes of the crystal, then ρ_1, ρ_2, and ρ_3 define the dependence of the electric field along one of the $\langle 100 \rangle$ crystal axes on the current in the same direction. ρ_4, ρ_5, and ρ_6 are cross-resistivities, relating the electric field along one axis to the current in a perpendicular direction. For an isotropic conductor, for example unstressed silicon, $\rho_1 = \rho_2 = \rho_3 = \rho$, and ρ_4, ρ_5, and ρ_6 are equal to zero.

In a piezoresistive material, these six resistivity components depend on the stress in the material, which can also be decomposed into six components: three normal stresses σ_1, σ_2, and σ_3, along the cubic crystal axes, and three shear stresses τ_1, τ_2, and τ_3, as defined in Fig. 3, where the stresses are represented

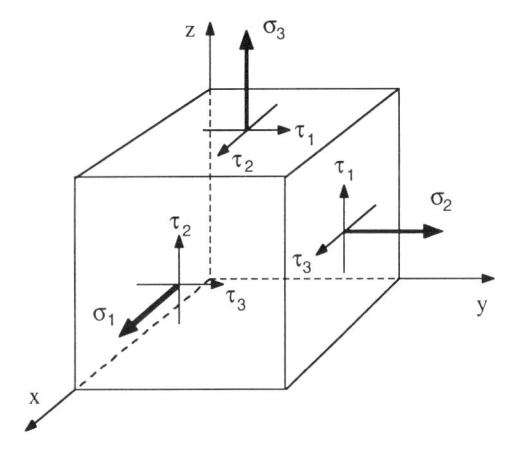

Fig. 3 Definition of the normal stresses σ_i and the shear stresses τ_i ($i=1, 2, 3$).

as acting on a cube of infinitesimal dimensions dx, dy, and dz. If we reference the resistivities to the isotropic unstressed case, then we can write the six resistivity components as:

$$\begin{bmatrix} \rho_1 \\ \rho_2 \\ \rho_3 \\ \rho_4 \\ \rho_5 \\ \rho_6 \end{bmatrix} = \begin{bmatrix} \rho \\ \rho \\ \rho \\ 0 \\ 0 \\ 0 \end{bmatrix} + \begin{bmatrix} \Delta\rho_1 \\ \Delta\rho_2 \\ \Delta\rho_3 \\ \Delta\rho_4 \\ \Delta\rho_5 \\ \Delta\rho_6 \end{bmatrix}. \tag{2}$$

The piezoresistance effect can now be described by relating each of the six fractional resistivity changes $\Delta\rho_i/\rho$ to each of the six stress components. Mathematically this yields a matrix of 36 coefficients. By definition, the elements of this matrix are called piezoresistance coefficients, π_{ij}, expressed in Pa^{-1}.

In order to define the matrix, it would be necessary to carry out 36 independent measurements. However, this task can be greatly simplified for a crystalline material. Since the matrix represents properties of a crystal, it must be invariant under the symmetry operations of the crystal lattice under study. Hence, the form of the matrix can be found, theoretically, for each class of crystals belonging to the same crystallographic point group. The symmetry conditions lead to certain relations between the different matrix components, which reduce the number of independent, non-vanishing components to considerably less than 36. For the cubic crystal structure of silicon and germanium, three different coefficients remain, π_{11}, π_{12}, and π_{44}, and the matrix takes the following form:

$$\frac{1}{\rho}\begin{bmatrix} \Delta\rho_1 \\ \Delta\rho_2 \\ \Delta\rho_3 \\ \Delta\rho_4 \\ \Delta\rho_5 \\ \Delta\rho_6 \end{bmatrix} = \begin{bmatrix} \pi_{11} & \pi_{12} & \pi_{12} & 0 & 0 & 0 \\ \pi_{12} & \pi_{11} & \pi_{12} & 0 & 0 & 0 \\ \pi_{12} & \pi_{12} & \pi_{11} & 0 & 0 & 0 \\ 0 & 0 & 0 & \pi_{44} & 0 & 0 \\ 0 & 0 & 0 & 0 & \pi_{44} & 0 \\ 0 & 0 & 0 & 0 & 0 & \pi_{44} \end{bmatrix} \cdot \begin{bmatrix} \sigma_1 \\ \sigma_2 \\ \sigma_3 \\ \tau_1 \\ \tau_2 \\ \tau_3 \end{bmatrix}. \quad (3)$$

Combining Eqs. 1, 2, and 3, we obtain an expression for the electric field in a cubic crystal lattice under stress:

$$\mathscr{E}_1 = \rho i_1 + \rho\pi_{11}\sigma_1 i_1 + \rho\pi_{12}(\sigma_2+\sigma_3)i_1 + \rho\pi_{44}(i_2\tau_3+i_3\tau_2)$$
$$\mathscr{E}_2 = \rho i_2 + \rho\pi_{11}\sigma_2 i_2 + \rho\pi_{12}(\sigma_1+\sigma_3)i_2 + \rho\pi_{44}(i_1\tau_3+i_3\tau_1) \quad (4)$$
$$\mathscr{E}_3 = \rho i_3 + \rho\pi_{11}\sigma_3 i_3 + \rho\pi_{12}(\sigma_1+\sigma_2)i_3 + \rho\pi_{44}(i_1\tau_2+i_2\tau_1).$$

The first term in Eq. 4 is the contribution of the unstressed conduction. The second term, containing π_{11}, represents the piezoresistance effect as it is known from wire and foil gages; it is the effect of a stress in the direction of current flow, on the potential drop in that direction. The other terms reflect the more complicated piezoresistive behavior of the stressed semiconductor lattice. These coefficients are properties of the material and vary from one material to another. Note that the expressions of Eq. 4 are only valid for uniform bulk material. For devices with finite dimensions, the influence of dimension changes ought to be added, although, in general, it is negligible compared with the larger effect due to the change in resistivity. It is also important to notice that the piezoresistance coefficients can be either negative or positive, and that, in general, they vary with doping concentration and temperature, as will be explained in more detail later. For nonuniform materials, such as the diffused or implanted silicon strain gages that will be discussed in Section 4.2.5, the equations have to be integrated over the complete structure.

Transformation of Axes. With the knowledge of the value of three parameters π_{11}, π_{12}, and π_{44} (which can be measured; see Section 4.2.2) that were defined in a coordinate system aligned to the $\langle 100 \rangle$ axes of the silicon crystal, all the piezoresistance properties of silicon can be calculated. In order to calculate the stresses and the electric field expressed in an arbitrary Cartesian system, the $\langle 100 \rangle$ axes can be transformed into the given coordinate system. If e_1, e_2, and e_3 are the unity vectors of the $\langle 100 \rangle$ axes system, and u_1, u_2, and u_3 are the unity vectors of the new axes system then the relations between the two systems can be written as:

$$e_1 = l_1u_1 + l_2u_2 + l_3u_3$$
$$e_2 = m_1u_1 + m_2u_2 + m_3u_3 \quad (5)$$
$$e_3 = n_1u_1 + n_2u_2 + n_3u_3.$$

In other words, (l_1, l_2, l_3), (m_1, m_2, m_3) and (n_1, n_2, n_3) are the coordinates in the new system of the unity vectors of the $\langle 100 \rangle$ system.

A vector (x, y, z) referred to the crystal axes is then transformed into a vector (x^*, y^*, z^*) in the new system using:

$$\begin{bmatrix} x^* \\ y^* \\ z^* \end{bmatrix} = \begin{bmatrix} l_1 & m_1 & n_1 \\ l_2 & m_2 & n_2 \\ l_3 & m_3 & n_3 \end{bmatrix} \cdot \begin{bmatrix} x \\ y \\ z \end{bmatrix}. \quad (6)$$

4.2.3 Longitudinal and Transverse Piezoresistance Coefficients

Of all possible orientations that can be calculated by means of Eq. 5, only two will be examined in more detail here, since they represent the most common situations for piezoresistive sensor devices: they are represented schematically in Fig. 4. The first one concerns a uniaxial state of stress σ^*, electric field $\mathscr{E}^*$ and current i^*, all in the same direction, but not necessarily along a crystal axis (Fig. 4a). In this case, the relation between stress and change of resistivity is called the longitudinal piezoresistance coefficient, noted π_l. In order to calculate π_l as a function of the three piezoresistance coefficients in the $\langle 100 \rangle$ axis system,

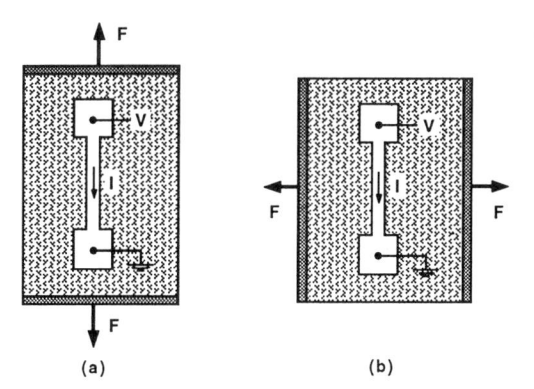

Fig. 4 Schematic representation of the stress/current situations that are ruled by (a) the longitudinal piezoresistance coefficient and (b) the transverse piezoresistance coefficient. F represents a force applied to the sample. Note that, for clarity, resistors are represented here, but the theory discussed in the text is for bulk material.

the axis transformation of Eq. 5 is applied to Eq. 4. The result is:

$$\mathscr{E}^* = \rho i^* + \rho i^* [\pi_{11} + 2(\pi_{44} + \pi_{12} - \pi_{11})(l_1^2 m_1^2 + l_1^2 n_1^2 + m_1^2 n_1^2)] \tag{7}$$

and hence, the longitudinal piezoresistance coefficient can be written as:

$$\pi_l = \pi_{11} + 2(\pi_{44} + \pi_{12} - \pi_{11})(l_1^2 m_1^2 + l_1^2 n_1^2 + m_1^2 n_1^2). \tag{8}$$

In another commonly used embodiment (very often combined with the previous case), the electric field and current are colinear, and the uniaxial stress is perpendicular to both (Fig. 4b), giving rise to a so-called transverse piezoresistance coefficient, π_t, which is calculated similarly. The result is:

$$\pi_t = \pi_{12} - (\pi_{44} + \pi_{12} - \pi_{11})(l_1^2 l_2^2 + m_1^2 m_2^2 + n_1^2 n_2^2). \tag{9}$$

It is easily shown that the factor $(l_1^2 m_1^2 + l_1^2 n_1^2 + m_1^2 n_1^2)$ is a maximum in the directions making equal angles with the crystal axes, i.e., in the $\langle 111 \rangle$ directions. It follows that, if $(\pi_{44} + \pi_{12} - \pi_{11}) \neq 0$, the longitudinal coefficient has either a maximum or a minimum in the $\langle 111 \rangle$ directions, depending on the relative magnitudes of π_{11}, π_{12} and π_{44}. Materials with a minimum in the $\langle 111 \rangle$ direction have their maximum π_l along the crystal axes. The value of the $\langle 111 \rangle$ π_l is obtained from Eq. 8 by setting $l_1^2 = m_1^2 = n_1^2 = 1/3$:

$$(\pi_l)_{\langle 111 \rangle} = \tfrac{1}{3}(\pi_{11} + 2\pi_{12} + 2\pi_{44}). \tag{10}$$

Calculated similarly, using Eqs. 8 and 9, Table 1 lists longitudinal and transverse piezoresistance coefficients for various practical directions in cubic crystals.

Note that, when piezoresistive sensors are fabricated using diffused resistors, as will be explained later, careful alignment of the diffusion mask to the crystal axes is required to realize maximum stress sensitivity, because of the high anisotropy of the piezoresistance coefficients. The piezoresistance coefficients would be independent of orientation for a semiconductor with

TABLE 1 Longitudinal and Transverse Piezoresistance Coefficients for Various Combinations of Directions in Cubic Crystals

Longitudinal Direction	π_l	Transverse Direction	π_t
(1 0 0)	π_{11}	(0 1 0)	π_{12}
(0 0 1)	π_{11}	(1 1 0)	π_{12}
(1 1 1)	$\tfrac{1}{3}(\pi_{11} + 2\pi_{12} + 2\pi_{44})$	(1 $\bar{1}$ 0)	$\tfrac{1}{3}(\pi_{11} + 2\pi_{12} - \pi_{44})$
(1 1 $\bar{0}$)	$\tfrac{1}{2}(\pi_{11} + \pi_{12} + \pi_{44})$	(1 1 1)	$\tfrac{1}{3}(\pi_{11} + 2\pi_{12} - \pi_{44})$
(1 1 $\bar{0}$)	$\tfrac{1}{2}(\pi_{11} + \pi_{12} + \pi_{44})$	(0 0 1)	π_{12}
(1 1 0)	$\tfrac{1}{2}(\pi_{11} + \pi_{12} + \pi_{44})$	(1 $\bar{1}$ 0)	$\tfrac{1}{2}(\pi_{11} + \pi_{12} - \pi_{44})$

$(\pi_{11}-\pi_{12}-\pi_{44})=0.^{46}$ In principle, it is possible to meet this requirement by an alloy of the silicon–germanium system, with still a usuably large piezoresistance effect. It can be calculated that the coefficients of this isotropic alloy would be:

$$\pi_{11}=-49 \quad \pi_{12}=+21 \quad \pi_{44}=-70 \quad (10^{-11}\text{ Pa}^{-1}).$$

However interesting this idea is from a theoretical point of view, a description of a device based on this alloy has never been reported.

4.2.4 Measurement of the Piezoresistance Coefficients

In order to measure the piezoresistance coefficients, a small known stress has to be applied and the resistivity change measured. Two types of stress can easily be applied to a solid: hydrostatic pressure and uniaxial tension or compression. The effect of pure hydrostatic pressure p is obtained by setting $\sigma_1=\sigma_2=\sigma_3=-p$ and $\tau_1=\tau_2=\tau_3=0$ in Eq. 4, which gives:

$$\mathscr{E}_i=\rho i_i[1-p(\pi_{11}+2\pi_{12})] \quad (i=1,2,3). \tag{11}$$

Two other independent measurements are required to determine the three coefficients, which will give three linear combinations of the coefficients. Smith,[2] who was the first to measure the coefficients π_{11}, π_{12} and π_{44} for different doping concentrations at room temperature, used the three independent arrangements presented in Fig. 5. He applied a uniaxial tensile stress to a

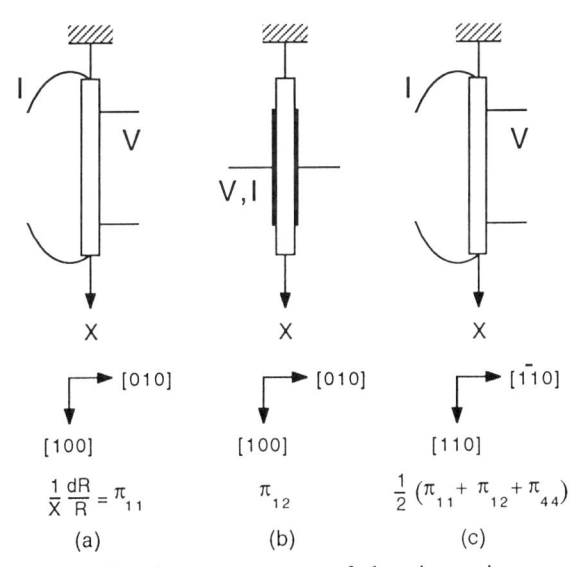

Fig. 5 Schematic set-up for the measurement of the piezoresistance coefficients by Smith,[2] and (combination of) the coefficients that are obtained. (See text for details.)

single-crystal rod by hanging a weight on a string, and measured the voltage drop. The first set-up (Fig. 5a) performs a longitudinal measurement on a [100] sample, yielding a value of π_{11}. The second (Fig. 5b) is a transversal measurement on the same sample, giving π_{12}. Finally a longitudinal measurement on a [110] sample (Fig. 5c) determines, in accordance with Table 1, a value for $0.5(\pi_{11}+\pi_{12}+\pi_{44})$. If the length of the specimens is large compared to the transverse dimensions, there will be a region in which the stress field is not perturbed by the grips used at the ends to apply the stress. The directly observed quantities, dR/R, have to be corrected for dimensional changes to obtain the required $d\rho/\rho$, although these corrections are small. In general, the measurements are performed under adiabatic conditions, that is, the measurement is completed in a short time compared to the time required for thermal equilibrium between the sample and its environment to be established after the stress is applied. The correction that would allow conversion of the measured constant to an isothermal constant is negligible.[43]

Table 2 lists the results that Smith published in 1954. Obviously, the values are considerably larger than the typical values of 2 to 4 for commercially available strain gages of other types. The largest coefficients for silicon are π_{11} in n-type silicon and π_{44} in p-type silicon, about -102×10^{-11} Pa^{-1} and 138×10^{-11} Pa^{-1}, respectively.

With the values of Table 2, π_l and π_t can now be calculated numerically for any orientation, using Eqs. 8 and 9. It was mentioned before that the maximum longitudinal piezoresistance coefficient π_l occurs either in the $\langle 111 \rangle$ or in the $\langle 100 \rangle$ directions, depending on the relative magnitudes of the main piezoresistance coefficients. The data of Table 2 show that p-type silicon and both types of germanium have maxima in the $\langle 111 \rangle$ directions, whereas n-type silicon has a maximum longitudinal effect along the crystal axes. The expressions of Table 1 give as maximum values for low doping concentrations and at room

TABLE 2 Adiabatic Piezoresistance Coefficients at Room Temperature (After Ref. 2)

Material	ρ (Ω cm)	π_{11}	π_{12}	π_{44}
		\multicolumn{3}{c}{(10^{-12} cm^2/dyne or 10^{-11} Pa^{-1})}		
Silicon				
(p-type)	7.8	+6.6	−1.1	+138.1
(n-type)	11.7	−102.2	+53.4	−13.6
Germanium				
(p-type)	1.1	−3.7	+3.2	+96.7
	15.0	−10.6	+5.0	+46.5
(n-type)	1.5	−2.3	−3.2	−138.1
	5.7	−2.7	−3.9	−136.8
	9.9	−4.7	−5.0	−137.9
	16.6	−5.2	−5.5	−138.7

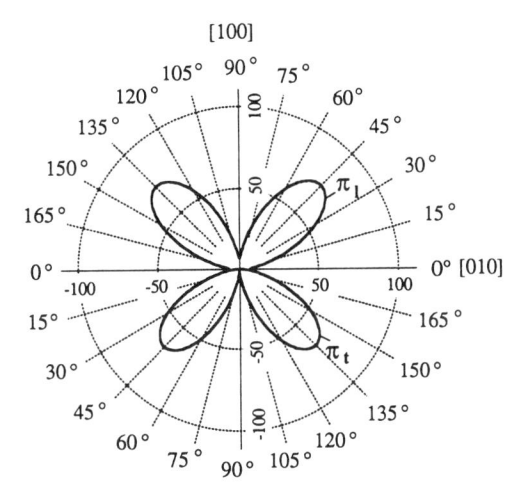

Fig. 6 Longitudinal and transverse piezoresistance coefficients in a $\langle 100 \rangle$ plane, as function of orientation for p-type silicon.

temperature:

$$p\text{-Si:} \quad \pi_{l\langle 111 \rangle} = 93.5 \times 10^{-11} \text{ Pa}^{-1}$$

$$n\text{-Si:} \quad \pi_{l\langle 100 \rangle} = -102.2 \times 10^{-11} \text{ Pa}^{-1}. \tag{12}$$

Figure 6 shows a graphical plot of Eqs. 8 and 9 in a $\langle 100 \rangle$ plane for p-type silicon, with the sample orientation as a parameter.[47] Although Eq. 12 indicates that a higher piezoresistivity can be obtained with n-type silicon, piezoresistors used for sensors are generally of p-type because of orientation limitations due to anisotropic etching of silicon. This will be explained in more detail in Section 4.3.

4.2.5 Quantum-Physical Explanation

This section briefly discusses the relation between electron or hole conductivity and crystal-lattice orientation in general terms. For a more elaborate description the reader is referred to the literature (Refs. 14, 42, 44, 48, and 49). We start with n-type silicon and consider the energy state of an electron in or above the conduction band. The theory of quantum mechanics attributes separate wave numbers k_1, k_2, and k_3 to the components of the motion of the electron in each of the directions 1, 2, and 3. In some materials, such as a silicon lattice, an electron can have the minimum energy it needs to remain in the conduction band, by different combinations of k_1, k_2, and k_3. These combinations are called band-edge points, since they represent lower limits for the energy required for a free electron. Figure 7 schematically represents band-edge points in k-space for n-type silicon, where this space is related to the crystal axes. An electron

170 MECHANICAL SENSORS

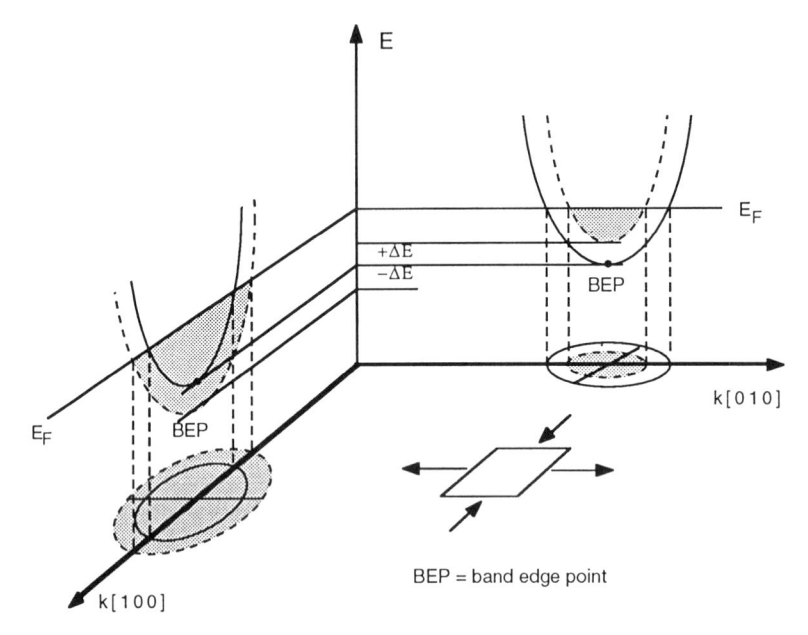

Fig. 7 Two n-type silicon valleys in k-space, aligned with the [100] axes. E_F is the Fermi level. $+\Delta E$ is an energy increase and $-\Delta E$ an energy decrease.

with more energy than is required at a band-edge point may possess such energy by a variety of combinations in (k_1, k_2, k_3) that describe a constant energy surface around the band-edge point.

A family of such surfaces, centered on a band-edge point, describes a so-called energy valley in k-space. In the case of silicon, these families consist of prolate ellipsoids of revolution that are aligned with the crystal axes. They are the projections in k-space of the energy-band edge of the first Brillouin zone, as also illustrated in Fig. 7. Since there are several band-edge points, the model is referred to as the many-valley model. Because the valleys can be transformed into one another, they are identical except for orientation. Figure 7 shows two of the six valleys aligned with the six $\langle 100 \rangle$ directions. The fact that the constant-energy surfaces possess principal axes of unequal lengths, may be interpreted to mean that the components of effective mass and mobility, μ_1, μ_2, and μ_3, of an electron in such a valley are different in the three principal directions. In Fig. 7 the mobility is lowest in the valley direction (i.e., $\langle 100 \rangle$) and highest normal to that direction. Consequently, these electrons make anisotropic contributions to the total conductivity of the lattice. If, however, all ellipsoids have the same properties and all valleys are equally populated with electrons, which is the case for silicon in the unstressed state, then the over-all conductivity of the lattice will be isotropic.

The application of an anisotropic stress condition now changes the relative energies, and, hence, changes the populations in these valleys. Traction in a valley direction removes electrons from that valley and transfers them to valleys lying normal to the traction. Compression has the opposite effect. In Fig. 7 the stress-induced shifts of the band-edge energies (ΔE and $-\Delta E$) are illustrated by dashed lines, for the case of traction in the [010] direction or compression in the [100] direction. The energy decreases with a value ΔE in the [100] direction, hence, more electrons have enough energy to enter the conduction band. In the [010] direction, the minimum required energy increases by ΔE and less electrons satisfy that condition. Thus, the average mobility becomes lower in the direction of traction (longitudinal effect) and higher in directions transverse to the traction axis (transverse effect).

The more the stress on the lattice destroys the symmetry of the valley structure, the larger the piezoresistance effect produced. In n-type silicon the valleys are aligned to the $\langle 100 \rangle$ axes, which explains why π_{11} is the largest coefficient for n-type silicon, since stress in a $\langle 100 \rangle$ direction significantly disturbs the symmetry. If, on the other hand, the crystal is stressed in a $\langle 111 \rangle$ direction, or if the resistance change is measured in a $\langle 111 \rangle$ direction, there is a negligibly small effect, because this direction is symmetrical to the three valleys.

Based on the many-valley model, the piezoresistance coefficients can be calculated explicitly.[38,43,50] For each ellipsoid, the conductivity is expressed as a function of the population, given by the Fermi distribution function, and the relaxation time. The influence of stress on these two functions is calculated, and the total conductivity change is obtained by adding the contribution of each ellipsoid. The mathematical treatment and the resulting expressions for the piezoresistance coefficients yield relative magnitudes that are essentially in agreement with measured values. Also, piezoresistance measurements have been useful in obtaining more detailed information on the band structure and scattering processes of semiconductors, and to verify the theoretical models.

For p-type silicon, the many-valley model is found to be less accurate. Still, some general conclusions can be drawn using the model. The symmetry of the piezoresistance coefficients is assumed to be that of a $\langle 111 \rangle$-valley material. They are imagined as lying in four valleys, one along each $\langle 111 \rangle$ direction. Stress in this direction will have an important impact on the symmetry of the valleys, resulting in a large coefficient π_{44}.

The mathematical expressions that are obtained by quantum physics contain terms that depend on temperature and on the doping concentration. Consequently, the influence of these two parameters can be studied. This is discussed in the next section.

4.2.6 Concentration and Temperature Dependence of Piezoresistivity

Doping Concentration Dependence. For the low-doped silicon samples that he used (about 10^{15} cm^{-3} for the values listed in Table 2), Smith[2] observed no dependence on zero-stress resistivity in his longitudinal measurements and concluded that the piezoresistance coefficients were independent of impurity

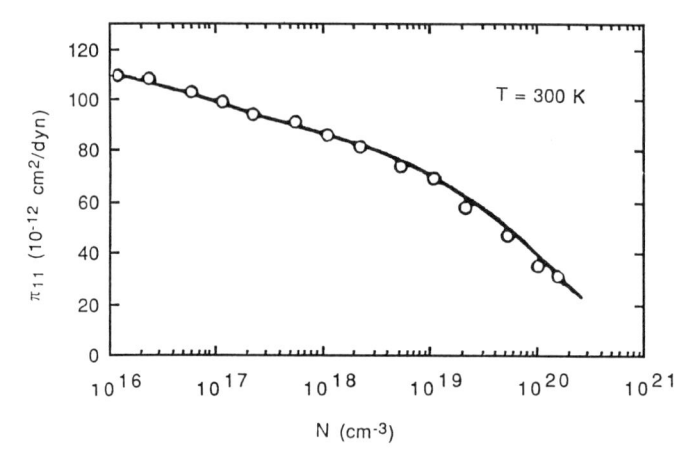

Fig. 8 π_{11} versus impurity concentration in n-type silicon, calculated (———) and measured values (○). (After Ref. 52)

concentration. For germanium, on the other hand, he did find a doping concentration dependence: the piezoresistance coefficients decrease with increasing impurity concentration (see Table 2). Later measurements showed that at higher impurity concentrations, this decrease is also observed for silicon. Based on quantum physics the doping concentration dependence can be calculated. As an example,[52] Fig. 8 shows calculated and measured values for π_{11} for n-type silicon at 27°C and for impurity concentrations ranging from 10^{16} to 10^{20} at/cm³.

It is clear that for practical use, the doping concentration should not be chosen too high in order to keep a reasonably high gage factor.

Temperature Dependence. Mathematical calculations based on the many-valley model predict a decrease of the piezoresistance effect with increasing temperature,[38,39,43] as illustrated schematically in Fig. 9, where the logarithm for any piezoresistance coefficient is plotted against the logarithm of temperature. At very low temperatures, from absolute zero, this relation is linear with slope -1. In other words, the piezoresistance coefficients increase linearly with the inverse of the temperature. At these temperatures, the scattering from electrons from one valley to another is greatly reduced. With increasing temperature, inter-valley scattering becomes more and more important. Since inter-valley scattering produces a larger asymmetry than intra-valley scattering, a larger piezoresistance is associated with the latter. Hence, the piezoresistance-reducing effect of increasing temperature is partly canceled by the enhancing effect of increased inter-valley scattering and the slope in Fig. 9 is less steep in this transition area. At higher temperatures (including room temperature), the $1/T$ behavior is restored again.

The $1/T$ relation predicted by the many-valley model has been measured.[39]

4.2 PIEZORESISTIVITY   173

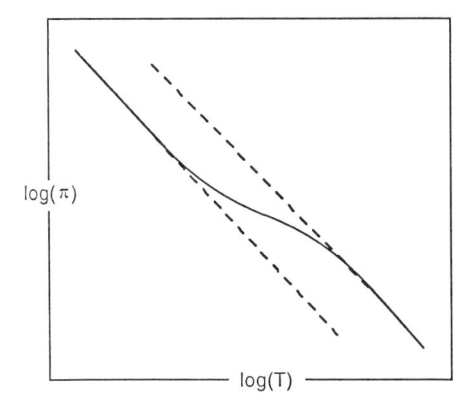

Fig. 9 Schematic variation of any piezoresistance coefficient with temperature.

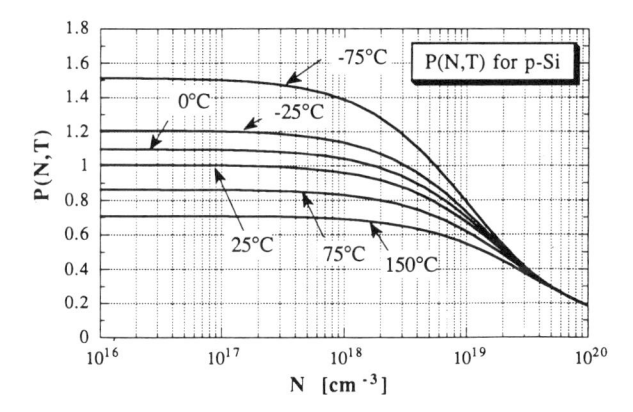

Fig. 10 Piezoresistance factor $P(N, T)$ as a function of impurity concentration and temperature for p-type silicon. (After Ref. 49)

It was shown that for n-type silicon, linearity was followed for a large temperature range, about -200 to $80°C$ (the highest temperature used in these measurements). For p-type silicon, the relation was observed to be valid in a more restricted range, from -100 to $80°C$.

In general, any piezoresistance coefficient can be expressed by its low-doped room-temperature value, referred to as π_0, multiplied by a dimensionless factor that is a function of doping concentration (N) and temperature (T):[50]

$$\pi(N, T) = \pi_0 \cdot P(N, T). \tag{13}$$

The piezoresistance factor $P(N, T)$ based on mathematical calculations, is shown for p-type silicon in Fig. 10. This figure graphically summarizes the discussion of this section: the piezoresistance decreases with increasing doping concentration and with increasing temperature. However, the decreasing distance between the curves indicates that the temperature coefficient of the piezoresistivity also decreases with increasing doping concentration. In practice, sensitivity is often sacrificed to obtain a lower temperature coefficient, one of the major inconveniences of piezoresistive sensors. Note that Fig. 10 is only valid for uniform bulk material.

4.3 PIEZORESISTIVE SENSORS

4.3.1 Piezoresistivity in Diffused Resistors

An important advantage of semiconductors for sensor applications is that very thin layers can be fabricated by standard microelectronic technology to limit the current to the surface of the element under bending or torsion, where stresses are maximum. Moreover, the geometry of the piezoresistors can easily be defined. If doping elements are either deposited on, or implanted into a silicon substrate of opposite doping concentration and then diffused, a piezoresistive layer of, typically 0.5 to 3 μm is obtained. In this section it will be assumed that the diffused layers are thin compared to the substrate thickness, so that the stresses in the layer can be considered independent of depth.

Effective Diffused Piezoresistive Coefficient. In the previous sections, the physical properties of the piezoresistivity in silicon were described for uniform bulk material. It was shown that the piezoresistance coefficients decrease with increasing impurity concentration (Figs. 8 and 10). In diffused or implanted resistors, the impurity concentration decreases with depth, hence, the piezoresistance coefficients show an increasing profile. If the (unstressed) impurity profile, as a function of depth (z), is known, the piezoresistance profile $\pi(z)$ can be determined. (The Z-axis points from the surface into the silicon bulk.) For each piezoresistance coefficient, an average value $\bar{\pi}$ can be defined as an effective coefficient that would yield the same electromechanical behavior for the given doping profile. Obviously, a higher contribution to the average coefficient has to be given to layers where the current flow is higher. Hence, the local coefficient $\pi(z)$ is weighted by the conductivity $\sigma(z)$, which is, again, a function of the doping profile:

$$\bar{\pi} = \int_0^j \pi(z)\sigma(z)\,dz \bigg/ \int_0^j \sigma(z)\,dz \qquad (14)$$

where j is the junction depth. It is assumed that the current distribution does not change significantly with applied stress. Although the deeper layers of the resistor are less doped, and thus exhibit a higher piezoresistivity, their contribution is limited by the higher resistivity and resulting lower current flow. Figure 11a schematically presents the piezoresistivity as a function of depth

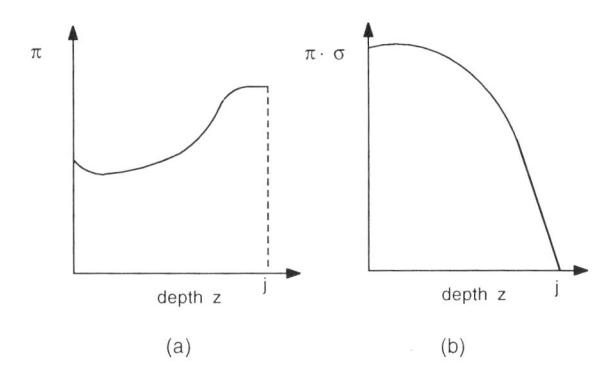

Fig. 11 (a) Piezoresistivity π as a function of depth for a diffused resistor; (b) effective piezoresistivity, obtained by multiplying $\pi(z)$ with the conductivity $\sigma(z)$ as a weighting function.

and Fig. 11b sketches a typical profile of the contribution of each resistor layer to the effective piezoresistance coefficient of Eq. 14.

It is evident that this piezoresistivity profile has to be multiplied by the stress profile, if it is not constant as was assumed, that is if the resistor depth is not negligible compared to the thickness of the substrate under stress in which it is diffused. For very thin silicon membranes or cantilever beams used in piezoresistive sensors, the effect of the stress profile may not be negligible.

Diffused Piezoresistance Coefficient as a Function of Doping Concentration. For a given deposition process, the shape of the dopant distribution function does not change significantly, even if the total amount of diffused dopants and the diffusion depth may be quite different. Based on this observation, it can be shown[51] that, for a given mathematical form of the impurity distribution function, the piezoresistance coefficients defined in Eq. 14 depend only on the surface concentration of impurities and not on the junction depth. Hence, the piezoresistance coefficient can be expressed as a function of the surface concentration only. The results of calculations for a complementary error, erfc, profile (diffusion from a source that maintains a constant surface concentration) and for a Gaussian profile (diffusion from a thin planar deposit of impurities) for p-type silicon with $\langle 111 \rangle$ orientation[51] are presented in Fig. 12. For comparison, the figure gives the concentration dependence of uniformly doped material. The diffused coefficients are systematically higher than the uniform coefficient due to the (small) contribution of deeper layers with higher piezoresistance. The longitudinal piezoresistance coefficient in p-type silicon has been measured for different surface concentrations for doped layers that have an erfc profile.[47] The measured values, also set out in Fig. 12, are in good agreement with the calculated coefficients.

176 MECHANICAL SENSORS

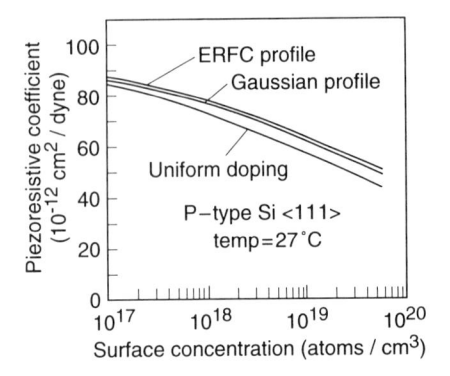

Fig. 12 Longitudinal piezoresistance coefficient of diffused *p*-type silicon with (111) orientation versus impurity surface concentration for complementary error function (erfc) and Gaussian profiles (after Ref. 48), and measured piezoresistance coefficient for erfc profile (after Ref. 49).

4.3.2 General Structure of Piezoresistive Sensors

In the previous sections, the conversion from mechanical stress to electrical signal was discussed. To make a sensor for a specific application, we now have to design structures in which mechanical stress appears when the effect to be measured is applied. Preferably the stress should be proportional to the quantity of the effect, since the resistance change was seen to be proportional to mechanical stress. The two main classes for piezoresistive sensors are membrane-type structures (typically pressure and flow sensors) and cantilever beam-type sensors (typically acceleration sensors).

Membrane-Type Sensors. In its most general embodiment, a piezoresistive membrane sensor consists of a thin monocrystalline silicon membrane supported by a thicker silicon rim. The membrane is fabricated by etching away the bulk silicon on a defined region until the required thickness is reached. Piezoresistors are integrated on the membrane, typically close to the edges.

Figure 13 presents a schematic cross-section of a piezoresistive pressure sensor. When a pressure difference is applied across the device, the thin

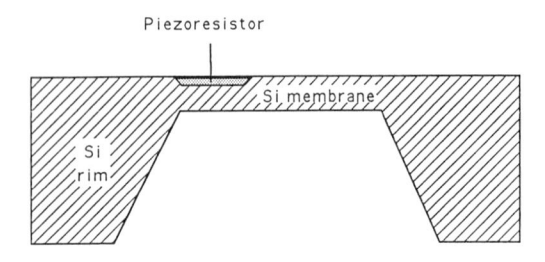

Fig. 13 Schematic cross-section of a piezoresistive pressure sensor.

4.3 PIEZORESISTIVE SENSORS 177

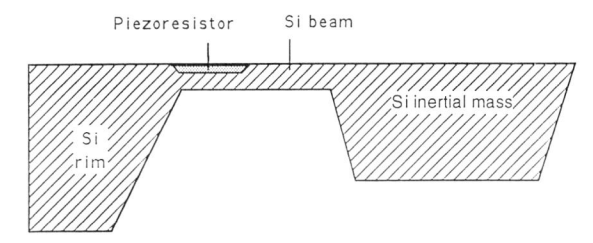

Fig. 14 Schematic cross-section of a piezoresistive acceleration sensor.

membrane will bend downward or upward, inducing traction or compression on the resistor. The resistance change caused by this stress can easily be measured.

Historically, sensors with monocrystalline silicon membranes were developed first. Monocrystalline silicon is strong (Young's modulus: 10^{11} Pa) with very little creep or hysteresis. In addition, the membranes are very easy to fabricate with isotropic or anisotropic etching. With the development of surface micromachining, polycrystalline membranes have been designed, in which resistors are diffused. Although the mechanical properties are somewhat inferior, advantages such as precise thickness control of thin membranes and small sensor dimensions can be realized. Finally, polycrystalline piezoresistors can be deposited and patterned on membranes of other material, for example a dielectric. This configuration is particularly useful for high-temperature applications, since the p–n-junctions, which are the only electrical insulation in the sensors discussed above, will have high leakage currents at higher temperatures. Evidently, these leakage currents are nonexistent in electrically-insulated polycrystalline piezoresistors.

Beam-Type Sensors. The general configuration of a beam-type accelerometer is shown in Fig. 14. The stress caused by deflection of the inertial mass under acceleration is concentrated on the surface of the beam. Piezoresistors are usually placed on the beam close to the rim, where the stress is maximal. The technologies required to fabricate silicon membrane and beam sensors are very similar. However, the beam process is more complicated because the silicon wafer is completely etched through, whereas for membrane sensors the surface of the wafer, which contains the piezoresistors, can be protected more easily from the etch solution.

4.3.3 Resistance Change as a Function of Stress

Based on the theory and data of Section 4.2, the resistance change can be calculated as a function of the membrane or cantilever beam stress. It was shown that there is a contribution to resistance change from stresses that are longitudinal (σ_l and transverse (σ_t) with respect to the current flow. Assuming that the mechanical stresses are constant over the resistors, the total resistance change ΔR is given by

$$\frac{\Delta R}{R} = \sigma_l \pi_l + \sigma_t \pi_t \qquad (15)$$

where π_l and π_t are the longitudinal and transverse piezoresistance coefficients, respectively. Note that dimensional changes are not taken into account in Eq. 15.

The orientation of the membrane or beam is determined by its anisotropic fabrication. The surface of the silicon wafer is usually a $\langle 100 \rangle$ plane and the edges of etched structures are intersections of $\langle 100 \rangle$ and $\langle 111 \rangle$ planes and are thus (110) directions. Therefore, the orientation of the piezoresistors with respect to the silicon crystal is also (110). Table 1 tells us that the longitudinal piezoresistive coefficient in the (110) direction is $\pi_l = 1/2(\pi_{11} + \pi_{12} + \pi_{44})$ and the corresponding transverse coefficient is $\pi_t = 1/2(\pi_{11} + \pi_{12} - \pi_{44})$. From Table 2 we know that for p-type resistors π_{44} is more important than the other two coefficients. Equation 15 is thus approximated for p-type resistors by:

$$\frac{\Delta R}{R} = \frac{\pi_{44}}{2}(\sigma_l - \sigma_t). \qquad (16)$$

For n-type resistors, π_{44} can be neglected, and we obtain:

$$\frac{\Delta R}{R} = \frac{\pi_{11} + \pi_{12}}{2}(\sigma_l + \sigma_t). \qquad (17)$$

It is noted that Eqs. 16 and 17 are only valid for uniform stress fields or if the resistor dimensions are small compared to the membrane or beam size. For small sensors, the stresses will vary across the resistors and have to be integrated. That can be done most conveniently by computer simulation programs.

Considering the values of the piezoresistive coefficients (Table 2), it is easily calculated from Eqs. 16 and 17, or seen in Fig. 6, that for the crystal orientations (110), which are imposed by the membrane or beam fabrication, a two to three times higher pressure sensitivity is obtained with p-type than with n-type resistors. For low doping concentrations and at room temperature, the exact values are 72 and 31 respectively for the longitudinal piezoresistance coefficients, and -66 and -18 respectively for the transverse coefficients (all coefficients are expressed in 10^{-11} Pa^{-1}). In spite of the fact that the maximum longitudinal coefficient in the $\langle 100 \rangle$ plane (Fig. 6) is larger for n-type than for p-type silicon, p-type resistors are preferable since their coefficients have a maximum in the (110) direction, whereas the n-type coefficients have a minimum in that direction.

Wheatstone Bridge Configuration. In general, four piezoresistors are used, as shown in Fig. 15. Two resistors are oriented so that they sense stress in the

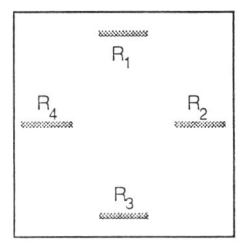

Fig. 15 Schematic representation of the basic position of four piezoresistors on a membrane.

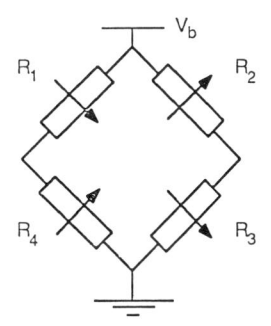

Fig. 16 Wheatstone-bridge configuration of the four piezoresistors shown in Fig. 15. The arrows indicate resistance changes when the membrane is bent downward.

direction of their current axes and two are placed to sense stress perpendicular to their current flow. Therefore, the resistance change of the first two piezoresistors will always be opposite to that of the other two. For instance, for membrane sensors, two piezoresistors can be placed parallel to opposite edges of the membrane, and the other two perpendicular to the other two edges. When the membrane is bent downwards, causing tensile stress on the membrane surface at the edges, the parallel resistors are under lateral stress and show a decrease in resistance while the perpendicular ones are under longitudinal stress and show an increase. If the resistors are correctly positioned with respect to the stress field over the membrane or beam, the absolute value of the four resistance changes can be made equal. The resistors are connected in a Wheatstone bridge, as shown schematically in Fig. 16. The arrows in the figure represent resistance changes as discussed before. Equally positioned resistors form opposite arms of the bridge so that, under applied pressure, the left and right output nodes of the bridge deviate from their zero-pressure voltage with opposite signs.

The Wheatstone-bridge configuration has some distinct advantages. It converts the resistance change directly to a voltage signal. It is easily calculated that the differential output voltage (ΔV) of an ideally balanced bridge with assumed identical (but opposite in sign) resistance changes ΔR, in response to a differential pressure change ΔP on a membrane sensor, is given by:

$$\Delta V = \frac{\Delta R}{R} V_b \qquad (18)$$

where R is the zero-stress resistance and V_b the bridge supply voltage. The pressure sensitivity (S) is then defined as the relative change of output voltage per unit of applied differential pressure (expressed in mV/V-bar):

$$S = \frac{\Delta V}{\Delta P} \frac{1}{V_b} = \frac{\Delta R}{\Delta P} \frac{1}{R}. \qquad (19)$$

Here, an important advantage of the Wheatstone-bridge configuration becomes apparent: the output voltage is, in first order, independent of the absolute value of the piezoresistors, but is determined by the relative resistance change and the bridge voltage.

If a constant bridge current (I_b) is applied, then the pressure sensitivity is defined as the change of differential output voltage per unit pressure and per unit bridge current (expressed in mV/mA-bar):

$$S = \frac{\Delta V}{I_b} \frac{1}{\Delta P} = \frac{\Delta R}{\Delta P}. \qquad (20)$$

In the ideal case, the total resistance of each half-bridge and, thus, of the total bridge is independent of pressure since the resistance changes cancel one another. Moreover common-mode effects, in particular temperature influences, are not felt at the differential bridge output. Indeed, a temperature rise increases the resistance of all piezoresistors equally, so that the output of the bridge remains zero. Note that this is the case only for a perfectly balanced bridge. It is also interesting to notice that at constant bridge voltage, the total current will vary with the temperature or, more practical to measure, for a constant-current bridge supply, the total bridge voltage will vary. In effect, a built-in temperature sensor is available for further compensation of temperature effects, as discussed in more detail later.

Geometrical Design of the Piezoresistors. In this section, the dimensions of the piezoresistors and their position on the thin silicon membrane or beam will be discussed. The discussion focuses on piezoresistive membrane sensors, but most conclusions remain valid for beam sensors. The dimensions and position of the piezoresistors will be chosen as a compromise between maximum pressure sensitivity and other important requirements such as the expected fabrication reproducibility.

Most sensors with square membranes have four piezoresistors disposed at the four edges of the membrane, as shown schematically in Fig. 15. However,

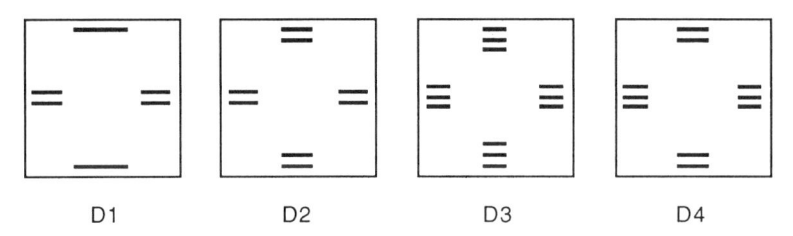

Fig. 17 Alternative layouts of the piezoresistors.

the exact layout varies. The first design rule is evidently to locate the resistors as close as possible to the center of the membrane edges, since that is where the stresses are maximal. From this central point, stress decreases more rapidly towards the center of the membrane than towards the corners, so that the perpendicular resistors are likely to be less pressure sensitive than the parallel ones. To preserve the Wheatstone-bridge symmetry, the parallel resistors have to be moved away from the edge until equal sensitivity is obtained, giving rise to a certain sensitivity loss. For very small membranes, the sensitivity loss can be quite important. In that case, it is advisable to cut the perpendicular resistors in two parts, as shown in Fig. 17 (D1). Alternative layouts are possible where both the perpendicular and the parallel resistors consist of two or three parts (Fig. 17 D2, D3 and D4).

A second design consideration is the minimum allowable distance between the resistors and the membrane edge. This parameter is limited by the fabrication reproducibility of the membrane. Normal commercially available three-inch silicon wafers have a certain thickness reproducibility guaranteed by the manufacturer, for example $\pm 7\,\mu$m. All membranes or beams are etched from the back-side to exactly the same thickness. However, because of wafer thickness variations, the etched holes will be deeper for some wafers than for others, and the pyramidal structure formed by the $\langle 111 \rangle$ slopes of the holes will be more or less closed, depending on the etched depth. The $\langle 111 \rangle$ slopes form angles of 54.74° with the $\langle 100 \rangle$ surface. Therefore, the variation of the sides of the membranes is $2/\tan(54.74°)$ times the thickness variation of the wafers. Moreover, alignment errors of the etch mask with respect to the silicon crystal orientation must also be taken into account. These two factors make it necessary to design the nominal distance between membrane edge and piezoresistor with enough margin, especially for the piezoresistors that are placed parallel to the membrane edge.

4.3.4 Temperature Coefficient of Piezoresistive Sensors

Temperature sensitivity is a major concern for piezoresistive sensors, since it was shown in Section 4.2 that the piezoresistance effect is inherently temperature dependent. Therefore, these types of sensors often require active temperature-compensation circuitry. However, some passive temperature compensation techniques are also available. The following paragraphs discuss the effect of temperature on the offset of the sensor and on its pressure sensitivity. The effects are described by the temperature coefficient of offset (TCO) and the temperature coefficient of sensitivity (TCS), respectively.

Diffused Piezoresistance Coefficients as a Function of Temperature. In the previous section it was shown that, for bulk material, the temperature coefficients of the piezoresistance coefficients decrease with increasing doping concentration. In order to calculate the average temperature dependence of the piezoresistance of diffused resistors, integration over the impurity profile as a function of depth is, again, required. Hence, a relatively high surface concentration does not necessarily result in a very low temperature dependence, as Figs. 9 and 10 (which are valid for uniform bulk material) would suggest, since the deeper, and thus lower, doped layers exhibit a higher temperature coefficient. For this reason, shallow implanted resistors with a very sharp doping-concentration decline can reach lower temperature dependencies than deep-diffused resistors with equal sheet resistance.

As was the case for the doping concentration dependence, it is possible to express the temperature behavior of resistors with equal concentration profiles, as a function of only the surface concentration.[47] Figure 18 plots the temperature dependence of two different p-type resistors with the same surface concentrations, but different depths. As for bulk material, the diffused piezoresistance coefficient is seen to decrease with increasing temperature. Note that the surface concentration of the samples that Tufte used was very high $(2 \times 10^{21}$ cm$^{-3})$, so that the temperature coefficient was very low: about $-0.05\%/°C$ estimated from Fig. 18.

Temperature Coefficient of Offset (TCO). To reduce the TCO, the Wheatstone-bridge configuration was shown to be effective, since temperature changes result only in common-mode effects, that is all resistors in the bridge

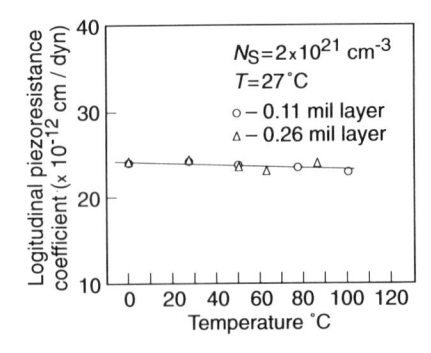

Fig. 18 Longitudinal piezoresistance coefficient versus temperature for two p-type diffused layers having the same mathematical impurity profile function and the same surface concentration but different layer thicknesses. (After Ref. 47)

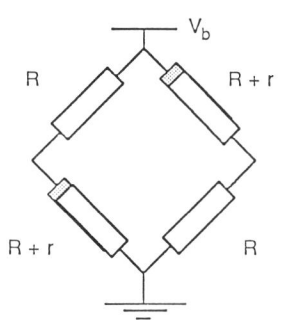

Fig. 19 Wheatstone bridge with symmetrical mismatch of the resistors.

change equally, and the output voltage of the bridge does not change, at least as long as the resistance changes are symmetrical. This is evident in the ideal case where the four resistors have equal values, in which case the offset is zero. In reality it is often found that resistors in opposite arms of the Wheatstone bridge are equal, but are different from resistors in touching arms by a value r, because their layout is slightly different (parallel and perpendicular to the edges of the membrane or the beams). In this case, illustrated in Fig. 19, the TCO is theoretically zero as long as the temperature coefficients of the resistors are equal. Indeed, for this configuration, the offset (V_0) per volt applied to the bridge (V_b) is:

$$O = \frac{V_0}{V_b} = \frac{r}{2R+r}. \tag{21}$$

The temperature dependence is easily calculated to be:

$$\frac{\partial O}{\partial T} = \frac{2(\dot{r}R - r\dot{R})}{(2R+r)^2} \tag{22}$$

where a dot on the variable denotes the derivative with respect to temperature. Now, if the resistors have equal temperature coefficients, then

$$\frac{\dot{r}}{r} = \frac{\dot{R}}{R} \tag{23}$$

and Eq. 22 becomes zero, that is the offset is insensitive to temperature changes, no matter how large the mismatch (r) is.

Consequently, the temperature coefficient of the piezoresistance coefficients is not a major player in the TCO game. Much more important for the offset of piezoresistive sensors and, thus, also for the TCO is the so-called pre-stress condition and its temperature dependence. This refers to the residual stresses

184   MECHANICAL SENSORS

on the resistors when no external pressure or force is applied. Origins of stress are typically passivation layers over the resistors and packaging stress. Both can be very dependent on temperature, depending on the materials used. Only careful design and the introduction of stress-releasing packaging configurations can reduce the TCO.

Temperature Coefficient of Sensitivity (TCS). Unlike the TCO, the temperature dependence of the piezoresistance coefficients discussed in Section 4.2 does have a great influence on the TCS. However, a simple compensation technique is available when a constant current is applied to the Wheatstone bridge instead of a constant voltage. Indeed, at constant bridge current, the bridge voltage increases with temperature due to the positive temperature coefficient (TC) of the resistors. This effect enhances the pressure sensitivity and, as a result, compensates for the loss of sensitivity due to the negative TC of the piezoresistance coefficients. At constant voltage, this internal negative feedback does not occur. The temperature coefficients for constant bridge voltage (TCS_v) and for constant bridge current (TCS_i) can be derived as follows:

For constant bridge voltage. Combining Eqs. 16 and 19 gives the following expression for the pressure sensitivity of the sensors sensitivity:

$$S = \frac{1}{2\Delta P} \pi_{44}(\sigma_l - \sigma_t) \tag{24}$$

temperature dependence of S:

$$\frac{\partial S}{\partial T} = \frac{\sigma_l - \sigma_t}{2\Delta P} \frac{\partial \pi_{44}}{\partial T} + \frac{\pi_{44}}{2\Delta P} \frac{\partial(\sigma_l - \sigma_t)}{\partial T} \tag{25}$$

hence:

$$TCS_v = \frac{1}{S} \frac{\partial S}{\partial T} = \frac{1}{\pi_{44}} \frac{\partial \pi_{44}}{\partial T} + \frac{1}{\sigma_l - \sigma_t} \frac{\partial(\sigma_l - \sigma_t)}{\partial T}. \tag{26}$$

Equation 26 says that the TC of the pressure sensitivity is essentially the same as the TC of π_{44}, apart from the temperature dependence of the membrane stress. In practice, the influence of the membrane stress on TCS is less important than for the TCO. However, the temperature coefficient of π_{44} can be high, especially for low doping concentrations, as explained in Section 4.2.

For constant bridge current. For a constant current, the pressure sensitivity was defined as the change of output voltage per unit bridge current and per unit pressure. Equations 16 and 20 are then combined to:

$$S = \frac{1}{2\Delta P} R\pi_{44}(\sigma_l - \sigma_t). \tag{27}$$

Equation 27 contains R, and thus the temperature coefficient of the piezoresistors will occur in the expression for TCS_i:

$$\text{TCS}_i = \frac{1}{S}\frac{\partial S}{\partial T} = \frac{1}{\pi_{44}}\frac{\partial \pi_{44}}{\partial T} + \frac{1}{R}\frac{\partial R}{\partial T} + \frac{1}{\sigma_l - \sigma_t}\frac{\partial(\sigma_l - \sigma_t)}{\partial T}. \tag{28}$$

The first and the second term in Eq. 28 are respectively negative and positive, so that they compensate each other. The effect of the compensation depends on the relative magnitudes of the terms. A low TCS can be realized by designing the piezoresistors to achieve a good matching of the two temperature coefficients. As discussed in Section 4.2, the temperature coefficient of π_{44} can be controlled by selecting the appropriate doping concentration of the piezoresistors.

4.4 CAPACITIVE SENSORS

Capacitive sensors convert a change in measurand into a change of capacitance. Since a capacitor consists basically of two electrodes separated by a dielectric, the capacitance change can be caused either by motion of one of the electrodes with respect to the other electrode, or by the changes in the dielectric between two fixed electrodes. In general, mechanical sensors rely on the first principle.

As opposed to piezoresistive sensors, the theory behind capacitive sensors is relatively simple. The capacitance between two parallel plates with surface area S, separated by a distance d by a dielectric with dielectric permittivity ε, is given by

$$C_0 = \varepsilon \frac{S}{d}. \tag{29}$$

When one electrode is displaced by a small distance Δd, while staying parallel to the other electrode, the capacitance changes by a value ΔC. If Δd is much smaller than d, the sensitivity to such an event is given by:

$$\frac{\Delta C}{\Delta d} = -\varepsilon \frac{S}{d^2}. \tag{30}$$

Therefore, to design capacitive sensors with a high sensitivity, it would suffice to make the plate area large and the gap distance narrow. However, other technological factors limit both values. These factors include sensor dimensions, fabrication accuracy and reproducibility, and damping of the movement of the electrode if the gap is filled with gas or liquid.

If the electrode movement is not parallel, or if the capacitance change is caused by deflection of part of the electrode, then the capacitance change should be calculated by integrating over the entire deformed dielectric space:

$$\Delta C = C_0 - \iint \frac{\varepsilon}{d - w(x,y)} dx\, dy \tag{31}$$

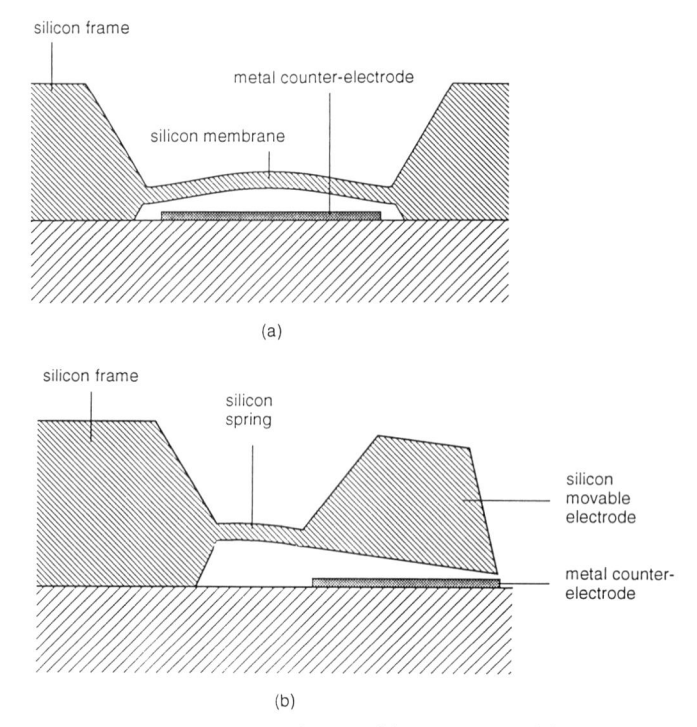

Fig. 20 Schematic cross-sections of capacitive sensors; (a) pressure sensor, (b) accelerometer.

where the two-dimensional function $w(x, y)$ gives the displacement from the original position of every point (x, y) of the capacitor plate.

For semiconductor sensors, the movable electrode is often a thin, square silicon membrane that is deflected, for instance, by a uniform pressure applied to it, as schematically represented in Fig. 20a. In this case, the deflection function $w(x, y)$ cannot be calculated easily. In the section on piezoresistive pressure sensors, a calculation method will be given.

Another common sensor construction is a stiff, square silicon plate suspended by flexible silicon beams, shown in Fig. 20b. Assuming a rotation θ around an axis parallel to one of the edges of the plate, the capacitance of Eq. 31 can be calculated as a function of θ (l is the length of the mass):

$$C(\theta) = \frac{\varepsilon S}{l\theta} \ln\left(\frac{2d + l\theta}{2d + l\theta}\right) \tag{32}$$

where d is the distance from the center line of the inclined plate to the fixed plate. This situation is common for capacitive accelerometers. In the case of a stiff plate suspended by flexible beams, the position d of the center line of the plate should be calculated by the elastic theory of beams.

For capacitive sensors, it is advisable to have the read-out electronics close to the sensor to reduce the stray capacitance of the leads. Therefore, the complete sensing element is often more complex than the piezoresistive accelerometer.

4.5 APPLICATIONS

In this section, the three most common applications of mechanical sensors will be discussed: pressure sensors, accelerometers, and flow sensors. Both piezoresistive and capacitive devices will be discussed. For the pressure sensor, the piezoresistive read-out will be given in more detail, since the calculation of the stress of the membrane under pressure is not trivial. For the accelerometer, the calculation of stress in a simple deflected beam is straightforward, and we will focus on some specific problems of the capacitive read-out technique.

4.5.1 Pressure Sensors

Piezoresistive Pressure Sensors. In Sections 4.2 and 4.3, piezoresistivity, piezoresistors, and piezoresistive pressure sensors were introduced. It was shown that changes in stress produce a sensor output voltage proportional to the applied pressure. Therefore, a particular problem for piezoresistive pressure sensors is the calculation of stress in a membrane as a function of applied pressure. For round membranes, the calculations are elementary. Therefore, this section will focus on square membranes to which a uniform pressure is applied. Square membranes are commonly used for piezoresistive pressure sensors, because they can be fabricated easily and accurately by anisotropic etching of silicon.

The standard reference for this purpose is the book from Timoshenko and Woinowsky-Kreiger "Theory of Plates and Shells".[53] Although hardly used for hand-calculations anymore, the equations are briefly reviewed here, since they form the basis for mechanical-simulation programs for sensor CAD packages.

Timoshenko and Woinowsky-Kreiger's book includes the treatment of pure bending of uniform plates under lateral loads, for example pressure. If the deflections are very small compared to the membrane thickness, the problem can be solved analytically for some situations. The following assumptions have to be adopted:

1. There is no deformation in the middle plane of the plate. This plane remains neutral during bending.
2. Points of the plate lying initially on a normal-to-the-middle plane of the plate remain on the normal-to-the-surface after bending. In other words, the effect of shear forces on the deflection of the plate are not considered.
3. The normal stresses in the direction transverse to the plate can be disregarded.

188   MECHANICAL SENSORS

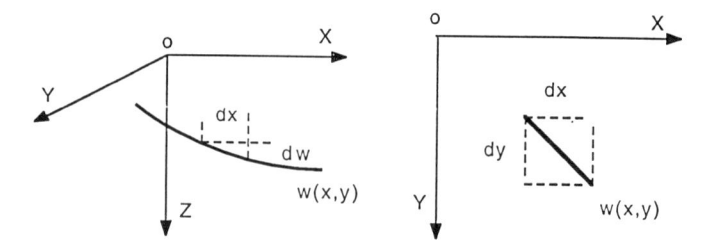

Fig. 21 Definition of the differentials used in the calculation of the deflection of a membrane.

The solution of the problem starts with the calculation of the deflection of the plate as a function of the position on the membrane $w(x, y)$, under a given pressure p. Based on Hooke's law, the differential equation describing this situation can be derived for an infinitesimal part of the membrane, which is shown in Fig. 21. The result is:

$$\frac{\partial^4 w}{\partial x^4} + \frac{\partial^4 w}{\partial x^2 \partial y^2} + \frac{\partial^4 w}{\partial y^4} = \frac{p}{D} \qquad (33)$$

where D is the rigidity defined as

$$D = \frac{Eh^3}{12(1-v^2)} \qquad (34)$$

E is Young's modulus, h is the membrane thickness and v is Poisson's ratio.

The first step is thus to solve Eq. 33 for the boundary conditions that are imposed by the configuration under study.

The second step is the calculation of the bending moments (M) based on the expression for the deflection calculated in the first step:

$$M_x = -D\left(\frac{\partial^2 w}{\partial x^2} + v\frac{\partial^2 w}{\partial y^2}\right) \quad M_y = -D\left(\frac{\partial^2 w}{\partial y^2} + v\frac{\partial^2 w}{\partial x^2}\right). \qquad (35)$$

Finally, the knowledge of the bending moments allows us to calculate the stress distribution. The stress profile in the z-direction is triangular: stress is zero at the middle plane and rises linearly to its maximum value at the surface. This maximum value is calculated for each position as:

$$(\sigma_x)_{max} = \frac{6M_x}{h^2} \quad (\sigma_y)_{max} = \frac{6M_y}{h^2}. \qquad (36)$$

For any given situation, the stresses can be calculated by applying these three steps:

1. membrane deflection,
2. bending moments,
3. stress.

We are interested in the calculations for a rectangular membrane with built-in edges. This is a rather complicated problem, which is usually solved by computer software based on the finite-element method, and we will not go into further detail here.

Timoshenko and Woinowsky-Krieger[53] derived approximate analytical expressions for the relation between pressure and stress in a square membrane, by developing the equations in double trigonometric series. They showed that the maximum of the absolute value of the bending moments appears at the center of the sides of the membrane and decreases towards the corners and towards the center of the membrane. They calculated the maximum bending moment as:

$$M_{max} = \mu_m p a^2 \tag{37}$$

where a is the membrane side length and μ_m depends on the number of series terms that are taken into account.

Combining Eqs. 36 and 37, it is found that the surface stress in the middle of the sides of the membrane is

$$(\sigma_x)_{max} = 0.31 p \frac{a^2}{h^2} \tag{38}$$

where the term 0.31 is calculated by taking seven series terms.

An important conclusion from Eq. 38 is that the stress, and hence the pressure sensitivity of piezoresistive membrane sensors, are proportional to the square of the ratio of the membrane side to the membrane thickness.

In practice, a maximum limit to a is imposed by the cost of surface area, and the minimum limit for h is given by fabrication limitations and thickness-reproducibility requirements. Silicon membranes are generally fabricated by anisotropic etching of bulk silicon using chemical-etch products such as potassium hydroxide,[54,55] ethyelenediamine/pyrocatechol[56,57] and hydrazine.[58,59] To produce membranes of 10 μm or less, while still respecting fabrication specifications, an automatic etch-stop is often required to make the membranes. As mentioned in Section 4.1, a number of etch-stop methods exist. However, membranes for piezoresistive sensors need a low doping concentration, to allow diffusion or implantation of the piezoresistors. Therefore, an electrochemical etch-stop is the most convenient technique to assist etching of membranes for piezoresistive pressure sensors. This method is briefly introduced here.

The basic idea is to make the membrane of n-type silicon, usually an epitaxial layer, and the substrate of p-type silicon, and then to make sure that p-type silicon is etched and n-type silicon is not. This is done by applying different electrochemical potentials to the two silicon types while the wafers are etched in the anisotropic etch solution. The electrochemical etch-stop technique combines the anodic passivation characteristics of silicon[60,61] with a reverse-biased p-n junction to provide a high etching selectivity of p-type silicon over n-type. The dimensional definition of micromachined structures can be controlled precisely by taking advantage of standard silicon-diode fabrication technologies. This technique was first proposed by Waggener[62] in 1970 and has, since, been successfully applied to the fabrication of several different microstructures.[63-65] The result is that etching can be stopped at a well-defined p-n junction. To achieve this, a positive voltage is applied directly to the n-type silicon via an ohmic electrical contact while the electrical contact to the p-type silicon is made via the etch solution with an appropriate counter electrode. Under sufficient anodic bias, silicon passivates as a result of anodic oxide formation and etching stops. Since the majority of the potential drop is across the reverse-biased p-n junction, the p-type silicon remains essentially at the open-circuit potential and etches. With the complete removal of the p-type silicon, the diode is destroyed and the n-type silicon becomes directly exposed to the etch solution. The positive potential applied to the n-type silicon then passivates it and etching terminates. Silicon membranes and other microstructures may be fabricated this way by selectively etching away p-type silicon and leaving the n-type silicon passivated. The definition of the microstructure morphology is determined precisely by the definition of the n-type silicon sections under anodic bias.

In its simplest embodiment, the required passivating potential can be applied between the epixial silicon and a single, inert-metal electrode in the etch solution. However, the solution potential is ill-defined and current dependent, which results in the lack of precise control over the fabrication parameters. Therefore, the use of a single metal electrode is not a viable process technique.

A three-electrode configuration, shown in Fig. 22, overcomes the aforementioned limitations of a two-electrode configuration and is the preferred electrochemical arrangement. A potentiostat, in conjunction with a reference electrode (RE), and an inert counter-electrode, maintains a constant and reproducible solution potential with respect to the reference electrode. In order to establish the required voltage between the silicon wafer, and the RE, the voltage between the working electrode and the RE is measured and compared to the required voltage, and the current through the counter-electrode is adjusted so as to null the readings. As a result, the reference electrode remains currentless (high-impedance input) and its interface potential with the solution is stable. By using three electrodes, the n-type epitaxial layer is kept at a well-defined passivation potential, but the substrate potential is still not under direct electrical

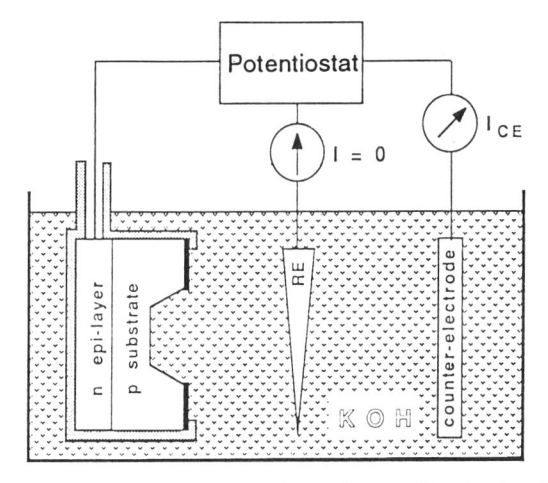

Fig. 22 Standard three-electrode system for etch-stop. RE is the Standard Calomel Reference Electrode.

control. In the ideal case it is etched, but, in practice, leakage currents will often prevent the *p*-type substrate from etching. The method is further refined by using four electrodes: an electrical connection is added to the substrate.[66]

4.5.2 Accelerometers

Deflection Detection. Most accelerometers consist of a mass–spring system. For a constant acceleration, the mass will deviate from its zero-acceleration position, until the force exercised by the spring exactly equals the force required to accelerate the mass. If, in first order, the force acting on the spring is proportional to its deflection, then the deflection is directly proportional to the acceleration.

The most common methods to detect the deflection of the inertial mass are:

  capacitive: gap between a movable and a fixed electrode
  piezoelectric: compression of the spring
  piezoresistive: stress in the spring.

Table 3 summarizes some of the characteristics of the three types.[67] Piezoresistive semiconductor accelerometers (Fig. 14) have been concieved and developed for more than a decade.[68] For these sensors, piezoresistors are diffused or implanted on the spring beams. The deflection and stress in the suspension springs can easily be calculated by the elastic theory of beams, and the output of the sensor is further derived with the data in the section on piezoresistivity. As for the pressure sensors, temperature sensitivity of the piezoresistance effect is a critical issue, and therefore a Wheatstone bridge is advisable. Special care

TABLE 3 Comparison of Some Characteristics of Three Common Sensing Technologies for Accelerometers (Adapted from Ref. 67)

	Capacitive	Piezoelectric	Piezoresistive
Impedance	High	High	Low
Size	Medium	Small	Medium
Temperature range	Very wide	Wide	Medium
Linearity error (sensor only)	High	Medium	Low
DC response	Yes	No	Yes
AC response	Wide	Wide	Medium
Damping available	Yes	No	Yes
Sensitivity	High	Medium	Medium
Zero shifts due to shock	No	Yes	No
Electronics required	Yes	Yes	No
Cost	Medium	High	Low

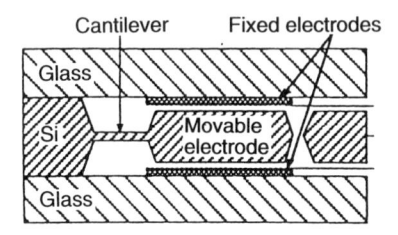

Fig. 23 Cross-section of a symmetrical accelerometer, where the movable electrode is suspended between two fixed electrodes.

must be taken to position the piezoresistors in the stress fields on the suspension beams.[69]

Capacitive semiconductor accelerometers came into use soon after the piezoresistive accelerometers.[70,71] A cross-section of a typical capacitive accelerometer is shown in Fig. 23. The inertial mass is suspended between two plates of Pyrex glass[72,73] or silicon,[74] with a narrow gap between them. The counter-plates contain fixed electrodes, and the inertial mass constitutes a movable electrode. Movement of the mass due to acceleration of the device will change the capacitance with the two fixed electrodes.

An important advantage of capacitive accelerometers is that, as opposed to the piezoresistive effect, there is no inherent temperature sensitivity. The only thermal effect is a change in the capacitance due to thermal expansion of the constituent elements. A symmetrical sensor design reduces the effects of thermal expansion to a minimum, so that these sensors often do not need any active temperature compensation.

The third read-out technique is based on the piezoelectric effect.[75] In many crystalline materials, a mechanical stress produces an electric polarization and, reciprocally, an applied electric field generates a mechanical strain. Quartz has a significant piezoelectric effect, but silicon does not. However, thin piezoelectric films, such as zinc oxide, can be deposited on silicon to convert the mechanical signal to an electric one. Under constant strain, the charges slowly leak away, and the electric field disappears. Therefore, piezoelectric accelerometers cannot be used for low-frequency applications. Although many classical vibration sensors are of the piezoelectric type, the principle is used less frequently in semiconductor accelerometers.

Damping of the Accelerometers. An important issue of accelerometers is the damping of the movable electrode.[76,77] The spring–mass structure is a second-order system, and the highest frequency bandwidth can be achieved when the movement is critically damped. A convenient way to damp the mass is to enclose it in a small cavity. This construction is already required for operation of capacitive sensors. The inertia of the viscous liquid or gas trapped in the gaps will damp the movement of the mass. This is called the squeeze-film effect.[78,79] Damping is determined by the dimensions of the gaps and by the viscosity or pressure of the medium.

For piezeresistive accelerometers, the gap between the movable electrode and other parts of the structure has no influence on the functioning of the sensor, although it should not be too wide in order to limit the excessive movement of the mass when shocks or over-range accelerations are applied. However, for capacitive accelerometers, the capacitor gap should be narrow to obtain high sensitivity, so that damping, even in air, can be very high, due to the squeeze-film effect. High damping results in a slow frequency response but air flow channels or low-pressure encapsulation can improve the frequency characteristics.

4.5.3 Flow Sensors

One of the classical methods to measure gas flow is to put a hot wire in the gas stream with constant heating power, and measure the temperature of the wire as a function of the flow (open loop), or to keep the wire at a constant temperature and measure the power required to do so (closed loop). Alternatively, a differential approach is possible by monitoring the temperatures of the gas upstream and downstream of the hot wire, as schematically represented in Fig. 24. When there is no gas stream, the temperature profile is symmetrical around the wire. The symmetry is disturbed when gas flows, since the upstream detector is cooled by the oncoming gas, while the gas is heated by the hot wire and then, in turn, passes heat to the downstream detector. Methods have been developed to calculate the flow from the temperature measurements. The differential measurement eliminates zero-point offsets due to factors such as ambient pressure and temperature.

The hot-wire system can easily be miniaturized and integrated into silicon. A simple differential sensor could consist of a diffused or thin-film resistor on silicon as the heating element, and junction diodes or resistors as temperature

194 MECHANICAL SENSORS

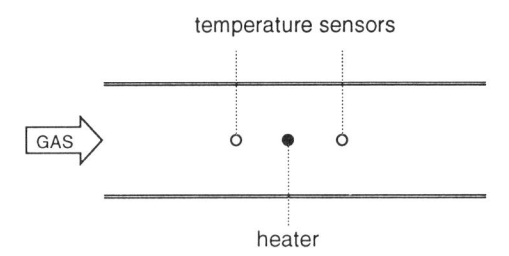

Fig. 24 Principle of a flow meter with a central heating wire and symmetrically placed temperature sensors.

sensors at both sides of the heater. However, the silicon micromachining tools that are available allow far-reaching refinements with respect to power consumption and sensitivity of the flow sensors, by using microbridges made of monocrystalline silicon, silicon nitride or polysilicon.[80-83].

Power consumption can be reduced drastically by thermally isolating the heating element from the substrate so that most power passes into the gas, and as little as possible is lost in the substrate. One possibility is to place the heater and the temperature sensors on free-standing silicon bridges, similar to the suspension springs of the piezoresistive accelerometers. These sensors can be robust, but the thermal conductivity of silicon is high, and an appreciable part of the heating power is still lost in the substrate.

For better thermal insulation, the heating and temperature-sensing elements can be placed on free-standing silicon nitride bridges. Figure 25 shows a top view and a cross-section of such a flow sensor.[81] The nitride bridges are less than one micrometer thick, and the thermal conductivity of silicon nitride is lower than that of silicon, resulting in a high thermal decoupling. The sensor consists of two silicon nitride microbridges, containing a central heating resistor divided equally between the two bridges, and two identical resistors with relatively-large temperature coefficients, placed symmetrically with respect to the heater, which serve as temperature detectors.

The temperature difference between the silicon substrate and the nitride bridge is 100 to 200°C, and the thermal efficiency is in the order of 15°C/mW, that is it takes less than 10 mW to increase the temperature of the heater by 100°C. The small mass of the nitride bridges results in a very small heat capacity, and therefore the time constant of the measurement is very small.

4.6 SUMMARY AND FUTURE TRENDS

In this chapter, the two most common working principles for mechanical semiconductor sensors were introduced, namely capacitive and piezoresistive transduction. A lot of attention was given to the piezoresistance effect in silicon,

4.6 SUMMARY AND FUTURE TRENDS **195**

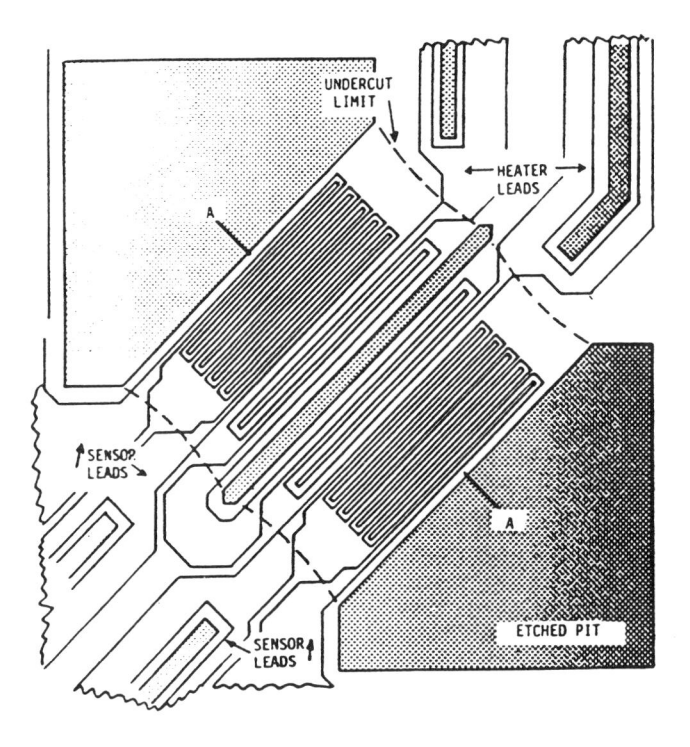

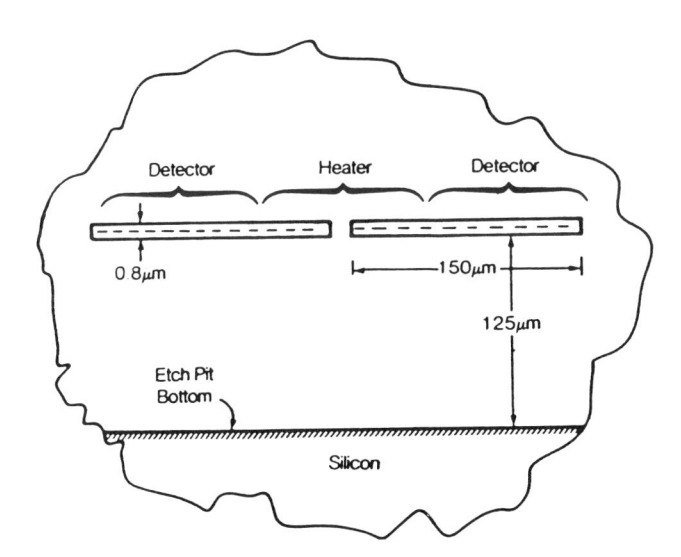

Fig. 25 Top: schematic view of a flow sensor with heating and temperature sensing resistors on a thin silicon nitride membrane (the bridges are free-standing until the dotted line marked "undercut limit"). Bottom: cross-section AA. (After Ref. 81)

since a knowledge and understanding of its characteristics are very important for designing semiconductor sensors.

The piezoresistance effect in monocrystalline silicon was seen to be highly anisotropic. The phenomenological description was supported by a mathematical model to derive expressions for all crystal directions. A physical explanation was given, based on the asymmetry of the three-dimensional band structure of silicon as described by the many-valley model. This model allows the depencence of piezoresistance on doping concentration and temperature to be calculated. The piezoresistance effect decreases with increasing doping concentration and with increasing temperature. Moreover, at higher doping concentrations, the temperature coefficient at room temperature decreases, and this effect is usually more significant than the loss of stress sensitivity due to the higher doping concentration.

The calculations and measurements were first discussed for homogeneous silicon layers of constant doping concentration. To apply them to diffused resistors, they had to be integrated over the doping profile of the resistor. It was seen that the described tendencies remain valid, but they are averaged out over layers with different doping concentrations. For very steep junctions (obtained by ion implantation), the contribution of deeper layers with low concentration becomes less important, and the resistors approach the characteristics of bulk material.

The layout of piezoresistive sensors was presented: basically the sensors consist of four piezoresistors disposed at the edges of a square silicon membrane or on a cantilever beam. It was shown how the mechanical stress can be calculated. An important conclusion can be drawn for membrane pressure sensors, which is that the stress in the middle of the membrane edge is proportional to the square of the membrane side (a) and inversely proportional to the square of the membrane thickness (h). Hence, for maximum pressure sensitivity, the ratio a/h should be chosen as high as possible.

Temperature influence is a major concern for piezoresistive sensors, because of the temperature sensitivity of the piezoresistance coefficient. It was shown that a Wheatstone-bridge configuration reduces the temperature coefficient of the offset of the sensors. The temperature coefficient of the sensitivity, on the other hand, can be deduced by designing the piezoresistors so that the effect of the decreasing value of the piezoresistance coefficients is compensated by the increasing value of the resistance.

Whereas the theory behind piezoresistive sensors is relatively complex, the read-out of the devices is easy: the output signal is available as a voltage difference at two nodes of the Wheatstone bridge. The main requirement for the read-out device is sufficient common-mode rejection.

The principle of the capacitive read-out is formulated simply by the expression for capacitance as a function of the dimensions of the capacitor and the dielectric constant of the medium. In general, the gap distance of the capacitor is the parameter that varies with the measurand. It was pointed out that the advantages of capacitive sensors include relatively simple construction and low inherent temperature coefficient. Although the principle is straightforward, care must be taken to avoid stray capacitances that vary with time, in particular that of leads. Therefore an electronic circuit is often required close to the sensor to convert the capacitance variation to a low-impedance signal, which complicates its construction and packaging.

Three semiconductor mechanical sensors were discussed in more detail, namely pressure sensors, accelerometers and flow sensors. At present, these sensors are the most common in industrial applications.

Most commercially available sensors are micromachined in bulk silicon. In the future, surface micromachining may take an increasingly important place, with polysilicon as the material for moving parts. Although the mechanical properties of polysilicon are somewhat inferior to those of monocrystalline silicon, significant advantages are better process compatibility with electronic-circuit fabrication and smaller dimensions and mass than similar monocrystalline devices. Polysilicon microstructures have been used for pressure sensors,[11] accelerometers,[71] and flow sensors.[83]

An on-going discussion in sensor technology is whether or not to integrate electronic circuitry on the same chip as the sensor device (i.e., the integrated sensor). The advantages include the reduction of packaging costs and the increase of device reliability because the number of bond wires is reduced. However, the processes for the electronic circuitry and the sensor device may not be compatible, and the fabrication yield may suffer. Most sensor processes are compatible with IC processes, but the total process is more complicated, which has its impact on the fabrication yield. For bulk-micromachined sensors it is generally felt unwise to integrate more than a resistive network on-chip. As mentioned, surface micromachining offers better process compatibility, and sensors with complete electric circuits may be expected in the near future. Complex integrated circuits with micromechanical components that take less chip surface than the electronic components have been demonstrated.[84]

Micromechanical sensor applications include the process industry, automotive electronics, medical devices and equipment, and household appliances. Closed-loop control systems for these applications require reliable and stable sensors to monitor the processes and actions that take place. With increasing safety and comfort requirements, the automotive industry is probably the fastest growing sensor market for applications such as air-bags, active suspension control, antilock brake systems, gas injection and combustion control, tire pressure monitoring, and others.

Research and development are slowly moving into the age of microsystems, which requires the integration of different technologies such as sensors, electronics, actuators, and optics in miniature hybrid systems. In the future we may even see the appearance of monolithic microsystems.

PROBLEMS

1. Most piezoresistive pressure sensors consist of a Wheatstone bridge with four piezoresistive elements on a thin silicon membrane. In another

198 MECHANICAL SENSORS

configuration, developed by Motorola, the Wheatstone bridge is replaced by a cross-shaped piezoresistive element with four connections, as shown in the figure below. (A plan view of Motorola "X-ducer" silicon pressure transducer.) Assuming that current only flows in the directions of the cross, and based on the piezoresistive theory, find the current/voltage relations for this element.

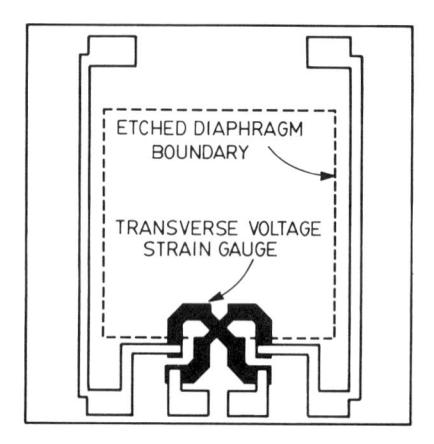

2. Take an accelerometer with an inertial mass suspended by one silicon beam. The dimensions are shown in the figure below. Assume a rectangular cross-section of the beam. Calculate the mass of the movable electrode and the stress on the top surface of the suspension beam, as a function of x, when an acceleration of 1 m/s^2 is applied. Assume there is no gravity, and neglect the mass of the beam. (Density of silicon: 2332 kg/m^3, Young's modulus of silicon: 1.3×10^{11} Pa.)

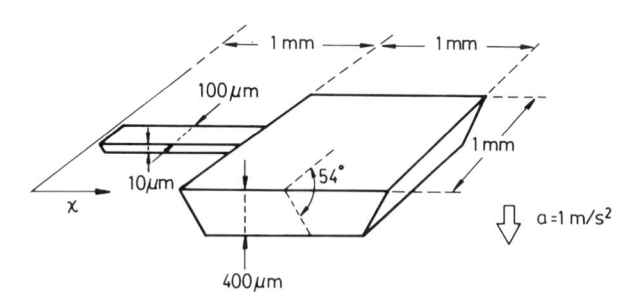

3. Comment on the best position for strain gages on the beam of the accelerometer of Problem 2 for highest acceleration sensitivity.

4. The suspension beam of Problem 2 has a (100) surface and is aligned to the $\langle 110 \rangle$ direction of the silicon crystal. A piezoresistor of 100 μm long, 10 μm wide and 1 μm deep is diffused into the beam, at 10 μm from the rim. The

sheet resistivity of the piezoresistor is 1000 Ω per square. Determine the piezoresistance coefficient required for this situation, and calculate it.

5. Calculate the zero-stress value of the piezoresistor of Problem 4, and the resistance change for an acceleration of 1 m/s^2, assuming constant stress and resistivity over the depth of the resistor.

6. The accelerometer of Problem 4 is placed horizontally in the earth's gravitational field, and rotated (slowly) around an axis in the top surface plane, perpendicular to the longitudinal direction of the beam. Calculate and plot the resistance change as a function of rotation angle.

7. Assuming Smith measured the piezoresistance coefficients at 25°C, and using Fig. 10, calculate (approximately) the temperature coefficient for the resistor of Problem 4 around 25°C, in Ω/°C.

8. Again using Fig. 10, calculate (approximately) the temperature increase and decrease equivalent to an acceleration of ±1 m/s^2 for the sensor of Problem 4.

9. Let us now use a similar structure as before to make a capacitive accelerometer. As shown in the figure below, an insulation layer of 1 μm prevents short-circuits when the two electrodes touch. The dimensions of beam and mass are the same as before. Calculate the nominal gap between the electrodes so that the tip of the movable electrode just touches the counterelectrode at an acceleration of 1 g (9.8 m/s^2). Calculate also the capacitance in that case, and compare with the zero-acceleration capacitance (neglect electrostatic forces).

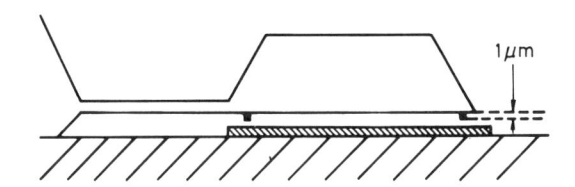

REFERENCES

1. K. E. Petersen, "Silicon as a mechanical material," *Proc. IEEE* **70**, 420–57 (1982).
2. C. S. Smith, Piezoresistance effect in germanium and silicon, *Physical Rev.* **94**, 42–9 (1954).
3. S. M. Sze, *Semiconductor Devices, Physics and Technology*, Wiley, New York, 1985.
4. H. Robbins and B. Schwartz, "Chemical etching of silicon, II. the system HF, HNO$_3$, H$_2$O," *J. Electrochem. Soc.* **107**, 108 (1960).
5. H. Robins and B. Schwartz, "Chemical etching of silicon, III. a temperature study in the acid system," *J. Electrochem. Soc.* **108**, 365 (1961).

6. M. J. Theunissen, J. A. Appels, and W. H. C. G. Verkuylen, Application of electrochemical etching of silicon to semiconductor device technology, *J. Electrochem. Soc.* **117**, 959 (1970).
7. R. L. Meek, "Electrochemically thinned n/n+ epitaxial silicon—method and applications," *J. Electrochem. Soc.* **118**, 1240 (1971).
8. A. C. M. Gieles and G. H. J. Somers, "Miniature pressure transducers with a silicon diaphragm," *Philips Tech. Rev.* **33**, 14–20 (1973).
9. N. F. Raley, Y. Sugiyama, and T. van Duzer, "(100) Silicon etch-rate dependence on boron concentration in ethylenediamine-pyrocatechol-water solutions," *J. Electrochem. Soc.* **131**, 162 (1984).
10. H. Seidel, "The mechanism of anisotropic silicon etching and its relevance for micromachining," *Transducers 87 Digest of Technical Papers, June 1987, Tokyo, Japan*, p. 120.
11. R. S. Hijab and R. S. Muller, "Micromechanical thin-film cavity structures for low pressure and acoustic transducer applications," *Transducers 85 Digest of Technical Papers, June 1985, Philadelphia, USA*, pp. 178–81.
12. R. T. Howe and R. S. Muller, "Polycrystalline silicon micromechanical beams," *J. Electrochem. Soc.* **130**, 1420–3 (1983).
13. H. C. G. Ligtenberg, "Miniaturization of indwelling biomedical pressure sensors," in *Sensors & Actuators, Microtechnology for Transducers*, J. C. Lodder, Ed., Kluwer, Deventer, The Netherlands, 1986, pp. 257–64.
14. W. G. Pfann and R. N. Thurston, "Semiconducting stress transducers utilizing the transverse and shear piezoresistance effects," *J. Appl. Phys.* **32**, 2008–19 (1961).
15. Y. Kanda and A. Yasukawa, "Hall-effect devices as strain and pressure sensors", *Sens. Actuators* **2**, 283–96 (1982).
16. J. E. Gragg, W. E. McCulley, W. B. Newton, and C. E. Derrington, "Compensation and calibration of a monolythic four terminal silicon pressure transducer," *Tech. Dig. IEEE Solid-State Sensors Workshop, Hilton Head Island, SC, June 1984*, pp. 21–7.
17. R. J. Veen, "Piezojunction effect of a planar *n-p-n* transistor for transducer aims," *Electron. Lett.* **15**, 333 (1979).
18. B. Puers and W. Sansen, "New mechanical sensors in silicon by micromachining piezojunction transistors," *Transducers 87 Digest of Techical Papers, June 1987, Tokyo, Japan*, pp. 324–7.
19. C. S. Sander, J. W. Knutti, and J. D. Meindl, "A monolithic capacitive pressure sensor with pulse-period output," *IEEE Trans. Electron Devices* **ED-17**, 927–30 (1980).
20. Y. S. Lee and K. D. Wise, "A batch-fabricated silicon capacitive pressure transducer with low temperature sensitivity," *IEEE Trans. Electron Devices* **ED-29**, 42–8 (1982).
21. W. H. Ko, M.-H. Bao, and Y.-D. Hong, "A high-sensitive integrated-circuit capacitive pressure transducer," *IEEE Trans. Electron Devices* **ED-29**, 48–56 (1982).
22. A. Jornod and F. Rudolf, "High precision capacitive absolute pressure sensor," *Tech. Dig. Eurosensors II, 4th Symp. on Sensors and Actuators, November 1988, Enschede, The Netherlands*, p. 186.
23. D. J. Warkentin, J. H. Haritonidis, M. Mehregany, and S. D. Senturia, "A micro-machined microphone with optical interference readout," *Transcucers 87 Dig. of Tech. Papers, June 1987, Tokyo, Japan*, pp. 60–3.

24. A. M. Young, J. E. Goldsberry, J. H. Haritonidis, R. L. Smith, and S. D. Senturia, "A twin-interferometer fiber-optic readout for diaphragm pressure transducers," *Tech. Dig. IEEE Solid-State Sensor and Actuator Workshop, Hilton Head Island, SC, June 1988*, pp. 19–22.

25. T. S. J. Lammerink and W. Wlodarski, "Integrated thermally excited resonant diaphragm pressure sensor," *Transducers 85 Dig. of Tech. Papers, June 1985, Philadelphia, USA*, pp. 97–100.

26. J. G. Smits, H. A. C. Tilmans, and T. S. J. Lammerink, "Pressure dependence of resonant diaphragm pressure sensors," *Transducers 85 Dig. of Tech. Papers, June 1985, Philadelphia, USA*, pp. 93–6.

27. P. J. French and A. G. R. Evans, Polysilicon strain sensors using shear piezoresistance, *Sens. Actuators* **15**, 257–72 (1988).

28. J. Detry, D. Koneval, and S. Blackstone, "A comparison of piezoresistance in polysilicon, laser recrystallized polysilicon and single crystal silicon," *Transducers 85 Dig. of Tech. Papers, June 1985, Philadelphia, USA*, pp. 278–80.

29. R. A. Buser and N. F. de Rooij, "Resonant silicon structures," *Sens. Actuators* **17**, 145–53 (1989).

30. T. A. Knecht, "Bonding techniques for solid-state pressure sensors," *Transducers 85 Dig. of Tech. Papers, June 1985, Philadelphia, USA*, pp. 95–8.

31. T. R. Anthony, *J. Appl. Phys.* **54**, 2419 (1983).

32. T. R. Anthony, "Dielectric isolation of silicon by anodic bonding," *J. Appl. Phys.* **58**, 1240–7 (1985).

33. J. Ohura, T. Tsukakoshi, K. Fukuda, M. Shimbo, and H. Ohashi, "A dielectrically isolated photodiode array by silicon-wafer direct bonding," *IEEE Electron Device Lett.* **EDL-8**, 454–6 (1987).

34. H. Li, G.-L. Sun, J. Zhan, and Q.-Y. Tong, "Some material structural properties of SOI substrates produced by SDB technology," *Appl. Surf. Sci.* (1987).

35. M. Shimbo, K. Furakawa, K. Fukuda, and L. Tanzawa, "Silicon-to-silicon direct bonding method," *J. Appl. Phys.* **60**, 2987–9 (1986).

36. K. Petersen, Ph. Barth, J. Poydock, J. Brown, J. Mallon, and J. Bryzek, "Silicon fusion bonding for pressure sensors," *Tech. Dig. IEEE Solid-State Sensor and Actuator Workshop, Hilton Head Island, SC, June 1999*, pp. 144–7.

37. P. W. Bridgman, "The effect of homogeneous mechanical stress on the electrical resistance of crystals," *Phys. Rev.* **42**, 858–63 (1932).

38. C. Herring, "Transport properties of a many-valley semiconductor", *Bell System Tech. J.* **34**, 237–90 (1955).

39. F. J. Morin, T. H. Geballe, and C. Herring, "Temperature dependence of the piezoresistance of high-purity silicon and germanium," *Phys. Rev.* **105**(2) (1957) pp. 525–39.

40. W. P. Mason and R. N. Thurston, "Use of piezoresistive materials in the measurement of displacement, force and torque," *J. Acoust. Soc. Am.* **11**, 1096–101 (1957).

41. L. E. Hollander, G. L. Vick, and T. J. Diesel, "The piezoresistive effect and its applications," *Rev. Sci. Instrum.* **31**, 323–7 (1960).

42. F. T. Geyling and J. J. Frost, "Semiconductor strain transducers," *Bell System Tech. J.* **39**, 705–31 (1960).
43. R. W. Keyes, "The effects of elastic deformation on the electrical conductivity of semiconductors," *Solid State Phys.* **11**, 149–221 (1960).
44. G. E. Pikus and G. L. Bir, "Effect of the deformation of the hole energy spectrum of germanium and silicon," *Sov. Phys.—Solid State* **1**, 1502–16 (1960).
45. O. N. Tufte, P. W. Chapman, and D. Long, "Silicon diffused-element piezoresistive diaphragms," *J. Appl. Phys.* **33**, 3322–7 (1962).
46. W. G. Pfann, "Isotropically piezoresistive semiconductor," *J. Appl. Phys.* **33**, 1618–19 (1962).
47. O. N. Tufte and E. L. Stelzer, "Piezoresistive properties of silicon diffused layers," *J. Appl. Phys.* **34**, 313–18 (1963).
48. G. Nuzillat and H. Helioui, "Transducteurs piézo-FET analogiques," *Revue Technique Thomsom-CSF* **5**, 49–80 (1973).
49. O. Jäntsch, *Piezowiderstandseffekte, in Halbleiter-Elektronik, Bank 17 Sensorik*, W. Heywang, Ed., Springer-Verlag, Berlin/Heidelberg, 1884, pp. 114–34.
50. Y. Kanda, "A graphical representation of the piezoresistance coefficients in silicon," *IEEE Trans. Electron Devices* **ED-29**, 64–70 (1982).
51. D. R. Kerr and A. G. Milnes, "Piezoresistance of diffused layers in cubic semiconductors," *J. Appl. Phys.* **34**, 727–31 (1963).
52. W. Pietrenko, "Einfluss von Temperatur und Störstellenkonzentration auf den Piezowiderstandseffekt in *n*-Silizium," *Phys. Status Solidi—Section A Appl. Res.* **41**, 197–205 (1977).
53. S. P. Timosheko and S. Woinowsky-Krieger, *Theory of Plates and Shells*, 2nd ed. McGraw–Hill, New York, 1970.
54. J. B. Price, "Anisotropic etching of silicon with KOH–H$_2$O–isopropyl alcohol," *Semicond. Silicon*, 339–53 (1973).
55. D. L. Kendall, "On etching very narrow grooves in silicon," *Appl. Phys. Lett.* **26**, 195–8 (1975).
56. R. M. Finne and D. L. Klein, "A water-amine-complexing agent system for etching silicon," *J. Electrochem. Soc.* **114**, 965–70 (1967).
57. E. Bassous, "Fabrication of novel three-dimensional microstructures by anisotropic etching of (110) and (110) silicon," *IEEE Trans Electron Devices* **ED-25**, 1178–85 (1978).
58. D. B. Lee, "Anisotropic etching of silicon," *J. Appl. Phys.* **40**, 4569–74 (1969).
59. M. J. Declercq, L. Gerzberg, and J. D. Meindl, "Optimization of the hydrazine-water solution for anisotropic etching of silicon in integrated circuit technology," *J. Electrochem. Soc.* **122**, 545–52 (1975).
60. R. L. Smith, B. Kloeck, N. F. de Rooij, and S. D. Collins, "The potential dependence of silicon anisotropic etching in KOH at 60°C," *J. Electroanalyt. Chem.* **238**, 103 (1987).
61. O. J. Glembocki, R. E. Stahlbush, and M. Tomkiewicz, "Bias-dependent etching of silicon in aqueous KOH," *J. Electrochem. Soc.* **132**, 145 (1985).
62. H. A. Waggener, "Electrochemically controlled thinning of silicon," *Bell System Tech. J.* **49**, 473 (1970).

63. T. N. Jackson, M. A. Tischler, and K. D. Wise, "An electrochemical *p–n* junction etch-stop for the formation of silicon microstructures," *IEEE Electron. Device Lett.* **EDL-2**, 44 (1981).
64. M. Hirata, S. Suwazono, and H. Tanigawa, "Diaphragm thickness control in silicon pressure sensors using an anodic oxidation etch-stop," *J. Electrochem. Soc.* **134**, 2037 (1987).
65. E. D. Palik, O. J. Glembocki, and R. E. Stahlbush, "Fabrication and characterization of Si membranes," *J. Electrochem. Soc.* **135**, 3126–34 (1988).
66. B. Kloeck, S. D. Collins, N. F. de Rooij, and R. L. Smith, "Study of electrochemical etch-stop for high precision thickness control of silicon membranes," *IEEE Trans. Electron Devices* **ED-36**, 663–9 (1989).
67. *Brochure on Accelerometers of IC Sensors*, Fremont, CA, USA.
68. L. M. Roylance and J. B. Angell, "A batch-fabricated silicon accelerometer," *IEEE Trans. Electron Devices* **ED-26**, 1911–17 (1979).
69. H. Sandmaier, K. Kühl, and E. Obermeier, "A silicon based micromechanical accelerometer with cross acceleration sensitivity compensation," *Transducers '87, Dig. of Tech. Papers, June 1987, Tokyo, Japan*, 1987, pp. 399–402.
70. F. Rudolf, "A micromechanical capacitive accelerometer with a two-point inertial-mass suspension," *Sens. Actuators* **4**, 191–8 (1982).
71. K. E. Petersen, A. Shartel, and N. F. Raley, "Micromechanical accelerometer integrated with MOS detection circuitry," *IEEE Trans. Electron Devices* **ED-29**, 23–6 (1982).
72. S. Suzuki, *et al.*, "Semiconductor capacitance-type accelerometer with PWM electrostatic servo technique," *Sens. Actuators* **A21–A23**, 316–19 (1990).
73. H. Seidel, *et al.*, "Capacitive silicon accelerometer with highly symmetrical design," *Sens. Actuators* **A21–A23**, 312–15 (1990).
74. F. Rudolf, A. Jornod, and P. Bencze, "Silicon microaccelerometer," *Transducers '87, Dig. of Tech. Papers, June 1987, Tokyo, Japan*, 1967, pp. 395–8.
75. P. L. Chen, *et al.*, "Integrated silicon microbeam PI-FET accelerometer," *IEEE Trans. Electron Devices* **ED-29**, 27–33 (1982).
76. T. Tschan, N. de Rooij, and A. Bezinge, "Damping of piezoresistive silicon accelerometers," *Sens. Actuators* **32**, 567–71 (1990).
77. S. Terry, "A miniature silicon accelerometer with built-in damping," *Tech. Dig. of IEEE Solid-State Sensor and Actuator Workshop, Hilton Head, SC*, 114–16 (1988).
78. J. Blech, "On isothermal squeeze films," *J. Lubrication Technol.* **105**, 615–20 (1983).
79. J. Starr, "Squeeze-film damping in solid-state accelerometers," *Tech. Dig. of IEEE Solid-State Sensor and Actuator Workshop, Hilton Head Island, SC* 1990, pp. 44–7.
80. K. E. Petersen, J. Brown, and W. Renken, "High-precision, high performance mass flow sensor with integral laminar flow micro-channels," *Transducers '85 Dig. of Tech. Papers, June 1985, Philadelphia, USA*, 1985, pp. 361–3.
81. R. G. Johnson and R. E. Higashi, "A highly sensitive chip microtransducer for air flow and differential pressure sensing applications," *Sens. Actuators* **11**, 63–72 (1987).
82. M. Esashi, S. Eoh, T. Matsuo, and S. Choi, "The fabrication of integrated mass flow controllers," *Transducers '87, Dig. of Tech. Papers, June 1987, Tokyo, Japan*, 1987, pp. 830–3.

83. Y. C. Tai and R. S. Muller, "Lightly-doped polysilicon bridge as a flow meter," *Sens. Actuators* **15**, 63–75 (1988).
84. "Analog devices combines micromachining and BiCMOS," *Semicond. Int.* **17**, 17 (1991).

5 Magnetic Sensors

H. BALTES and R. CASTAGNETTI
ETH Zurich
Zurich, Switzerland

5.1 INTRODUCTION

The expression magnet originates from the region Magnesia in Greece where magnetite (Fe_3O_4) is found. Certain animals are able to sense the magnetic field of the earth and can use this ability for orientation. Man relies on the magnetic compass, which helped Columbus on his transatlantic voyages. The compass can be traced back more than 4000 years to the Chinese and may be considered as the first sensor-like device.

In modern terms, a magnetic sensor is a transducer that converts a magnetic field into an electrical signal. Semiconductor magnetic sensors exploit galvanomagnetic effects due to the Lorentz force on charge carriers in semiconductors. Integrated semiconductor, notably silicon, magnetic sensors are manufactured using integrated circuit technologies, which allow batch fabrication and on-chip signal conditioning circuitry.

Galvanomagnetic effects occur when a material carrying an electric current is exposed to a magnetic field. The best known of these is the Hall effect, which produces an electric field perpendicular to the magnetic induction vector and the original current direction. It was discovered by Edwin Hall in 1879 at the University of Baltimore. The counterpart of the Hall effect is the Lorentz deflection or carrier deflection, producing a current component perpendicular to the magnetic induction vector and the original current direction. The effect is named after the Dutch physicist H. A. Lorentz (1853–1928) who was the first to fully explain the movement of charge carriers in a magnetic field.

Other galvanomagnetic effects include the magnetoresistance and the magnetoconcentration effect. Magnetoresistance, the modulation of the

Semiconductor Sensors, Edited by S. M. Sze.
ISBN 0-471-54609-7 © 1994 John Wiley & Sons, Inc.

electrical resistance by a magnetic field, was discovered by William Thomson Kelvin at the University of Glasgow in 1856. The magnetoconcentration or Suhl effect produces a gradient of the carrier concentration perpendicular to the magnetic inductor vector and the original current direction. The effect was discovered by Harry Suhl at Bell Laboratories in 1949.

High mobility and low carrier concentration favor galvanomagnetic effects. That is why semiconductors rather than metals are the preferred materials for galvanomagnetic sensors. Pearson[1] proposed a germanium Hall-effect sensor in 1948. Twenty years later Bosch[2] proposed to incorporate the Hall sensor in a silicon integrated circuit. Since then, Hall sensors have been mass produced.

5.1.1 Applications and Ranges

A magnetic sensor is capable of converting a magnetic field into a useful electrical signal. A magnetic sensor is also needed whenever a nonmagnetic signal is detected by means of an intermediary conversion into a magnetic signal in a so-called tandem transducer. Examples are the detection of a current through its magnetic field or the mechanical displacement of a magnet. Thus, we can distinguish two groups of direct and indirect magnetic-sensor applications.[3]

In *direct* applications, the magnetic sensor is part of a magnetometer. Examples are the measurement of the geomagnetic field, the reading of magnetic data storage media, the identification of magnetic patterns in cards or banknotes, and the control of magnetic apparatus.

In *indirect* applications, the magnetic field is used as an intermediary carrier for detecting nonmagnetic signals. Examples are potential-free current detection for overload protection, integrated watt-hour meters, and contactless linear or angular position, displacement, or velocity detection using a permanent magnet.

These applications require the detection of magnetic fields in the micro- and millitesla range, which can be achieved by integrated semiconductor sensors.

Contactless switching for keyboards or collectorless DC motor control, displacement detection for proximity switches or crankshaft position sensors, and current detection seem to comprise most of the large-scale applications of magnetic sensors. It is for these large-scale applications that inexpensive batch-fabricated semiconductor magnetic sensors are highly desirable. It is unlikely that integrated silicon magnetic sensors will ever replace nuclear magnetic resonance (NMR) magnetometry with resolution in the nanotesla region, let alone the superconducting quantum interference devices (SQUID) resolving picotesla fields occurring in biomagnetometry.

With respect to the above ranges of magnetic resolution, we recall the following magnetic units. As a measure for the magnetic field strength H we use the related magnetic induction B, whose unit is 1 tesla $= 1$ V-s/m^2. This is the inverse of the unit of carrier mobility, namely 1 m^2/V-s $= 10^4$ cm^2/V-s $= 1$ T^{-1}. The product of magnetic induction and mobility is a dimensionless number which controls the strength of the galvanomagnetic effects.

5.1 INTRODUCTION

Semiconductor magnetic sensors including integrated silicon and GaAs sensors are useful in the range between 1 μT and 1 T. Here are some examples of magnetic induction within that range:

- geomagnetic field   30–60 μT
- magnetic storage media   about 1 mT
- permanent magnets in switches   5–100 mT
- conductor carrying a 10 A current   1 mT
- superconducting coils   10–20 T

5.1.2 Magnetic-Sensor Families

Figure 1 presents the family tree of some magnetic sensor materials and effects. Most magnetic sensors exploit the Lorentz force

$$F = -q\mathbf{v} \times \mathbf{B} \qquad (1)$$

on moving electrons in a metal, semiconductor, or an insulator in one way or another. Here, q denotes the absolute value of the electron charge, $\mathbf{v}$ the electron velocity, $\mathbf{B}$ the magnetic induction vector, and $\times$ the vector product. While the magnetic field vector $\mathbf{H}$ is the measurand, it is the magnetic induction $\mathbf{B}$ that acts on the mobile carriers and determines the sensor response.

In view of the relation $\mathbf{B} = \mu\mu_0\mathbf{H}$, where μ_0 denotes the permeability of free space and μ the relative permeability of the sensor material, the sensor response is boosted by high relative permeability. Thus, we can readily distinguish two major classes of magnetic sensors, as shown in Fig. 2.[3,4]

Magnetic sensors using high-permeability (ferro- or ferrimagnetic) materials, where $\mu \gg 1$, bring about a corresponding enhancement of sensitivity. Examples are sensors based on the magnetoresistance of NiFe thin films, the

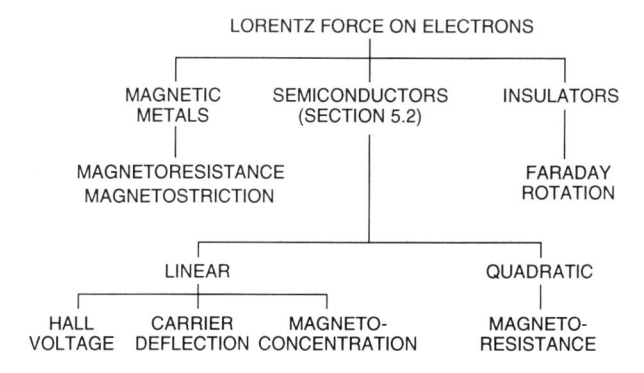

Fig. 1 Family tree of magnetic-sensor materials and effects.

208 MAGNETIC SENSORS

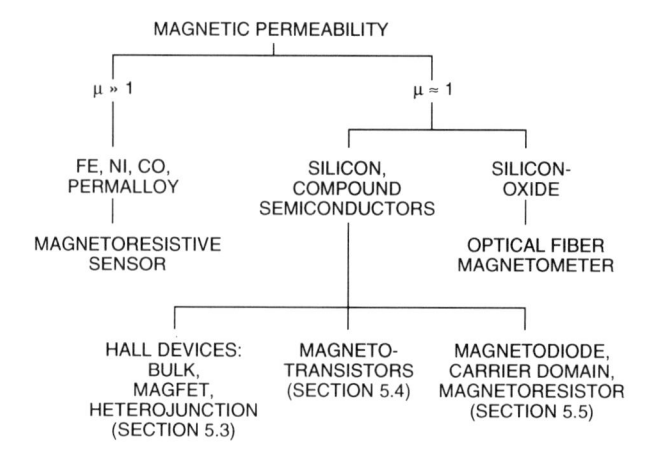

Fig. 2 Family tree of magnetic-sensor structures.

magnetostriction of the nickel cladding of an optical fiber, or the magneto-optic effects in garnets, and any sensor combined with a flux-concentrating device.

Magnetic sensors using low-permeability (dia- or paramagnetic) materials, where $\mu \approx 1$, do not provide appreciable amplification by the factor μ. All sensors based on galvanomagnetic effects in semiconductors belong to this class.

Thin Metal Film Magnetic Sensors. These sensors are based on ferromagnetic materials.[5] The low-magnetostriction alloy $Ni_{81}Fe_{19}$ (permalloy) is preferably used for thin-film magnetic sensors. The most successful sensor effect with such materials is the magnetoresistive (MR) switching of anisotropic NiFe or NiCo films. Another application of permalloy films is the recent miniaturization of the fluxgate magnetometer, reaching nanotesla resolution.[6]

Semiconductor Magnetic Sensors. These sensors and the underlying galvanomagnetic effects are the topic of this chapter.[3,4,7-13] Semiconductor magnetic sensors are flexible in design and application, small in size, rugged, and they provide a direct electronic signal output. They are currently fabricated with silicon or III–V compound semiconductor materials. Silicon magnetic sensors are by far the least expensive because they can draw on integrated circuit (IC) technology. Fabricated devices include bulk and inversion-layer Hall elements, magnetotransistors, magnetodiodes, and carrier-domain magnetometers. Certain III–V sensors show higher magnetic resolution than comparable silicon devices because of the higher carrier mobility. The III–V materials are used in the form of Hall and magnetoresistive devices.[4,14]

Optoelectronic Magnetic Sensors. These sensors use light as an intermediary signal carrier.[3,4] Magneto-optic sensors are based on the Faraday rotation of the polarization plane of linearly polarized light due to the Lorentz force on bound electrons in insulators. Magnetic sensors can be realized by using optical fiber coils providing a long light path and an accordingly large rotation per unit magnetic field. Magneto-optic current sensors for high-voltage transmission lines have been realized in this way. Minimum detectable fields of less than one nanotesla have been reported for optical fibers covered with magnetostrictive materials such as nickel. The strain transferred to the fiber from the magnetostrictive material results in a change of the optical path length, leading to a phase shift that is detected with an optical-fiber interferometer.

Superconductor Magnetic Sensors. These sensors, for example the SQUID and the supermagnetoresistor, do not fit in the above groups.[15] The SQUID is a high-resolution magnetometer for the picotesla range. It exploits quantum mechanical galvanomagnetic effects occurring in Josephson junctions between superconducting materials, for example $Nb/Al_2O_3/Nb$, at sufficiently low temperatures (below 20 K). Using thin-film techniques, the SQUID can be integrated together with a superconducting niobium input coil and signal conversion circuitry on a single chip. Since the recent discovery of high-temperature superconductors, a magnetoresistive sensor using the ceramic superconducting Y–Ba–Cu–O compound material has been demonstrated, which works at the temperature of liquid nitrogen.[16]

5.1.3 Specifications

Each magnetic-sensor application comes with its specific requirements, such as magnetic field sensitivity and resolution.[3] Switching and displacement-detection applications involve permanent magnets with fields in the millitesla range, whereas the stray fields of magnetic domains in recording media are between a few microtesla and a few millitesla.

The selection of the appropriate sensor design and technology depends on a number of further specifications that may vary widely from one application to another. Spatial resolution, for example, is crucial for reading high-density binary magnetic recordings while linearity is not. A tentative list of specifications may include the following items:

- availability and cost of technology
- application environment:
  temperature, humidity and chemical stress, vibrations and mechanical stress
- geometry:
  field vector parallel or perpendicular to chip surface
- sensitivity, signal level
- magnetic field resolution, signal-to-noise ratio:
  accuracy (absolute) and precision (relative)
- spatial resolution
- time resolution, frequency response

210   MAGNETIC SENSORS

- linearity of magnetic response
- temperature coefficient of sensitivity
- offset and its temperature dependence
- power consumption, size, weight
- electrical input and output impedance
- stability, reliability, lifetime
- testability of above specifications.

Many of these specifications are shared by sensors for other measurands and also by analog integrated circuits.

5.1.4 Scope and Literature

From here on, the subject of this chapter is limited to miniaturized semiconductor magnetic sensors based on classical galvanomagnetic effects.† Emphasis is on integrated semiconductor magnetic sensors and their physical principles, design, materials, technology, performance, and modeling. Quantum Hall effect devices[17] are beyond the scope of this chapter.

The development of a new sensor starts with the choice of the physical effect that converts the measurand in question into an electrical signal. Next, comes the choice of the material that shows this effect in a sufficiently strong way. Then, a sensor structure is designed and a technology to manufacture this structure is established. Finally, the performance of the sensor functions is tested. Section 5.2, therefore, deals with the galvanomagnetic effects basic to semiconductor magnetic sensors and the pertinent semiconductor materials, technologies, and figures of merit. The discrete Hall plate is used for illustration and for the definition of the figures of merit.

Section 5.3 on integrated Hall sensors presents bulk Hall devices as well as magnetic-field sensitive MOSFET and heterojunction structures. The different types of bipolar magnetotransistors are the topic of Section 5.4. Numerical modeling results are included in both Sections 5.3 and 5.4. Section 5.5 reviews other semiconductor magnetic sensors such as magnetodiodes (MD), current domain magnetometers (CDM), and magnetoresistive (MR) sensors. The relative merits of various magnetic sensors are compared in Section 5.6.

We conclude the introduction with a summary of the more recent review-type literature. Semiconductor magnetic sensors are the topic of several review articles in technical journals[3,7,9,11] and chapters in monographs.[4,10] For further reading we recommend the elaborate treatise by Popovic.[12] Semiconductor magnetic

---

†The text of this chapter grew out of lecture notes for a graduate course on solid-state sensors given at the Univeristy of Alberta, Edmonton, Canada, in 1985 and 1987 and a review article[3] published in 1986. The notes were updated for a fourth-year class on microsensors taught at ETH Zurich in 1989 and 1991. For this book, we revised this material, in particular the sections in advanced galvanomagnetic effects, bipolar magnetotransistors, and numerical modeling.

sensor modeling is covered in a number of comprehensive papers.[18-22] For non-semiconductor magnetic sensors we refer to the excellent handbook edited by Boll and Overshott.[5,15] We finally mention two previous textbook chapters on magnetic sensors.[8,13]

5.2 EFFECTS AND MATERIALS

In this section we introduce the galvanomagnetic phenomena responsible for semiconductor magnetic-sensor action and its side effects, including noise, offset, temperature coefficient, and nonlinearity. We present the current transport equations in the presence of a magnetic field and discuss the Hall plate and its figures of merit as a model magnetic sensor. Semiconductor magnetic-sensor materials and technologies are reviewed including silicon IC technologies and compound semiconductors.

5.2.1 Basic Galvanomagnetic Effects

The action of the Lorentz force, Eq. 1, manifests itself in the carrier transport.[23-25] Let us assume an isotropic, non-degenerate n-type semiconducting material with zero temperature gradient. We denote the electron current density for zero magnetic induction ($\boldsymbol{B}=0$) by $\boldsymbol{J}_n(0)$. The drift-diffusion approximation of the Boltzmann transport equation leads to the current density equations[26]

$$\boldsymbol{J}_n(0) = \sigma_n \boldsymbol{\mathscr{E}} + q D_n \nabla n, \tag{2}$$

where $\sigma_n = q\mu_n n$ denotes the electronic conductivity for $\boldsymbol{B}=0$, $\boldsymbol{\mathscr{E}} = (\mathscr{E}_x, \mathscr{E}_y, \mathscr{E}_z)$ the electrical field, $D_n = \mu_n kT/q$ the diffusion coefficient, n the electron density, $\mu_n = v_n/|\boldsymbol{\mathscr{E}}|$ the electron drift mobility, v_n the average electron velocity, and ∇ the gradient.

Let us try to describe the action of the Lorentz force, Eq. 1, in a quick heuristic way. We neglect the statistical velocity distribution of the electrons and assume all electrons to have just the average drift-diffusion velocity v_n. We simply replace v_n involved in Eq. 2 by $v_n - v_n \times (\mu_n \boldsymbol{B})$ and, thus, produce the approximation[3,12]

$$\boldsymbol{J}_n(\boldsymbol{B}) \approx \boldsymbol{J}_n(0) - \mu_n \boldsymbol{J}_n(\boldsymbol{B}) \times \boldsymbol{B} \tag{3}$$

for the current density $\boldsymbol{J}_n(\boldsymbol{B})$ under non-zero magnetic induction. Solving Eq. 3 with respect to $\boldsymbol{J}_n(\boldsymbol{B})$ leads to

$$\boldsymbol{J}_n(\boldsymbol{B}) \approx [\boldsymbol{J}_n(0) + \mu_n \boldsymbol{B} \times \boldsymbol{J}_n(0) + (\mu_n)^2 \boldsymbol{B} \cdot \boldsymbol{J}_n(0) \boldsymbol{B}][1 + (\mu_n \boldsymbol{B})^2]^{-1} \tag{4}$$

with · denoting the scalar product. Such an expression can indeed be derived in the framework of formal transport theory for a strongly degenerate electron gas (metal), where all electrons have the same energy.[24]

For nondegenerate semiconductors, however, the energy distribution of the electrons and, hence, the energy dependence of scattering mechanisms, cannot be neglected. This effect can be described in terms of the scattering factor r_n for electrons,

$$r_n = \langle \tau^2 \rangle / \langle \tau \rangle^2 \tag{5}$$

where τ denotes the (energy-dependent) relaxation time or mean free transit time between two successive electron collisions, and the brackets $\langle \rangle$ denote the average over the electron energy distribution. The theoretical values $r_n = 1.18$ for phonon scattering and $r_n = 1.93$ for impurity scattering can be derived from Boltzmann's distribution for nondengerate semiconductors with a spherical constant-energy surface.[26] For n-type silicon (an anisotropic many-valley semiconductor) the empirical value of r_n is about 1.15 at room temperature for low donor concentration.[3] To first order in B, the energy-dependent scattering can be accounted for by replacing the drift mobility μ_n by the Hall mobility

$$\mu_n^* = r_n \mu_n. \tag{6}$$

Unlike in Eq. 4, the electron current density $J_n(B)$ in an isotropic nondegenerate n-type semiconductor exposed to the magnetic induction B is approximately described by the equation[12,24]

$$J_n(B) = [J_n(0) + \mu_n^* B \times J_n(0) + K(\mu_n^*)^2 B \cdot J_n(0) B][1 + (\mu_n^* B)^2]^{-1}. \tag{7}$$

The factor K summarizes effects of higher power averages such as $\langle \tau^3 \rangle$ on the third term of Eq. 7. This and other corrections are further pursued in Section 5.2.5. Equation 7 describes only isothermal ($\nabla T = 0$) galvanomagnetic effects for electrons. It includes thermal effects only through the temperature dependence of carrier concentration, diffusion coefficient, and mobility, but does not include thermomagnetic or thermoelectric cross effects. If carrier concentration gradients ∇n are negligible, as in a semiconductor slab with ohmic contacts, we can replace Eq. 7 by

$$J_n(B) = \sigma_{nB}[\mathscr{E} + \mu_n^* B \times \mathscr{E} + K(\mu_n^*)^2 (B \cdot \mathscr{E}) B] \tag{8}$$

with the magnetic-field dependent conductivity

$$\sigma_{nB} = \sigma_n [1 + (\mu_n^* B)^2]^{-1}. \tag{9}$$

If B is parallel to $\mathscr{E}$, then $(B \times \mathscr{E}) = 0$ and the current density is reduced to $J_n(B) = \sigma_{nB}[1 + K(\mu_n^* B)^2] \mathscr{E}$ showing a longitudinal magnetoresistance effect (see also Section 5.2.5). This effect occurs in nondegenerate semiconductors, but vanishes in isotropic metals where $K = 1$.

Let us now consider the more important case where the magnetic induction vector $\boldsymbol{B}$ is perpendicular to the electric field $\mathscr{E}$. In this case we have $(\boldsymbol{B}\cdot\mathscr{E})=0$ and, hence, the electron current density

$$\boldsymbol{J}_n(\boldsymbol{B})=\sigma_{nB}(\mathscr{E}+\mu_n^*\boldsymbol{B}\times\mathscr{E}). \tag{10}$$

This equation describes transverse galvanomagnetic effects for electrons in the case of negligible diffusion. The term $\mu_n^*(\boldsymbol{B}\times\mathscr{E})=-(\boldsymbol{v}_n\times\boldsymbol{B})$ is often interpreted as a transverse electric field counterbalancing the deflection of the carrier path under the Lorentz force in long samples. Let us now assume a specific geometric configuration where the vectors $\mathscr{E}$ and $\boldsymbol{J}_n(\boldsymbol{B})$ are in the x–y plane and the magnetic induction vector $\boldsymbol{B}$ is parallel to the z direction. Then in terms of $\mathscr{E}=(\mathscr{E}_x,\mathscr{E}_y,0)$, $\boldsymbol{J}_n(\boldsymbol{B})=(J_{nx},J_{ny},0)$, and $\boldsymbol{B}=(0,0,B)$, Eq. 10 reads.

$$J_{nx}=\sigma_{nB}(\mathscr{E}_x-\mu_n^*B\mathscr{E}_y) \qquad J_{ny}=\sigma_{nB}(\mathscr{E}_y+\mu_n^*B\mathscr{E}_x). \tag{11}$$

We discuss the two limiting cases of an infinitely long and an infinitely wide semiconductor slab.

Hall Field. It is assumed that the current density has only an x-component, while $J_{ny}=0$. This condition can be approximated by a long sample (length L) with narrow cross section (width $W\ll L$) and current electrodes at the small faces (see Fig. 3). Then the Hall field resulting from Eq. 11 is

$$\mathscr{E}_y=-\mu_n^*B\mathscr{E}_x=R_HJ_{nx}B, \tag{12}$$

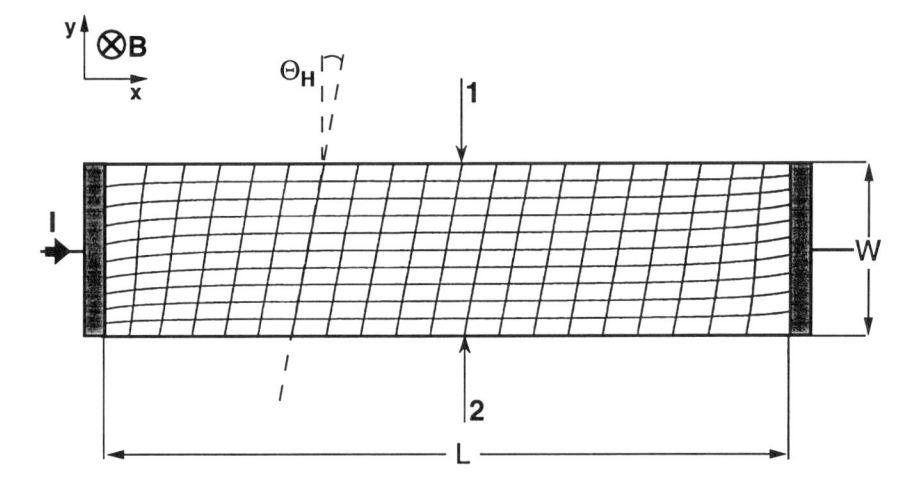

Fig. 3 Hall effect in a long semiconductor plate ($L=4W$). Current lines connect (shaded) ohmic contacts. Equipotential lines are deflected from vertical direction by Hall angle θ_H. The Hall voltage appears between border locations 1 and 2. The curves originate from numerical modeling with $\mu_n^*B=0.21$. (After Ref. 3)

where

$$R_H = -\mu_n^*/\sigma_n = -r_n/qn \qquad (13)$$

denotes the Hall coefficient. Its value depends on the temperature and doping of the semiconductor material. The presence of the Hall field results in a rotation of the equipotential lines by the Hall angle θ_H with

$$\tan\theta_H = \mathscr{E}_y/\mathscr{E}_x = -\mu_n^* B = \sigma_n R_H B. \qquad (14)$$

From Eq. 13 we learn that low carrier concentration produces a large Hall coefficient R_H. This explains why semiconductors are more useful here than metals are. From Eq. 14 we note that high mobility favors a large Hall angle θ_H and Hall field $\mathscr{E}_y$. Across a long, thin plate of thickness t carrying a current I, the Hall field, Eq. 12, produces the corresponding Hall voltage

$$V_H = R_H I B/t. \qquad (15)$$

The corresponding relative sensitivity, that is the Hall voltage per unit current and magnetic induction, is

$$V_H/IB = R_H/t = -r_n/qnt. \qquad (16)$$

High relative sensitivity, thus, requires small carrier concentration and small plate thickness. We notice that the relative sensitivity with respect to current does not depend on the carrier mobility.

Carrier Deflection and Magnetoresistance. Let us now make the contrary assumption of zero Hall field, $\mathscr{E}_y = 0$. This condition is approximately realized by a short sample with wide cross-section ($L \ll W$) and current electrodes at the large faces (see Fig. 4). Under these conditions, Eq. 11 leads to a lateral current-density component, J_{ny}, which produces a rotation of the current lines described by the ratio

$$-J_{ny}/J_{nx} = \mu_n^* B = \tan\theta_L. \qquad (17)$$

This is the carrier or current or Lorentz deflection. The deflection angle of the current lines ("Lorentz angle") is $\theta_L = \arctan(\mu_n^* B)$.

Figure 5 compares the Hall and Lorentz modes of operation of semiconductor plates. In the case of a wide plate with a split contact carrying the total current I, the deflection produces a current imbalance I_L between the left and right side of the contact, so that each side carries the current $(I+I_L)/2$ or $(I-I_L)/2$, respectively. The resulting Lorentz current I_L is described by

$$I_L = \mu_n^*(L/W)IB, \qquad (18)$$

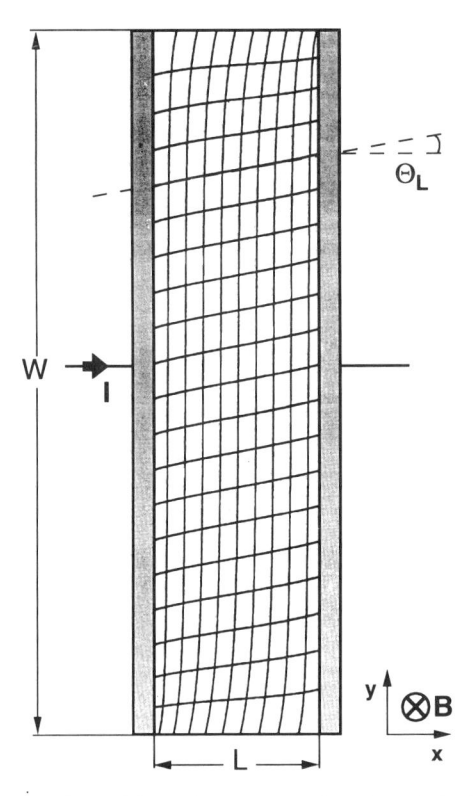

Fig. 4 Lorentz deflection in a wide semiconductor plate ($W = 4L$). Current lines connect contacts and are deflected from the horizontal direction by the angle θ_L. Equipotential lines are nearly parallel to the contacts. The curves originate from modeling with $\mu_n^* B = 0.21$. (After Ref. 3)

provided that the separation between the two contacts is very small. The related relative sensitivity with respect to the current, I_L/IB, is proportional to the carrier mobility.

The longer carrier-drift path brought about by the deflection leads to the (transversal) geometric magnetoresistance effect,

$$(\rho_{nB} - \rho_n)/\rho_n = (\mu_n^* B)^2. \tag{19}$$

Here $\rho_n = 1/\sigma_n$ denotes the electrical resistivity for zero magnetic induction, and $\rho_{nB} = \mathcal{E}_x/J_{nx} = 1/\sigma_{nB}$ the resistivity enhanced by the magnetic field. Besides this geometrical magnetoresistance, there is also a physical magnetoresistance due to the averaging over different velocity-dependent coefficients.[12,24] In any case, the relative change of resistivity increases with the square of the mobility-induction product. Hence, this effect is very small for silicon, for example $\rho_{nB} \approx 1.02 \rho_n$ for B as high as 1 T. Sensors based on this effect require

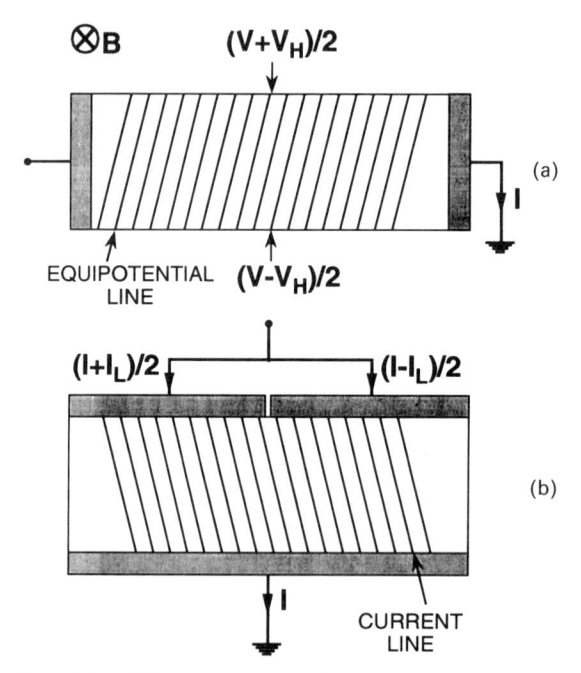

Fig. 5 Hall voltage (a) and Lorentz current (b) modes of operation of a semiconductor plate in perpendicular magnetic induction producing the Hall voltage V_H and the Lorentz current I_L, respectively.

high-mobility III–V compounds such as InSb or InAs. For bulk InSb, a 2% resistivity enhancement is obtained with a magnetic induction of only about 18 mT.

For p-type semiconductor material characterized by drift and Hall mobilities μ_p and $\mu_p^* = r_p \mu_p$ we have to pursue the above analysis starting from the opposite sign of the Lorentz force. Instead of Eq. 10, the hole current density $J_n(B)$ for B perpendicular to $\mathscr{E}$ is given by

$$J_p(B) = \sigma_{pB}(\mathscr{E} - \mu_p^* B \times \mathscr{E}) \quad \text{with } \sigma_{pB} = \sigma_p[1 + (\mu_p^* B)^2]^{-1}, \tag{20}$$

where $\sigma_p = q\mu_p p$ denotes the hole conductivity at zero magnetic induction. Correspondingly, the Hall coefficient for holes is

$$R_H = \mu_p^*/\sigma_p = r_p/qp \tag{21}$$

and the Hall angle for holes is given by

$$\tan \theta_H = \mu_p^* B. \tag{22}$$

The magnetoresistance effect for holes is described by $\rho_{pB}/\rho_p = 1 + (\mu_{np}^* B)^2$ with

$\rho_p = 1/\sigma_p$. For mixed n- and p-conduction, the Hall coefficient has the general form

$$R_H = -[r_n(\mu_n/\mu_p)^2 n - r_p p]/q[\mu_n/\mu_p)n + p]^2. \tag{23}$$

Strictly speaking, the relations derived for the Hall fields, Eqs. 12–14 and Eqs. 21 and 22, or the carrier deflection, Eqs. 17 and 18, for semiconductor plates are valid only in the limit of infinite and zero length-to-width ratio L/W, respectively. Numerical modeling, however, demonstrates that the limiting cases are good approximations for $L/W \geqslant 4$ and $L/W \leqslant \frac{1}{4}$, respectively.[4,27]

A way of realizing a finite sample showing just carrier deflection without any Hall electric field is the *Corbino disc*. This is a round semiconductor plate with one electrode at the center and the other around the circumference. Without magnetic field the current flow is radial. In a perpendicular magnetic field the current is tilted by the Lorentz angle, Eq. 17, relative to the electric field. The electric field is always radial irrespective of the applied magnetic induction.

The magnetoconcentration effect or Suhl effect is a variation in carrier concentration in a current-carrying sample exposed to a magnetic field.[28] In its simplest form, this effect appears in a semiconductor slab of nearly intrinsic conduction.[4,27] Since the sign of the Lorentz force is the same for electrons and holes drifting in opposite directions, both types of carriers are pushed towards the same insulating boundary, where their concentration builds up. This leads to a localized region of increased conductivity and current crowding at the expense of another region where carriers and current lines are depleted. Figure 6 demonstrates the magnetoconcentration effect for a quadratic slab of nearly

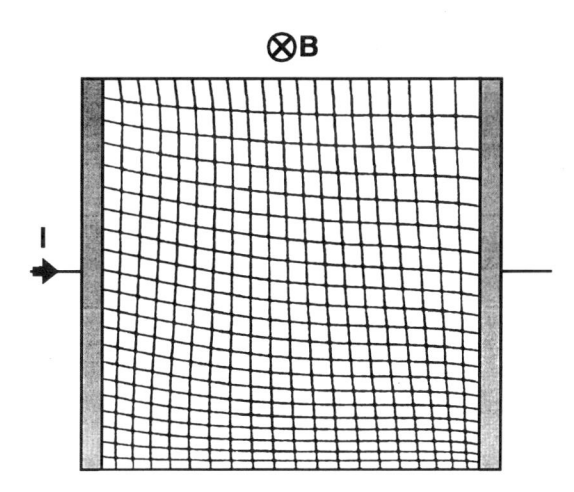

Fig. 6 Magnetoconcentration in a nearly intrinsic ($T = 500$ K) bulk silicon plate. Current lines connect ohmic contacts and crowd near bottom. Equipotential lines are approximately parallel to the contacts. The curves originate from modeling with $\mu_n^* B = 0.21$ and $\mu_p^* B = 0.07$. (After Ref. 3)

intrinsic material. The current lines connect the left- and right-hand contact of the slab, and the equipotential lines end at the upper and lower borders of the quadratic slab. In the presence of a magnetic induction perpendicular to the drawing plane, the current lines are crowded towards the lower right corner of the slab.

Another form of magnetoconcentration effect occurs when a long semiconductor slab of intrinsic conduction is prepared such that the two faces of the sample have two very different surface-recombination velocities.[12] If the direction of the magnetic induction vector is such that both types of carriers are pushed towards the high recombination surface, the carriers recombine there, but their concentration at that surface stays close to equilibrium. Since the generation rate in the rest of the sample is small, it cannot compensate for the loss of carriers at the highly-recombining surface and the resistance of the sample increases: *vice versa*, if the carriers are deflected towards the low-recombining surface, resistance decreases.

Magnetoconcentration plays a role in magnetodiodes and bipolar magnetotransistors under the condition of high injection of both electrons and holes. In principle, the effect may also appear as a parasitic effect in Hall devices, but it is negligible in modern Hall devices made of highly-doped extrinsic material.

5.2.2 Ideal Hall Plates

The Hall plate is a thin, usually rectangular plate of relatively high-resistivity material provided with four ohmic contacts, as shown in Fig. 7. The Hall voltage V_H appears between the sensor contacts SC1 and SC2. It is proportional to the perpendicular magnetic induction B and the bias current I supplied via the current contacts CC1 and CC2 by an applied voltage V. Its absolute value is

$$V_H = R_H G t^{-1} I B = -GIBr_n(qnt)^{-1}. \qquad (24)$$

The geometric correction factor G describes the shape effects of the plate. The factor G depends not only on the length L and the width W of the Hall plate, but also on the contact size s, the position of the sensor contacts, and on the

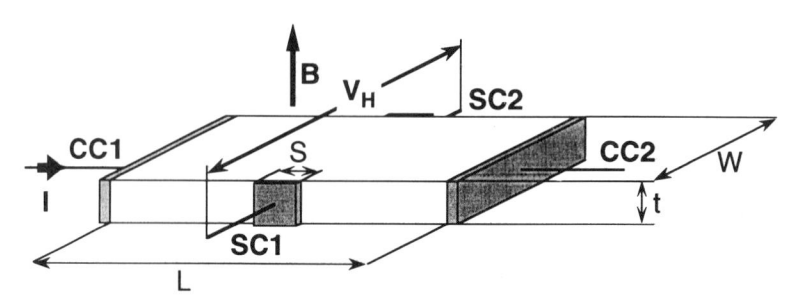

Fig. 7 Rectangular Hall plate. The symbols are explained in the text. (After Ref. 3)

Hall angle θ_H. The factor G approaches unity in the limit of the infinitely long Hall plate described by Eq. 15; otherwise, it accounts for the short-circuiting effects between the current and sensor contacts.

Let us for now dispense with the shape effects and consider a quantitative example.[12] As Hall plate material, we choose a uniform n-type silicon with $1\,\Omega\,\text{cm}$ resistivity at room temperature, which corresponds to a carrier concentration of $n \approx 4.5 \times 10^{15}\,\text{cm}^{-3}$. Hence, the Hall coefficient R_H is about $-1.4 \times 10^{-3}\,\text{C}^{-1}\,\text{m}^3$. Assuming $L = 300\,\mu\text{m}$, $W = 100\,\mu\text{m}$, and $t = 20\,\mu\text{m}$ the device resistance turns out to be $1.5 \times 10^3\,\Omega$. For 10 V biasing voltage, this leads to a current of about 6.7 mA. With magnetic induction as large as 1 T, the resulting Hall voltage is about 0.46 V. Looking up the electron mobility in a table,[26] we find $\mu_n \simeq 0.14\,\text{m}^2/\text{V-s}$. Neglecting the scattering coefficient, this leads to $\mu_n^* B \approx 0.14$ and a Hall angle $\theta_H \approx 8$ degrees. The relative sensitivity V_H/IB of our example device is about 70 V/A-T. We notice that with half the thickness and five times less doping, we would achieve 700 V/A-T.

A large variety of analytical formulae and numerical results for rectangular, circular, octagonal, and cross-shaped Hall plates are available.[3,12] As a first example we present the case of a relatively long rectangular plate with $L/W > 1.5$ and small sensor contact size $s/W < 0.18$, where

$$G \approx [1 - (16/\pi^2) \exp(-\pi L/2W)\theta_H/\tan\theta_H](1 - 2s\theta_H/\pi W \tan\theta_H) \qquad (25)$$

for small Hall angles. The first term in square brackets represents the influence of the current contacts due to the finite length L, while the second term accounts for the finite size s of the sensor contacts. Within these limits, the accuracy of the approximation is better than 4%. It approaches unity for $L/W > 3$ and $s/W < 1/20$. An opposite example is the short rectangular plate ($L < W$) with point sensor contacts located at $y = L/2$ and small Hall angle, where the factor G can be approximated by

$$G \approx 0.74 L/W. \qquad (26)$$

Inspection of Eqs. 24 and 25 seems to suggest that the Hall plate should be as long, as thin, and as low doped as possible for efficient use of the available bias current. On the other hand, the voltage drop

$$V = (I/qn\mu_n^*)(L/Wt) \qquad (27)$$

along the plate may become unacceptably high. By inserting Eq. 26 into Eq. 24 we can express the Hall voltage in terms of the supply voltage V rather than the supply current I,

$$V_H = \mu_n^*(GW/L)BV. \qquad (28)$$

In contrast to Eq. 16, this relation shows the value of high carrier mobility for

Hall plate efficiency. According to Eq. 28, the Hall voltage of a plate with given bias voltage seems to increase with the aspect ratio W/L. There is, however, a limiting value revealed by combining Eqs. 26 and 28. Thus, the maximum Hall voltage for short samples under constant bias voltage V is

$$V_H \approx 0.74 \mu_n^* BV. \tag{29}$$

Combining Eqs. 24 and 28, we obtain the Hall voltage in terms of the power $P = IV$ dissipated in the plate,

$$V_H = G(W/L)^{1/2} r_n (\mu_n/qnt)^{1/2} P^{1/2} B. \tag{30}$$

The product $G(W/L)^{1/2}$ shows a broad maximum of about 0.7 at $L/W \approx 1.35$. We can summarize these results in terms of the following rules to obtain the highest Hall voltage using rectangular plates with small contacts.[12]

- For given bias current, use long plate with $L/W > 3$ yielding $G \approx 1$
- For given bias voltage, use short plate with $L/W < 1$ yielding $G < 0.7$
- For given power dissipation, use plate with $L/W \approx 1.35$ yielding $G \approx 0.7$

In principle, a thin semiconductor plate with any shape and four contacts at any position will still produce some Hall voltage, though possibly with low efficiency. Nevertheless, some shapes are more desirable than others. Diamond-shaped plates (with contacts at the four corners) and Hall crosses (with contacts at the end of each cross bar) are used to reduce the short-circuit effect. For integrated Hall plates, limitations of shape are imposed by photolighography. Small rectangular plates with $s/L < 0.1$ are difficult to fabricate, but the same geometrical factor can be achieved by a cross-shaped configuration.

In Fig. 5 we compared the Hall-voltage and Lorentz-current modes of operation. The Hall-voltage operation, biased with a constant current source, is preferred in modern Hall devices. Figure 8 shows the preferable circuit for biasing a Hall sensor.[3,12] With the help of an operational amplifier (OA) the

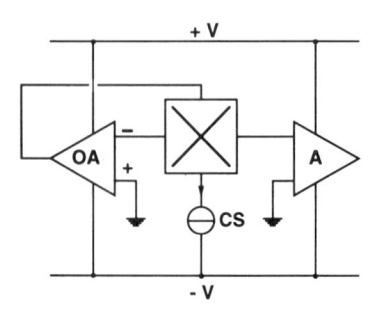

Fig. 8   Preferable biasing and amplification circuitry for Hall plate. (After Ref. 3)

left sensor contact is virtually grounded and the full Hall voltage appears at the right sensor contact. Otherwise, without the operational amplifier, a large common-mode voltage would appear at the amplifier (A) input.

Sensitivity. Sensitivity is the most important figure of merit of a sensor. Three different sensitivities for ideal Hall plates under perpendicular magnetic induction have been defined. The absolute sensitivity of a Hall magnetic sensor is its transduction ratio for large signals:

$$S_A = |V_H/B| \quad \text{with } V_H = S_A B. \tag{31}$$

The ratio of the absolute sensitivity of a modulating transducer and its bias quantity yields a relative sensitivity. For Hall-plate sensors, the supply-current related sensitivity S_I (also known as product sensitivity) and the supply-voltage related sensivitity S_V can be distinguished. We first discuss

$$S_I = S_A/I = |V_H/BI| \quad \text{with } V_H = S_I IB. \tag{32}$$

The units of S_I are V/A-T (volts per ampere and tesla). From Eqs. 13 and 24,

$$S_I = R_H G/t = G r_n/qnt \tag{33}$$

for a plate of extrinsic *n*-type material. If the Hall plate is nonuniformly doped across its thickness, then the product nt should be replaced by the surface-charge carrier density N_s. The resulting denominator qN_s equals the charge carried by free electrons per unit plate area.

The current-related sensitivity hardly depends on the plate material, since r_n is close to unity for all relevant materials. Typical values range between 70 (our numerical example) and several hundred V/A-T. The highest value reported so far[29] seems to be 3100 V/A-T corresponding to $N_s = 2 \times 10^{11}$ cm^{-2}. Low values of the charge density in the active layer of an integrated Hall sensor, however, come with the penalty of a strong junction field effect (see Sec. 5.3.1).

The supply-voltage related sensitivity is defined as

$$S_V = S_A/V = |V_H/BV| \quad \text{with } V_H = S_V VB. \tag{34}$$

Its unit is V/V-T = T^{-1} (per tesla, often given in % per tesla). From Eq. 28 we find for an extrinsic *n*-type semiconductor,

$$S_V = \mu_n^*(GW/L). \tag{35}$$

The material plays a decisive role in view of the mobility dependence of S_V. Typical values are 0.07 T^{-1} or 7%/T for silicon Hall plates and 0.20 T^{-1} for GaAs plates.[30] The upper physical limit set by Eq. 29 is

$$S_{V\max} = 0.74 \, \mu_n^*. \tag{36}$$

This leads to upper limits of 0.12 T^{-1} and 0.67 T^{-1} for low-doped n-type silicon and GaAs, respectively. A voltage-related sensitivity as high as 0.48 T^{-1} has been reached for a III–V heterojunction Hall device with 0.72 T^{-1} mobility at room temperature.[31]

Finally, the power-related sensitivity can be defined as

$$S_P = S_A/P = G(W/L)^{1/2} r_n (\mu_n/qnt)^{1/2}/P^{1/2}. \tag{37}$$

This quantity is proportional to the power-related figure of merit[31]

$$M = (S_I \mu_n^*) = \mu_n^* (\rho/t)^{1/2}. \tag{38}$$

5.2.3 Hall-Plate Limiting Effects

Before we can discuss the proper choice of semiconductor sensor materials and fabrication technologies, we have to introduce further figures of merit. They characterize side effects causing deviations from ideal Hall-plate behavior and concern noise, offset, temperature coefficient, and nonlinearity.

Noise Phenomena. Noise phenomena severely limit the performance of any sensor.[3,12] The signal-to-noise ratio (SNR) determines the resolution of the measurand and is the dominant figure of merit for many applications. Voltages induced in the Hall circuit by the time variation of magnetic induction can virtually be eliminated by careful design. Once this effect has been taken care of, the output voltage of a Hall device can be written as

$$V_{out} = V_H(I, B) + V_N(t). \tag{39}$$

Here $V_H(I, B)$ denotes the voltage between the sensor contacts, which depends on the bias current and the magnetic induction in a deterministic way, and $V_N(t)$ denotes the noise voltage. The noise voltage at the terminals of a Hall device can be due to thermal noise (Johnson–Nyquist noise), generation–recombination (GR) noise (fluctuation of number of carriers), and $1/f$ noise (flicker noise). In comparison with $1/f$ noise, GR noise is often negligible. Going to the Fourier domain (frequency f) representation, we, therefore, keep the voltage-noise spectral density

$$S_V(f) = S_{V\alpha}(f) + S_{VT} \tag{40}$$

across the Hall sensor contacts, with $S_{V\alpha}(f)$ denoting the voltage-noise spectral density due to $1/f$ noise and S_{VT} the thermal voltage-noise spectral density. For a rectangular Hall plate $S_{V\alpha}(f)$ can be approximated by

$$S_{V\alpha}(f) \simeq \alpha (V/L)^2 (2\pi ntf)^{-1} \ln(W/s) \tag{41}$$

with α denoting the Hooge $1/f$ noise parameter.[32] The thermal noise of a resistance R at absolute temperature T is generally described by the Nyquist relation

$$S_{VT} = 4kTR \qquad (42)$$

with k denoting Boltzmann's constant. From Eq. 42 and the output resistance[12]

$$R_{out} \approx 2(\rho_{nB}/\pi t) \ln(W/s) \qquad (43)$$

of a long rectangular Hall plate with small sensor contacts we obtain

$$S_{VT} \approx 8kT(\rho_{nB}/\pi t) \ln(W/s). \qquad (44)$$

The SNR in a narrow bandwidth Δf around f is given by

$$\text{SNR} = V_H(I, B)(S_V(f) \Delta f)^{-1/2}. \qquad (45)$$

At low frequencies, where $1/f$ noise is dominant, we obtain from Eqs. 24, 41, and 43

$$\text{SNR} \approx \mu_n^* B(2\pi n t L W/\alpha \ln(W/s)^{1/2} G(W/L)^{1/2} (f/\Delta f)^{1/2}. \qquad (46)$$

This approximation holds for frequencies that are much smaller than the corner frequency f_c, defined as the frequency where $S_{V\alpha}(f)$ equals $S_{VT}(f)$,

$$f_c = (q\mu_n^* \alpha/16kT)(V/L)^2. \qquad (47)$$

Equation 46 demonstrates how the mobility-induction product $\mu_n^* B$ controls the low-frequency SNR.

Thermal noise dominates at high frequencies, $f \gg f_c$. The SNR increases with higher bias current and is limited by the maximum acceptable power dissipation P_{max}. Using Eqs. 30 and 44 we obtain

$$\text{SNR} \approx 0.44 \, \mu_n^* B[P_{max}/kT \Delta f \ln(W/s)]^{1/2}. \qquad (48)$$

From Eq. 48 we recognize the role of the mobility-induction product $\mu_n^* B$ for the high-frequency SNR.

The spectral density of the equivalent magnetic-induction noise is defined as

$$S_B(f) = S_V(f)/S_A^2 \qquad (49)$$

and is measured in T^2/Hz. Typical values for silicon Hall sensors are $3 \times 10^{-13} \, T^2/Hz$ at 100 Hz (flicker noise prevailing) and $10^{-15} \, T^2/Hz$ at 100 kHz (thermal noise). The smallest detectable equivalent magnetic induction,

B_N, in a given frequency band of width Δf is defined as the square root of the band integral over $S_B(f)$; this definition corresponds to a SNR of unity. The spectral noise density B_N for the bandwidth $\Delta f = 1$ Hz is called the smallest detectable magnetic induction, B_{min}. These values correspond to $B_{min} = 500$ nT and $B_{min} = 30$ nT, respectively. In this way, the detection limit of an InSb Hall plate with flicker noise has been assessed theoretically as $B_{min} = 40$ pT.

Offset Voltage. Offset voltage is a static or very slowly varying output voltage V_0 of a Hall sensor at zero magnetic induction. The offset can be characterized by an equivalent magnetic induction $B_0 = V_0/S_A$. The major causes of offset in integrated Hall sensors are imperfections of the fabrication process such as misalignment, notably of the sensor contacts, and nonuniformity of material resistivity and thickness. The piezoresistance effect activated by mechanical stress such as encapsulation stress can also produce offset.[33] Through the magnetoresistance effect, the offset voltage may also depend on the magnetic induction.

Temperature Coefficient. The temperature coefficient of the current-related sensitivity, TC_I, is defined as the relative partial derivative with respect to temperature,

$$TC_I = (S_I)^{-1}(\partial S_I/\partial T). \tag{50}$$

Let us consider strongly extrinsic material in the temperature range, where n equals the donor concentration N_D. Thus, in Eq. 33, $S_I = Gr_n/qnt$, only the scattering factor r_n depends on temperature, whence

$$TC_I = (r_n)^{-1}(\partial r_n/\partial T). \tag{51}$$

For low-doped silicon around room temperature, a typical value is $10^{-3}/K$. The temperature coefficient of the voltage-related sensitivity, (TC_V) is dominated by the strong temperature dependence of the drift mobility, since

$$TC_V = (\mu_n^*)^{-1}(\partial \mu_n^*/\partial T). \tag{52}$$

In low-doped silicon, μ_n is proportional to $T^{-2.4}$ and the corresponding TC_V is about $-8 \times 10^{-3}/K$.

Nonlinearity. Nonlinearity (NL) has material and geometrical causes; the Hall coefficient and the geometry factor G depend on the magnetic induction. The joint effect is described in terms of

$$NL = (V_H(I, B) - V_{H0})/V_{H0}, \tag{53}$$

where $V_H(I, B)$ denotes the measured Hall voltage at bias current I and V_{H0} is the best linear fit. Experimentally, a quadratic NL proportional to B^2, is usually found. The material NL is approximately equal to $-\beta(\mu_n^* B)^2$ with $\beta > 0$ denoting the material NL coefficient. For low-doped n-type silicon and GaAs at room temperature, β is about 0.5 and 0.1, respectively. The geometrical NL can be approximated by $\gamma(\mu_n^* B)^2$ with the geometrical NL coefficient $\gamma > 0$. The coefficient γ vanishes as the geometric correction factor G gets close to unity and approaches the maximum value 0.604 for very small G.[12]

5.2.4 Materials and Technologies

An appropriate motto for this section is the following citation from the book by Popovic:[12] "A great majority of contemporary Hall devices for magnetic sensor applications are made of silicon and GaAs, in spite of the fact that these materials, in particular silicon, have rather low carrier mobilities. Their large band gaps and the possibility of using microelectronics technology outweigh the drawback of small carrier mobility."

The choice of a semiconductor magnetic sensor material starts from key figures of merit such as output sensitivity and signal-to-noise ratio, temperature range and coefficient, as well as input/output impedance for matching to circuit components.[3,4] As we have seen, the magnitude of the underlying galvanomagnetic effects and the resulting SNR for Hall plates are governed by the mobility-induction product $\mu_n^* B$. This seems to favor materials with high mobility (the values are given at room temperature in units of $m^2/V\,s = T^{-1}$). Thus, InSb (about 8 in bulk and about 6 in thin film) and InAs (about 3) seem to be far superior to GaAs (about 0.7 in low-doped bulk material and in the two-dimensional electron gas at the AlGaAs/GaAs hetero-interface), let alone Si (up to 0.14 in bulk and 0.07 in n-channel MOSFET inversion layers). Generally, n-type semiconductor material is superior to p-type material because of the much lower hole mobility (0.05 for bulk Si, 0.04 for GaAs, 0.14 for InSb).

Figure 9 offers a systematic comparison of bulk silicon with different III–V technologies[31] in terms of the Hall mobility μ_n^* (proportional to S_V and SNR) and the current related sensitivity $S_I = |V_H/BI|$. The figure locates the materials Si, GaAs (epi and implanted), InSb (bulk and film), and the two-dimensional electron gas (2DEG) near the AlGaAs/GaAs heterojunction, at room temperature and 77 K, in a logarithmic-scale coordinate system formed by μ_n^* (unit $m^2/V\,s = T^{-1}$) and S_I (unit: V/A-T). For example, we gather from Fig. 9 that the 2DEG at 77 K shows a Hall mobility of about $15\,m^2/V\,s$ and a sensitivity of about 1000 V/A T. A more complex chart including the power-related figure of merit $M = \mu_n^*(\rho/t)^{1/2}$ and the sheet resistance ρ/t (relevant for the impedance matching) is given in the literature.[4,31]

The bandgap is another important material figure of merit related to the feasibility of high-temperature operation, which is possibly disturbed by increasing numbers of thermally-excited minority carriers. Although InSb and InAs show high electron mobility, the drawback with these materials is the small bandgap (0.2 eV for InSb and 0.4 eV for InAs). At room-temperature

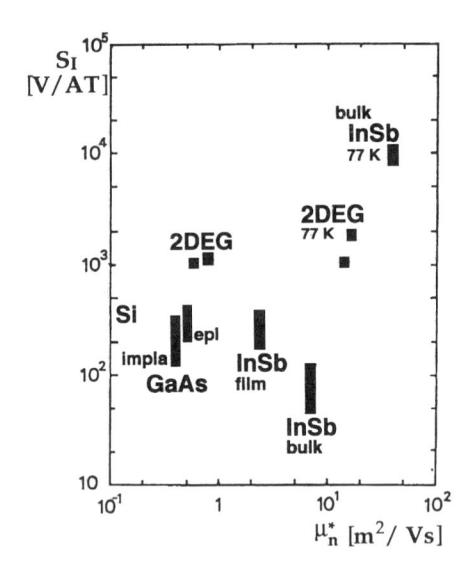

Fig. 9 Comparison of Hall sensors fabricated with different materials and technologies in terms of current-related sensitivity S_I and Hall mobility μ_n^*. The materials are explained in the text. (After Ref. 31)

operation, intrinsic behavior prevails, which excludes the use of such materials other than for magnetoresistive applications. In this respect, Si (1.12 eV) and GaAs (1.42 eV) are outstanding materials with temperature coefficients that are far superior to InSb or InAs bulk and thin-film devices. The larger bandgap of GaAs permits device operation up to 250°C, while corresponding devices in Si stop working at about 150°C. Also, the temperature coefficient of GaAs (bulk or implanted) Hall plates seems to be slightly lower than that of comparable silicon devices, although this point may need further clarification. Minority exclusion by using n^+n or p^+p junctions has been proposed recently[34] as a way to extend the temperature range of operation for silicon Hall sensors up to 300°C.

As we saw above, the SNR of the Hall voltage with respect to $1/f$ noise is proportional to the mobility. More important, however, is the control of flicker noise by the Hooge parameter α. This parameter can generally be defined in terms of the squared relative fluctuation of a two-pole of resistance R,

$$\langle \Delta R^2/R^2 \rangle = (\alpha/n)(\Delta f/f), \tag{54}$$

where n denotes the total number of free electrons (assuming n-type material) in the resistance block. Depending on material, technology, and geometry, values of α between 10^{-3} and 10^{-9} have been reported.[3,12] For Hall elements, appropriate choice of material, processing technology, and design geometry (e.g., a buried Hall layer reminiscent of the junction field effect transistor) that reduces the Hooge parameter can provide a much larger improvement in SNR than that achievable by a material with higher mobility. At high frequencies, thermal noise dominates and the corresponding SNR is proportional to the mobility, thus, favoring high-mobility materials.

Silicon offers the unique advantage of inexpensive batch-fabrication and allows the integration of basic sensor elements together with support and signal-processing circuitry in IC technologies of established reliability, such as bipolar or CMOS technology. Indeed, a large variety of integrated silicon magnetic sensors have been investigated following the design rules of IC processes offered by custom-chip manufacturers or university laboratories. Using silicon IC technology, *in situ* compensation of nonlinearity, offset, and temperature dependence can be attempted by appropriate circuitry. Integrated silicon Hall devices are currently sold in large quantities by a number of component manufacturers. The ever-advancing silicon IC technologies and, in particular, the proliferation of the BiCMOS process, will continue to offer further sensor-design opportunities.

Magnetic-field-sensor development outside established mainstream IC technologies has to face the extra cost of developing specific manufacturing technologies and tools for mass production as well as appropriate test and reliability procedures. This development cost may surpass the financial resources of small and medium size companies. Such investment may become justifiable when the applicability of silicon magnetic sensors can be clearly ruled out as in the case of higher operating temperature or resolution requirements. Integrated GaAs magnetic sensors seem to be the next proper choice if operating temperatures above 150°C are required. Successful examples of GaAs Hall IC chips have indeed been described.[30,35]

The constituent parts of the well-established bipolar and CMOS technologies include: computer-aided design (CAD) with pertinent CAD tools; device models for the simulation of device and circuitry functions; mask and chip fabrication; assembly, including packaging; and test and quality procedures. By definition, the integration of semiconductor magnetic sensors means a design of such devices for a given IC chip fabrication process as offered by an IC chip manufacturer. Several nonstandard procedures are required by magnetic-sensor integration. Device structures unconventional by the standards of IC designers, but compatible with the fabrication process, have to be designed. Adequate CAD tools must include device models that allow the simulation of magnetic sensors. Unconventional packaging that minimizes mechanical stress and avoids ferromagnetic materials is crucial. Characterization and test procedures must include the specific sensor functions, for example by comparison with a calibrated magnetic probe. For component reliability, the mature quality of the IC manufacturing process can usually be taken for granted. This is another important advantage offered by integrated semiconductor magnetic sensors.

Advanced III–V semiconductor technologies are beginning to make their impact on magnetic sensors. By means of molecular beam epitaxy (MBE), Hall devices have been fabricated which use the 2DEG confined to a very thin (about 10 nm) layer close to a heterojunction as the active sensor layer. The

AlGaAs/GaAs heterojunction and the AlAs/GaAs heterojunction in a modulation-doped AlAs/GaAs superlattice structure have been investigated for this purpose.[31,36] A quite different technique is the maskless implantation of Si^{2+} ions into semi-insulating GaAs using focused ion-beam technology. It has been employed in the fabrication of miniature Hall elements with active regions of submicron dimensions.[37] This technique requires a minimum number of process steps, provides accurate pattern definition (crucial, e.g., for minimizing offset) and allows variation in the implanted profile to optimize magnetic sensitivity and device impedance for on-chip integration with circuitry.

High temperature superconductors such as ceramic Y-Ba-Cu-O are outside the realm of semiconductors, but should be mentioned as a further advanced technology for magnetoresistive microsensors operating at the temperature of liquid nitrogen.

5.2.5 Advanced Galvanomagnetic Effects

A more accurate analysis of the galvanomagnetic effects is based on the current transport equation including the full transport coefficients resulting from the relaxation-time approximation of the Boltzmann equation.[12,23,24] Here we have room only to compile some key results. Let us consider the case of uniform electron concentration ($\nabla n = 0$) in an n-type semiconductor. The current density under magnetic induction $\boldsymbol{B}$ is

$$\boldsymbol{J}_n(\boldsymbol{B}) = q^2 K_1 \boldsymbol{\mathscr{E}} + (q^3/m^*) K_2 \boldsymbol{B} \times \boldsymbol{\mathscr{E}} + (q^4/m^{*2}) K_3 (\boldsymbol{B} \cdot \boldsymbol{\mathscr{E}}) \boldsymbol{B}. \tag{55}$$

This is the more general expression comprising Eq. 8 as a special approximation. The kinetic coefficients K_1, K_2, and K_3 are energy-weighted averages over the term $\tau^s[1+(\mu_n B)^2]^{-1}$ with energy-dependent collision time τ and drift mobility $\mu_n = q\tau/m^*$,

$$K_s = (n/m^*)\langle \tau^s[1+(\mu_n B)^2]^{-1} \rangle \qquad s = 1, 2, 3. \tag{56}$$

Here m^* denotes the effective mass of the electron and n the electron density. For zero magnetic induction, we have

$$K_s(0) = (n/m^*)\langle \tau^s \rangle. \tag{57}$$

We readily recognize the first term as Ohm's law, since

$$q^2 K_1(0) = qn(q\langle \tau \rangle/m^*) = qn\mu_n = \sigma_n. \tag{58}$$

For the magnetic induction perpendicular to the electric field, $\boldsymbol{B} \cdot \boldsymbol{\mathscr{E}} = 0$, Eq. 55 leads to an expression similar to Eq. 10,

$$\boldsymbol{J}_n(\boldsymbol{B}) = \sigma_n(B)\boldsymbol{\mathscr{E}} + \sigma_n(B)\mu_n^*(B)\boldsymbol{B} \times \boldsymbol{\mathscr{E}}, \tag{59}$$

but in contrast to Eq. 10, the first term on the right-hand side of Eq. 59 exhibits a more general electrical conducticity $\sigma_n(B)$ which now includes the (transversal) physical magnetoresistance effect. This effect is distinct from the geometrical magnetoresistance Eq. 19. Moreover, the generalized Hall coefficient

$$R_H(B) = \mu_n^*(B)/\sigma_n(B) \tag{60}$$

involved in the resulting electrical field

$$\mathscr{E} = \mathbf{J}_n/\sigma_n(B) - R_H(B)\mathbf{B} \times \mathscr{E} \tag{61}$$

can also depend on the magnetic induction. For weak fields B with $(\mu_n B)^2 \ll 1$, we use the series expansion

$$[1 + (\mu_n B)^2]^{-1} = \sum_{s=0}^{\infty} (-1)^s (\mu_n B)^{2s}. \tag{62}$$

With $\tau = m^* \mu_n/q$ this leads to the corresponding expansion for the kinetic coefficients. Keeping only the lowest terms in $\mu_n B$, we obtain the following approximations valid for nondegenerate carriers:

$$K_1 \approx (n/q)[\langle \mu_n \rangle + \langle \mu_n^3 \rangle B^2] \tag{63}$$

$$K_2 \approx (nm^*/q^2)[\langle \mu_n^2 \rangle + \langle \mu_n^4 \rangle B^2] \tag{64}$$

$$K_3 \approx (nm^{*2}/q^3)[\langle \mu_n^3 \rangle + \langle \mu_n^5 \rangle B^2]. \tag{65}$$

An analogous expansion exists for high magnetic fields with $(\mu_n B)^2 \gg 1$.

For various collision processes the mean free transit time is proportional to some power p of the electron energy. For example, for semiconductors with spherical constant energy surface, $p = -\frac{1}{2}$ for phonon scattering and $p = \frac{3}{2}$ for ionized impurity scattering. The calculation of the sth power of the mobility averaged over the electron energy distribution yields

$$\langle \mu_n^s \rangle \propto (q/m^*)(kT)^{ps} \Gamma(\tfrac{5}{2} + sp)/\Gamma(\tfrac{5}{2}), \tag{66}$$

where Γ denotes Euler's gamma function. From Eq. 66, the scattering coefficients of order s,

$$r_s = \langle \tau^s \rangle / \langle \tau \rangle^s = \langle \mu_n^s \rangle / \langle \mu_n \rangle^s = \Gamma(\tfrac{5}{2} + sp)[\Gamma(\tfrac{5}{2})]^{s-1}[\Gamma(\tfrac{5}{2} + p)]^{-s}, \tag{67}$$

are readily obtained. For $s = 2$ we find an expression for the scattering coefficient introduced in Eq. 5,

$$r_n = r_2 = \langle \tau^2 \rangle / \langle \tau \rangle^2 = \langle \mu_n^2 \rangle / \langle \mu_n \rangle^2 = \Gamma(\tfrac{5}{2} + 2p)\Gamma(\tfrac{5}{2})[\Gamma(\tfrac{5}{2} + p)]^{-2}. \tag{68}$$

The temperature dependence of r_n is crucial for the temperature coefficient of the current-related sensitivity; see Eqs. 50 and 51. Measurements of r_n versus temperature with low-doped n-type silicon result in a decrease with lower temperature down to 100 K, with $r_n = 1.15 \pm 0.3$ at 300 K and $r_n = 0.92 \pm 0.3$ at 100 K.

Comparing Eqs. 55 and 59, we recognize that the Hall mobility can be expressed in terms of the kinetic coefficients as

$$\mu_n^*(B) = (q/m^*) K_2/K_1. \tag{69}$$

For low magnetic induction and nondegenerate electrons, Eqs. 63 and 64 lead to

$$\mu_n^*(B) \approx \mu_n^*(0)[1 - \beta(\mu_n B)^2], \tag{70}$$

where $\beta = r_4/r_2 - r_3$ is the nonlinearity coefficient of the Hall mobility in terms of the scattering coefficients, Eqs. 67 and 68. This affects the Hall coefficient and contributes the nonlinear component due to the material itself, Eq. 53.

So far we have assumed the simple model of a single energy minimum in an energy band with isotropic effective mass. This is inadequate for silicon, whose conduction band shows several energy minima or valleys with anisotropic effective mass. The anisotropy leads to transversal and longitudinal mobilities, μ_T and μ_L, effective masses, and relaxation times. The anisotropic scattering factor for low induction can be approximated by $r_n = 3\mu_T(2\mu_L + \mu_T)(\mu_L + 2\mu_T)^{-2}$. In comparison with Eqs. 5 or 58 based on the average drift mobility $\mu_n = (2\mu_T + \mu_L)/3$, this leads to a 13% reduction of the scattering coefficient.[12]

Finally, we identify the correction factor K introduced in Eq. 8. By comparison with Eq. 55 we find $K(\mu_n^*)^2 = (q^4/m^{*2})K_3$. This term is related to the case where the current density is collinear with the electric field and, thus, describes the longitudinal magnetoresistance effect. For nondegenerate electrons in an isotropic band and low induction, the approximation $K(\mu_n^*)^2 \approx \sigma_n \mu_n^2 r_3$ is found.[12]

Piezoresistance and the Piezo-Hall Effect. These occur upon application of mechanical forces. The former is the change of the electrical resistivity, $\Delta\rho/\rho$, and the latter the change of the Hall coefficient $\Delta R_H/R_H$. Piezoresistive material is characterized by the piezoresistive coefficient Π occurring in the expression $\Delta\rho/\rho = \Pi X$ with X denoting the mechanical stress. Likewise, the piezo-Hall effect is described by $\Delta R_H/R_H = \mathscr{P} X$ with $\mathscr{P}$ denoting the piezo-Hall coefficient. In silicon, longitudinal and transverse piezoresistive and piezo-Hall coefficients can be distinguished. The coefficients depend on the crystal orientation and the direction of the current-density vector. For (100) silicon plates, the piezoresistive coefficients are minimal if the current is parallel to the (010) direction.[12]

Thermomagnetic Effects. These effects involve electric current, magnetic induction, and heat flow.[38] As a side effect of the Hall voltage, a temperature

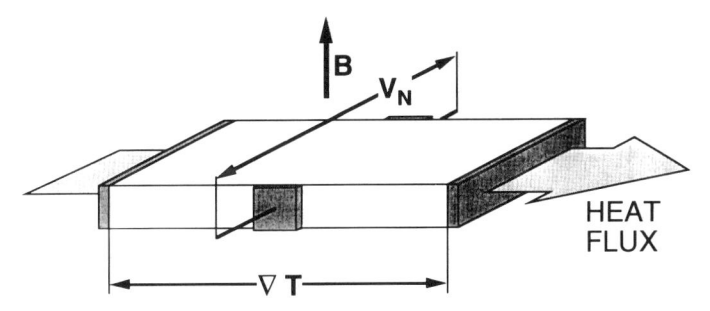

Fig. 10 Ettinghausen–Nernst effect producing the Nernst voltage V_N across a semiconductor plate with heat flow exposed to a perpendicular magnetic induction.

gradient $\nabla T \propto J \times B$ occurs across the Hall plate carrying the current density J, where ∇T is parallel to the Hall field and perpendicular to the applied fields $\mathscr{E}$ and B. This is called the Ettinghausen effect. Another example is the Ettinghausen–Nernst effect, also known as the isothermal Nernst effect, where an electric field $\mathscr{E}_N \propto \nabla T \times B$ is generated perpendicular to a heat flow (caused by a corresponding temperature gradient) and a magnetic induction, as shown in Fig. 10. Thus, a temperature gradient parallel to the current density in a Hall plate produces a Nernst voltage V_N superimposed on the Hall voltage. This Nernst voltage appears as a thermally generated offset. This effect is unlikely to trouble integrated magnetic sensors, since it is small in extrinsic material, bulk silicon is a good thermal conductor, and the sensors are small in size.

A thermal counterpart of the Hall plate is the infrared radiation sensor based on the Ettinghausen–Nernst plate.[39] A thin bismuth-tin layer with $p \approx n$ produces a signal of about 1 mV/W-T when exposed to a magnetic field and a temperature gradient between the radiation absorbing surface and a heat sink. The Ettinghausen–Nernst coefficient of n-doped polysilicon made by CMOS IC technology has been measured recently.[40] A thin polysilicon beam structure of 200 μm length is subjected to a temperature gradient of 100 K. Under 1 T magnetic induction this produces a Nernst voltage of about 1 μV across the beam width of 30 μm. This results in an Ettinghausen–Nernst coefficient of approximately 7×10^{-8} V/T-K.

5.3 INTEGRATED HALL SENSORS

In this section we present integrated bulk and interface Hall devices and their Lorentz counterparts. Before discussing the Hall sensors, we first compile a list of integrated semiconductor magnetic-sensor structures with their common abbreviations. We distinguish two groups of structures by their response to a magnetic-induction vector either perpendicular or parallel to the chip surface. The section, where the specific type of sensor is described, is shown in parentheses.

***B* Perpendicular to Chip Surface:**

- Integrated Bulk Hall plates (5.3)
- Differential amplification magnetic sensor DAMS (5.3)
- Magnetic field effect transitor MAGFET (5.3)
- Magnetic heterojunction device (5.3)
- Lateral magnetotransistor LMT (5.4)
- Carrier domain magnetometer CDM (5.5)

***B* Parallel to Chip Surface:**

- Vertical Hall device VHD (5.3)
- Vertical magnetotransistor VMT (5.4)
- Lateral magnetotransistor LMT (5.4)
- Magnetodiode MD (5.5)

5.3.1 Integrated Bulk Hall Devices

The bulk Hall plate of the type shown in Fig. 7 can be readily incorporated into a silicon integrated circuit (IC) chip, as was proposed by Bosch more than two decades ago.[2] Figure 11 shows a horizontally integrated Hall plate realized as part of the *n*-type epitaxial (collector) layer grown on a *p*-type substrate in a silicon bipolar IC process. The lateral boundary of the plate is defined by a deep *p*-diffusion, that is the usual isolation diffusion. The current and the sensor contacts (CC and SC) are made by n^+-diffusions, which correspond to the usual emitter diffusions and are used as intermediate layers providing ohmic contact to the metal layer. The isolation of the Hall plate from the rest of the chip is

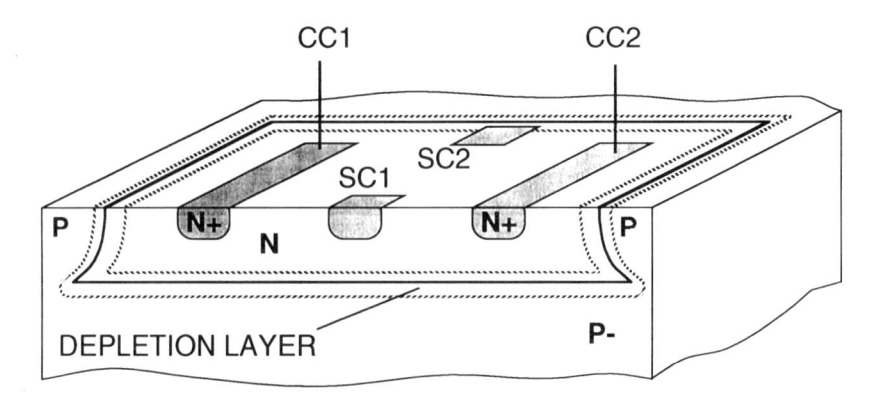

Fig. 11 Structure of integrated Hall plate in bipolar IC technology. The symbols are explained in the text. (After Ref. 3)

achieved by reverse-biasing the p–n junction surrounding the plate. Typical integrated Hall plates are 400 μm long and 200 μm wide.

Preferably, n-type material is used as the active region in integrated Hall devices because of its higher voltage-related sensitivity, Eq. 35. Incidentally, the typical epitaxial-layer doping and thickness used in bipolar IC technology are nearly optimal for Hall plates. Typical values are $N_D \approx 10^{15}$ cm^{-3} to 10^{16} cm^{-3} and $t = 5$ μm to 10 μm. Assuming that all dopants are ionized, these values lead to an nt product (see Eq. 16) in the range 5×10^{11} to 10^{13} cm^{-2}. If the doping density in the epitaxial layer is insufficient for the Hall plate, it can be upgraded by ion implantation. In that case, the carrier density $n(z)$ depends on the depth z, and the nt value must be replaced by an effective value determined by the integral over $n(z)$.

The upper surface of the integrated Hall plate is usually covered by silicon dioxide ("oxide"). The oxide–silicon interface involves carrier scattering, which is detrimental to the Hall device performance. This effect is avoided in a refined, buried structure, where the top of the integrated Hall plate is covered with a shallow p-layer that shields the carriers from the interface.[9,12]

The junction field effect is another problem of junction-isolated Hall plates. Because of this effect, the width of the depletion layer of the p–n junction depends on the reverse bias. The effective thickness of the integrated Hall plate is, therefore, smaller than the metallurgical thickness and depends on the bias voltage. As a consequence, the Hall voltage and the corresponding voltage drop across the plate depend on the applied junction bias. Due to the voltage drop, the effective junction bias is position-dependent and so is the effective plate thickness. This effect results in a modulation of the device thickness by the Hall voltage and, hence, in a nonlinearity of the sensor response.

Thus, the junction field effect is a further source of nonlinearity besides the material and geometric nonlinearities discussed in Section 5.2.3. In this situation the equations presented in Section 5.2.2 describing Hall plates with uniform thickness are not rigorously valid. However, in a bulk Hall plate with an nt product of more than 10^{12} cm^{-2}, the junction field effect is rather weak and these equations are good approximations. There is an obvious trade-off between sensitivity and linearity. A low nt product favors high sensitivity, but produces a strong junction field effect. High-sensitivity junction-isolated Hall sensors can show a pronounced junction field effect, which may be the dominant nonlinear effect (as much as 1% nonlinearity at $B = 1$ T). It can be compensated by a circuit using the very same effect, by making the bias of the isolation diffusion dependent on the Hall voltage so that the effective plate thickness remains almost constant.[29]

We recall that the nonlinearity due to the material as well as the geometrical nonlinearity are proportional to the square of the magnetic induction, but have opposite signs (see Section 5.2.3). Cancelation of these two effects can be attempted by choosing a material whose nonlinearity coefficient matches that of the geometrical nonlinearity coefficients. Another way to compensate the geometrical nonlinearity is to load the Hall device output with a resistor.[12]

Offset reduction is another important issue. Laser trimming of an external resistor is often used for offset adjustment. Prior to such adjustment, equivalent offset magnetic induction of 10 to 100 mT is not unusual, mainly due to mechanical stress and geometric tolerances. In order to avoid costly trimming procedures, on-chip circuits for offset cancelation have been devised. One offset reduction concept exploits the matching properties of integrated devices. In Hall devices with a shape symmetrical with respect to rotation by $\pi/2$, current and sensor contacts can be interchanged (e.g., in a symmetrical Hall cross). Two such Hall devices of identical design integrated side by side on a chip have very similar properties. When they are interconnected so that the resulting Hall voltages are parallel, but the two current directions orthogonal, their asymmetries tend to cancel and the offset is reduced accordingly.[13] An interesting new offset reduction method[41] for a single integrated Hall plate uses a plate shape with 16 contacts that is symmetrical with respect to rotation by $\pi/8$. The direction of the current is made to spin by contact commutation. Averaging of the consecutive Hall voltages reduces the offset by a factor 100.

Commercial general purpose Hall devices are available from several companies including Honeywell, Siemens, Sprague Electric, and Texas Instruments.[7] Besides the integrated Hall sensor, these silicon Hall ICs contain on-chip amplification and stabilization circuitry. Further commercial products are Hall switches for applications where only a binary output signal is needed. These devices also contain a Schmitt trigger to control an output stage. A picture of a Hall IC chip is shown in Fig. 12. The photo shows a section of a whole wafer with alignment marks and scribe lines which define each chip. In the center of the chip we distinguish four diamond-shaped Hall plates, which are probably connected in a bridge configuration and are surrounded by biasing and signal-conditioning circuitry.

GaAs IC technology has also been used for the fabrication of Hall devices. The electron mobility of GaAs is more than five times higher than that of silicon. More important, GaAs integrated circuits can operate at higher temperatures than silicon ICs. The active Hall-plate region can be made in an epitaxial layer or by ion implantation.[30,35] The cross-shaped active layer of a commercially available GaAs Hall device[30] has an effective thickness of about 0.4 μm. It is formed by local implantation of silicon ions into a substrate of semi-insulating GaAs. For another implanted GaAs Hall cross, a nonlinearity of 300 ppm at room temperature for magnetic induction below 1 T has been achieved.[35]

A submicron Hall device has been made by maskless implantation of silicon (Si^{2+}) ions into semi-insulating GaAs using a focused ion beam.[37] Apart from its application potential in the semiconductor IC industry, maskless ion implantation is a promising sensor technology, particularly for the design of magnetic sensors with high spatial resolution. The fabricated submicron Hall device has the shape of a cross, whose center is the active sensor region. The cross bars are approximately 0.3 μm in diameter with a sheet carrier concentration of the order of 10^{13} cm^{-2}. The relative sensitivity of the device

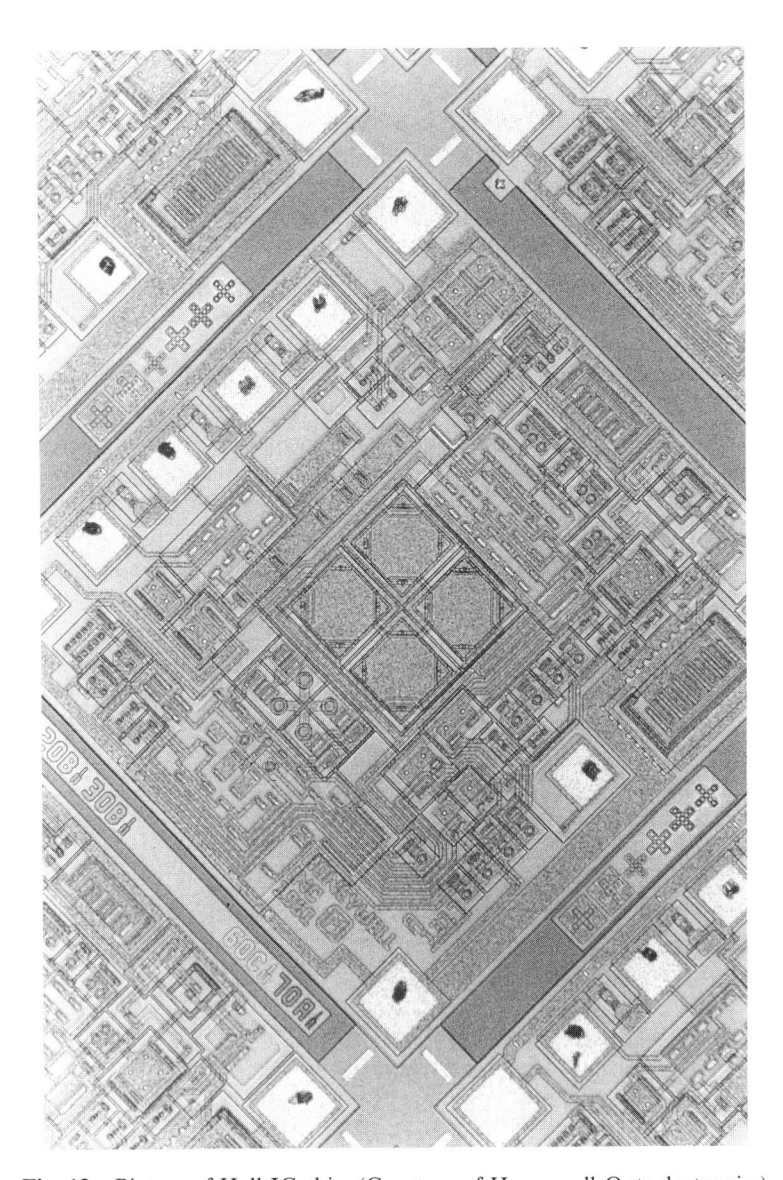

Fig. 12 Picture of Hall IC chip. (Courtesy of Honeywell Optoelectronics)

is below 100 V/A-T. The minimum width of the implanted region is limited by the diameter of the focused ion beam.

Vertical Hall Devices. All Hall sensors described so far have the form of a plate merged in the chip surface. Thus, they are sensitive to a magnetic-induction vector perpendicular to the chip surface and could be referred to as "horizontal" Hall devices. On the other hand, integrated Hall sensors responding to a magnetic-induction vector parallel to the chip plane are preferred for some applications. Such an integrated vertical Hall device (VHD) was proposed by Popovic in 1984.[42] Its structure is illustrated in Fig. 13. Understanding of its operating principle is eased by the modeling results[43] shown in Fig. 14. The general plate shape is chosen so that all necessary contacts are located at the top of the chip surface. The vertical plate geometry is defined by a deep ring-shaped p-diffusion isolation. The active device region is part of the n-type substrate material. It is reminiscent of a plate placed vertically with respect to the chip surface.

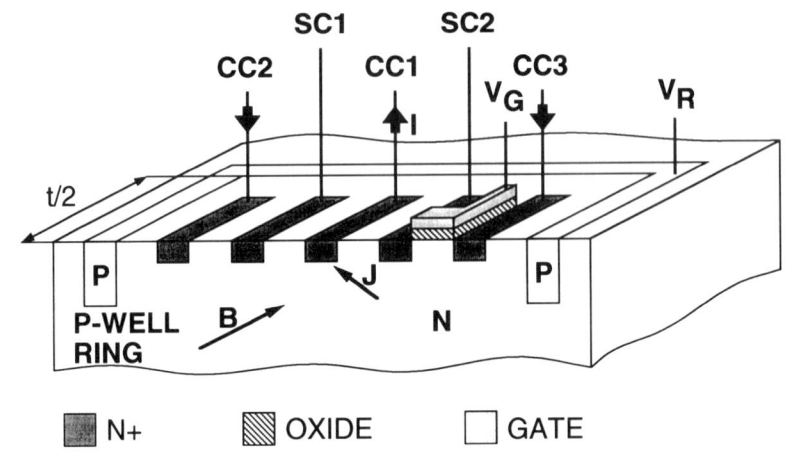

Fig. 13 Structure of a vertical Hall device (VHD) in CMOS technology. The symbols are explained in the text. (After Ref. 3)

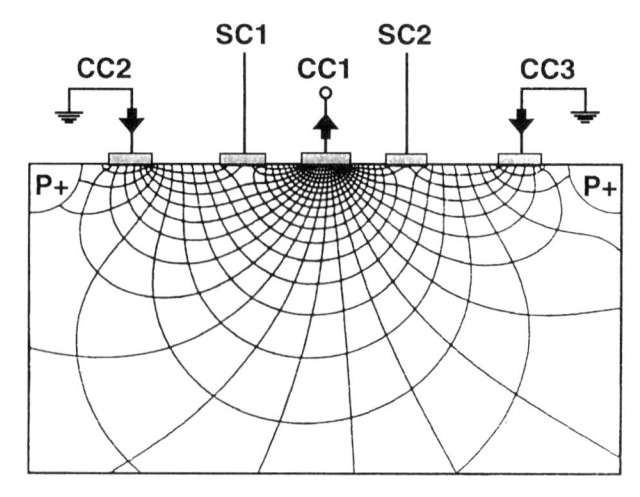

Fig. 14 Operation of a vertical Hall device (VHD). Current and equipotential lines are shown for $\mu_n^* B = 0.21$. The sumbols are explained in the text. (After Ref. 3)

The operation of the VHD is demonstrated in Fig. 14 showing its cross section with the current and equipotential lines for $\mu_n^* B = 0.21$ resulting from two-dimensional numerical modeling. In Fig. 14, the current lines connect the central current contact CC1 to the two grounded current contacts CC2 and CC3. The equipotential lines are approximately normal to the current lines. The asymmetry of the lines is due to the magnetic induction vector pointing perpendicular to the drawing plane. A Hall-type potential difference can be recognized in Fig. 14 between the two sensor contacts SC1 and SC2 because of the asymmetry of the equipotential lines.

An early version of this device was fabricated using standard bulk CMOS technology, where the isolation ring was realized by a *p*-well diffusion depth of 12 μm.[42] Apart from the bottom, the boundaries of the vertical Hall plate are defined by the reverse-biased *p–n* junction ring and the silicon/silicon-dioxide interface at the chip surface. The doping $N_D \approx 10^{15}$ cm^{-3} of the substrate and the effective width $t \approx 12$ μm of the active Hall plate region result in an *nt* product of 1.2×10^{12} cm^{-2}. As we would expect from the discussion following Eq. 30, the operation of the VHD is essentially unaffected by the unusual geometry. Its current-related sensitivity is about 600 V/A-T. In a later version of the VHD, a shallow *p*-layer was used to obtain a buried structure.[12] An advanced VHD chip, assembled between magnetic flux concentrators, has become part of a commercial product of Landis and Gyr.[9] Popovic's VHD is probably the most important single innovation in integrated silicon magnetic sensors over the last decade.

Differential Amplification Magnetic Sensor (DAMS). The DAMS, devised by Takamiya and Fujikawa,[44] merges a horizontal *n*-type Hall plate with two bipolar *p–n–p* transistors forming part of a differential amplifier.[3,7,12] The device cross-section is shown in Fig. 15. The *n*-doped top silicon layer simultaneously acts as a Hall plate and a shared base region of the two transistors. Two

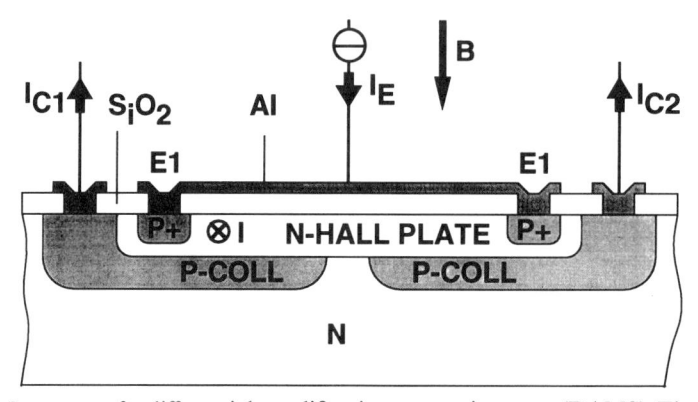

Fig. 15 Structure of a differential amplification magnetic sensor (DAMS). The symbols are explained in the text. (After Ref. 3)

p^+-emitters, E1 and E2, are diffused into the Hall plate instead of the usual n^+-sensor contacts. Two separate p-collectors underneath the Hall plate almost isolate it from the n-substrate. The direction of the bias current applied to the Hall plate is perpendicular to the chip surface, the Hall voltage appears across the base region, the Hall effect being negligible in the heavily doped p^+-emitter region. If the two emitters are kept at the same potential, the Hall voltage acts as the differential emitter–base voltage of the transistor pair. Under proper bias conditions, this results in a corresponding collector-current difference, which can be converted into a final voltage difference by load resistors.

The DAMS operation can be interpreted in terms of the minority-carrier injection being modulated by the Hall field produced by a majority-carrier current in the base region. This is similar to the injection modulation occurring in certain magnetotransistors to be discussed in Section 5.4.

5.3.2 Modeling of Hall Devices

Device modeling reduces the amount of trial-and-error and provides unique insight into the device interior not easily accessible to experiments, as illustrated by the current and equipotential distributions shown in Figs. 3–5 and 14. Simple analytical models such as Hall field and Lorentz deflection (see Sections 5.2.1 and 5.2.2) are the first choice and may serve as heuristic design tools. The validity of such models is, however, limited to special geometry, material, and operating conditions. For instance, magnetic-sensor modeling can be reduced to the Laplace equation for the electric potential only in the case of uniform magnetic induction and uniform extrinsic plate material. In more general situations, we have to solve the partial differential equations that describe the carrier transport in semiconductors in the presence of a magnetic field. Magnetic sensors require at least two-dimensional modeling in view of the vector character of the magnetic induction. Milestones of numerical magnetic sensor modeling were the modeling of magnetoconcentration[27] (Fig. 6), the vertical Hall device[43] (Fig. 14), horizontal Hall plates exposed to nonuniform induction[18,19] (this section), and magnetotransistors[19-22] (Section 5.4).

The system of partial differential equations describing semiconductors in the presence of magnetic induction includes the Poisson equation

$$\mathrm{div}(\varepsilon_s \mathrm{\ grad\ } \psi) = q(n - p + N_A - N_D) \tag{71}$$

and the continuity equations[23]

$$\mathrm{div\ } \boldsymbol{J}_n(\boldsymbol{B}) - q\ \partial n/\partial t = qR \tag{72}$$

$$\mathrm{div\ } \boldsymbol{J}_p(\boldsymbol{B}) + q\ \partial p/\partial t = -qR. \tag{73}$$

Here, ε_s denotes the material permittivity and ψ the electrostatic potential, N_D and N_A denote the fully ionized donor and acceptor impurity distributions, and R stands for the net generation–recombination rate of electrons and holes. The electron and hole current densities $J_n(B)$ and $J_p(B)$ depend on the magnetic induction B and are given by the current transport relations (cf. Eq. 7, but with $B \cdot J_n(0) = 0$)

$$J_n(B) = [J_n(0) + \mu_n^* B \times J_n(0)][1 + (\mu_n^* B)^2]^{-1} \quad (74)$$

$$J_p(B) = -[J_p(0) + \mu_p^* B \times J_p(0)][1 + (\mu_p^* B)^2]^{-1}. \quad (75)$$

Here, $J_n(0)$ is given by Eq. 2, $J_p(0) = \sigma_p \mathcal{E} - qD_p \nabla p$ with D_p denoting the hole diffusion coefficient. Under the assumption of nondegenerate carrier statistics, n and p may be expressed in terms of the electric potential ψ and the quasi-Fermi potentials φ_n and φ_p as

$$n = n_{ien} \exp[q(\psi - \varphi_n)/kT] \quad (76)$$

$$p = n_{iep} \exp[q(\varphi_p - \psi)/kT], \quad (77)$$

where T is the lattice temperature and n_{ien} and n_{iep} are the effective intrinsic numbers for electrons and holes. The effective intrinsic numbers differ from the usual intrinsic concentration n_i by factor which account for high-injection effects (Fermi–Dirac statistics) and the position dependence of the band edges due to, for example, bandgap narrowing.

Equations 71–73 have to be solved subject to proper boundary conditions. The electric potential and carrier densities at ideal ohmic contacts are not affected by the magnetic induction and are prescribed by the Dirichlet conditions[22]

$$\psi = V_a + (kT/q) \ln[(N/(2n_{ien})] \quad (78)$$

$$\psi = V_a - (kT/q) \ln[N/(2n_{iep})] \quad (79)$$

$$n = (N^2/4 + n_{ien} n_{iep})^{1/2} + N/2 \quad (80)$$

$$p = (N^2/4 + n_{ien} n_{iep})^{1/2} - N/2 \quad (81)$$

resulting from the assumption of thermal equilibrium and charge neutrality along the contact. Here, V_a denotes the applied voltage and $N = N_D - N_A$ the net ionized impurity concentration. The boundary conditions along noncontacted outer semiconductor surfaces is affected by the magnetic field. The current densities at insulating boundaries are determined by the recombination at the interface. For the technicalities of the numerical solution we refer to the original literature.[18–22] Only a few selected results are presented here and in Sections 5.3.3 and 5.4.3.

Figures 16 and 17 show equipotential and current line distributions computed for horizontal sensors operating in the Hall and Lorentz mode, respectively. The current lines connect the upper and lower current contacts and the equipotential lines end at the left- and right-hand edges of the structures. The

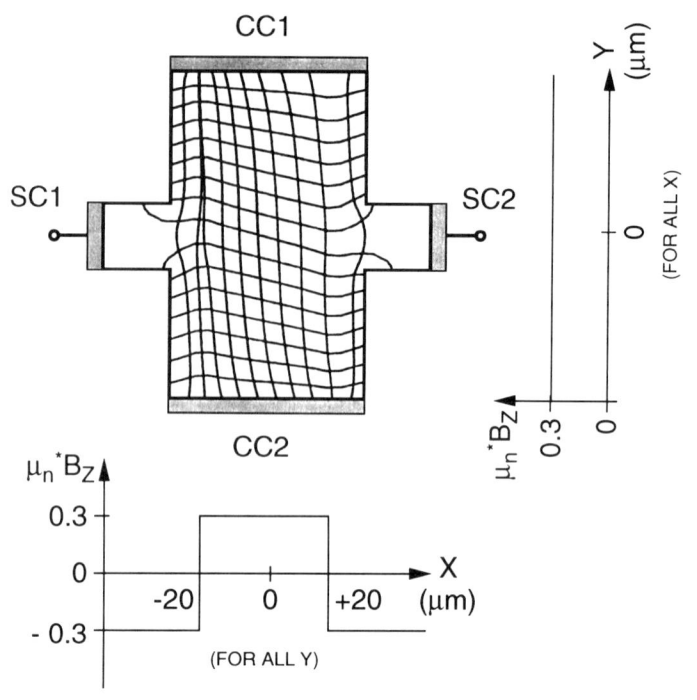

Fig. 16 Equipotential and current lines (Hall mode) for a Hall-cross detecting a longitudinal strip domain. The inserts show the spatial distribution of the nonuniform $\mu_n^* B(x, y)$. (After Ref. 18)

nonuniform magnetic induction applied perpendicular to the device surfaces is plotted on both figures as a function of $\mu_n^* B_z$ versus x and y, respectively.[18] The magnetic induction is defined by a one-dimensional, locally-inverted strip domain. In Fig. 16 the strip domain has a constant value in the y-direction of the current flow between the two current contacts CC1 and CC2, whereas the product $\mu_n^* B_z$ changes stepwise from -0.3 to 0.3 in the x direction between the two sensor contacts SC1 and SC2. We call this domain, therefore, longitudinal. Figure 17 shows the situation of a transversal domain. The magnetic induction is constant along the equipotential lines (x direction) and changes stepwise in direction of the current (y direction). A distribution reminiscent of the Hall field with the current flow more or less parallel to the inversion boundaries is observed in Fig. 16 for the Hall plate exposed to the longitudinal domain. In the case of the transversal domain applied to the plate operating in the Lorentz mode, Fig. 17 shows current lines that are skewed locally in a way reminiscent of the Lorentz deflection.

5.3.3 Magnetic Field-Effect Transistors (MAGFET)

In 1966, Gallagher and Corak recognized that the surface inversion layer or channel of a MOSFET (metal-oxide-semiconductor field-effect transistor) can

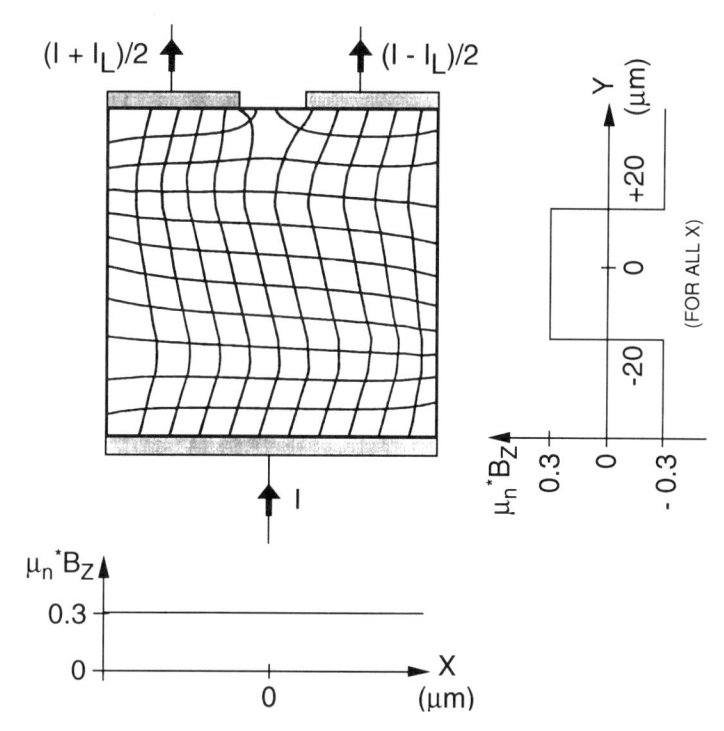

Fig. 17 Equipotential and current lines (Lorentz mode) for a split-electrode device exposed to a transverse strip domain. The inserts show the spatial distribution of the nonuniform $\mu_n^* B(x, y)$. (After Ref. 18)

be used as the active region of a Hall sensor.[45] Such a sensor is compatible with MOS bias and signal-conditioning circuitry. The acronym MAGFET is used to refer to the family of MOSFET magnetic sensors. These devices exploit the Hall effect or the Lorentz deflection of carriers in the inversion layer. The most serious drawback of the MAGFET is the $1/f$ noise, which is usually several orders of magnitude higher than that of a comparable bulk device. Further disadvantages are the low channel mobility (about half of the bulk mobility) and possible surface instability.

Hall MAGFET. Figure 18 shows the structure of a rectangular *n*-channel Hall MAGFET.[3] The channel (Ch) under the gate G is used as an extremely thin Hall plate. The channel charge density is $Q_{ch} = C_{OX}(V_G - V_T)$ with the gate-oxide capacitance per area C_{OX}, the gate voltage V_G, and threshold voltage V_T. Source (S) and drain (D) act as current contacts. Two additional n^+-diffusions (SC1 and SC2) provide the sensing contacts, which are fabricated together with the source and drain. The channel length is denoted by L and its width by W. The position of the sensor contacts is characterized by the distance y from the source region. A MOSFET working in the linear region (drain voltage $V_D \ll V_G - V_T$)

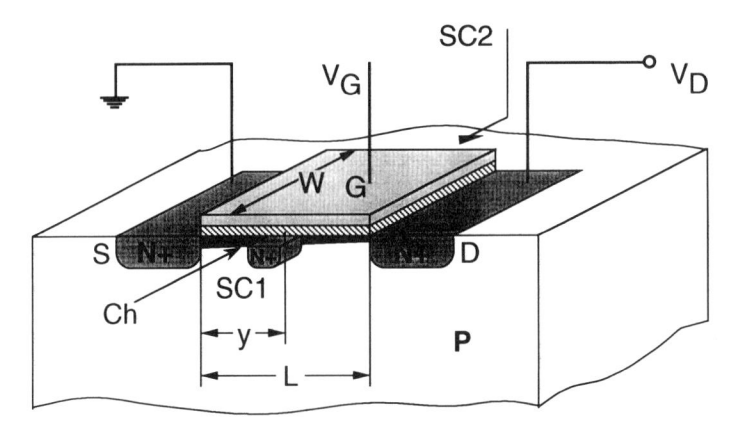

Fig. 18 Structure of a Hall MAGFET. The symbols are explained in the text. (After Ref. 3)

is similar to the bulk Hall plate described by Eq. 24. We only have to replace the product qnt by the charge density Q_{ch}, the bias current I by the drain current I_D, and the bulk Hall scattering factor r_n by the channel scattering factor r_{nch}. Then, the absolute value of the Hall voltage of the Hall MAGFET is

$$V_H = GI_D Br_{nch}/Q_{ch}. \tag{82}$$

The n-channel Hall scattering factor r_{nch} ranges between 1.1 and 1.4, depending on the operating conditions (gate voltage V_G).[4] By analogy with Eq. 33 the current-related sensitivity of the "linear" Hall MAGFET is given by

$$S_I = Gr_{nch}/Q_{ch} = Gr_{nch}/(V_G - V_T). \tag{83}$$

A current-related sensitivity of several hundred V/A-T is easily achieved, but the supply-voltage-related relative sensitivity V_H/BV_D is proportional to the channel Hall mobility μ_{nch}^* and is, hence, lower than that of bulk Hall plates.[3] Moreover, linear operating conditions limit the drain voltage and current to rather small values and accordingly small Hall voltage. It is, therefore, preferable to operate the Hall MAGFET slightly below pinch-off or even in saturation. The Hall voltage increases with larger y/L because the channel is thinner and the resistivity higher in the vicinity of the drain region. On the other hand, offset due to alignment tolerances increases as y approaches L. The practical optimum position y of the sensor contacts is between $0.7L$ and $0.8L$. With respect to shape, the optimal Hall MAGFET is characterized by $W \approx 1.2L$ and is shorter than the optimal bulk Hall plate.

An exemplary Hall MAGFET[46] has been designed for a noncontact keyboard switch. The n-channel depletion type Hall device is made by polysilicon-gate NMOS technology. Offset due to piezoresistance is reduced by annealing. Offset, due to sensor-contact misalignment, is kept small by choosing a large device size with $L=600$ μm. Empirical values for optimal device parameters and operating conditions are as follows: $W=1.2L$, $y=0.7L$, and $V_G=V_D=5$ V, resulting in $I_D=0.5$ mA and a 640 V/A T current-related sensitivity with about 14 mT offset-equivalent magnetic induction.

Dual-Drain MAGFET. The MOSFET can also be modified for Lorentz-current rather than Hall-voltage operation (cf. Fig. 7). To this end, two separate drain regions are introduced, which share the drain current. This leads to the dual-drain or split-drain MOSFET. Its structure is shown in Fig. 19 with the two drains labeled D1 and D2. A magnetic induction vector perpendicular to the inversion layer produces a current imbalance $\Delta I_D = I_{D1} - I_{D2}$ between the two drains. By analogy with Eq. 18, in the linear MOSFET range with the drains kept at the same drain voltage, V_D, the current difference is

$$\Delta I_D = G\mu_{nch}^*(L/W)BI_D. \tag{84}$$

Here, I_D denotes the sum of the two drain currents at zero magnetic induction and G the geometric correction factor of the device, where $G=1$ for $L \ll W$ and a narrow separation between the two drains. Modeling results[3,47] for the distribution of current, potential, and surface charge are shown in Figs. 20 and 21. Figure 20 shows equipotential and current line density distributions for a split-drain MOSFET. Current lines are drawn from the source S to the drains $D1$ and $D2$, respectively. The channel length is L and its width is W. The current line density is increased on drain $D2$ with respect to drain $D1$ in the presence of a magnetic induction perpendicular to the drawing plane. Figure 21 shows

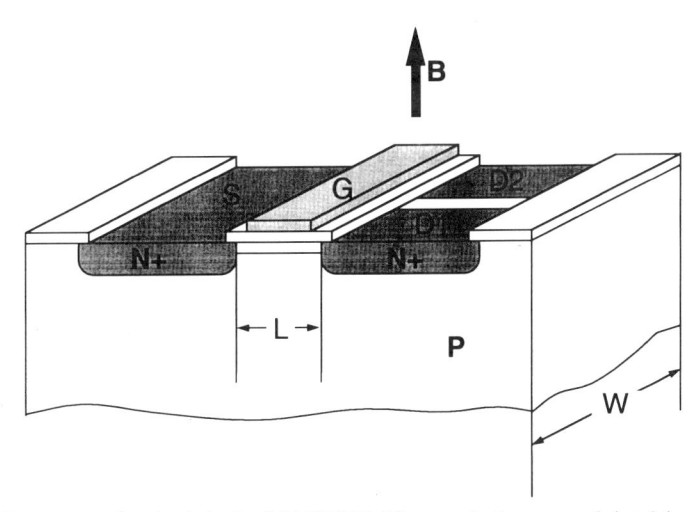

Fig. 19 Structure of a dual-drain MAGFET. The symbols are explained in the text. (After Ref. 3)

244 MAGNETIC SENSORS

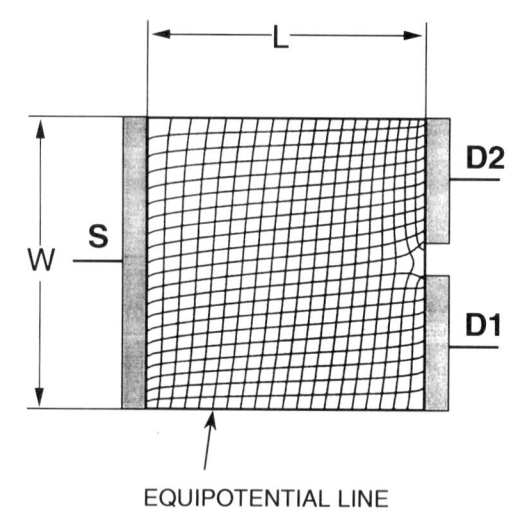

Fig. 20 Operation of the dual-drain MAGFET in the linear range. Equipotential and current lines are shown. Current lines ending at the left-hand source and right-hand split drain are deflected towards the top, resulting in an increased current in the upper drain at expense of the lower drain. (After Ref. 3)

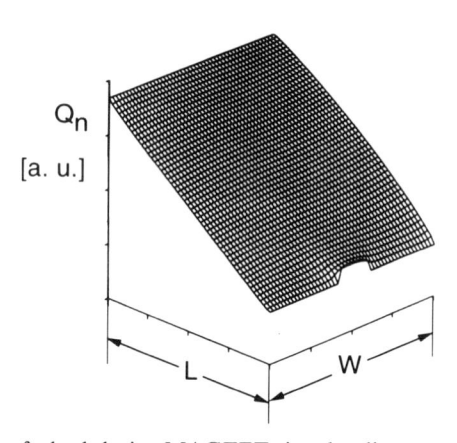

Fig. 21 Operation of dual-drain MAGFET in the linear range. Surface charge distribution was obtained from numerical modeling. (After Ref. 3)

the distribution of electrons along the channel. The electron density is high in the vicinity of the source and decreases along the channel towards the two drains. The separation of the two drains can easily be distinguished.

The sensitivity of the dual-drain device is usually defined as the relative drain current imbalance per magnetic induction,

$$S = |\Delta I / I_D B|. \tag{85}$$

In silicon MAGFETs, S is typically[48] about 0.05/T. The maximum sensitivity

$$S_{max} = 0.74 \mu_{nch}^* \qquad (86)$$

is expected in the limit of a wide, short channel, $W \gg L$, and a narrow separation between the two drains, in analogy to Eqs. 29 and 36. From sensitivity measurements and modeling of the factor G, the channel Hall mobility μ_{nch}^* and, hence, the Hall scattering coefficient r_{nch} can be determined.[4,48] For a variety of device geometries and operating conditions that ensure strong inversion in the linear regime, μ_{nch}^* has been found to range between 0.6 and 0.85 m^2/V-s.

Measurements of the $1/f$ noise spectral density of dual-drain MAGFETs reveal a strong anticorrelation between the two drain currents, since the noise spectral density of the differential signal $I_{D1} - I_{D2}$ is larger than that of the single-ended signals I_{D1} or I_{D2}.[49] The drain-current output signal of dual-drain MAGFETs can be efficiently converted to a voltage or frequency by co-integration with CMOS circuitry. A pair of complementary dual-drain MOSFETs in a CMOS differential amplifier-like configuration has been demonstrated to produce a voltage response of 1.2 V/T with a 10 V supply and 100 μA current consumption.[3,12,50] A dual-drain MAGFET integrated with a current-controlled oscillator forms a magnetically-controlled oscillator (MCO). Its frequency varies linearly with magnetic induction around the center frequency of 3.29 kHz with a modulation of 230 Hz/T.[51]

5.3.4 Magnetic Heterojunction Devices

Cross-shaped Hall devices have been fabricated by means of molecular beam epitaxy (MBE).[4,31] Their active region is a very thin rectangular layer ($t \approx 0.01$ μm, $L = 346$ μm, $W = 200$ μm) located at an (AlGa)As/GaAs heterojunction containing a two-dimensional electron gas (2DEG). At liquid nitrogen temperature (77 K), the electron mobility is as high as 15 m^2/V-s. Thus, the voltage-related sensitivity can reach about 7/T = 0.7%/mT. The current-related sensitivity, however, is only about 1000 V/A-T. According to Eq. 33, it is determined by the nt product and cannot be increased by a higher mobility. The sheet carrier concentration (i.e., the effective nt product) is 5.8×10^{11} cm^{-2} at 77 K.

At room temperature, the electron mobility is 0.72 m^2/V-s, with a sheet carrier concentration of 4.9×10^{11} cm^{-2}, and a current-related sensitivity 1000 V/A-T. Around room temperature, the mobility exhibits a T$^{-2.4}$ temperature dependence and the temperature coefficient of the sensitivity is about -0.7%/K. From noise measurements for the Al$_{0.3}$Ga$_{0.7}$As/GaAs heterojunction Hall element, a minimum detectable magnetic field of about 2 nT at 1 kHz is claimed.[31]

The sensor performance is improved by using an AlAs/GaAs heterojunction in a modulation-doped AlAs/GaAs superlattice structure. A current-related sensitivity of 1200 V/A-T with only $-0.1\%/\text{K}$ temperature coefficient is achieved in this way.[36] The Lorentz-current operation for this type of active region has been investigated by designing a split-contact structure reminiscent of the dual-drain MAGFET.[4] The electron mobility of the underlying active region is 0.64 m^2/V-s at room temperature and leads to the sensivitity $S = 48\%/\text{T}$ with S defined analogously to Eq. 85. This is one order of magnitude larger than the sensitivity of a comparable dual-drain silicon MAGFET with only about 0.07 m^2/V-s n-channel Hall mobility, as one would expect from Eqs. 84 to 86.

5.4 MAGNETOTRANSISTORS

Magnetotransistors (MTs) are bipolar transistors designed in such a way that the collector current is modulated by a magnetic field. Depending on the MT geometry, magnetic fields parallel or perpendicular to the chip plane can be detected. Most MTs have a dual-collector structure. At zero magnetic field, their operation is symmetrical with respect to the two collectors, which show identical currents, $I_{C1} = I_{C2}$. In the presence of a magnetic field, the Lorentz force creates an asymmetry in the potential and current distribution, which translates into a collector current imbalance, $\Delta I = I_{C1} - I_{C2}$. The influence of magnetic fields on transistors was first studied[28] as early as 1949. A wide variety of transistors designed for the detection of magnetic fields have been proposed over the last 30 years.[3] While intrinsically bipolar in function, some types of MTs can also be made using CMOS technology.

The concept of the dual-collector MT is reminiscent not only of the Lorentz operation of the dual-drain MAGFET (Section 5.3.3), but also of the Hall-voltage operation of the DAMS (Section 5.3.1). From these diverse roots, two competing MT models have been proposed: (1) Lorentz deflection: the Lorentz force deflects minority carriers towards one collector and away from the other collector; (2) Injection modulation: the magnetic induction acting on the majority carriers moving in the base region creates a Hall voltage, which modulates the emitter-base voltage and, thus, creates an asymmetry in the minority-carrier injection. Moreover, magnetoconcentration (Section 5.2.1) is thought to be involved in the MT operation. Which effect plays a major or minor role, depends on the MT structure and operating conditions.

The MT family tree is shown in Fig. 22. The terms vertical MT (VMT) and lateral MT (LMT) are used ambiguously in the literature. Here, we use them with respect to the prevailing direction of motion of the minority carriers in the base at zero magnetic field, not with respect to the direction of the magnetic-induction vector. A selection of fundamental MT structures is presented in Section 5.4.1. MTs designed for a given IC process, their noise properties, and their circuitry are the topics of Section 5.4.2. In Section 5.4.3 we discuss MT models in the light of numerical simulation results.

5.4 MAGNETOTRANSISTORS   247

```
                        MAGNETOTRANSISTORS
                    ┌───────────┴───────────┐
                1 D PROBES              VECTOR PROBES
              ┌─────┴─────┐             ┌─────┴─────┐
             B⊥          B∥           2 D, B∥        3 D
              │      ┌────┼────┐
             LMT    VMT  LMT  SSIMT
```

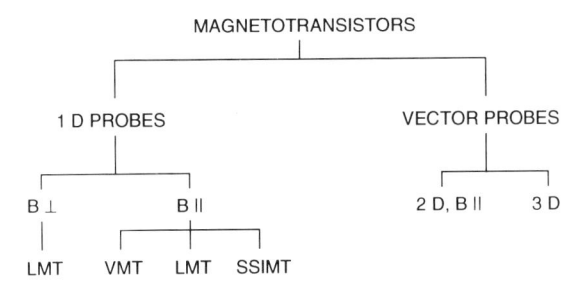

Fig. 22 Family tree of magnetotransistors. One-dimensional (1D) probes measure the magnitude of the magnetic field perpendicular ($B_\perp$) or parallel ($B_\parallel$) to the chip plane. Vector probes measure the direction of the magnetic-induction vector in two (2D) or three (3D) dimensions. The SSIMT is the suppressed sidewall injection magnetotransistor.

5.4.1 Vertical and Lateral Magnetotransistors

In this section we describe the basic VMT and LMT structures. A classification of the many versions of each kind, differing in geometry and operating conditions, is found in a recent review.[53] All of these can be fabricated using either bipolar or CMOS IC technologies.

Vertical Magnetotransistors. Figure 23 shows an example of a VMT.[54] It was fabricated using a bipolar IC process and consists of two n–p–n transistors coupled by a common emitter and a common base. The vertical device geometry is imposed by the two buried layers reaching underneath the base region. They are connected to the two lateral collector contacts by deep n^+ diffusions.

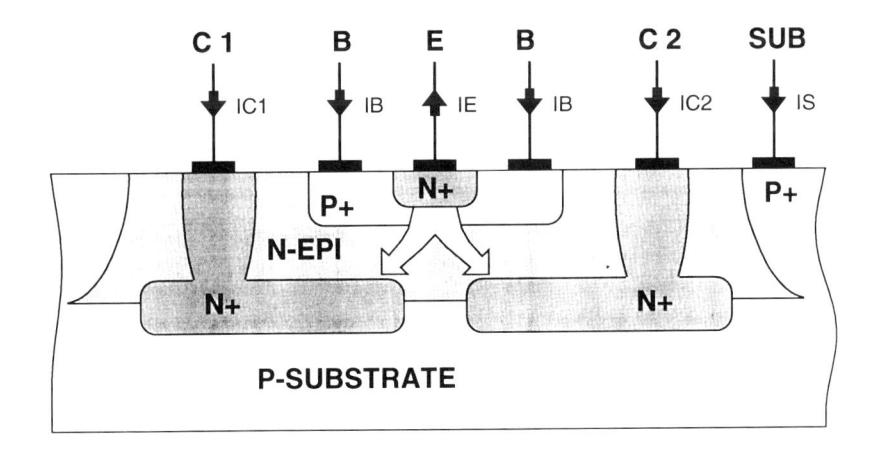

Fig. 23 Dual-collector vertical magnetotransistor (VMT). (After Ref. 3)

Short-circuiting of the two collectors is avoided by the gap between the buried layers. This gap defines the ohmic coupling of the two collectors through a high-resistivity path in the epi layer. The magnetic induction B has to be perpendicular to the minority carrier flow and, thus, parallel to the chip plane. The Lorentz force must have a direction, such that the device symmetry is disturbed. Hence, B has to be perpendicular to the drawing plane.

The minority carriers are injected from the highly-doped emitter region, pass through a shallow base region, and reach the lightly-doped epitaxial collector layer, where they become majority carriers. Here, the current path is split into two portions, each of which reaches one of the two buried layers. With zero magnetic field and perfectly symmetric device structure, both paths carry the same amount of collector current, $I_{C1} = I_{C2} = I_{C0}/2$, with I_{C0} denoting the total collector current at zero magnetic induction.

Let us now switch on the magnetic field and assume the Lorentz deflection model. The Lorentz force deflects the injected carriers in the base and the subsequent epi layer causing a collector-current imbalance $\Delta I_C = I_{C1} - I_{C2}$. Because of the short path in the base region, we have to deal mainly with the deflection of majority carriers in the epi layer. This model is reminiscent of the dual-drain MAGFET discussed in Section 5.3.2. Therefore, the expression for the collector current difference is analogous to Eq. 18 and is

$$\Delta I_C = G\mu_n^*(L/W_E)I_{C0}B. \tag{87}$$

Here, L denotes the emitter–collector distance, W_E the emitter width, and G a geometry factor analogous to Eqs. 18 and 84. The relative sensitivitity S can be defined in analogy with Eq. 85,

$$S = |\Delta I_C/I_{C0}B|. \tag{88}$$

By inserting Eq. 87 in Eq. 88 we obtain

$$S = G\mu_n^*(L/W_E). \tag{89}$$

In general, however, the absolute and the relative MT sensitivities, S_A and S, are defined as derivatives at zero magnetic induction,

$$S_A \equiv |\partial V_C/\partial B|_{B=0} = R_C^{-1}|\partial I_C/\partial B|_{B=0} \tag{90}$$

$$S \equiv V_C^{-1}|\partial V_C/\partial B|_{B=0} = I_C^{-1}|\partial I_C/\partial B|_{B=0}. \tag{91}$$

The right-hand side of Eq. 91 holds only if the collector differential resistance $R_C = \partial V_C/\partial I_C$ is much higher than the load resistance.[3]

Values for the relative sensitivity in this device range from 0.03/T to 0.05/T, which are similar to the values found for the dual-drain MAGFET. These results, however, are 2.5 times smaller than predicted by Eq. 89. This discrepancy may be explained as follows. Equation 89 does not hold for all values of L/W_E because the electron spread invalidates the underlying model. In view of the assumptions basic to Eq. 18, we would expect Lorentz deflection to prevail only if $L < W_E$.

Lateral Magnetotransistors. Lateral magnetotransistors (LMTs) are characterized by a large base region where the minority carriers flow laterally from the emitter to the collectors. In contrast to the above VMT, we have to deal mainly with the deflection of minority carriers in the base region. Figure 24 shows a lateral MT structure, the so-called drift-aided lateral dual-collector *p–n–p* transistor fabricated using a bipolar IC process.[11] The emitter and the two collectors are embedded in the epitaxial layer, which serves as the base region. The two n^+ base contacts are used to create an accelerating field across the base region. Due to the accelerating voltage, most minority carriers injected from the emitter are directed towards the two collectors and only a small amount flows into the substrate. This device is sensitive to magnetic fields perpendicular to the chip surface.

An LMT sensitive to a magnetic field parallel to the chip surface, as shown in Fig. 25, can be made by CMOS IC technology.[3] In this case the transistor is of the *n–p–n* type. Emitter and collector are embedded in the *p*-well, which serves as the base region. This device has only one collector and uses the substrate as a second collector. The minority carriers flowing laterally through the base region are deflected either towards the collector or towards the substrate, depending on the direction of the Lorentz force. Thus, the ratio I_C/I_S between collector current and substrate current I_S is modulated by the magnetic field. Sensitivity, as defined by Eq. 88, is between 0.07/T and 1.5/T.

Suppressed-Sidewall-Injection MT. The suppressed-sidewall-injection MT (SSIMT) is a refined version of the LMT.[55] Figure 26 shows its cross-section

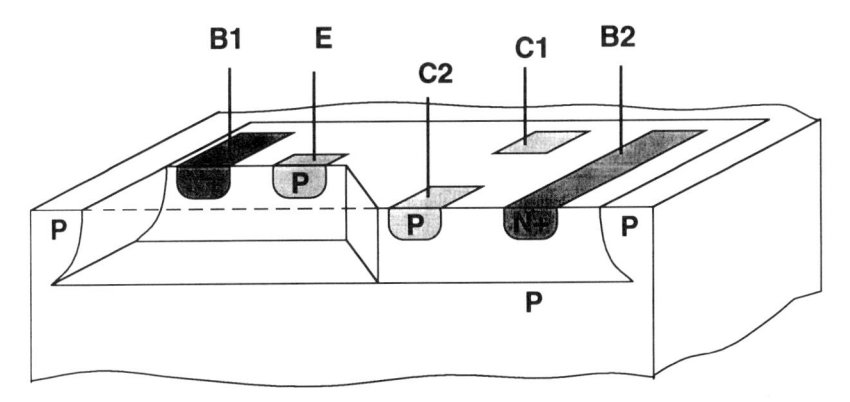

Fig. 24 Dual-collector lateral magnetotransistor (LMT) in bipolar IC technology, sensitive to field perpendicular to the chip surface.

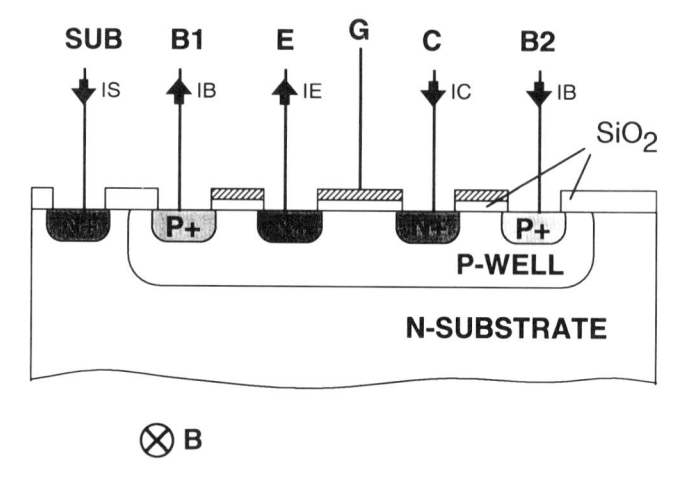

Fig. 25 Dual-collector lateral magnetotransistor (LMT) sensitive to a field parallel to the chip surface. (After Ref. 3)

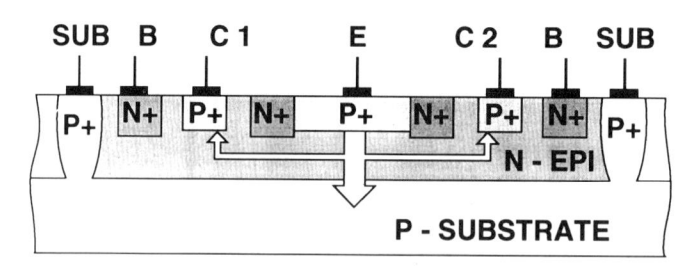

Fig. 26 Suppressed sidewall injection magnetotransistor (SSIMT) in bipolar IC technology with sensor and substrate current for zero magnetic induction.

as designed for bipolar technology. Two p–n–p transistors share the same emitter and base. The collectors are located at opposite sides of the emitter. The base region is defined by the epi layer. A highly-doped n^+ guard ring surrounding the emitter prevents the lateral injection of minority carriers from the emitter into the base. This makes the device more sensitive to magnetic fields parallel to the chip surface and perpendicular to the drawing plane. By reverse biasing the emitter–guard junction, the carrier injection can be confined to the center of the emitter–base junction. An accelerating field is simultaneously formed between the guard and the base contacts. All these effects boost the magnetic response.

In terms of Lorentz deflection, an intuitive explanation of the SSIMT operation can be attempted as follows. Minority carriers (holes) are injected from the emitter into the base region. Here, the current flow splits into three parts. At zero magnetic induction, an equal amount of minority carriers is collected by the left- and right-hand side collector. Other carriers are collected by the substrate, which acts as an additional collector. With a magnetic field, the following double deflection occurs. The substrate current is deflected by the Lorentz force. Depending on the field direction, it contributes to the left- or right-hand collector current. The laterally-flowing collector currents are deflected towards the substrate or the collector. Both types of deflection cooperate. Sensitivities up to 1.5/T have been found for such p–n–p structures fabricated in bipolar technology. Up to 30/T have been reported for the corresponding n–p–n device made by CMOS technology. The latter SSIMT consists of two n–p–n transistors, where the n^+-doped emitter and collectors are embedded in a p-well, serving as the base region.

Sensitivitity Revisited. All LMT structures are driven at high base currents of several mA; their current gain β is very small, $\beta \ll 1$, for three reasons:

1. The large effective base width exceeds the diffusion length of the minority carriers and fewer carriers reach the collector.
2. The substrate acts as an additional collector. The vertical dimension of the LMT is defined by the thickness of the base region, which is much smaller than its lateral extension. Thus, a vertical transistor between emitter, base, and substrate is much more efficient than a lateral one.
3. In the case of the SSIMT, the highly-doped guard ring around the emitter reduces the emitter efficiency and, hence, the current gain.[22]

The reported high relative sensitivities of the LMT and, in particular, the SSIMT appear to be futile, since Eq. 88 neglects the substrate current. Sensitivity values above $1/T$ and the large sensitivity spread can be attributed to the magnetic modulation of the ratio of the substrate current and collector current. Thus, we have to revise the figure of merit and replace the relative sensitivity S of Eq. 88 by the supply-current related sensitivity

$$S_T \equiv I_T^{-1} |\partial I_C / \partial B|_{B=0} \tag{92}$$

with $I_T = I_E - I_B = I_C + I_S$ denoting the total supply current of the transistor. Equation 88 is still appropriate for the VMT of Fig. 23 with current gain $\beta \gg 1$ and negligible substrate current I_S, so that $I_C \simeq I_E$ and $S \simeq S_I$. But Eq. 92 is preferable for LMTs with $\beta \ll 1$ and $S_T \ll S$. In terms of S_T, the sensitivity values of LMTs and VMTs are of the same order. The substrate current is further discussed in Section 5.4.2.

Vector Probes. A unique feature of MTs is the possibility to detect two, or even all three, components of the magnetic field by using several VMT or LMT structures in an appropriate design geometry. For example, a two-dimensional vector probe can be designed by choosing some MT sensitive to a field parallel to the chip surface. Then the same structure is duplicated and rotated by 90 degrees with respect to the first one. With the VMT of Fig. 23 this procedure leads to a four-collector VMT.

5.4.2 Noise, Substrate Current, and Circuits

Low-frequency noise ($f < 1$ MHz) limits the detection of the magnetic field. For a single MT collector, say collector no. 1, with average output-signal current I_{C1} the related signal-to-noise ratio is

$$\text{SNR} = I_{C1} \bigg/ \left[\int_{BW} S_{11}(f)\, df \right]^{1/2} \tag{93}$$

where $S_{11}(f)$ denotes the power spectral density of the collector-current fluctuation. We recall that $S_{11}(f)$ is defined as the Fourier transform of the time autocorrelation function $C_{11}(t) = \langle I_{C1}(t') I_{C1}(t'+t) \rangle$ of the fluctuating collector current $I_{C1}(t)$ with the brackets $\langle \rangle$ denoting the average over the time t'. The integral is performed over the bandwidth (BW) of interest. For dual-collector MTs, Eq. 93 translates into

$$\text{SNR} = \Delta I_C \bigg/ \left[\int_{BW} S_D(f)\, df \right]^{1/2}, \tag{94}$$

where ΔI_C denotes the average collector-current imbalance and $S_D(f)$ the differential current power spectral density. The latter is the Fourier transform of the time autocorrelation of the fluctuating current difference $\Delta I_C(t) = I_{C1}(t) - I_{C2}(t)$. From Eqs. 88 and 94 we can express the magnetic resolution in terms of the minimum detectable induction $B_{\min}$ estimated by taking SNR = 1:

$$B_{\min} = \left[\int_{BW} S_D(f)\, df \right]^{1/2} \bigg/ I_{C0} S. \tag{95}$$

Types of noise sources found in bipolar MTs are thermal or Nyquist noise, shot noise, $1/f$ or flicker noise, and generation–recombination (GR) noise. For thermal noise, we express Eq. 42 in terms of current and obtain the spectral density

$$S_{IT}(f) = 4kT/R(f). \tag{96}$$

The resistance $R(f)$ is constant at low frequencies. When a current flows across the emitter–base or the base–collector junction it exhibits shot noise, whose power spectral density is

$$S_{IS}(f) = 2qI. \tag{97}$$

Here, I denotes the current passing through the junction. The most important noise source at low frequencies is $1/f$ noise. Its spectral density depends on the Hooge parameter α (Sections 5.2.3 and 5.2.4) and is proportional to the square of the current. GR noise may be relevant in LMTs with a large base region. All these types of noise contribute to the power spectral density at the MT output.

Noise Correlation. In dual-collector devices the amount of common information hidden in the fluctuations at each collector can be measured in terms of the degree of correlation or coherence $\gamma_{12}(f)$ defined as the normalized cross-spectral density between the two output currents,

$$\gamma_{12}(f) \equiv S_{12}(f)/[S_{11}(f)S_{22}(f)]^{1/2}. \tag{98}$$

The cross spectral density $S_{12}(f)$ is the Fourier transform of the cross correlation function $C_{12}(t) = \langle I_{C1}(t')I_{C2}(t'+t) \rangle$ of the two collector noise currents, while $S_{11}(f)$ and $S_{22}(f)$ are the single power spectral densities at each collector (cf. Eq. 93). With this definition the degree of correlation ranges between 1 (perfect correlation) and -1 (anticorrelation). The value 0 corresponds to no correlation at all, that is the noise of two output currents is statistically independent. Values between 0 and 1 correspond to some partial correlation. In view of the symmetry of most dual-collector devices, we assume that both power spectral densities are approximately equal, $S_{11}(f) = S_{22}(f)$. Then the degree of coherence can be expressed in terms of the differential power spectral density S_D:

$$\gamma_{12}(f) = 1 - 0.5[S_D(f)/S_{11}(f)]. \tag{99}$$

Since $\gamma_{12}(f)$ ranges between -1 and 1, $S_D(f)$ tends to zero when $\gamma_{12}(f)$ approaches 1. For $\gamma_{12}(f) \approx -1$ we find $S_D(f) \approx 4S_{11}(f)$. By substituting Eq. 99 into Eq. 95 we express the minimum detectable magnetic induction in terms of $\gamma_{12}(f)$,

$$B_{\min} = \left[2 \int_{BW} (1 - \gamma_{12}(f))S_{11}(f)\, df \right]^{1/2} \bigg/ I_{C0}S. \tag{100}$$

The degree of correlation has been measured for various MT structures and operating conditions.[56] The single power spectral densities always show a $1/f$ type dependence for frequencies below 100 kHz. A strong positive correlation, $\gamma \approx 1$, is found between 1 Hz and 100 kHz for VMTs operated in the active regime below high injection. Thus, we would expect that the noise nearly cancels if we register the difference of both signals. Indeed, the differential power spectral density is lowered by more than 60 dB with respect to the single power spectra of the collectors. This yields a very large SNR even if the sensitivity of these devices is low. A resolution of 189 nT/Hz$^{1/2}$ at 10 Hz has been reported.[56]

The strong positive correlation is lost in SSIMT structures operated at large base currents and low current gain. The differential power spectral density becomes comparable to that of a single collector (partial correlation) or can even be twice as high (zero correlation). The measured resolution for a typical SSIMT made in CMOS technology is 14 μT/Hz$^{1/2}$.

Thus, there is a trade-off between high sensitivity and high resolution. The high-resolution VMTs produce small output signals and, therefore, the subsequent amplifier stage rather than the MT may become the limiting factor for magnetic resolution. The LMTs show large output signals at the expense of large base and substrate currents, which deteriorate the noise performance. Not only is the correlation lost, but also the single power spectral densities are increased because of the current dependence of the noise.

Reduction of Substrate Current. Different ways to reduce the substrate current in SSIMTs have been investigated.[57] In bipolar IC technology, a buried layer can be used to suppress the vertical transistor formed by emitter, base, and substrate. This structure increases the ratio between collector and substrate current by more than one order of magnitude. The sensitivity, however, decreases by an order of magnitude, because the high doping of the buried layer forms a low resistive path in the base region reducing the voltage drop and, hence, the current deflection.

Another possibility is to shorten the lateral base width to increase the collector current at the expense of the substrate current. Indeed, the current gain, and the collector current are inversely proportional to the square of the base width,[23] but the sensitivity decreases linearly with decreasing lateral dimensions, as one would expect from Eq. 89. Thus, optimization for both high sensitivity and low substrate current seems to be futile. Still, by shrinking the device using minimum design rules one keeps a reasonable sensitivity while increasing I_C/I_S. A further attempt to prevent the presence of the vertical transistor is to introduce a buried insulating SiO_2 layer.[58]

MT Circuitry. Low total current (and, hence, power consumption) is not only beneficial for the SNR, but also eases its cointegration with biasing and signal-conditioning circuitry. In contrast to the obvious integration of the MT together with a conventional amplifier, an alternative strategy uses the "active" function of the bipolar transistors inherent in all MT structures. The sensor element becomes part of the amplifying circuit itself, where it works as the first input stage. The sensor thus amplifies its own signal; the resulting circuit is called a merged amplifier. Merged structures require less total supply current and produce lower input equivalent noise for a given supply current.[3]

An example is the so-called magneto-operational amplifier (MOP),[11] whose block diagram is shown in Fig. 27. It is a feedback amplifier similar to a conventional operational amplifier (OA) and contains a differential amplifier stage including an SSIMT as magnetic detector, a high-gain amplifier, an output stage, and a temperature compensation circuit. Like the OA, the MOP can

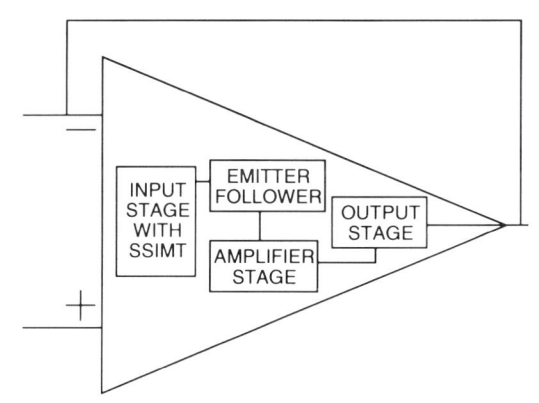

Fig. 27 Block diagram of the magneto-operational amplifier (MOP).

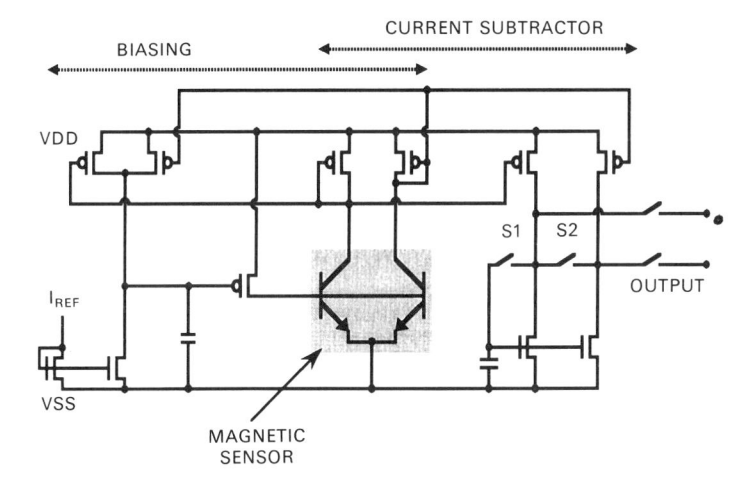

Fig. 28 Schematic of the biasing and signal-conditioning circuit for SSIMT in CMOS technology. (After Ref. 60)

perform diverse operations; it can serve as a linear magnetic sensor or magnetic switch or filter. The system sensivitity is 96 mV/T.

A magnetic-field dependent frequency output can be obtained by merging the MT with a current-controlled multivibrator.[59] The MT is part of the multivibrator whose oscillation frequency is modulated by the MT collector current. The circuit includes an SSIMT, an adaptor circuit to reduce the large bias currents present in the SSIMT, a current-steering circuit, and a Schmitt trigger as a threshold circuit. The oscillation frequency depends linearly in the collector current, which can range between $1\,\mu A$ and $100\,\mu A$. The center frequency ($B=0$) of the oscillator can be adjusted by an external capacitor.

Cointegration of analog and digital circuitry is attractive in CMOS technology. Immediate conversion into a digital signal is advisable for small and noisy signals. The circuit shown in Fig. 28 biases the MT, subtracts the collector currents, and feeds the difference into a differential pair that serves as the input stage of a Sigma-Delta A/D (analog-to-digital) converter.[60] Two current mirrors replicate the collector currents on a current mirror composed of matched transistors, a capacitor, and two switches S_1 and S_2. The circuit is externally clocked at a frequency of 5 MHz. It works in two phases so that the differential component of the sensor output currents is extracted and the common mode is eliminated. The total collector current is defined externally to allow MT operation at different biasing conditions. The circuit then produces the required base drive current. Convergence is achieved through frequency compensation of the circuit.

5.4.3 Modeling of Magnetotransistors

In Section 5.3.2 we assessed the validity of simple analytical models for Hall devices and the benefits of numerical modeling. There are even more, and sometimes controversial, analytical models of the MT operation, which reflect the intricacies of the bipolar action in the presence of a magnetic field. Numerical modeling of MTs requires a large computational effort in at least two dimensions and has been broached only recently.[19-22] The calculations are based on Eqs. 71 to 75 with proper boundary conditions. The results allow us to associate MT operating principles with device structures.

Figure 29 shows the center of the VMT structure with the n^+ emitter diffused inside the p^+ base region. The p–n junctions present in the structure are defined by dark gray lines. The underlying n-epi layer acts as collector region. Two n^+ collectors (not shown) are located at each side of the structure. There is no buried layer. The shape of the base region is due to the emitter push effect.[26] When the device is operated in the active mode, a considerable amount of hole (majority) current flows from the base contacts, B, towards the emitter. Due to the large extension of the emitter and the bottleneck shape of the base profile,

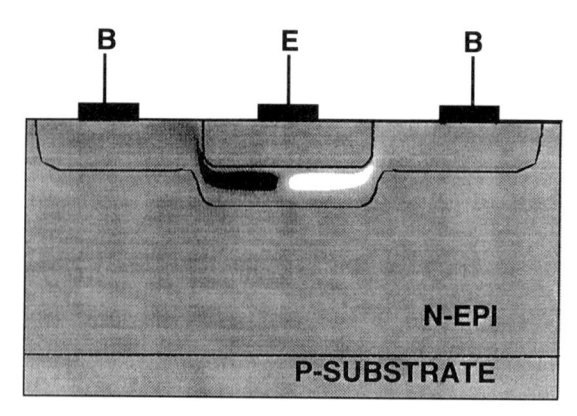

Fig. 29 Numerical simulation of the VMT. Shades are explained in the text. (After Ref. 22)

the hole current is concentrated at the edges of the emitter. Therefore, the minority-carrier injection is split into a left- and right-hand side component already at the emitter–base junction, in contrast to the assumption made in Section 5.4.1.

For zero magnetic field the symmetry of the device geometry leads to a symmetric distribution of the currents. With a magnetic field perpendicular to the drawing plane, the hole currents flowing from the base contacts are affected by the Lorentz force pushing holes up towards the emitter–base junction or dragging them down. This effect slightly increases the hole current beneath the one half of the emitter and decreases it beneath the other. This modulation of the hole current appears in Fig. 29, where the difference in hole current density between $B = 0.5$ T and $B = 0$ T is displayed. Gray shading means no change, white means maximum increase and black means maximum decrease in hole current density.

The emitter responds to the asymmetry in the hole current density along the emitter–base junction by injecting more electrons at the right part of the base than into the left. This leads to an asymmetric current flow in the collector region and finally an increased right- and decreased left-hand side collector current. The asymmetry of the current flow is enhanced by the action of majority-carrier deflection in the collector region, where the carrier transport is governed by drift in the electric field. This effect is, however, partly canceled by the Hall field built-up across the epi layer.

Thus, for the VMT of Fig. 29, the emitter-injection modulation is confirmed by the simulation, but its interpretation is somewhat different from the previous intuitive concept. It is not the (small) Hall voltage across the emitter–base junction that causes the asymmetric minority-carrier injection, but rather the large lateral extension of the emitter and the peculiar shape of the base region. The magnetic sensor action occurs in a small portion of the device; this explains the relatively low sensivitity.

From a corresponding simulation we can gain insight into the operation of the SSIMT of Fig. 26. Figure 30 shows the difference in minority-carrier (hole) current density between 0.5 T and zero magnetic induction. The p–n junctions appearing in this structure are again defined by dark gray lines. Again, gray means no difference, black maximum decrease, and white maximum increase of the minority-carrier current density. From Fig. 30 we deduce that the deflection of minority carriers in the base region is the dominant effect, as conjectured in Section 5.4.1. Moreover, we see that the minority carriers injected from the emitter reach the collectors either directly, or are reflected at the base–substrate junction. The deflection of the portion of minority carriers reaching the substrate leads to an increased substrate current on the left side and a decreased substrate current on the right side. Majority-carrier deflection, emitter-injection modulation, and magnetoconcentration play only a minor role.

Table 1 shows simulated and measured sensor parameters of two SSIMT structures differing in emitter–collector distance and made by two different bipolar IC processes.[22] The trend to about two times better sensitivity, S, when

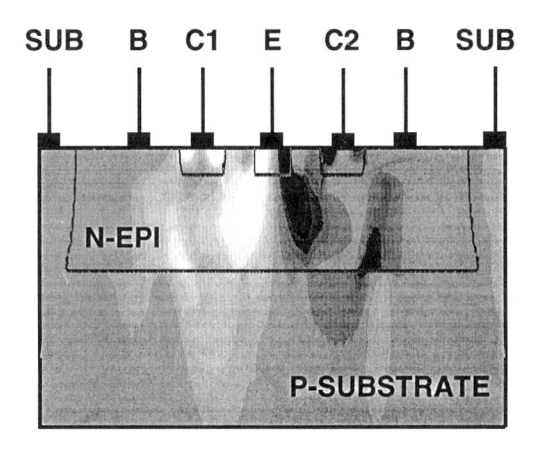

Fig. 30 Numerical simulation of the SSIMT. Shades are explained in the text. (After Ref. 22)

TABLE 1 Simulation and Experimental Results for Two SSIMT Structures[22]

	Simulation	Experiment
Emitter collector distance: 16 μm		
I_C (mA)	1.106	0.975
I_S (mA)	7.303	7.446
S (1/T)	0.444	0.11
Emitter collector distance: 32 μm		
I_C (mA)	0.672	0.124
I_S (mA)	7.749	8.002
S (1/T)	0.875	0.24

going from 16 μm to 32 μm emitter–collector distance is reproduced by the simulations. The discrepancies in the absolute values of S may be due to the difference between the ideal, symmetrical device and the real, measured sensor with offset due to mask misalignment and stress.

5.5 OTHER MAGNETIC SENSORS

In this section we present other types of semiconductor magnetic sensors: magnetodiodes (MDs) and carrier-domain magnetometers (CDMs). Their description is adapted mainly from Reference 3. In addition, a non-semiconductor magnetic sensor is outlined: the supermagnetoresistor (SMR) based on superconducting Y–Ba–Cu–O films.

5.5.1 Magnetodiodes

We recall that the carrier concentration in a nearly intrinsic, current-carrying semiconductor slab exposed to a magnetic field is modulated by the magnetoconcentration effect. Magnetodiodes (MDs) use this effect in combination with double-injection and surface recombination phenomena. Both electrons and holes injected by the end contacts of a semiconductor slab are deflected by the Lorentz force against the same surface. The recombination process at this surface is responsible for the carrier-density variation. The resulting transverse carried gradient affects the diode current. In view of the involved carrier concentration variation, the MD is much slower than the sensors discussed above. Its sensitivity drops for magnetic-field frequencies above 10 kHz.

The general MD structure is shown in Fig. 31. Carriers are injected from the n^+ and p^+ regions into the low-doped semiconductor slab of length L, width W, and thickness t, where they drift under the action of the electric field $\mathscr{E}$. The semiconductor slab is supposed to have a low- and a high-recombining surface, S1 and S2, respectively, whose recombination rates are s_1, s_2. Depending on the direction of the magnetic induction ($\boldsymbol{B}$ or $-\boldsymbol{B}$) perpendicular to $\mathscr{E}$, both electrons and holes are deflected towards either the surface S1 of the surface S2 of the slab. This leads to a carrier-concentration gradient perpendicular to $\mathscr{E}$ and $\boldsymbol{B}$ and finally to a magnetic-field modulation of the diode current–voltage characteristics. Crucial requirements for the MD operation are the difference between the recombination rates s_1 and s_2, the slab geometry (the thickness t should be of the order of the ambipolar diffusion length), and high-injection current in order to achieve the quasi-intrinsic condition.

Discrete MDs have been realized using Ge, Si, or GaAs. The difference in surface-recombination rate is achieved through a difference in surface roughness by grinding one surface and polishing and etching the other. The first integrated MDs were developed using SOS (silicon-on-sapphire) IC technology. They rely on the high-recombination rate of the Si/Al$_2$O$_3$ interface and the low-recombination rate of the Si/SiO$_2$ interface.

It is more desirable to exploit the MD effect with sensors that can be made

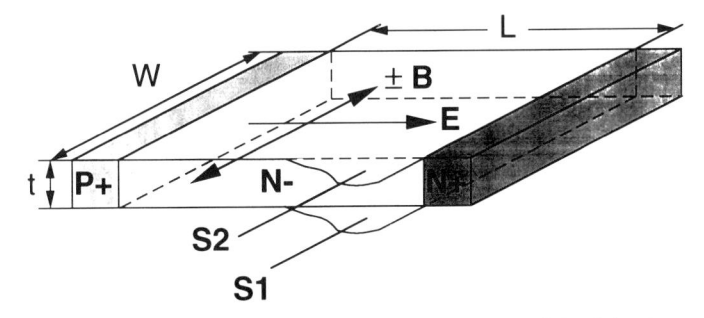

Fig. 31 Basic magnetodiode structure. The symbols are explained in the text. (After Ref. 3)

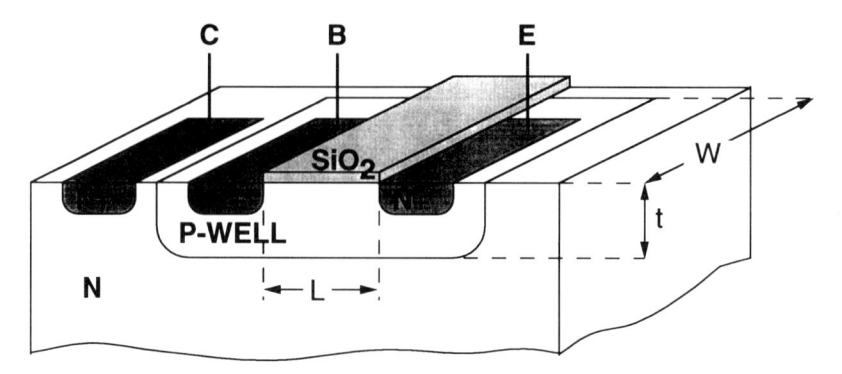

Fig. 32 Integrated magnetodiode made by a *p*-well CMOS IC process. The symbols are explained in the text. (After Ref. 3)

using a standard IC process like CMOS. This can be realized by a structure reminiscent of a bipolar transistor, but whose operation is essentially that of a magnetodiode.[61] A reverse-biased *p–n* (collector) junction functions as the highly-recombining interface, recombination being replaced by collection. The magnetic sensitivity of such CMOS MDs is about 5 V/mA-T with bias currents of a few mA.

Figure 32 shows the basic device geometry in the case of a *p*-well process. The *p*-well is used as the base active region, the substrate serves as the collector region, and n^+ and p^+ source and drain diffusions are used to create the emitter, base, and collector contacts, E, B, and C, respectively. The Si/SiO$_2$ interface on top of the *p*-well between the B and E contacts provides the low-recombining surface. The effective length, width, and thickness of the MD region is denoted by *L*, *W*, and *t*, respectively. The corresponding *n*-well devices show higher sensitivity, in particular those with a long base in excess of 100 μm. The device is operated under a high-injection condition. The EB junction is forward-biased and the BC junction is reverse-biased. In contrast to bipolar transistor operation, the useful signal is obtained between the E and B terminals, and the magnetic-field sensitive region is the lightly-doped base region between the chip surface, the E and B contacts, and the collector junction.

5.5.2 Carrier-Domain Magnetometers

A carrier domain is a region of high, nonequilibrium carrier density. The domain consists of an electron–hole plasma, since the concentrations of both types of carriers are approximately equal because of the charge-neutrality condition. Localization of the domain is enforced by a suitable potential distribution or some positive feedback. Carrier-domain magnetometers (CDMs) exploit the action of the Lorentz force on the charge carriers moving in the domain. This force moves the entire carrier domain through the semiconductor or modulates a domain migration caused by some other effect. In each case, detection of the domain motion provides information on the magnetic field.

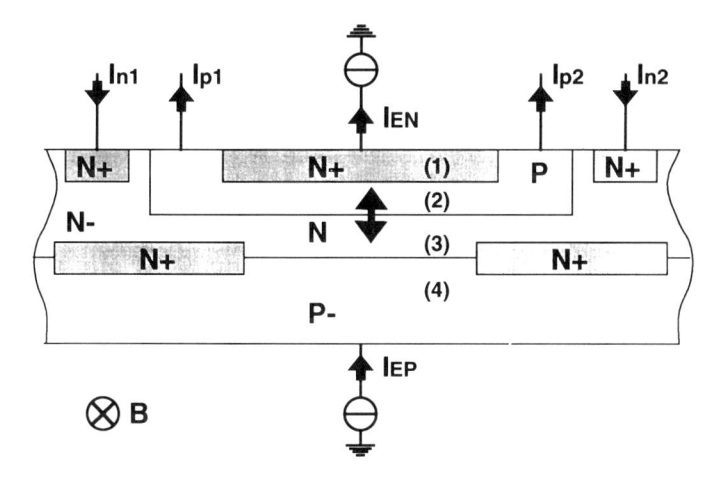

Fig. 33 Vertical four-layer carrier-domain magnetometer (CDM). The vertical two-pointed arrow indicates the carrier domain, which moves to the left or to the right under a magnetic field perpendicular to the drawing plane. The other symbols are explained in the text. (After Ref. 3)

Vertical Four-Layer CDM. The vertical four-layer CDM shown in Fig. 33 can be fabricated using bipolar IC technology.[62] Its n–p–n–p structure (layers numbered (1) to (4) in Fig. 33) can be perceived as a combination of n–p–n and p–n–p transistors that share a common base–collector junction (between layers 2 and 3). In operation, this junction is always reverse-biased. Both transistors are biased by emitter current sources providing the currents I_{EN} and I_{EP} and operate in the forward active region. Electrons are injected into the base of the n–p–n transistor (layer 2) and collected in the p–n–p base (layer 3), which simultaneously serves as the n–p–n collector. Similarly, holes are injected from the p–n–p emitter (layer 4) and collected in the n–p–n base (layer 2). Due to the lateral voltage drops in both bases, the carrier injection occurs at two opposite small spots of the base–emitter junctions, and thus forms a carrier domain. The domain consists of electrons and holes moving vertically in opposite directions.

A magnetic field perpendicular to the drawing plane of Fig. 33 produces a displacement of the domain. If, for example, the domain moves to the right, both right-hand base currents, I_{p2} and I_{n2} increase, while both left-hand base currents, I_{p1} and I_{n1}, decrease. This current imbalance indicates the domain displacement, and hence the presence of the magnetic field. The asymmetry produced by the domain displacement eventually brings about a restoring force preventing further displacement of the domain. At low magnetic induction, the current imbalance $(I_{n2} - I_{n1})/T$ is 3 mA/T with a 10 mA current drive. This corresponds to the high relative sensitivity of 0.3/T. The equivalent noise magnetic field is 5 μT. The high temperature coefficient (about 3%/K) is a disadvantage of this device.

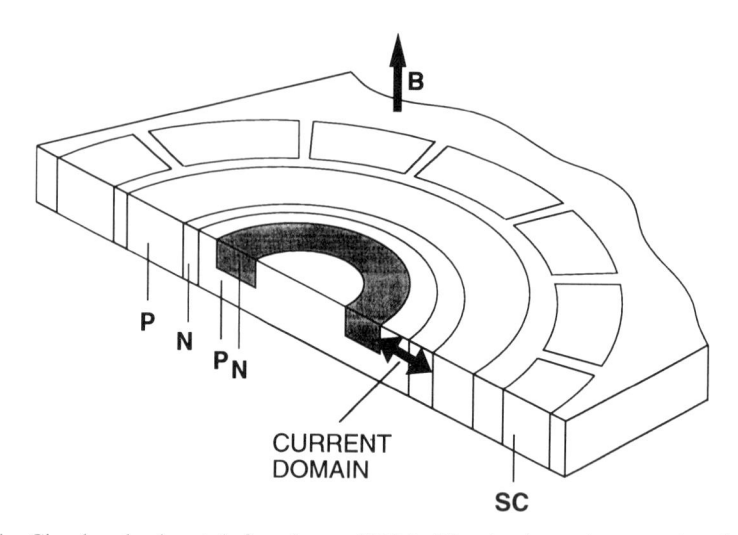

Fig. 34 Circular horizontal four-layer CDM. The horizontal two-pointed arrow indicates the carrier domain, which rotates under a magnetic induction perpendicular to the device surface. (After Ref. 3)

Circular, Horizontal Four-Layer CDM. The circular, horizontal four-layer CDM shown in Fig. 34 avoids the restoring forces that would limit the migration of the domain.[63] The device basically consists of a ring-shaped lateral four-layer n–p–n–p structure. The four layers in the circular structure alternate in the radial rather than in the vertical direction. Otherwise, carrier-domain formation and the action of the magnetic field are the same as in the vertical four-layer CDM. A horizontal, radial domain is formed. Its actual position is detected by monitoring currents at the segmented outer contacts SC. Under the action of the magnetic induction (perpendicular to the chip surface), the domain travels around the circumference of the structure. The frequency of this rotation is proportional to the applied magnetic induction. This generation of a frequency output is a unique feature of the circular CDM.

Sensitivities are between 10 and 100 kHz/T at about 10 mA supply current. The proportionality between output frequency and magnetic induction holds within 1% accuracy provided that $B \geqslant 0.2$ T. The domain rotation stops if B drops below a threshold value, which is between 0.1 and 0.4 T. It is believed that the domain is trapped at some preferred location because of process tolerances. Besides the high threshold field, the large temperature coefficient is another drawback of this device. The rotation frequency is proportional to $T^{-4.4}$ and the temperature coefficient is about $-1.5\%/\text{K}$. This is about twice the TC of the silicon Hall plates.

Circular, Horizontal Three-Layer CDM. The circular, horizontal three-layer CDM shown in Fig. 35 has the structure of a ring-shaped lateral n–p–n transistor (with B, E, and C denoting the base, emitter, and collector contacts) surrounded

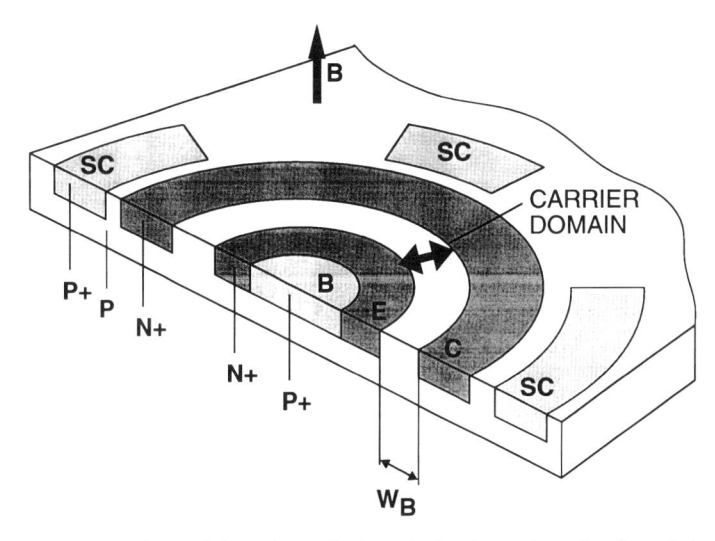

Fig. 35 Circular horizontal three-layer CDM. The horizontal carrier domain (see arrow) rotates under a magnetic induction perpendicular to the device surface. The other symbols are explained in the text. (After Ref. 3)

by four voltage probing contacts.[64] The diameter of the collector is 500 μm and the width W_B of the base regions is 8 μm. The device operates in the collector–emitter breakdown regime with short-circuited emitter and base contacts. The carrier domain appears for sufficiently large supply current (about 30 mA) provided by an external current source. The internal feedback involved in the transistor breakdown mechanism confines the current domain to a narrow sector of the base region.

In contrast to the circular four-layer CDM, no threshold magnetic induction is required. The domain rotates spontaneously with $f = 280$ kHz at $B = 0$, probably driven by thermal effects. The direction (left or right) of the spontaneous rotation is incidental. The domain rotation is detected from voltage pulses appearing at the contacts SC whenever the domain passes by. The angular frequency of the carrier domain rotation is modulated by a magnetic field perpendicular to the chip surface. The rotation frequency is raised or lowered, depending on the relative directions of B and the domain rotation. The sensitivity is about 250 kHz/T at low induction. The rotation is reversed when the frequency drops to about 150 kHz. Because of the high current, the device needs cooling. Another drawback is that the breakdown voltage is not a precise IC fabrication parameter.

5.5.3 Supermagnetoresistor

The supermagnetoresistor is the most recent addition to the realm of magnetic sensor technology.[16] The sensor operates at the temperature of liquid nitrogen

(77 K) and responds to small fields (magnetic induction below 10 mT). The device exploits the effect that a weak magnetic field disrupts the superconductivity of a granular Y–Ba–Cu–O ceramic sample by increasing the resistance between the superconducting grains. This leads to an abrupt change in the resistance of the sample with magnetic field. Meander-shaped ceramic films are prepared by spray pyrolysis and contacted by evaporated titanium films. The resistance modulation is up to 23 Ω/mT at 10 mA current and leads to a voltage response of up to 230 mV/mT. A magnetic resolution of 0.2 nT/Hz$^{1/2}$ is achieved at 100 Hz in the range between 10 μT and 10 mT magnetic induction.

5.6 SUMMARY AND FUTURE TRENDS

In order to put this chapter into perspective, let us remember that the range of magnetic induction of scientific or technical interest spreads over many orders of magnitude between the 10 fT of weak biomagnetic fields and the 10 T of magnets with superconducting coils. Out of these 15 orders of magnitude, only the "upper crust" of magnetic inductions above 10 μT or so are resolved by the integrated silicon magnetic sensors, unless additional flux concentration devices or expensive external measurement circuitry are employed. More demanding methods, such as the flux-gate magnetometer, nuclear magnetic resonance, and the superconducting quantum interference device, must be invoked for the weak and the very weak magnetic fields.

Nevertheless, integrated silicon magnetic sensors constitute the main topic of this chapter. One justification is that they are adequate for a number of important applications requiring vast numbers of inexpensive sensors such as the detection of displacement (keyboards and automation), current, or magnetic recording. Another important point is the unique opportunity of using silicon IC manufacturing offering inexpensive, reliable batch-fabrication and the cointegration of signal-conditioning circuitry.

There seem to be almost as many different types of magnetic sensors as there are researchers in the field. Even if we restrict our attention to integrated semiconductor magnetic sensors, there is still an overwhelming variety, out of which only a selection could be presented in this chapter. Out of the many integrated silicon magnetic sensors, which are the most useful ones today and the most promising ones for the tomorrow? While the answer to such a question always depends on which is required for a specific application, we may still attempt some general statements.

By and large, integrated bulk Hall plates constitute the most mature family of semiconductor magnetic sensors.[3,12] The Hall IC chip with integrated circuitry is currently widely utilized. One useful research direction is to improve its performance by inventive new structures and circuits for better offset and temperature properties.[34,41] Next in line, magnetotransistors seem to be a valid research topic. Apart from the intellectual challenge to understand the bipolar transistor action (including noise) in the presence of a magnetic field, they seem to have potential as part of a magnetic-field sensitive circuit. In this respect, it is important to remember that some bipolar magnetic-sensor structures can be manufactured by bipolar as well as by CMOS IC technology. A recent promising approach is to combine bipolar IC technology with anisotropic etching, removing a portion of the substrate in order to increase the sensor efficiency.[65]

The MAGFET is inherently noisy. Incorporated in low-power CMOS circuitry, it may find applications where very low current consumption is of paramount importance.[3] But it may have to yield to the competition of the less noisy lateral magnetotransistor, which also can be incorporated into CMOS circuitry. On the other hand the sensitivity of split-drain MAGFETs has been increased recently (0.185%/T) by using multiple gates, which establish a longitudinal accelerating field in the channel.[66] Our outlook for the other types of silicon integrated magnetic sensors is less optimistic.

ACKNOWLEDGMENTS

The authors would like to thank Prof. A. Nathan, University of Waterloo, Canada, and Prof. Dr. R. S. Popovic, EPF, Lausanne, Switzerland, for a critical reading of the manuscript, Dr. J. R. Biard, Honeywell Optoelectronics, Richardson, USA, for arranging the use of Fig. 12, and Ms. C. Riccobene, ETH Zurich, Switzerland for providing Figs. 29 and 30.

PROBLEMS

1. Compile a list of types of magnetic field sensors with their magnetic field range and possible applications.

2. Derive the expression for the Hall coefficient R_H for ambipolar current flow.

3. An Si plate is doped with phosphorus and boron. $N_D = 4 \times 10^{14}$ cm^{-3}, $N_A = 4.001 \times 10^{14}$ cm^{-3}, $r_n = 1.15$, $r_p = 0.7$, $\mu_p = 0.047$ T^{-1}, $\mu_n = 0.138$ T^{-1}. What is the value for R_H?

4. Describe three galvanomagnetic and two thermomagnetic effects.

5. A Hall plate is integrated using a standard bipolar IC process (Fig. 11). The epi layer defining the plate has a thickness of 10 μm and a sheet resistance $\rho_s = 1000$ Ω/sq. Assume $L = 600$ μm, and $W = 200$ μm. The supply current is $I = 10$ mA and the presence of a magnetic induction $B = 100$ Gauss. Calculate:
   a) the Hall coefficient, R_H
   b) the Hall voltage, V_H
   c) the Hall angle, θ_H
   d) the supply-voltage related sensitivity, S_V.

266 MAGNETIC SENSORS

6. What is the resolution of the Hall plate described in Problem 4, assuming only thermal noise, sensor contacts $s = 5$ μmm and a bandwidth of 1 Hz?

7. Calculate the Lorentz current I_L for the same structure with interchanged L and W dimensions.

8. Using either bipolar or CMOS IC technologies design a magnetic field sensor sensitive to all three components of the magnetic induction vector.

9. A random signal for which the power spectral density equals the Fourier transform of the autocorrelation function is called ergodic and stationary. Stationarity means that the mean value and the mean square value of the signal are independent of time. With this additional information derive the expression for the degree of correlation in dual collector magnetotransistors, $\gamma_{12}(f) = 1 - 0.5[S_D(f)/S_{11}(f)]$ starting from Eq. 98.

10. Design a simple bias circuit for a dual-collector magnetotransistor.

REFERENCES

1. G. L. Pearson, "A magnetic field strength meter employing the Hall effect in germanium," *Rev. Sci. Instrum.* **19**, 263 (1948).
2. G. Bosch, "A Hall device in an integrated circuit," *Solid-State Electron.* **11**, 712 (1968).
3. H. Baltes and R. S. Popovic, "Integrated semiconductor magnetic field sensors," *Proc. IEEE* **74**, 1107 (1986).
4. H. Baltes and A. Nathan, "Integrated magnetic sensors," in *Sensors 1: Fundamentals*, T. Grandke and W. Ko, Eds., VCH Verlagsgesellschaft, Weinheim, 1989, Ch. 7, p. 195.
5. U. Dibbern, "Magnetoresistive sensors," in *Sensors 5: Magnetic Sensors*, K. Boll and K. J. Overshott, Eds., VCH Verlagsgesellschaft, Weinheim, 1989, Ch. 9, p. 341.
6. T. Seitz, "Flux gate sensor in planar microtechnology," *Sens. Actuators* **A21–A23**, 799 (1990).
7. S. Kordic, "Integrated silicon magnetic-field sensors," *Sens. Actuators* **10**, 347 (1986).
8. S. Middelhoek and S. A. Audet, *Silicon Sensors*, Academic Press, London 1989, Ch. 5, p. 201.
9. R. S. Popovic, "Hall effect devices," *Sens. Actuators* **17**, 39 (1989).
10. R. S. Popovic and W. Heidenreich, "Magnetogalvanic scensors," in *Sensors 5: Magnetic Sensors*, K. Boll and K. J. Overshott, Eds., VCH Verlagsgesellschaft, Weinheim, 1989, Ch. 3, p. 43.
11. T. Nakamura and K. Maenaka, "Integrated magnetic sensors," *Sens. Actuators* **A21–A23**, 762 (1990).
12. R. S. Popovic, *Hall Effect Devices*, Adam Hilger, Bristol, 1991.
13. J. T. Maupin and M. L. Geske, "The Hall effect in silicon circuits," in *The Hall Effect and its Applications*, C. L. Chien and C. R. Westgate, Eds., Plenum Press, New York, 1980, p. 421.
14. Y. Sugiyama, "Fundamental research on Hall effects in inhomogeneous magnetic fields," in *Res. Electrotech. Lab.*, No. 838, Electrotechnical Laboratory, Tokyo, 1983.

15. H. Koch, "SQUID Sensors," in *Sensors 5: Magnetic Sensors*, K. Boll and K. J. Overshott, Eds., VCH Verlagsgesellschaft, Weinheim, 1989, Ch. 10, p. 381.
16. H. Nojima, H. Shinkatu, M. Nagata, and S. Kataoka, "Fundamental characteristics of a magnetoresistive sensor using $Y_1Ba_2Cu_3O_{7-x}$ ceramic superconducting films," *IEEE Trans. Electron Devices* **ED-39**, 576 (1992).
17. D. R. Yennie, "Integral quantum Hall effect for non-specialists," *Rev. Mod. Phys.* **59**, 781 (1987).
18. A. Nathan, W. Allegretto, H. Baltes, and Y. Sugiyama, "Carrier transport in semiconductor detectors of magnetic domains," *IEEE Trans. Electron Devices* **ED-34**, 2077 (1987).
19. A. Nathan and H. Baltes, "Sensor modeling," in *Sensors 1: Fundamentals*, T. Grandke and W. Ko, Eds., VCH Verlagsgesellschaft, Weinheim, 1989, Ch. 3, p. 45.
20. A. Nathan, H. Baltes, and W. Allegretto, "Review of physical models for numerical simulation of semiconductor microsensors," *IEEE Trans. Computer-Aided Design* **CAD-9**, 1198 (1990).
21. W. Allegretto, A. Nathan, and H. Baltes, "Numerical analysis of magnetic field-sensitive bipolar devices," *IEEE Trans. Computer-Aided Design* **CAD-10**, 1198 (1990).
22. C. Riccobene, G. Wachutka, H. Baltes, and J. Bürgler, "Operating principle of dual collector magnetotransistors studied by two-dimensional simulation," *IEEE Trans. Electron Devices*, **41** (1994).
23. O. Madelung, "Halbleiter," (in German), in *Handbuch der Physik XX: Elektrische Leitungsphänomene II*. S. Flügge, Ed., Springer-Verlag, Berlin, 1957, pp. 1–245.
24. O. Madelung, *Introduction to Solid State Theory*, Springer-Verlag, Berlin, 1978, Ch. 4.
25. K. Seeger, *Semiconductor Physics*, Springer-Verlag, Berlin, 1982, Ch. 4.
26. S. M. Sze, *Physics of Semiconductor Devices*, 2nd ed., Wiley, New York, 1981.
27. H. Baltes, L. Andor, A. Nathan, and H. G. Schmidt-Weinmar, "Two-dimensional numerical analysis of a silicon magnetic field sensor," *IEEE Trans. Electron Devices* **ED-31**, 996 (1984).
28. H. Suhl and W. Shockley, "Concentrating holes and electrons by magnetic fields," *Phys. Rev.* **75**, 1617 (1949).
29. R. S. Popovic and B. Hälg, "Nonlinearity in Hall devices and its compensation," *Solid-State Electron.* **31**, 681 (1988).
30. T. T. Hara, M. Mihara, N. Toyoda, and M. Zama, "Highly linear GaAs Hall devices fabricated by ion implantation," *IEEE Trans. Electron Devices* **ED-29**, 78 (1982).
31. Y. Sugiyama, T. Taguchi, and M. Tacano, "Highly sensitive magnetic sensor made of AlGaAs/GaAs heterojunction semiconductor," *Proc. 6th Sensor Symp.*, IEE Japan, Tokyo, 1986, p. 55.
32. T. G. M. Kleinpenning and L. K. J. Vandamme, "Comment on transverse $1/f$ noise in InSb thin films and the SNR of related Hall elements," *J. Appl. Phys.* **50**, 5547 (1979).
33. B. Hälg and R. S. Popovic, "How to liberate integrated sensors from encapsulation stress," *Sens. Actuators* **A21–A23**, 908 (1990).
34. S. R. in't Hout and S. Middelhoek, "High temperature silicon Hall sensor," *Sens. Actuators*, **A37-38**, 26 (1993).

35. E. Pettenpaul, W. Flossmann, J. Huber, U. v. Borke, and H. Weidlich, "GaAs Hall devices produced by local ion implantation," *Solid-State Electron.* **24**, 781 (1981).
36. Y. Sugiyama, T. Taguchi, and M. Tacano, "Two-dimensional electron gas magnetic field sensor," *Transducers 87 Dig. of Tech. Papers,* S. Kataoka, Ed., IEE Japan, Tokyo, 1987, p. 547.
37. T. Kanayama, H. Hiroshima, and M. Komura, "A quarter-micron Hall sensor fabricated with maskless ion implantation," *J. Vac. Sci. Technol.* **136**, 1010 (1988).
38. H. B. Callen, *Thermodynamics,* Wiley, New York, 1960.
39. T. Elbel, "Ettinghausen-Nernst-Detektoren zum Nachweis thermischer Strahlung," *Feingerätetechnik* **28**, 243 (1979).
40. O. Paul and H. Baltes, "Measuring thermogalvanomagnetic properties of polysilicon for the optionization of CMOS sensors," *Transducer '93 Dig. of Tech. Papers, IEE Japan*, Yokohama, 1993, p. 606.
41. P. J. A. Munter, "A low-offset spinning-current Hall plate," *Sens. Actuators* **A21–A23**, 743 (1990).
42. R. S. Popovic, "The vertical Hall-effect device," *IEEE Electron Device Lett.* **EDL-5**, 357 (1984).
43. A. M. J. Huiser and H. Baltes, "Numerical modeling of vertical Hall-effect devices," *IEEE Electron Device Lett.* **EDL-5**, 482 (1984).
44. S. Takamiya and K. Fujikawa, "Differential amplification magnetic sensor," *IEEE Trans. Electron Devices* **ED-19**, 1085 (1972).
45. R. C. Gallagher and W. S. Corak, "A metal-oxide-semiconductor (MOS) Hall element," *Solid-State Electron.* **9**, 571 (1966).
46. M. Hirata and S. Suzuki, "Integrated magnetic sensor," *Proc. 1st Sensor Symposium,* S. Kataoka, Ed., IEE Japan, Tokyo, 1982, p. 37.
47. A. Nathan, A. M. J. Huiser, and H. Baltes, "Two-dimensional numerical modeling of magnetic field sensors in CMOS technology," *IEEE Trans. Electron Devices* **ED-32**, 1212 (1985).
48. D. R. Briglio, A. Nathan, and H. Baltes, "Measurement of Hall mobility in n-channel silicon inversion layer," *Can. J. Phys.* **65**, 842 (1987).
49. A. Nathan, H. Baltes, D. R. Briglio, and My T. Doan, "Noise correlation in dual-collector magnetotransistors," *IEEE Trans. Electron Devices* **ED-36**, 1073 (1989).
50. R. S. Popovic and H. Baltes, "A CMOS magnetic field sensor," *IEEE J. Solid-State Circuits* **SC-18**, 426 (1983).
51. A. Nathan, I. A. Kay, I. M. Filanovsky, and H. Baltes, "Design of a CMOS oscillator with magnetic-field frequency modulation," *IEEE J. Solid-State Circuits* **SC-22**, 230 (1987).
52. S. M. Sze, Ed., *High-Speed Semiconductor Devices,* Wiley, New York, 1990.
53. Ch. S. Roumenin, "Bipolar magnetotransistor sensors. An invited review," *Sens. Actuators* **A24**, 83 (1990).
54. V. Zieren and B. P. M. Duyndam, "Magnetic-field-sensitive multicollector n-p-n transistors," *IEEE Trans. Electron Devices* **ED-19**, 83 (1982).
55. L. Ristic, H. Baltes, T. Smy, and I. Filanovsky, "Suppressed sidewall injection magnetotransistor with focused injection and carrier double deflection," *IEEE Electron Device Lett.* **EDL-8**, 395 (1977).

56. A. Nathan and H. Baltes, "Integrated silicon magnetotransistors- high sensivity or high resolution?," *Sens. Actuators* **A21–A23**, 780 (1990).
57. R. Castagnetti, H. Baltes, A. Bezinge, and N. C. Bui, "PNP bipolar magnetotransistors for sensor applications," *Transducers '91 Dig. Tech. Papers,* IEEE, New York, 1991, p. 1065.
58. R. Gottfried-Gottfried, G. Zimmer, and W. Mokwa, "CMOS-compatible magnetic field sensors fabricated in standard and in silicon on insulator technologies," *Sens. Actuators* **A25–A27**, 753 (1991).
59. I. Filanovsky and H. Baltes, "A CMOS current-controlled multivibrator," *Int. J. Electronics* **73**, 333 (1992).
60. R. Castagnetti, C. Azeredo Leme, and H. Baltes, "Dual collector magnetotransistors with on-chip bias and signal conditioning circuitry," *Sens. Actuators* **A37–38**, 638 (1993).
61. R. S. Popovic, H. Baltes, and F. Rudolf, "An integrated silicon magnetic field sensor using the magnetodiode principle," *IEEE Trans. Electron Devices* **ED-31**, 286 (1984).
62. J. I. Goicolea, R. S. Muller, and J. E. Smith, "Highly sensitive silicon carrier domain magnetometer," *Sens. Actuators* **5**, 147 (1984).
63. M. H. Manley and G. G. Bloodworth, "The design and operation on a second-generation carrier-domain magnetometer device," *Radio Electron. Eng.* **53**, 125 (1983).
64. R. S. Popovic and H. Baltes, "A new carrier domain magnetometer," *Sens. Actuators* **4**, 229 (1983).
65. C. Riccobene, G. Wachutka, and H. Baltes, "Two-dimensional numerical analysis of novel magnetotransistors with partially removed substrate," *IEDM Tech. Digest,* IEEE, New York, 1992, p. 513.
66. F. J. Kub and C. S. Scott, "Multi-gate split-drain MOSFET magnetic-field sensing device and amplifier," *IEDM Tech. Digest,* IEEE, New York, 1992, p. 517.

6 Radiation Sensors

S. AUDET
Princeton Gamma-Tech, Inc.
1200 State Road
Princeton, NJ 08540, USA

J. STEIGERWALD
Analog Devices
804 Woburn Street
Wilmington, MA 01887, USA

6.1 INTRODUCTION

Radiation sensors transform incident radiant signals into standard electrical output signals to be used for data collection, processing, and storage. Radiant signals can be categorized into one of the following types: electromagnetic, neutrons, fast electrons, or heavy-charge particles. Electromagnetic radiation and neutrons are uncharged, while fast electrons and heavy-charged particles are charged-particulate radiation.[1,2] All radiant signals originate in atomic or nuclear processes, and similar techniques are used for their detection.[3] Sensors based on direct transduction will be discussed in this chapter. Tandem transducers such as thermal-based radiation-detectors that utilize intermediary signals in the conversion process are covered in Chapter 7.

The various types of electromagnetic radiation, displayed in Fig. 1, are distinguished by their characteristic frequencies and their origins.[2] Electromagnetic energy is carried by photons, which are the quantum particles involved in the exchange of radiant energy. Photons are generated when an atom or a nucleus relaxes following the absorption of energy. Infrared-, visible-, and ultraviolet-light photons are emitted in radiative-recombination processes involving the transition of an electron from the conduction band to the valence

---

Semiconductor Sensors, Edited by S. M. Sze.
ISBN 0-471-54609-7 © 1994 John Wiley & Sons, Inc.

272 RADIATION SENSORS

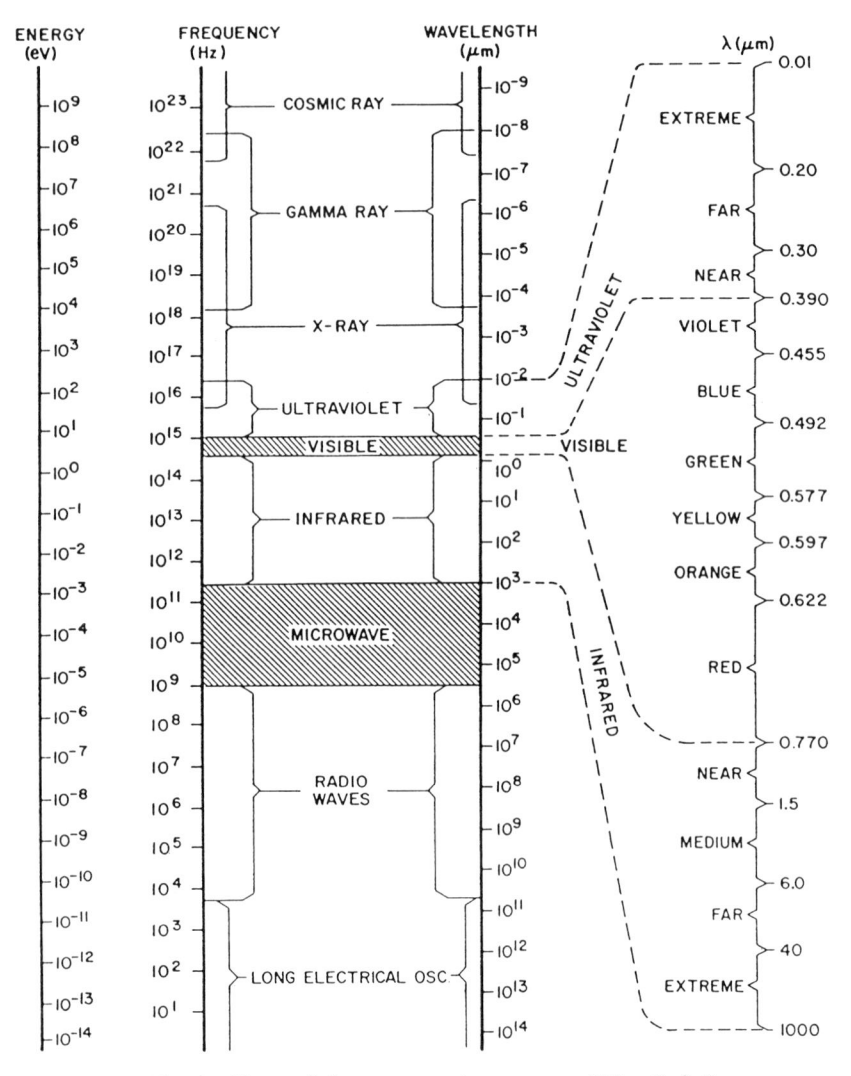

Fig. 1 Chart of electromagnetic spectrum. (After Ref. 4)

band of a semiconductor or when an electric discharge is passed through heated solids or gases. X-ray photons result from the transitions between different states of the orbital electrons as an atom returns to the ground state from an excited state, or may be produced by deceleration of an electron beam. Similarly, gamma photons are emitted in de-excitation processes of excited nuclei in their transition to lower, more stable energy levels. Gamma photons are also generated in nuclear reactions and in annihilation (pair-production) processes.[2] This chapter will cover electromagnetic radiation with energies ranging from

10 meV to 20 MeV. Energies greater than 100 eV will be called high-energy electromagnetic radiation.

Fast electrons have the mass of an electron and a negative or positive charge, corresponding to a beta particle or a positron, respectively. They may be emitted through natural or induced nuclear decay or may be produced by the acceleration of electrons in an electromagnetic field.[1] Heavy-charged particles consist of all atomic charged nuclei. The particles in this class may carry any charge from q to Zq, where q is the elemental charge and Z is the atomic number.[1] Charged helium nuclei, alpha particles, belong to this category and have a 2+ charge. Neutrons are neural particles generated in nuclear-decay processes.[1] Fast electrons, heavy-charged particles, and neutrons will collectively be termed nuclear-particle radiation. This chapter will cover nuclear-particle radiation with energies ranging from 10 eV to 20 MeV.

A review of the basic semiconductor physics and the applications of common devices involved in the detection of radiation is consolidated in Section 6.2. Infrared detectors, specifically HgCdTe photodiodes, are discussed in Section 6.3. Section 6.4 reviews contemporary and alternative color-sensing techniques. High-energy radiation detectors are covered in Sections 6.5 and 6.6, the latter being dedicated to silicon-drift chambers. The chapter is concluded with future trends in Section 6.7.

6.2 COMMON PHYSICS

6.2.1 Introduction

The requirements of radiation sensors include high sensitivity and linear performance over the intended energy range, fast response time, low noise, and acceptable reliability under the exposure conditions. Due to the large number of different applications, there are many variants of the radiation sensor. Different materials, geometric arrangements and different physical-detection techniques are used to optimize the trade-offs that exist among the requirements above.

Sections 6.2.2 and 6.2.3 provide a foundation for the discussion in Section 6.2.4 concerning the operation of specific radiation sensors. The characteristic properties of electromagnetic and nuclear-particle radiation are reviewed in Section 6.2.2 An understanding of the interactions of electromagnetic and nuclear-particle radiation with semiconductors is essential for the proper operation of the radiation sensor. Section 6.2.3 details these interactions concentrating on absorption, which is a function of the sensor material's energy-band construction. Radiation sensors played a critical role in the development of quantum mechanics, and thus the related energy-band theory has been studied in considerable detail.[4-6] Our coverage will be that required as a basis for further discussion.

6.2.2 Characteristics of Electromagnetic and Nuclear-Particle Radiation

A thorough discussion of the wave and particle properties of electromagnetic radiation can be found in contemporary physics textbooks.[5] Briefly, electromagnetic radiation consists of electric-field and magnetic-flux-density vectors that are perpendicular to each other with peak vector magnitudes of $\mathscr{E}_0$ and B_0, respectively. An electromagnetic wave is also distinguished with a characteristic frequency, v, wavelength, λ, velocity, $v = \lambda v$, and direction of propagation. The velocity of an electromagnetic wave in vacuum equals the speed of light, c, given by

$$c = \lambda_0 v = \mathscr{E}_0/B_0 = 1/\sqrt{\mu_0 \varepsilon_0} = 2.988 \times 10^8 \text{ m/s} \quad (1)$$

where $\varepsilon_0 = 8.854 \times 10^{-12}$ F/m and $\mu_0 = 1.258 \times 10^6$ H/m are the permittivity and permeability of free space, respectively. These two physical constants define the spatial extent of the electric and magnetic fields, respectively. In a medium other than free space, the radiation travels at a reduced velocity, $v = c/\bar{n} = \lambda v$, where $\bar{n} = (\mu_r \varepsilon_r)^{1/2}$ is the dimensionless index of refraction, and ε_r and μ_r are the dimensionless relative permittivity and permeability of a given material, respectively, and are both equal to unity for a vacuum and greater than unity for all other materials. Note that the wavelength is reduced while the frequency remains constant.

As determined by Einstein, electromagnetic energy is not distributed continuously, but quantized in small bundles called photons. The energy of a given photon, E, is constant and directly related to its frequency through Planck's constant, $h = 6.626 \times 10^{-34}$ J-s.

$$E = hv. \quad (2)$$

Since the refractive index of a given material affects the radiation wavelength but not its frequency, the photon energy is independent of the material in which the wave is traveling, as required for conservation of energy at material boundaries.† The electric-field and magnetic-flux-density magnitudes, $\mathscr{E}_0$ and B_0, are related to the energy flux of the wave as specified by the Ponyting vector S,

$$S = \frac{1}{\mu_0}(\mathscr{E} \times B) \quad (3)$$

which is the wave irradiance.‡ The Poynting vector has units of photon areal

---

†Maxwell's equations require that both the tangential components of $\mathscr{E}$ and H and the normal components of $D = \varepsilon_r \varepsilon_0 \mathscr{E}$ and $B = \mu_r \mu_0 H$ are continuous across boundaries, where H is the magnetic field, B is the inductance vector or magnetic-flux density, $\mathscr{E}$ is the electric field, and D is the displacement vector.

‡Irradiance is the internationally agreed term, however, intensity is commonly used.

density times photon energy per unit time or equivalently radiant power per unit area. Note that although the photon energy is constant, the irradiation varies with space and time as described by the cross product of the electric field and inductance vectors. The quantum probability of interaction between the photon- and electron-wave functions becomes a significant factor at higher radiant energies where the wavelength approaches the atomic spacing. However, the average power per unit area of an electromagnetic wave, derived using $v = \mathscr{E}_0/B_0$, Eq. 1, generally provides sufficient representation of the radiation.

$$P_{avg} = \mathscr{E}_0^2/(2\mu_0 c). \qquad (4)$$

Finally, a photon's momentum, p, can be determined realizing a photon has no rest mass, m_0, so the relativistic energy, $E = (m_0^2 c^4 + p^2 c^2)^{1/2}$, simplifies to $E = pc$ or $E = pv$ for radiation propagating through a material other than free space. Thus, the photon momentum is inversely proportional to the radiation wavelength.

$$p = E/v = h\nu/v = h/\lambda. \qquad (5)$$

Nuclear-particle radiation has been well characterized.[2,3] Fast electrons arise most commonly in nuclear-decay processes involving beta-minus emission. A vast number of radionuclides that decay by this mechanism are available for use as broad-energy, fast-electron sources because a continuum of energies is produced by any one source. Nearly monoenergetic electrons are emitted in internal conversion processes of an excited nucleus. Energy distribution diagrams are well documented.[2,3]

Atomic-charged nuclei or heavy-charged particles are produced by alpha decay. The particles produced by this type of decay are nearly monoenergetic with a fixed energy difference between an initial and a final state. Heavy charge particles can also be produced by spontaneous fission, which produces fission fragments with a distribution in mass numbers and energies.[2] Energy spectra can be found in the literature.[2,3] Neutrons are emitted by spontaneous fission or in nuclear reactions. A continuum of particle energies is produced in both cases.[2] Characteristics of fast-electron, heavy-charged particle, and neutron sources have been tabulated.[2,3]

6.2.3 Interactions of Electromagnetic and Nuclear-Particle Radiation with Semiconductors

Electromagnetic radiation interacts with semiconductors primarily through absorption processes. However, interference, diffraction, reflection, polarization, transmission, and refraction all play a role relative to the electromagnetic radiation's propagation through the various media separating the semiconductor and the radiation source. Absorption is defined as the relative decrease of the irradiance, Φ, per unit path length, $\delta\Phi(x)/\Phi = \alpha\,\delta x$, which has the solution

$$\Phi(x) = \Phi_0 \exp(-\alpha x) \tag{6}$$

where Φ_0 is the incident irradiance, α is the absorption coefficient and x is a path-length variable. The absorbed photon density, $1 - \Phi(x)$, reaches 63% of the incident value within one absorption length, $1/\alpha$. The total absorption coefficient is the sum of three mechanisms: the photoelectric effect, Compton scattering, and pair production. We shall consider these mechanisms in the following paragraphs.

The photoelectric effect dominates for low to moderate radiant energies, $hv < 50$ keV. Absorption by the photoelectric effect results in a complete transfer of the electromagnetic radiant energy from an incident photon to the interacting atom, which ejects a photoelectron. The minimum photon energy required is approximately equal to the semiconductor bandgap, for example the binding energy of a valence electron involved in an atomic bond. Photons with energy below this threshold may be absorbed by carriers already within the conduction band, which is called free-carrier absorption. However, the probability for this absorption mechanism is low in semiconductors because of small concentration of conduction-band electrons relative to the concentration in the valence band. For higher-energy radiation, the most probable origin of the electron is from the innermost atomic shell, that is the K-shell, Fig. 2a. The excited atom returns to its lowest energy state by filling the inner-shell electron vacancy through the capture of a free electron or an electron transition from an outer-shell to the inner shell, Fig. 2b. For the latter, a characteristic X-ray will be released with an energy equal to the difference between the electron in its initial and final stages, Fig. 2c.

The characteristic X-ray then undergoes one of two interactions: reabsorption by the photoelectric effect or Auger-electron production, the latter mechanism has a significantly smaller probability of occurrence than the former. The ejected photoelectron acquires a kinetic energy equal to the difference between the initial photon energy and the electron's binding energy. The photoelectron gains

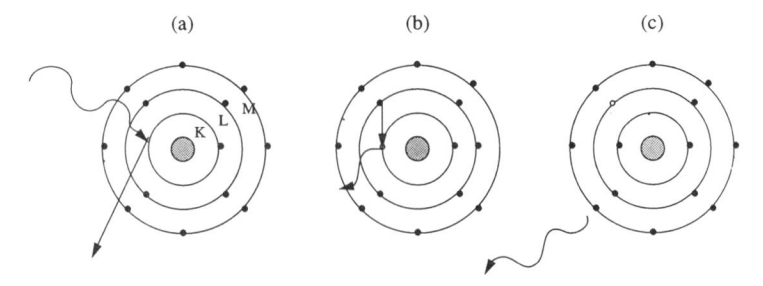

Fig. 2 (a) Incident photon interaction with a K-shell electron. (b) The de-excitation process in which an electron transition occurs from the outer L-shell to the inner K-shell. (c) Emission of a characteristic secondary X-ray photon.

additional kinetic energy as it settles to a lower potential energy of the $E(k)$ dispersion function of the conduction band. The dispersion functions for important crystal momentum vectors of silicon (Si), germanium (Ge), gallium arsenide (GaAs), mercury telluride (HgTe), and cadmium telluride (CdTe) are shown in Figs. 3a–e. The photohole can also gain kinetic energy if the electron–hole pair originates below the valence-band maximum. With sufficient kinetic energy, the photocarriers' periodic collisions within the lattice can produce additional electron–hole pairs through impact ionization. Through impact ionization and the release of characteristic X-rays, photoelectric absorption can produce an internal quantum efficiency, η, greater than unity even though the incident photon interacts with only one atom.

The characteristic length for the exponential decay of the electron density created by impact ionization depends on the photoelectron's kinetic energy. At initial energies of 10 and 100 keV, the characteristic decay lengths of electrons in silicon are approximately 4 and 40 μm, respectively.[2] Similarly, the characteristic length for the absorption of secondary photons depends on the energy of the radiation. To absorb 99% of 10 and 100 keV photons, a silicon substrate must be approximately 0.06 and 60 cm thick, respectively. Characteristic times for the exponential decay of both energetic electrons and photons are in the picosecond range.[2]

Photoelectric absorption is dependent on the density of occupied states at the initial energy, the density of available states at the final energy, and the transition probability. Although it is conceivable for an electron to be photo-excited from the valence band to a mid-gap state and then again photo-excited into the conduction band, the probability of such an occurrence is generally insignificant due to the low density of occupied states within the bandgap versus that within the valence band. Transforming the valence- and conduction-band density-of-state equations[4] from $N(E)\delta(E)$ to $N(h\nu)\delta(h\nu)$, the absorption coefficient for direct transitions is

$$\alpha(h\nu) = A^*(h\nu - E_g)^{3/2} \qquad (7)$$

where

$$A^* = q^2(2m_r)^{1.5}/(\bar{n}ch^2 m_e^*)$$

$$1/m_r = 1/m_e^* + 1/m_h^*$$

where m_e^*, m_h^* and m_r are the electron, hole and reduced effective masses, $\bar{n}$ is the dimensionless index of refraction, and E_g is the bandgap. In addition to describing absorption for direct semiconductors such as GaAs and HgCdTe, this equation is valid for direct transitions in silicon and germanium. The room-temperature dispersion diagrams, Fig. 3, show that direct transitions in silicon and germanium require approximately 3.4 and 0.8 eV, respectively, which explains the sharp increases of the absorption coefficient at these energies in Fig. 4.

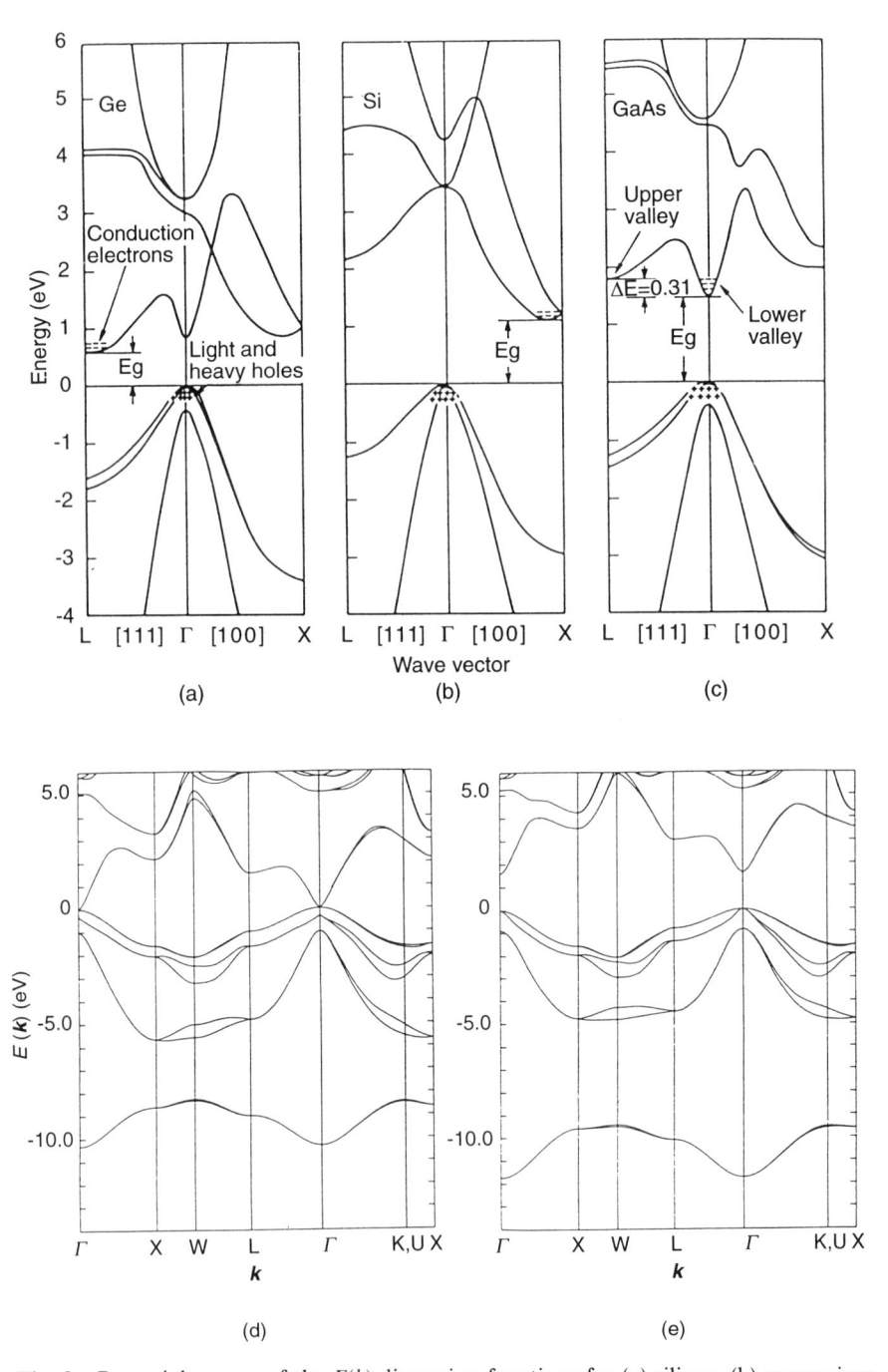

Fig. 3 Potential energy of the $E(k)$ dispersion functions for (a) silicon, (b) germanium, (c) gallium arsenide (After Ref. 4), (d) mercury telluride, (e) cadmium telluride. (After Ref. 31)

6.2 COMMON PHYSICS 279

(a)

(b)

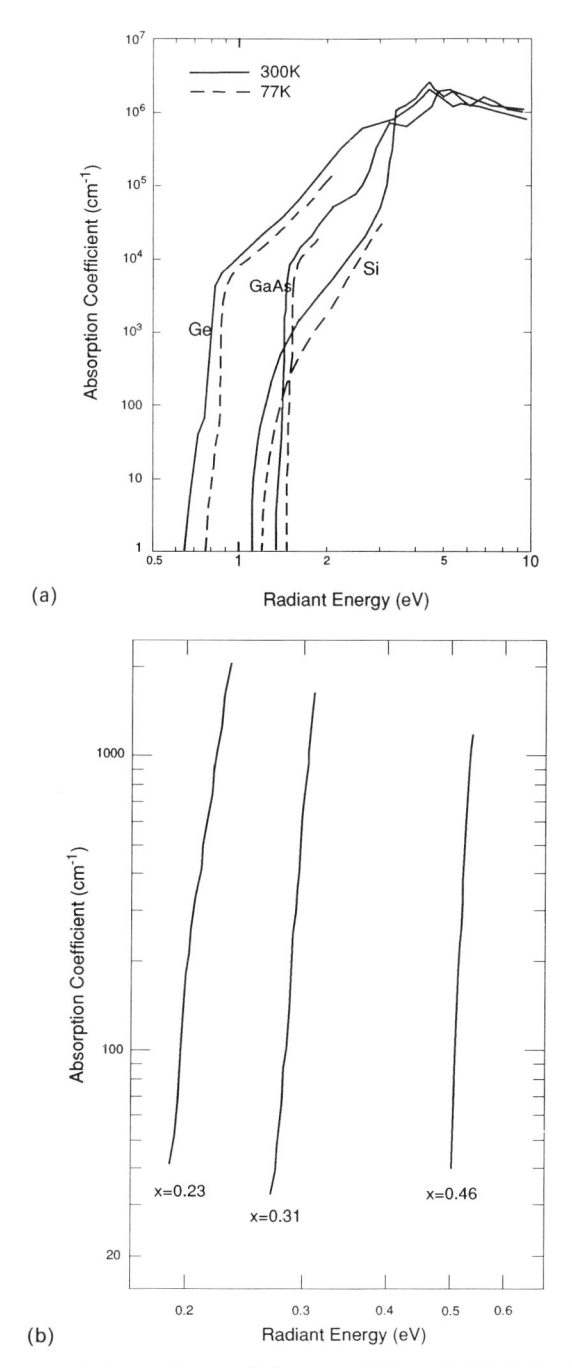

Fig. 4 (a) Measured absorption coefficients at 77 K and 300 K for Si, Ge and GaAs as a function of radiant energy. (After Ref. 4) (b) Measured absorption coefficients at 300 K for $Hg_{1-x}Cd_xTe$. (After Ref. 7)

A momentum change on the order of h/a, where a is the lattice constant, is required for bandgap minimum transitions in indirect semiconductors. Although a photon does carry a momentum of h/λ (Eq. 5) its value only begins to approach h/a at photon energies of 100 keV. This range of photon energies is well above the threshold of direct transition of all semiconductors. The momentum change is provided by a phonon, which is the energy quantum of lattice vibration storing the thermal energy of the semiconductor. Although phonons and photons have similar relationships with respect to frequency, wavelength, energy, and momentum, phonons travel with velocities in the acoustic as well as in the optical range, and, therefore, have much greater momentum for a given energy, $p = E/v$. The requirement of a momentum change causes indirect absorption to be dependent on the distribution of phonon energy, E_p. Phonon energy is described by Bose–Einstein statistics, $[\exp(E_p/kT) - 1]^{-1}$, where $k = 1.38 \times 10^{-23}$ J/K is the Boltzmann constant and T is the temperature in degrees Kelvin. Indirect absorption is also dependent on whether a phonon is absorbed or emitted, with absorption and emission coefficients of α_a and α_e, as shown in Eqs. 8a and b below, respectively.

$$\alpha_a(hv) = A'(hv - E_g + E_p)^2/(\exp(E_p/kT) - 1) \quad \text{for } hv > E_g - E_p \quad (8a)$$

$$\alpha_e(hv) = A'(hv - E_g - E_p)^2/(1 - \exp(-E_p/kT)) \quad \text{for } hv > E_g + E_p \quad (8b)$$

where A' is dependent on the effective masses and index of refraction similar to A^* of Eq. 7, but is generally assumed to be constant for a given energy gap[8] $A' = 5 \times 10^3/E_g^2/\text{cm eV}^2$. These relationships assume that indirect absorption is not limited by the density of mid-gap states with energy levels close to the conduction-band edge, and the same wave number, k, as the participating valence-band state. Although the pre-existing mid-gap state density is generally insufficient to support this assumption, for semiconductors with a bandgap greater than 0.5 eV, the mid-gap states are generally provided by excitons.[6] Excitons are electron–hole pairs coupled through Coulombic attraction and revolve with radii that are dependent on the electron- and hole-effective masses. Drawing a corollary between the exciton radius, using the reduced effective mass, m_r, defined in Eq. 7, and that of the Bohr atom, the exciton-binding energy is found to be comparable to that of shallow donors. Although excitons can have a strong dipole moment, above 70 K it is improbable that they will be bound to an ionized impurity atom, and, thus, are generally free to move through the lattice.

Equations 8a and b indicate that plotting the square root of the measured absorption coefficient versus photon energy will result in two straight-line segments. The slope is relatively shallow at low photon energies corresponding to the reduced photon energy dependence with phonon absorption, Eq. 8a. An increased dependence of the absorption coefficient on the photon energy exists at higher photon energies where phonon emission dominates, Eq. 8b. As can be expected, the probability of phonon absorption has a strong temperature dependence, whereas that for phonon emission is small. Consequently, the temperature dependence of α_a is dominated by the phonon energy distribution, the denominator of Eq. 8a, and is much greater than the temperature dependence of α_e which is primarily a function of the bandgap. The bandgap temperature dependencies are shown in Eq. 9a–d below.[4,9] Information concerning the semiconductor bandgap and the phonon energy distribution can be obtained from extrapolation of $\sqrt{\alpha}$ versus $h\nu$, which yields intercepts of $E_g - E_p$ for phonon absorption and $E_g + E_p$ for phonon emission.

$$\text{Si } E_g(T) = 1.170 - 4.730 \times 10^{-4} T^2/(T + 636 \text{ K}) \tag{9a}$$

$$\text{Ge } E_g(T) = 0.744 - 4.774 \times 10^{-4} T^2/(T + 235 \text{ K}) \tag{9b}$$

$$\text{GaAs } E_g(T) = 1.519 - 5.405 \times 10^{-4} T^2/(T + 204 \text{ K}) \tag{9c}$$

$$\text{HgCdTe } E_g(T) = -0.302 + 1.93x + 5.35 \times 10^{-4} T(1 - 2x) - 0.810x^2 + 0.832x^3. \tag{9d}$$

The probability of carrier–carrier scattering becomes significant at high carrier concentrations resulting in a reduced requirement of phonon-supplied momentum change for bandgap minimum, indirect transitions. High carrier concentration can be obtained through high irradiance or high dopant concentrations. For high dopant concentrations, the absorption coefficient will be increased further due to dopant-related bandgap narrowing.[10] However, in either case, the high probability of filled states near the conduction-band minimum or empty states near the valence-band maximum causes an effective increase of the bandgap known as the Burstein–Moss shift. Note that the Burstein–Moss shift will affect photon absorption but not photon emission, which allows this effect to be experimentally isolated from other bandgap dependencies. The absorption coefficient for high carrier concentrations is expressed below, where N is the number of scatterers, for example carriers, E_g is the bandgap, N_D is the dopant concentration, and ζ represents the Burstein–Moss shift

$$\alpha_{cc}(h\nu) = A' N [h\nu - E_g(N_D) - \zeta]^2. \tag{10}$$

As mentioned earlier, at high photon energies the photoelectric quantum efficiency can exceed unity through secondary X-ray emission and photocarrier-initiated impact ionization. However, at high photon energies the absorption coefficient for photoelectric absorption, α_{pe}, also begins to degrade as described by the empirical relationship

$$\alpha_{pe} = A_{pe}(h\nu)^{-7/2} \rho_d Z^5 \tag{11}$$

where A_{pe} is a constant and ρ_d and Z are the absorbing material's density and atomic number, respectively. The reduced absorption coefficient at higher photon energies is related to interaction between the different sub-bands of the

282 RADIATION SENSORS

valence and conduction bands. It is also related to the quantum uncertainty of high-energy photons and electrons of the interacting atom existing at a given time and space. This uncertainty becomes appreciable as the radiation wavelength approaches the atomic spacing, for example $\lambda = 5$ Å at $hv = hc/(\bar{n}\lambda) = 662$ eV for silicon, which is the lower extent of the X-ray spectrum. Figures 4 and 5 display the photoelectric absorption coefficients versus the incident photon energy of several different semiconductors, Figures 1.27, 13.5 and 13.6 of Ref. 4 display similar curves for additional materials.

At even higher photon energies, the photon itself can be scattered by an ionizing collision as identified by Comptom in 1923. Compton provided strong empirical support of Einstein's quantum theory of light. The scattered photon,

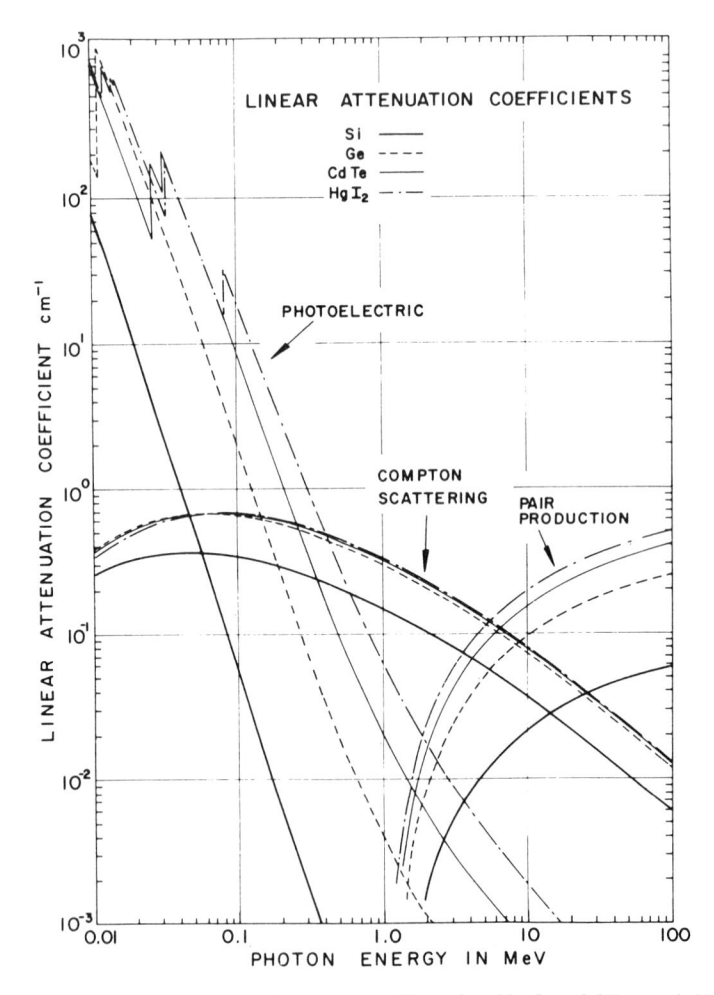

Fig. 5 Measured absorption coefficients at 300 K for Si, Ge, CdTe, and HgI_2 as a function of incident photon energy in the range of 0.01 to 100 MeV. (After Ref. 11)

which is actually a separate photon with reduced energy, can then participate in another Comptom or photoelectric event. Although Compton scattering exists for all radiant energies, its impact is proportional to the photon's wavelength shift, 0.05 Å maximum,[5] relative to its original value. Therefore, the Compton effect is only signficant for wavelengths less than 1 Å, which corresponds to the upper end of the X-ray spectrum, that is electromagnetic radiation with energies greater than 12.5 keV. However, since the photoelectric absorption coefficient decreases with shorter wavelengths, the Comptom effect begins to dominate for energies greater than 57 keV in silicon and 150 keV in germanium. The absorption coefficient for Compton scattering, α_{cs}, is independent of atomic number, and varies with radiant energy according to the empirical relationship:

$$\alpha_{cs} = A_{cs}\rho_d/(hv) \tag{12}$$

where A_{cs} is a constant. It is interesting to note that the Compton effect does not involve a threshold energy as does the photoelectric effect, and the following absorption mechanism—pair production. Therefore, it may be appreciable at low radiant energies in addition to free-carrier photoelectric absorption.

Pair production involves the conversion of electromagnetic energy into rest energy through a photon transformation into an electron and a positron. The energy requirement for pair production is calculated from the sum of the electron and positron rest mass energies, $E = 2m_0c^2 = 1.02$ MeV, which is at the highest levels of radiant energy. In addition, a nucleus must be nearby for conservation of momentum. A positron has the same mass but opposite charge of an electron and annihilates with an electron almost immediately releasing two 0.51 MeV X-rays that may be absorbed via Compton scattering. Empirically, the pair-production absorption coefficient, α_{pp}, is found to increase with radiant energy based on the relationship below where A_{pp} is a constant.

$$\alpha_{pp} = A_{pp} \ln(hv)p_d Z^2. \tag{13}$$

Unlike electromagnetic radiation, charged-particulate radiation carries an electric charge. Fast electrons and heavy-charged particles interact primarily with semiconductor materials through Coulombic processes.[2] Attractive (repulsive) Coulombic forces between the positive (negative) charge on the particle and the negative charge on the electrons of the semiconductor atoms are the primary mechanism for energy loss. In each interaction, Coulombic forces may cause an electron of the semiconductor to be excited to a higher electron shell or may cause the electron to be completely removed from the atom, thereby ionizing the semiconductor atom. If the semiconductor is ionized and the free electron is given minimal energy, there is a high probability that the ions will recombine to form neutral atoms. However, given sufficient kinetic energy, the free electron may cause further ionizations in the semiconductor. Free electrons that cause secondary ionizations are called delta rays.[2]

Interactions of fast electrons with the semiconductor absorber material also include radiative processes.[2] As a fast electron is passing through a semiconductor, the electron can be deflected and radiate energy in the form of a photon. The radiation is called Bremsstrahlung, which is translated as breaking radiation. Radiative processes occur with higher probability for high electron energies and high-atomic-number absorber materials. Energy loss from radiative processes are insignificant compared to the energy loss in the form of excitation and ionization of the semiconductor atoms. Interactions of heavy charged particles with the nuclei of the semiconductor also occur. This type of interaction can be maximized through orientation of the lattice to the incident beam and is the basis of the material-characterization technique called Rutherford backscattering.

With each interaction, energy is released by the fast electron or heavy-charged particle and the particle's velocity is decreased. Because the mass of a fast electron is equal to that of an atomic electron, a large part of its energy may be lost with each interaction and large-angle deflections may occur along its track. The angle of deflection may be so great that the fast electron emerges from the same surface it entered. This phenomenon is termed backscattering. Backscattering occurs with highest probability with electrons of low incident energy and absorber materials of high atomic numbers. The range of fast electrons is difficult to define because the electron is easily deflected and its total path length is much greater than its net penetration depth into the material. The mean range is defined as that thickness of the absorbing material required to reduce the particle count to one half its value in the absence of the material. Graphs of the electron ranges in several detector materials appear in Ref. 2.

The amount of energy released by a heavy-charged particle in an interaction with an absorbing material follows a Bragg curve.[2] The maximum energy released is given by $4E_k m_e/m_p$, where E_k is the initial kinetic energy of the particle, m_p is the mass of the charged particle, and m_e is the mass of the electron. Many interactions are necessary to bring the particle to rest. When the particle's energy is essentially dissipated, it will pick up electrons from the semiconductor and becomes a neutral atom. Because heavy-charged particles have masses greater than the electrons with which they are interacting, they are not significantly deflected. Heavy-charged particles can, therefore, be characterized with a definite range in a semiconductor. Compared to a fast electron of equivalent energy, the specific energy loss of a heavy-charged particle is much greater and its path length is hundreds of times smaller.[2]

Fast electrons traveling at relativistic velocities, called minimum-ionizing particles, generate approximately 85 electron–hole-pairs per μm through excitation and ionization. The response of semiconductor detectors to heavy-charged particles is more complicated than for fast electrons. Due to the low energy of heavy-charged particles, much of the incident energy can be lost in the entrance window of the detector. In addition, the energy-loss mechanisms of heavy-charged particles do not occur entirely through electron collisions. Nuclear collisions that are not detected electronically become important as the ion slows down. Also, recombinations of generated electrons and holes occur due to the dense plasma generated along the ionization track. A pulse-height deficit, therefore, occurs in the measured spectrum.

Because they are not charged, neutrons do not interact with semiconductors through Coulombic interactions. Interactions of neutrons occur with nuclei of the absorbing material. A complete transfer of the neutron's energy of the interacting atom may occur with the ejection of secondary radiation from the atom. The energy of the incident neutron may also be transferred only partially to the interacting atom. In this case, the energy and direction of the incident neutron change significantly, and, again, secondary radiation in the form of heavy-charged particles is emitted. Secondary radiation is primarily in the form of heavy-charged particles, which are produced as a result of neutron-induced nuclear interactions or may be the nuclei of the absorber material.[2,3]

Neutrons may travel long distances, tens of centimeters, in semiconductor materials before undergoing an interaction. Therefore, it is necessary to deposit a thin layer of a conversion material on the semiconductor surface to detect neutrons efficiently. The neutrons will be absorbed in this material and secondary charged-particulate radiation will exit. The detection mechanism is then the same as that for fast electrons and heavy-charged particles.[2,3]

6.2.4 Semiconductor Sensors for Detecting Electromagnetic and Nuclear-Particle Radiation

Unlike solar cells, detectors are designed to have short response times and high sensitivities to a relatively narrow band of the radiation spectrum. Although fiber optics and charge-coupled devices (CCDs) are commonly used in conjunction with radiation detectors, they are not detecting devices. Therefore, the specific design attributes of solar cells and fiber optics are beyond the scope of this chapter. The devices outlined below have been developed primarily for use in detecting electromagnetic radiation in the visible spectrum. In certain situations, which will be highlighted, the sensors have been modified for efficient detection of high-energy electromagnetic and/or nuclear-particle radiation. In addition, CCDs for high-energy radiation sensors will be discussed with regard to the unique attributes required for this application.

Photoconductors. Simple unipolar photoconductors subjected to a radiant power, P_{opt}, can provide high gain since the incident photocurrent modulates the device conductance, σ, and has a multiplicative effect on the device current. The incident photocurrent, I_{ph}, is defined as the product of incident photons per unit time, P_{opt}/hv, and the quantum efficiency, η, which is the number of electron–hole pairs generated per incident photon[4]

$$I_{ph} = q\eta P_{opt}/hv. \tag{14}$$

The quantum efficiency or yield, η, is an important figure of merit for all radiation detectors. The number of free charge carriers generated per incident photon is termed the intrinsic quantum yield. However, a detector is more properly characterized by the external quantum yield. This is equal to the number of charge carriers successfully transferred out of the device per incident photon and is always smaller than the intrinsic quantum yield due to various loss mechanisms. However, high-quality starting material and proper device processing can yield devices with external quantum yields close to the intrinsic value.[11]

The photoconductor current, I_p, is derived in Eq. 16 below using Ohm's law and steady-state generation rate, $G = n/\tau$, where n is the number of carriers per unit volume and τ is the carrier lifetime. The generation rate can be expressed in terms of the incident photocurrent assuming that the thickness in the direction of radiation travel, D, is much greater than the absorption length, $D \gg 1/\alpha$, where α is the absorption coefficient, and the optically-generated carrier concentration is much greater than that generated thermally, $n_{optical} \gg n_{thermal}$.

$$G = \frac{n}{\tau} = \frac{\eta(P_{opt}/h\nu)}{WLD} \tag{15}$$

where W and L are the photoconductor width and electrode spacing, respectively,

$$I_p = (\sigma\mathscr{E})WD = (q\mu_n n\varepsilon)WD = q\left(n\frac{P_{opt}}{h\nu}\right)\left(\frac{\mu_n \tau \mathscr{E}}{L}\right) = I_{ph}\frac{\tau}{t_r}. \tag{16}$$

Identifying the carrier transit time, $t_r = L/V_{drift} = L/(\mu_n \mathscr{E})$, where V_{drift} is the drift voltage, clarifies the requirement of long carrier lifetime relative to the transit time for high photoconductor gain I_p/I_{ph}. However, lifetime is a difficult property to control and a direct trade-off exists between the high gain and the slow response to modulation of the radiant signal that is associated with long lifetime. The carrier transit time is limited by the compromise with sensitive area (assuming the device is not illuminated through transparent electrodes), as well as by velocity saturation and space-charge effects. In addition, the equations above become inaccurate and the photoconductor response degrades if the assumptions of complete absorption and high-level illumination are not met.

The noise equivalent power, NEP, is the incident rms radiant power, $mP_{opt}/\sqrt{2}$, required to produce a unity signal-to-noise ratio in a 1 Hz bandwidth, B, where m is the modulation index.[4] In other words, a lower NEP corresponds to increased photodetector sensitivity. For thermal noise and the generation–recombination component of shot noise for a photoconductor illuminated by a radiant signal with a modulation index, m, and frequency, ω, the NEP is directly proportional to the radiant energy and inversely proportional to the modulation index and quantum efficiency, which is also a function of the radiant energy,

$$\text{NEP} = mP_{opt}/\sqrt{2} = \frac{4\sqrt{2}h\nu B}{\eta m}\left[1 + \frac{kT/q}{V_{applied}}(1+\omega^2\tau^2)\right]. \quad (17)$$

The second term inside the brackets may be neglected for modulation frequencies well below $1/\tau$ and for large ratios of applied voltage to the thermal voltage, kT/q. Under these conditions, the NEP is dominated by the quantum noise of the radiant signal. However, the NEP becomes directly proportional to lifetime as the modulation frequency increases. A comparison of several photoconductor detectivities, $D = \sqrt{\text{area}}/\text{NEP}$, is shown in Fig. 13.4 of Ref. 4.

A considerable reduction of the NEP, as well as of the response time, can be obtained with photodiodes. However, prior to the photodiode discussion, a brief review of noise in electronic materials will be presented since noise accounts for one of the primary limitations on radiation-sensor operation. A more detailed review may be found in Refs. 4 and 12.

Noise. Noise generated by solid-state devices involves the random behavior of individual carriers as opposed to the predictable device characteristics related to the mean of a large population of carriers. An understanding of noise components is important because a change in the measured output current related to the generation of a finite number of photocarriers may be obscured by the random nature of these and other carriers. Since random fluctuations will have a mean zero and exist with different frequencies, the output current is squared and filtered to a given frequency over a certain bandwidth, B, typically 1 Hz. Therefore, the spectral density of current fluctuations is reported as the mean square per unit bandwidth at a given frequency, $\langle\Delta I^2\text{-sec}\rangle$. Measurands such as voltage, power, resistance, etc., are treated in a similar manner.

Four predominant sources of noise exist in semiconductor devices; thermal, shot, flicker, and popcorn noise. Thermal noise, also known as Nyquist or Johnson noise, is related to a carrier's random motion at its thermal velocity in a conductor. As can be expected, it is independent of the modulation frequency, since carriers cannot respond faster than the thermal velocity, and it is directly proportional to the thermal energy kT. Thermal noise dominates the noise spectrum at high frequencies beyond the response time of other noise components. Relationships for the mean square current, $\langle i_{th}^2 \rangle$, and voltage, $\langle v_{th}^2 \rangle$, constituents are shown below where G and R are the device's conductance and resistance, respectively

$$\langle i_{th}^2 \rangle = 4kTBG \qquad \langle v_{th}^2 \rangle = 4kTBR. \quad (18a,b)$$

Shot noise is related to the probabilistic nature of carriers, which is typically dominated by generation and recombination modifying the carrier concentrations available for conduction. The thermal generation current is commonly referred to as dark current. The probability of thermal generation or recombination is described through the carrier lifetime, τ, which is dependent on temperature, carrier concentrations and direct versus indirect transitions. For indirect transitions, the lifetime is additionally dependent on surface- and bulk-trap concentrations, energy levels and capture cross-sections.[13] The dependence on carrier concentration leads to much shorter recombination versus generation lifetimes for indirect semiconductors. For both direct and indirect semiconductors, the lifetime is reduced with increased temperature due to more energetic carriers and narrower bandgaps. This effect limits the operating temperature of narrow-gap semiconductors used for infrared detectors. The mean square shot or dark-current noise, $\langle i_D^2 \rangle$, is shown below where I_D is the steady-state dark current and f is the modulation frequency of the incident radiation

$$\langle i_D^2 \rangle = \frac{2qI_D B}{1+(2\pi f \tau)^2}. \tag{19}$$

At frequencies greater than $1/\tau$, the generation–recombination process is unable to keep pace and the total noise current decreases with increased frequency until it levels off at the thermal-noise level. Longer carrier lifetime extends the spectrum of reduced generation–recombination noise to lower frequencies. At frequencies much less than $1/\tau$, the generation–recombination processes are in equilibrium with the radiant-signal modulation and the shot noise becomes frequency independent. In this range, I_D is often replaced by the total frequency-independent noise current related to probabilistic events, including the desired and background photocurrents, I_P and I_B, tunneling current, I_T, as well as I_D. The total frequency-independent shot noise is represented by

$$\langle i_s^2 \rangle = 2q(I_P + I_B + I_D + I_T)B. \tag{20}$$

Note that both moderate-frequency shot noise and thermal noise are frequency independent and thus linearly proportional to the bandwidth within which they are measured. This characteristic is referred to as white noise.

Flicker or $1/f$ noise has historically been explained as free-carrier exchange with traps at the semiconductor/passivation-layer interface, which causes a fluctuation of the carrier density. The model explains the near-linear increase of $1/f$ noise with decreasing frequency. The increase in noise is a result of a distribution of characteristic exchange times related to the traps' spatial distance into the oxide and/or energy difference from the semiconductor band edges.[14] However, the existence of $1/f$ noise in long-base diodes, which are not appreciably affected by surface recombination, and the inadequate modelling of $1/f$ noise in thermoelectric emf and Hall voltage, indicate that this mechanism is not solely responsible.[15] As pointed out by Hooge[16] and later expanded upon by Kleinpenning,[15] the observed resistance noise can be equally explained by mobility fluctuations, $n\,\delta R/\delta n = \mu\,\delta R/\delta \mu$, where R is resistance, n is the carrier concentration and μ is the carrier mobility. This explanation is not limited to a surface effect and suitably models the $1/f$ noise of thermoelectric emf, and Hall voltage. The mobility model predicts that the frequency at which the $1/f$ noise equals the shot noise, the knee frequency f_K, is dependent on long- or short-base diode geometry, the diode ideality factor, and illumination. Mobility fluctuations translate to noise in the minority-carrier diffusion current through the carrier diffusivity, D, and Einstein's relationship, $D=(kT/q)\mu$. For non-illuminated, long-base, ideal diodes with current flow much greater than the saturation current, I_0, for example when the diode is forward-biased, this knee frequency is inversely proportional to the recombination lifetime as shown below, where κ is an empirical constant on the order of 10^{-3}.

$$f_{K\text{dark}}(I \gg I_0) = \kappa/8\tau. \tag{21}$$

The illuminated knee frequency, again for a long-base, ideal diode with $I \gg I_0$, approaches the non-illuminated value when $\alpha X_J < 1$, where X_J is the junction depth. This condition corresponds to most of the radiation being absorbed before the junction, and therefore, not affecting the minority-carrier current in the diode base. However, when $\alpha X_J > 1$, the $1/f$ noise is reduced considerably because the current in the diode base is now dominated by the more stable supply of majority carriers,[15]

$$f_{K\text{illum}} = \frac{\kappa}{8\tau} \frac{\frac{1}{2}\exp(-\alpha\tau)}{1 - \frac{1}{2}\exp(-\alpha\tau)}. \tag{22}$$

However, this empirical mobility model has been found deficient for modeling $1/f$ noise in diodes dominated by generation–recombination and tunneling currents as well as in short-base photodiodes.[17,18] The quantum $1/f$ noise model was developed to account for several of these deficiencies.[19] However, this model also fails to explain fully the different dependence of $1/f$ noise on bulk versus surface-generation leakage and tunneling current. Reported differences in the surface versus bulk components and temperature coefficients corresponding to $1/n_i$ versus $1/n_i^2$ [15,17] suggest several physical mechanisms are responsible for $1/f$ noise.

Popcorn or burst noise is characteristic of a sizable current burst or quantized shift in conductance with frequencies ranging from 10^{-6} to 10^3 Hz. It occurs in only a small fraction of manufactured devices suggesting a defect-related mechanism. A strong interest exists in both $1/f$- and popcorn-noise components since devices are often expected to operate over a wide frequency range, and these low-frequency noise components impose an effective DC offset at high-frequency operation.

Photodiodes. Photodiodes may be operated in the photovoltaic mode where they are connected to a high impedance (near open circuit) and the developed potential across the diode has a logarithmic dependence on the photon flux.

However, the more typical mode of operation is the photoconductive mode in which the photodiode is short-circuited or reverse-biased and the photocurrent is measured. The output current in the photoconductive mode has a linear dependence on the flux for low-energy photons and charged-particle radiation and is directly proportional to the energy of high-energy photons. For most conditions, the photoconductive mode also provides faster response and a greater dynamic range. In contrast to the photoconductor, the direct measurement of the incident photocurrent by most photodiodes limits the gain to the external quantum efficiency, that is the number of carriers transferred out of the diode per incident photon or charged particle. The external quantum efficiency has a maximum of unity for low-energy radiation and equals hv/ε_i for high-energy radiation, where ε_i is the energy required for ionization from secondary high energy photocarriers or X-ray emission. Above unity gain can be obtained from low-energy radiation using avalanche photodiodes, which will be discussed shortly.

For low modulation frequencies of low-energy electromagnetic-radiant signals, the photocurrent generated in the depletion region (e.g. drift region) and that generated within a few diffusion lengths of the depletion region edge, J_{dr} and J_{diff} respectively, are nearly linerly dependent on the incident irradiance, Φ_0. The relationships are derived through Eq. 6[4] and are listed below for a P^+/N^- photodiode.

$$J_{dr} = [q\Phi_0(1 - e^{-\alpha W_d})] \qquad (23a)$$

$$J_{diff} = \left[q\Phi_0 \frac{\alpha L_p}{1 + \alpha L_p} e^{-\alpha W_d} + qp_{n0}\frac{D_p}{L_p}\right] \qquad (23b)$$

$$J_{total} = \left[q\Phi_0\left(1 - \frac{e^{-\alpha W_d}}{1 + \alpha L_p}\right) + qp_{n0}\frac{D_p}{L_p}\right] \qquad (23c)$$

where α is the absorption coefficient, W_d is the depletion width and L_p, D_p, and p_{n0} are, respectively, the hole diffusion length, diffusion coefficient, and equilibrium minority-carrier concentration in the n-type side of the diode. As the modulation frequency is increased, the measured rms photocurrent is reduced and a phase shift occurs. The phase shift results from the diffusion current and, eventually, the drift current not being able to keep pace, due to their respective transit times.

Electrons and holes generated from electromagnetic radiation that is absorbed within the depletion region will drift out of this region toward their respective contacts. The charge-carrier collection time is expressed as

$$t = \frac{1}{\mu}\int_x^d [\mathscr{E}(x)]^{-1}\, \partial x \qquad (24)$$

where μ is the carrier mobility, x is the point of generation and d is the depletion-layer edge.

For material with a low density of recombination centers, there is a high probability that all of these carriers will be collected and a true measure of the energy of the radiation can be made. If the bulk is not fully depleted, the charge created in the undepleted region will move by diffusion. The equivalent of the charge generated within one diffusion length, L, of the depleted region will be collected where $L=(D\tau)^{1/2}$ and D and τ are the carrier diffusivity and lifetime, respectively. However, one diffusion length is only an average and carriers generated as far as three or four diffusion lengths from the depletion-layer edges have a significant probability of being collected rather than recombining. Collection by diffusion is comparatively slow, and therefore, the collection time has to be lengthened considerably for the measurement to include this charge.

The electric field within the diode-depletion region separates the photo-carriers at velocities near their thermal limit compared to the drift velocity and recombination lifetime that regulate the response of photoconductors. Although the width of the depletion region has a direct impact on the photocarrier transit time, $t_r = W_d/v_{sat} \simeq W_d \times 10 \text{ ps}/\mu\text{m}$, the photodiode response is also limited by junction capacitance, which is inversely proportional to the depletion width. The optimal compromise between the transit time and capacitance-charging time, $R_{eq}C_J$, is one half the modulation period of the incident signal for detection of low-energy electromagnetic-radiation with a properly matched load resistance.[4] A wide depletion region, $W_d \approx 1/\alpha$, is required in indirect semiconductors for reasonable quantum efficiencies with low and very high radiant energies. Side illumination is an alternative. However, in addition to the requirement of special packaging, one dimension of the sensitive area is defined by the depletion width. As mentioned above, carriers absorbed beyond the depletion region but within a few diffusion lengths of its edge can contribute to the photocurrent, but with an additional transmit time. Estimating the diffusion transit time, τ_{diff}, reveals it is several orders of magnitude greater than that of the depletion region

$$\tau_{diff} \simeq L^2/D = \tau \tag{25}$$

where the carrier lifetime, τ, generally ranges from 1ns to 1ms. Direct semiconductors tend to have a sharp absorption cut-off, and thus wide depletion regions are normally not required for low radiant energies. However, a given material or alloy is only applicable for a limited range of photon energies.

The quantum efficiency is limited at very high radiant energies for the reasons discussed in Section 6.2.3. The high density of recombination centers at the semiconductor surface limit the quantum efficiency for moderately high radiant energies where the photocarriers are generated close to the surface and are likely to recombine before they are transported across the depletion region. A high density of bulk recombination centers will also exist for a heavily doped diffusion as well as a reduced absorption length because of the dopant related bandgap narrowing.[10]

To produce an external photocurrent, the photocarriers must completely transverse the depletion region and then recombine with majority carriers supplied by the ohmic contact. A carrier flow from the diode electrodes is not required when a photo-electron and photo-hole recombine either in the neutral regions, for example due to the high density of midgap states at the surface, or after they enter the depletion region from opposite ends. Shallow junctions are used to reduce the recombination and the diffusion time of carriers generated in the surface neutral region. However, shallow junctions must provide an ohmic contact and avoid depletion-region exposure of the high recombination velocity at the contact–metal interface. Shallow junctions must also be free of crystalline defects to minimize photocarrier recombination within the depletion region.

Improved response times can be obtained for radiant energies below that of the SiO_2 bandgap, 8.8 eV, using an oxide isolated diode to minimize the diffusion current. The response time can be further improved by geometrically altering the compromise between the depletion region transit time and junction capacitance. The interaction of depletion regions spreading from two or more directions will provide reduced junction capacitance and a longer absorption path-length with a smaller depletion region transit length relative to the planar diode. A cross-section of this type of diode is shown in Fig. 6, which can be integrated into a dielectric-isolated, complementary bipolar process. An epitaxial dopant concentration slightly below 10^{15} cm^{-3} with a ± 5 V power supply provides a 10 μm wide exposure window per cell of the photodiode.

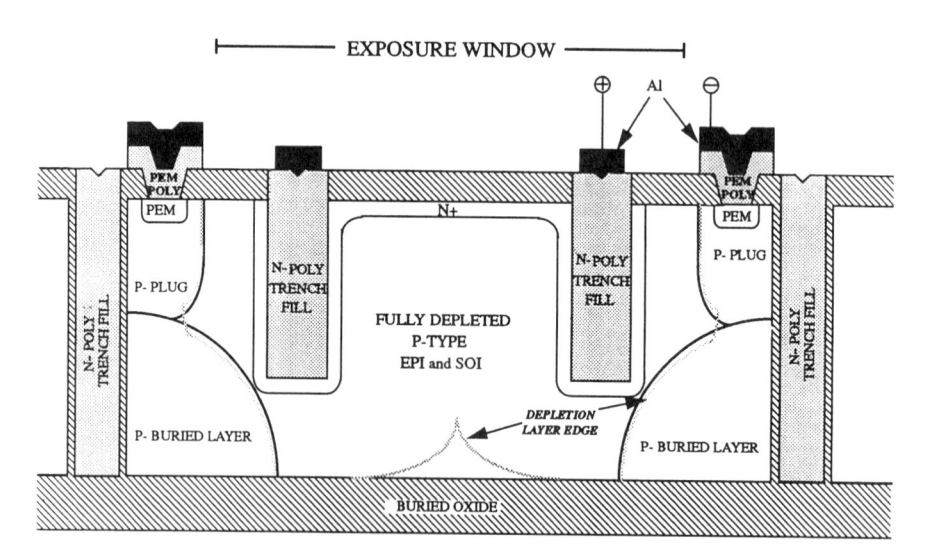

Fig. 6 One cell of a silicon-on-insulator (SOI) photodiode. The buried oxide minimizes the diffusion component of the photocurrent and the related slow response. The vertical and horizontal merging of the depletion regions below the exposure window improves the response time compromise between junction capacitance and depletion region transmit time.

The center p–epi region is depleted vertically from the surface N^+ region and horizontally from the sidewall n-type regions which are diffused from the poly of the shallow trenches. Although the sidewall n-type regions increase the junction area, and thus capacitance, the capacitance in the center region(s) is small due to the merged depletion regions. The reduced transit length applies even for carriers generated close to the exposure surface since the electrons do not need to travel the full absorption path-length to get around the shallow trench to the neutral p-type side of the diode. The absorption path-length can be adjusted through either the epi or SOI thickness to meet the quantum efficiency requirements at low radiant energies. This flexibility enables improved integration of the photodiode into a bipolar process.

This SOI diode could also be used as an avalanche photodiode utilizing the high electric field between the p-buried layer and the trench corner n-region. The separation of the multiplication region and the absorbing volume provides improved noise characteristics, which will be discussed later in this section. Alternative geometric arrangements are also possible such as a "side-illuminated" diode with one sidewall n-type and the other p-type.

The surface N^+ implant can be adjusted to optimize this region's characteristics in relation to bandgap narrowing, Auger recombination, and a deep junction resulting from higher implant doses; and a comparison can be made with the problems of interface-trap recombination and high series resistance that result from lower implant doses. The problem of high recombination velocity at the contact–metal interface as a result of complete depletion of the surface region can be avoided by contacting the n^+-poly of the shallow trenches. This structure's inherent capability for integration into the upcoming generation of high-speed bipolar processes would enable on-chip amplification, conversions, and digital processing at the frequencies of interest.

To further reduce the impact of surface recombination, silicon Schottky diodes with transparent 100 Å electrodes and anti-reflective coatings are commonly used for ultraviolet and soft X-ray detection. A low interface-recombination velocity is essential for such devices. In addition to enhancing the quantum efficiency of radiation with a high absorption coefficient, for example moderately high radiant energies, metal–semiconductor diodes can be used to detect radiant energies below the bandgap cut-off. For photon energies greater than the Schottky barrier, $q\phi_b$, but less than the semiconductor bandgap, free-electron absorption on the metal can provide sufficient energy for photoelectrons to surmount the Schottky barrier and pass into the semiconductor. This mode of operation has been used extensively for empirical determination of barrier heights, as well as in commercial applications, such as PtSi infrared detectors.

Heterojunction photodiodes are also employed to avoid the collection inefficiencies resulting from surface recombination associated with high absorption coefficients. Such photodiodes are illuminated through the large bandgap material, which is essentially transparent to the radiant signal, and generation occurs on the narrow-gap side of the junction. Provided the interface recombination velocity is low and the bulk lifetimes are sufficiently high,[4] photocurrent generated within the narrow-gap material can be efficiently collected. Superlattice structures, a variant of the heterojunction, consist of multiple narrow layers of different bandgap materials. These layers are as thin as 50 to 100 Å. This removes one degree of freedom of the carriers' wave function, which is especially significant for excitons whose Bohr radius is considerably greater than the lattice constant. Many novel devices are based on the unique properties of superlattices.[6] Well-defined transitions within the conduction band of ternary and quaternary semiconductor heterojunctions make superlattices a primary candidate of long-wavelength infrared detectors.

Matching of the lattice constants of two semiconductors presents the fundamental limitation on the heterojunction-interface characteristics. Near perfect matching and recent improvements in the epitaxial-growth techniques of GaAs-, InP- and HgCdTe-based alloys have provided manufacturable alternatives to silicon-based Schottky diodes for moderate radiant energies and alternatives to germanium diodes for radiant energies below the silicon limit of approximately 1 eV. However, the narrow-gap semiconductors used for infrared detection have inherently higher dark current and are limited to low reverse biases to avoid excessive tunneling and ionization currents. Low temperatures, for example 77 K, are commonly used to reduce the dark current. However, this results in increased ionization† and lower diffusion coefficients. Therefore, narrow-gap semiconductors are commonly short-circuited so that only the build-in field exists and minority carriers are not extracted from the neutral regions, thereby eliminating the diffusion photocurrent.

In addition to the collection efficiency, the external quantum efficiency and the related figure of merit, responsivity, are also dependent on the transmission efficiency of the incident radiation to the semiconductor. Proper protective oxide or coating thickness, for example $t_{ox} = \lambda_{ox}/2$ where t_{ox} is the oxide thickness and λ_{ox} is the radiation wavelength within the oxide, can provide some degree of anti-reflection properties through constructive interference of the reflected and refracted waves, whereas an improper thickness can cause destructive interference.[8] In addition, non-absorbing, anti-reflective coatings or elaborate inverted-pyramid topography[21] may be applied in addition to back-side metal to reflect back radiation transmitted through the full semiconductor thickness. Note that the responsivity, R, which is the photocurrent per unit radiant power, contains an inherent dependence on the radiant energy, decreasing linearly with increasing energy at a fixed quantum efficiency

$$R = I_p/P_{opt} = \eta\lambda \ (\mu m)/1.24 \ (A/W). \tag{26}$$

Diode structures with a near-intrinsic region separating the regions of p- and

---

†The ionization coefficient increases at lower temperatures because the reduction in lattice vibrations results in a longer carrier mean-free path, the distance over which the carrier gains energy from the local electric field.

n-type dopants, p–i–n diodes, are commonly used to control the width of the depletion region and provide a nearly uniform electric field. Although a moderately high-electric field is desired so that the carriers reach their saturation velocity, >1 V/μm for silicon, the electric field should not be so high as to cause appreciable ionization, <20 V/μm for silicon, unless the device is designed to operate in such a mode, for example an avalanche photodiode.

Relatively thick p–i–n diodes fabricated on nearly-intrinsic semiconductor substrates are used to detect high-energy electromagnetic and nuclear-particle radiation. Given that the magnitude of the reverse-bias voltage across the semiconductor diode is high enough to deplete the entire intrinsic region, the detector's sensitive volume will approximately equal the volume between the boundary electrodes. Fully depleting the substrate creates a relatively-high electric field that will ensure fast and complete collection of the charge generated within the detector. From Poisson's equation (Problem 1 provides a review exercise) the expression for the depletion width, W_d, of a one-sided abrupt p^+–n junction, appropriate for $N_A \gg N_D$ is[4]

$$W_d^2 = (2\varepsilon_0\varepsilon_s V_t)/(qN_D) \tag{27}$$

where W_d is the depletion layer width, N_A and N_D are the concentrations of acceptors and donors, respectively, ε_0 and ε_s are the permittivity of free space and the dielectric constant of the semiconductor, respectively, and V_t is the sum of the built-in and reverse-bias voltages. The breakdown voltage due to avalanche breakdown, appropriate for $N_A \gg N_D$ is[4]

$$V_B = (\varepsilon_0\varepsilon_s/2q)(\mathscr{E}_m^2/N_D) \tag{28}$$

where $\mathscr{E}_m$ is the electric field at the onset of avalanche. Since the breakdown phenomenon limits the reverse-bias voltage that can be applied, high-purity or compensated semiconductor material must be used to achieve large depletion layers.

Noise equivalent power, NEP, as defined previously (see Eq. 17), is the radiant power required to produce a signal-to-noise ratio of unity over a 1 Hz bandwidth, B. The NEP expression below accounts for the thermal and shot noise of a photodiode illuminated by an rms power, $mP_{opt}/\sqrt{2}$, with a modulation index, m, of 100% and a modulation frequency, ω.[4] The thermal noise involves the parallel combination of the junction-, load-, and amplifier-input resistances, R_{eq}. The shot noise includes the frequency-independent dark current, for example thermal generation current as well as the background- and incident-radiation current components, I_D, I_B, and I_P, respectively. Equation 14 is used to separate the noise of the incident radiation into power and energy components providing a distinction between the cases where NEP is dominated by the noise of the radiant signal rather than the system noise of the sensor. Reduction of the dark and background currents and maximizing the equivalent parallel resistances reduces system noise and can provide a substantial increase in the photodiode sensitivity, that is reduced NEP, only when the system noise is significant.

$$\text{NEP} = \frac{P_{opt}}{\sqrt{2}} = \frac{\sqrt{2h\nu B}}{\eta} \left[1 + \sqrt{1 + \frac{1}{qB}\left(I_B + I_D + \frac{2kT}{qR_{eq}}\right)} \right]. \quad (29)$$

As with the photoconductor, the photodiode NEP is reduced with higher quantum efficiency or lower radiant energy. However, unlike the photoconductor, the photodiode NEP is less sensitive to high modulation frequencies and reduced applied voltage since the photodiode signal is less dependent on these parameters given its faster response time. The photodiode is also less sensitive to recombination noise. The photocarriers are rapidly separated so that nearly all recombination occurs in the neutral regions. For diodes designed to minimize the widths of the neutral regions, the recombination occurs primarily at the ohmic contacts and further constrains the probable recombination events.

Avalanche photodiodes are frequency used because their internal gain can eliminate the need for separate amplification of the photocurrent. Although the signal multiplication, M, produces an inherent noise factor, $F(M) = \langle M^2 \rangle / M^2$, related to the probabilistic nature of impact ionization, the noise factor does not act on the thermal-noise component, which is dominated by temperature and the series and external resistances, following Eq. 18. Therefore, the signal-to-noise ratio of an avalanche photodiode can be increased at high frequencies where thermal noise dominates and the signal still experiences multiplication. In other words, the NEP can be reduced although an additional noise source has been added[4]

$$\text{NEP} = \frac{P_{opt}}{\sqrt{2}} = \frac{\sqrt{2h\nu B}}{\eta} \left[1 + \sqrt{1 + \frac{(I_B + I_D)F(M) + 2kT/qR_{eq}M^2}{qBF(M)}} \right]. \quad (30)$$

The possible variations of the ionization hierarchy are reduced if the feedback effects associated with nearly equivalent electron- and hole-ionization coefficients can be avoided. For example, a large increase in the number of possibilities exists in the case of an electron-initiated event with equal ionization rates since a generated hole is likely to produce electron–hole pairs upstream, sending additional electrons downstream. This in turn, may send more holes upstream. Therefore, the multiplication-noise factor is dependent on the ratio of ionization rates, $\bar{k}$, where the ionization rate of the initiation carrier is the denominator, for example $\bar{k} = \bar{\alpha}_p / \bar{\alpha}_n$ for electron initiation,[4] where $\bar{\alpha}_p$ and $\bar{\alpha}_n$ are the hole and electron ionization rate, respectively,

$$F(M) = \bar{k}M + (2 - 1/M)(1 - \bar{k}). \quad (31)$$

For silicon, the hole-to-electron ionization rate is below 5% for electric fields of 20 V/μm or less, but increases at higher fields as shown in Fig. 1.30 of Ref. 4. The hole-ionization rate dominates for germanium and GaAs, but to a lesser extent. Again a greater difference exists at lower fields. Therefore, it is advantageous to operate avalanche photodiodes at electric fields substantially below that corresponding to true avalanche, $M = \infty$, and to have electron-initiated multiplication in silicon diodes and hole-initiated multiplication in germanium and GaAs diodes. Separation of the absorption and ionization regions provides control over the carrier type that initiates the ionization, and is commonly obtained using p–i–n structures with a wide low-field region for absorption and a narrow high-field ionization region. Schottky[21] and heterojunction[23] avalanche photodiodes have also been reported.

The benefits of avalanche photodiodes comes with the expense of greater processing and operation control. Low dislocation and impurity densities are required to avoid localized high electric-field regions. Lightly-doped guardbands or beveling is also required to avoid the higher fields at the junction curvature and possibly at the surface. A low diode-series resistance and a very stable voltage supply across the range of output currents are required, since the multiplication factor has an exponential dependence on the voltage drop across the junction. As noted earlier, the ionization coefficient also has a considerable temperature coefficient. Control circuitry can be added to account for ambient temperature variations by adjusting the voltage source to maintain a constant current from a second, identical diode in parallel that is shielded from the radiation. However, differences in self-heating between the two diodes at high irradiances must be minimized through adequate heat sinking. Although a cost is associated with these additional requirements for avalanche photodiodes, it is often out-weighed by the elimination of an amplifier. A more detailed review of these important devices is provided in Refs. 4, 21, 22, and 23.

Phototransistor. Transistor action is an alternative means of obtaining more than unity gain from a photodetector. MOSFETs are typically used only for infrared and visible light because the gate-oxide is sensitive to higher energy radiation. Additionally, atypical processing is often required, such as the introduction of relatively-deep acceptor or donor levels that are photo-ionized, causing a threshold voltage shift. A bipolar phototransistor typically involves a large collector–base junction that is used as the photon-collection diode. When the device is biased with the base floating, the generated holes (electrons) raise (lower) the NPN (PNP) base potential, forward-biasing the emitter, which provides $h_{FE}I_{ph}$ carriers for every photocarrier, where h_{FE} is the common-emitter current gain. Thus, the total device current is $(1 + h_{FE})I_{ph}$, similar to the dark current ratio of I_{CEO} versus I_{CES}. However, processing complexity and reduced high-frequency response are the trade-offs for improved gain due to the increased parasitic resistances, specifically R_{base}, and the large storage capacitance associated with the forward-biased emitter junction.

6.3 HgCdTe INFRARED SENSORS

Infrared radiation encompasses the energy range between the microwave and visible spectra and is commonly divided into four regions as shown in Fig. 1: *near* is the range just below red visible light and has a maximum energy of 1.6 eV, *medium* covers energies from 830 to 200 meV, *far* spans from 200 to 30 meV, and *extreme* extends to 1.2 meV. Infrared sensors have been used for thermal imaging or "night vision" for several decades, but recent expectations of a commercial surge have arisen because fiber-optic cables with reduced attenuation for infrared radiation can now be manufactured.

Silicon photodiodes have been used in fiber-optic communication systems with wavelengths at 0.83 μm (i.e. 1.494 eV), but the fiber-optic signal attenuation in this wavelength range is sensitive to hydrogen exposure.[25] Germanium and InGaAs photodetectors can have high quantum efficiencies at wavelengths 1.3 and 1.55 μm (i.e., 954 and 800 meV, respectively). When operated for use at these wavelengths, the attenuation of the output signal is 5 to 10 times less than the attenuation at 0.83 μm. In addition, there is a considerable reduction in the fiber cable's sensitivity to hydrogen exposure at 1.3 and 1.55 μm and these wavelengths are modest distances from the fiber cable's OH-related absorption peaks at 1.24 and 1.40 μm. However, the present technological effort for optical communications is to first move to wavelengths between 3 and 5 μm and eventually to utilize the 8 to 14 μm range to further improve the performance and reliability aspects of fiber-cable attenuation. Although Schottky barrier diodes, such as PtSi or Pd_2Si built with SiGe alloys, as well as conventional silicon[21] and several narrow-gap semiconductors, InSb, PbS, PbSe, and PbTe,[4] are candidates for infrared radiation, the ternary-semiconductor alloy HgCdTe provides several unique advantages in these wavelength ranges.

The atomic structure of $Hg_{1-x}Cd_xTe$ consists of cation (Hg, Cd) and anion (Te) sublattices. Adjustment of the Cd percentage, x, on the cation sublattice can provide energy gaps ranging from 1.6 eV to overlapped conduction- and valence-band edges. Complete miscibility of HgCdTe, which means the absence of phase restrictions over the entire composition range, enables a continuous set of bandgaps extending the photodetector's active range well beyond that of the common III–V-based alloys as well as silicon and germanium. The HgCdTe system has been actively studied for over 30 years. Major technological advances have been made over the past ten years leading to improved manufacturability and reduced costs of HgCdTe relative to other IR-detector candidates.†

Material Properties. $Hg_{1-x}Cd_xTe$ alloys are direct-bandgap semiconductors with a 6.48 Å lattice constant approximately independent of composition. The

---

†Since 1981 this knowledge base has been disseminated at the U.S. Workshop on Physics and Chemistry of Mercury Cadmium Telluride, for which the proceedings of the last six workshops were published in the *Journal of Vacuum Science and Technology*; Jan-Feb '85, Sep-Oct '87, Jul-Aug '88, Mar-Apr '89, May-Jun '91 and July-Aug '92.

6.3 HgCdTe INFRARED SENSORS

bandgap dependence on composition is nearly linear ranging from -0.25 eV for the semi-metal HgTe to 1.6 eV for the ionic semiconductor CdTe at 77 K. Equation 32 is an empirical relationship describing the $Hg_{1-x}Cd_xTe$ bandgap dependence on composition, x, and temperature, T, which is often referred to as the HSC equation after its authors.[9] Compositions of $x = 0.20$ to 0.25 provide 77 K bandgaps from 83 to 163 meV, which have cut-off wavelengths of 14.9 to 7.6 μm and span the targeted range of 8 to 14 μm. However, gradual as well as the expected sharp absorption edges have been reported,[7,27,28] although only sharp transitions are shown in Fig. 4b. Variation between samples with comparable carrier concentrations indicates the residual absorption at longer wavelengths is a result of material inhomogeneities rather than free-carrier absorption.[7] The intrinsic-carrier concentration and parabolic effective electron masses have been derived, Eqs. 33 and 34, from Kane's non-parabolic band model. This model has been used for several years,[27,29] although it does not account for the metal d-band electron screening that recent theoretical studies have shown to be significant.[30,31] The common-anode band model is also typically applied which assumes a composition-independent energy of the valence-band edge, $\chi + E_g$. However, at the composition extremes, the HgTe valence-band edge has been shown to be 35 to 40 meV higher than that of CdTe.[30] The electron affinity, χ, was found from MOS studies to be 4.65 eV at 77 K for $Hg_{0.7}Cd_{0.3}Te$, which has a bandgap of 0.25 eV.[32] HgCdTe dispersion diagrams were shown previously in Fig. 3d,e.[31] The equations below, summarizing several decades of research in the field of HgCdTe solid-state physics,[29] provide a reasonably accurate description of the important material constants

$$E_g = -0.302 + 1.93x + (5.35 \times 10^{-4})T(1-2x) - 0.810x^2 + 0.832x^3 \quad (32)$$

$$n_i = 5.585 - 3.820x + (1.753 \times 10^{-3})T - (1.364 \times 10^{-3})xT -$$
$$[10^{14} E_g^{3/4} T^{3/2} \exp(-E_g/2kT)] \quad (33)$$

$$\frac{m_{ep}^*}{m_0} = \exp\left[\frac{4}{3} \ln\left(\frac{n_i}{3.136 \times 10^{15} T^{3/2} \exp(-E_g/(2kT))}\right)\right] \quad (34)$$

$$\mu_e(300 \text{ K}) = 1/(8.754 \times 10^{-4} x - 1.044 \times 10^{-4}). \quad (35)$$

Unlike silicon, germanium, GaAs, and even CdTe, the $Hg_{1-x}Cd_xTe$ bandgap decreases with decreasing temperature for $x < 0.5$ due to the metallic HgTe bond. The competing effects of the HgTe and CdTe bands on the bandgap provide a reduced temperature dependence of the sensor quantum efficiency. Other researchers[33] have found a slighly greater bandgap dependence on temperature than that shown in Eq. 29 through more elaborate experimental techniques. The conduction-band edge effective masses from these experiments were 0.008 and 0.018 at 77 K for $x = 0.224$ and $x = 310$, respectively, which are in relative agreement with Eq. 34. Most HgCdTe photodiodes are operated at liquid nitrogen temperature, 77 K, for three primary reasons: to reduce the dark currents, to increase the mobilities and to avoid the instabilities related to Hg vacancy diffusion, which is significant at room temperature.

Photocarrier velocities are often not saturated in HgCdTe photodiodes, since these photodiodes are commonly operated with zero-applied bias to minimize the dark current and eliminate the slow-diffusion component of the photocurrent resulting from minority-carrier gradients in the neutral regions. Therefore, high carrier mobility is of vital importance for short response times. The electron mobility relationship of Eq. 35 applies to high-quality material. Up to a factor of 200 times lower mobility has been reported for the case of large, ungettered, impurity concentrations.[34] The 77 K mobility is approximately 15 times that at 300 K for compositions between $x = 0.2$ to 0.3.[28,35] The 77 K hole mobility is approximately 200 cm^2/V-s for high-quality material,[36] however, 20 K values as high as 10^4 cm^2/V-s have been reported for supperlattices. Shockley–Read–Hall (SRH) and Auger lifetimes have been summarized in Refs. 37 and 38, respectively, the latter being dominant for carrier concentrations above 10^{16} cm^{-3}. High carrier concentration effects such as Burstein–Moss shift and band tailing, as described in the preceding common physics section, have also been reported.[39]

Material Defects. The reason why the advantageous optical properties of HgCdTe have not been put to greater commercial use has predominantly involved the defect issues related to the weak HgTe metallic bond, which is further weakened in the presence of the ionic CdTe bond. The low-binding energy of II–VI compounds makes them less stable than group IV and III–IV semiconductors, which principally contain stronger, covalent bonds. However, advances in the field of low-temperature processing, specifically epitaxy growth, are anticipated to enable HgCdTe to be a strong competitor in the 8 to 14 μm detector market.

The principle *p*-type dopant in HgCdTe has historically been the Hg vacancy, which produces deep level traps at $E_v + 0.4E_g$ and $+0.7E_g$ and a shallow acceptor.[40] Although optimized anneals can reduce the trap densities well below those of the shallow acceptors, the densities remain appreciable and contribute to dark currents through SRH generation and trap-assisted tunneling. The trap concentrations, energy levels, and capture cross-sections determined through deep-level transient spectroscopy (DLTS) were found to fit lifetime and noise measurements without any parameter adjustments.[40] Also, the Hg vacancy is normally doubly charged, which increases Coulombic scattering and causes degraded mobilities.[40] Hg vacancies are prevalent since Hg tends to out-diffuse at rates that are dependent on temperature, ambient partial pressures, composition, concentration of Hg vacancies already present, surface preparation, and even illumination. Annealing at temperatures between 250 and 500°C in a Hg overpressure is used to fill the Hg vacancies, converting the material to *n*-type with either residual impurities, Hg interstitials or Te antisites, intersitials, or vacancies acting as donors. A 1 hour anneal at 300°C in a Hg overpressure will result in a junction depth of several micrometers. Although both host and impurity-atom diffusion are believed to be dominated by an Hg vacancy-assisted mechanism, multiple defect structures play significant and coupled roles, and thus, junction-depth prediction and control are difficult. In addition, the excess Hg dissolves Te precipitants, which results in the growth of dislocations and the release of impurities previously gettered by the Te precipitants. Cross-sectional micrographs will typically display a high etch-pit density, 10^5 to 10^7 cm^{-2}, near the surface or the *skin* and a low etch-pit density in the bulk or *core*. The cleaved surface also tends to have ductile characteristics, as opposed to the brittle nature of most ionic and covalent materials. This is consistent with the presence of the metallic HgTe bond and the softening resulting from dislocation motion.

Dopants. Extrinsic dopants in HgCdTe behave primarily as expected based on the periodic table, Fig. 7, and generally provide improved carrier-concentration and junction-depth control over that obtained with intrinsic dopants such as Hg vacancies, Te interstitials, etc. Although a few group V exceptions exist for Te-rich material, which is common for the non-equilibrium epitaxy techniques to be discussed shortly,[41] the group Ia, Ib, and Va elements generally present shallow acceptor levels when substitutionally incorporated into the cation and anion sublattices, respectively. In the same respect, the group IIIa and VIIa elements act as donors. Group III and V elements have been found to be relatively stable during typical anneal cycles, and therefore, are the recommended replacements for their intrinsic counterparts.[24,40,41] Group Ia and Ib elements tend to be fast diffusers with a strong dependence on the Hg-vacancy concentration, and thus are typically gettered from the skin into the Hg vacancy and Te precipitant-rich core. The 270°C diffusivity of silver varies from 3×10^{-9} cm^2/s under Te saturation to 1×10^{-6} cm^2/s for Hg

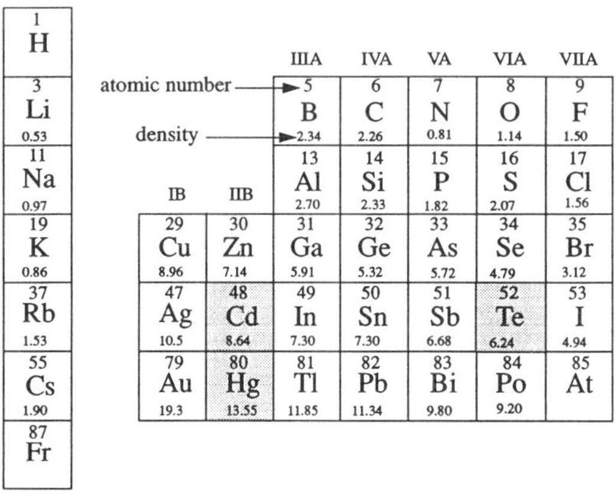

Fig. 7 Portion of the periodic table involved in the fabrication of HgCdTe photodetectors.

saturation. Also, similar to the Hg vacancy, group I elements typically present deep-level traps[40] and are susceptible to surface segregation under illumination.[34]

Arsenic is a particularly well-suited replacement for the Hg vacancy, since it is not known to present a midgap-energy level and is singularly charged. It, therefore, provides reduced dark currents, less noise, and higher mobilities. Extrinsic p-type doping decouples the lifetime from the carrier concentration enabling lifetimes as much as 10 times greater than that available with intrinsic doping.[40] Indium has been the most widely used n-type dopant. However, boron and iodine are also being studied to provide higher electrically-active concentrations. A set of mass action laws have been compiled describing the incorporation of dopants in HgCdTe and the related interactions with point defects.[24]

Processing. HgCdTe substrate growth using horizontal and vertical Bridgman techniques has improved remarkably over the past ten years. However, problems remain with regard to purity of materials, acceptable defect densities for large-area focal planes, material-strength limitation of boule diameter, and finally cost, which generally exceeds $500 for a 0.5 inch diameter wafer.

Incorporation of zinc into the substrate has provided greater material strength and reduced defect densities due to the increased Hg-vacancy formation energy. $Cd_{1-y}Zn_yTe$ has become the primary substrate for epitaxial growth for these reasons, but a trade-off results related to lattice mismatch—approximately 6.46 Å for $y=0.04$ versus 6.48 Å for CdTe. Although the lattice constant of CdZnTe can be adjusted through the zinc composition, the relatively high liquid-to-solid segregation coefficient of Zn, ~ 1.3, causes a greater than 50% concentration variation along the length of the boule. This variation can actually show up across an individual wafer since off-axis sawing is a common means of obtaining the desired orientation. The defect density of equilibrium-epitaxial techniques, such as the industry-standard liquid-phase epitaxy (LPE), are susceptible to lattice mismatch, resulting in increased dark currents and $1/f$ noise, as well as reduced diode yields. However, non-equilibrium techniques such as molecular-beam epitaxy (MBE) and metal-organic chemical-vapor deposition (MOCVD), are relatively independent of lattice mismatch at the expense of greater processing complexity. In fact, considerable attention is being directed to the use of MBE and MOCVD to grow HgCdTe epitaxy on sapphire, GaAs, and silicon substrates to avoid the high cost and small diameters of CdTe substrates.[42] Although its ductile properties make HgCdTe more amenable to mismatched substrates than covalent semiconductors, the quality of such films is still below that of epitaxy-grown films on CdTe or CdZnTe.

Improvements in epitaxial-growth techniques have also enabled the growth of controlled extrinsic doping, as well as composition changes within a few monolayers for heterojunctions and superlattice. MBE at 200°C and MOCVD near 400°C (200 to 250°C with pre-cracking) provide low-temperature alternatives to LPE, which requires temperatures in the range of 500 to 600°C.

Photo-assisted MBE and MOCVD processes currently under development offer the possibility of further temperature reduction. Lower growth temperatures increase the control over the HgCdTe stoichiometry, which can translate into greater composition and dopant control and the possibility of lower defect densities. Another advantage of MBE is the availability of *in situ* electron diffraction, Hall measurements, and ellipsometry, provided the capability of even greater processing control and early defect detection.

Implantation of boron, indium, arsenic and non-substitutional dopants, such as xenon, have been shown to provide reproducible results.[36,43] Without an anneal, the implant damage generates comparable donor concentrations for each species. The junction depths are considerably greater than the projected implant range, most likely due to propagation of dislocation loops. Following an anneal of 250 to 400°C, the In and B sheet resistances are reduced due to diffusion, the As implant is converted to *p*-type, and the dopant effects of the Xe implant are eliminated. These characteristics reveal the dominant extrinsic behavior of these implants. Indium presents several problems as a dopant in that it has a low electrical activity for concentrations above 10^{16} cm^{-3} and its diffusivity is strongly dependent on the Hg overpressure of the anneal and its own concentration when above 10^{21} cm^{-3}.[36,44] Boron and arsenic electrical activities as high as 3×10^{17} and 10^{18} cm^{-3}, respectively, have been demonstrated.[43] Low to moderate temperatures, 200 to 300°C, are commonly used with an Hg overpressure as the last step of an implant anneal to reduce the intrinsic-acceptor (Hg vacancy) concentration to an acceptable level. Although not an intrinsic dopant, oxygen is an inherent donor in HgCdTe, since Hg and Cd readily oxidize.[45] However, oxygen concentrations below 10^{14} cm^{-3} can be obtained through the use of high-purity Hg and Cd, careful processing, de-oxidation via hydrogen carrier gas, and the use of a carbon-coated quartz ampoule for annealing.

Ion-beam milling is used as an anisotropic alternative to bromine-based, isotropic wet etching of HgCdTe. However, the sputter etching generates beam damage converting the HgCdTe to *n*-type with two distinct regions; a thin surface region, <0.1 um, with high-carrier concentration, short lifetime and low mobility, is generated as well as a deep region with moderate concentration, 10^{16} cm^{-3}, long lifetime, and high mobility.[46] The lifetime and mobility of the deep region suggest Hg or Cd interstitials are driven into the substrate, reducing the Hg-vacancy concentration, as opposed to a compensation mechanism. The extent of the deep region, up to 10 μm after only a few minutes etch, is dependent on the substrate history and ion-beam dose, but is only weakly related to the ion species or energy. Propagation of extended defects from the surface region is believed to provide pipes through which the interstitials can quickly travel and then diffuse into the lattice. The degraded properties of the shallow surface region can be removed with an 80°C anneal, well below the temperature required for annealing ion implant damage, suggesting a fundamentally different damage mechanism.

Indium contacts to *n*-type HgCdTe produce lower contact resistance than platinum or gold due to the smaller work function difference and room-temperature diffusion of In into HgCdTe, which creates an n^+ region and promotes tunneling.[47] The In contact resistance is linearly correlated to the Cd composition and related variation in HgCdTe resistivity. The contact resistance varies from 2×10^{-5} Ω-cm^2 at $x = 0.3$ to 2.6×10^{-2} Ω-cm^2 at $x = 0.68$, above which the contacts become non-ohmic.

Electroless-gold deposition from AuCl$_3$ provides ohmic contacts to p-type HgCdTe with as-deposited contact resistances around 1 Ω-cm^2, which drops by an order of magnitude, with a comparable tightening of the distribution, after a 9-day, 100°C anneal.[48] The low solid-solubility of Au in HgCdTe, approximately 10^{16} cm^{-3}, prohibits a tunneling conduction mechanism for the contact. HgTe contacts to p-type HgCdTe have been suggested recently to eliminate the need for an interfacial oxide and improve the adhesion properties of the contact material.[62]

Oxides have been grown or deposited on HgCdTe by a large number of techniques. The surface preparation usually involves a 0.1 to 0.25 μm alumina polish followed by a bromine-based chemical etch. The details of the surface treatment have a large impact on the interface and oxide quality. Anodic oxides have been grown with thicknesses close to 1000 Å, however, they are typically of low quality. Oxides grown on an O$_2$ plasma provide a higher quality film and interface, but are generally less than 100 Å in thickness. A high-quality 10 to 20 Å TeO$_2$ layer will grow in air after a couple of days at room temperature, while an Hg-rich oxide will grow in ozone. PHOTOX, which is SiO$_2$ CVD from SiH$_4$ and O$_2$ at or below 100°C, can provide a thick oxide with low charge densities if the deposition is followed by a low-temperature vacuum bake to drive out moisture. An air-grown native oxide is required between the PHOTOX and the HgCdTe to prevent silicon reaction with Te. Fabrication of MIS capacitors from these oxides has been performed for surface investigations, as well as for a possible photodiode alternative in certain applications. Oxides have also been studied for use as the passivation layer, however, this function remains dominated by large bandgap, low absorption, II–VI compounds such as CdTe and ZnS because of the effect the passivation has on dark currents and noise. HF or HCl can be used for patterning native oxides, SiO$_2$, and ZnS layers without physically distrubing the HgCdTe surface.

Photoconductors, MIS capacitors, standard and heterojunction photodiodes, heterojunction superlattices, and dilute magnetic semiconductor devices have all been fabricated from HgCdTe-based compounds. The latter groups of devices provide improved optical properties for certain applications at the cost of greater processing complexity. Heterojunctions and superlattices offer higher quantum efficiency and reduced dark currents. Dilute magnetic semiconductors contain an additional magneto-element such as Mn or Fe that enhances the quantization of the conduction band into Landau levels in the presence of a magnetic field. Transitions between the Landau levels are used for sensing far infrared with energies below 10 meV. Similar devices are discussed further in Chapter 5 of this text. In conclusion, although several technological barriers must be overcome to ascertain the role of HgCdTe over the next decade, the many unique properties of HgCdTe infrared-radiation sensors suggest it will be advantageous to make the required effort.

6.4 VISIBLE-LIGHT COLOR SENSORS

Conventional color sensors consists of three photodiodes per pixel shielded by separately dyed polymer coatings that provide a set of primary or complementary color filters. Evaluation of the photocurrents from the three diodes can provide an unambiguous transduction of the incident average wavelength and irradiance. The average includes the spectral distribution, as well as any variation over the time and space of the collection. Silicon color sensors of this type are being used in CCD cameras with pixel counts over 1000 per side as well as less demanding applications such as feedback to manufacturing processes, which require sensitivity only over a portion of the visible spectrum and minimal spatial and temporal resolution. However, the requirement of three diodes per pixel increases the sensor cost for all applications and limits the spatial resolution of solid-state cameras. In addition, the deposition and patterning of the dyed polymers adds nonstandard silicon-processing steps. This section reviews an alternative silicon color-sensor design that reduces the photodiode count per pixel and can be readily integrated into a typical bipolar process. Integration of the color sensor with its control- and output-conditioning circuitry provides reduced noise and offset due to shorter interconnects and improved component and thermal matching, as well as reduced cost and complexity.

The color sensor to be discussed[8] is based on the resolvable wavelength dependence of the silicon absorption length within the visible range of the radiant spectrum as shown in Fig. 8. The absorption length, $1/\alpha$, corresponds to absorption of 63% of the incident photon flux, as in Eq. 6 in Section 2. This characteristic of silicon results from its indirect bandgap, that is the percentage of phonons capable of completing the energy-band transition increases as the photon energy is increased beyond the bandgap minimum, reducing the required momentum change. Therefore, it is impractical to construct this type of color sensor from most compound semiconductors consisting of direct bandgaps, or even from germanium since it direct bandgap is only modestly greater than its minimum gap, 0.8 versus 0.66 eV as compared to 3.4 versus 1.12 eV for silicon. Although the photocurrent from a single diode will have a wavelength dependence on reverse bias, it is not able to differentiate between wavelength and intensity variations. However, this inadequacy can be overcome through a set of vertical diodes similar to a bipolar transistor, but with both junctions reverse biased and a much larger base width.

A cross-section of the structure is shown in Fig. 9 with one of the investigated doping profiles.[8] Note that the depth of the dopant profile is plotted on a log axis providing a straight-line transformation of the exponential photon-density decay. Using the epitaxy for the primary absorption volume, this structure can be integrated into standard bipolar processes without atypical processing. The

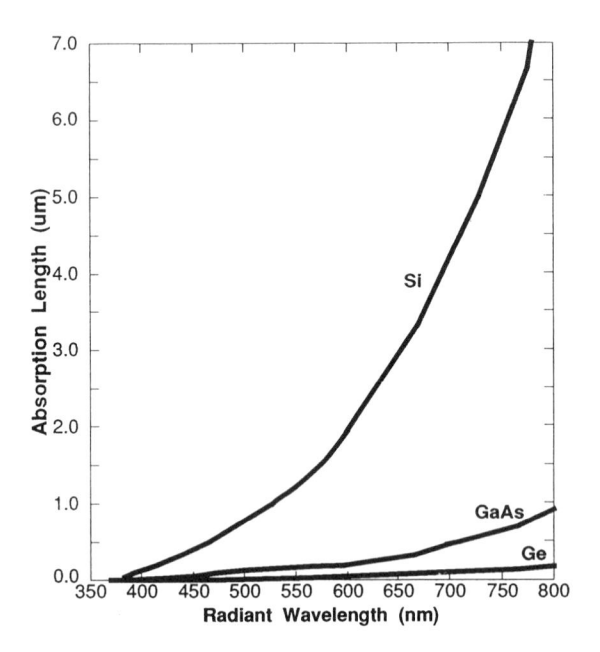

Fig. 8 Absorption lengths in Si, GaAs and Ge for radiant wavelengths spanning the visible range. Note the sensitivity of Si compared to GaAs and Ge making it a good candidate for color-sensing applications. (After Ref. 8)

wavelength and intensity of the incident optical signal is resolved by monitoring the ratio of the photocurrents from the upper and lower junctions at multiple sets of junction reverse biases. It is advantageous to have the upper and lower depletion regions touch to avoid photo-generation of the slow and less predictable diffusion current within the epitaxy. It is also desirable to minimize interaction between the two depletion regions to avoid control issues related to competing electric fields. This can be accomplished using the gate pinch-off voltage of a neighboring JFET channel with the same construction as the color sensor. Note that a JFET is not required for every pixel. In fact, if the biasing of all pixels is swept together, one JFET per die is sufficient assuming acceptable uniformity of the epitaxy thickness and dopant concentration.

The total absorbed photon density per unit time is equal to the sum of the upper and lower junction photocurrents. However, the signal from the upper junction corresponds to only those photons with absorption lengths less than the depth reached by the upper junction's depletion region, and greater than the minimum related to surface recombination. Thus, two short wavelength limits exist: that discussed in Section 2 where surface recombination limits the quantum efficiency to an insufficient level, and the second corresponding to photons absorbed within the upper diffusion's neutral region or equilibrium depletion width. Although a photocurrent is registered for the latter, the wavelength cannot be distinguished beyond that it is below the second limit.

6.4 VISIBLE-LIGHT COLOR SENSORS

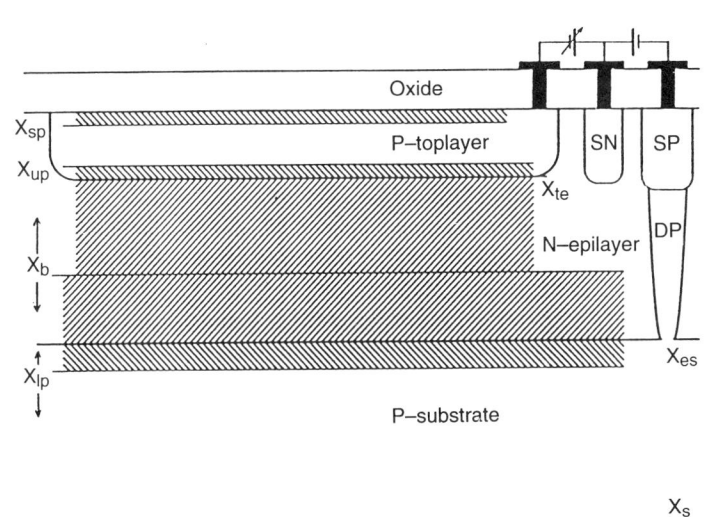

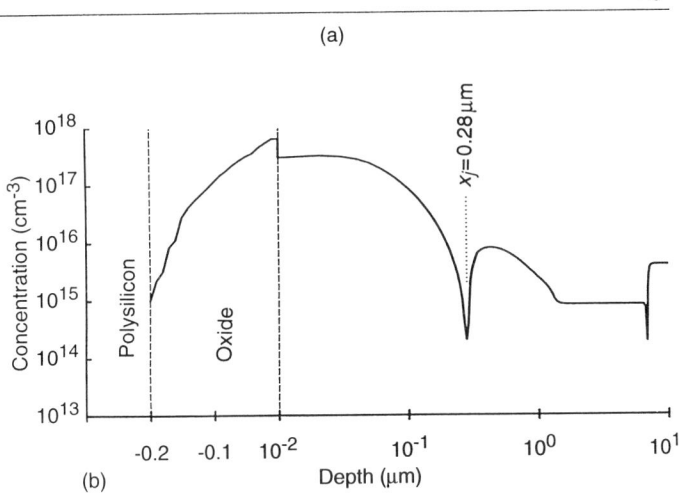

Fig. 9 (a) Schematic diagram of the basic color sensor in silicon. (b) SUPREM simulation of the doping profile in 9a. First a buffer layer is formed by a 150 keV/ 2×10^{12} cm^{-2} As$^+$ implantation followed by a drive-in at 1200°C for 60 minutes and concluded by a 60 keV/5×10^{12} cm^{-2} B$^+$ implantation and an anneal for 30 minutes at 800°C. (After Ref. 8)

The maximum and two minimum wavelength limits of the vertical color sensor have different physical origins and therefore can be addressed separately.

The variation of silicon absorption lengths for radiation within the visible range, 0.1 μm at $\lambda = 390$ nm to 7 μm at $\lambda = 770$ nm, dictates the need for a shallow upper junction and epitaxy thickness on the order of 10 μm. An epitaxy concentration of approximately 3×10^{14} cm^{-3} is required for a 10 μm wide depletion region extending from a single junction using a reasonable integrated-circuit operating-bias maximum of 20 volts. As mentioned above, low quantum efficiency is obtained for violet light due to the short characteristic-absorption length and the resulting surface recombination. Such effects can be minimized through the use of a dedicated, moderately-doped surface region, although care must be taken to avoid depletion-region exposure to the high-recombination velocity at the contact–metal interface. The short wavelength sensitivity can also be enhanced through use of a 1 kÅ poly gate to reduce surface-region recombination by providing an electric field that drives the generated minority carriers across the junction rather than allowing them to diffuse to the interface traps. Absorption within a 1 kÅ poly layer is generally acceptable, but will depend on deposition characteristics. Deposition of wide-gap amorphous silicon or SiC to form the upper junction would also improve the device sensitivity to violet light if recombination at the hetero-interface is minimized. An n-type implant, which increases the epitaxial concentration just below the surface junction, as shown in Fig. 9, reduces the built-in depletion width and thus the resolvable short-wavelength limit. Also the dependence of the depletion width on reverse bias becomes more linear improving the trade-off between long- and short-wavelength sensitivity and even provides improved long-wavelength sensitivity for depletion-region edges within the region of the concentration gradient.

At least three separate bias points are required to reproduce the incident color with algorithms similar to that using the dyed polymer sensors. Additional reverse-bias points provide an improved definition of the spectral distribution and a reduced sensitivity to noise. The time required to vary the reverse biases and interpret the photocurrent ratio is the principal trade-off of the vertical color sensor over that of the three diode design. Background infrared radiation absorbed deep in the neutral bulk generates a diffusion photocurrent limiting the frequency at which the reverse biases can be switched. Such infrared absorption could be eliminated using a derivative of the SOI diode in Fig. 6, or a blanket infrared filter over the entire sensor. However, the response time of the vertical sensor integrated into a standard bipolar process is sufficient for most applications and can provide improved spatial resolution, as well as a substantial cost reduction for large arrays.

6.5 HIGH-ENERGY PHOTODIODES

p–i–n diodes are commonly used to detect high-energy and nuclear-particle radiation, Fig. 10.[11] Ideally, the diode would be fabricated on an intrinsic semiconductor substrate, regions I and III would be highly doped to provide contacts to the substrate, an abrupt p–n junction would be formed at the interface between regions I and II, and a high–low junction would be formed at the interface between regions II and III. In practice, truly intrinsic material cannot be manufactured. The substrate will either be slightly p- or n-type, which are referred to as π or ν-type, respectively.

6.5 HIGH-ENERGY PHOTODIODES

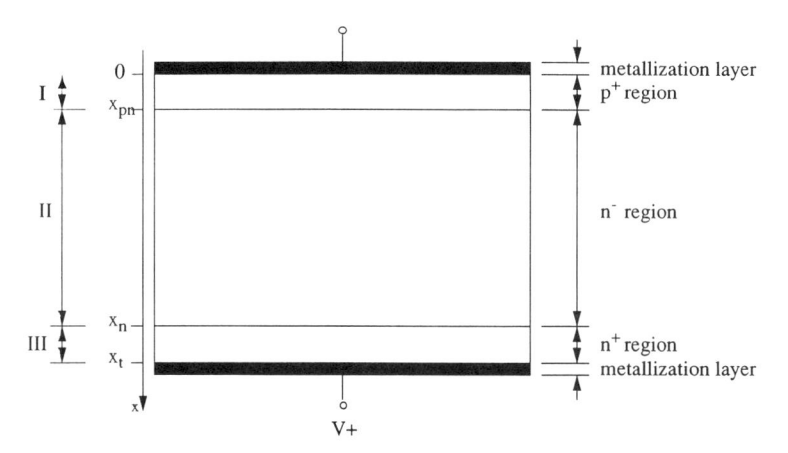

Fig. 10 Generalized structure of a p^+–n^-–n^+ diode indicating doping densities and regional divisions. (After Ref. 11)

A fully reverse-biased semiconductor diode has exceptional performance in the separation and collection of free-charge carriers created within the depletion region through ionizations by the absorption of high-energy electromagnetic or nuclear-particle radiation. High-resistivity or compensated semiconductor-substrate material is essential to provide the large sensitive volume necessary to absorb the incident radiation efficiently.

The sensitive volume required for the efficient detection of high-energy electromagnetic and nuclear-particle radiation depends on the substrate material used, the energy and the type of radiation to be detected, and the specific application. For high-energy radiation, Eqs. 6, 10, 11, 12, and 13 can be used to estimate the thickness the sensitive volume must be to absorb a sufficient percentage of the incident radiation for a particular applications. Fast electrons traveling at relativistic velocities are termed minimum-ionizing particles and generate approximately 85 electron–hole pairs per micron in silicon. This translates to an energy loss of 300 eV per micron in silicon. If the silicon detector were 300 μm thick, a signal of 4 fC would be generated. A silicon fast-electron detector must have a mimimum thickness of 300 μm to generate a detectable signal. Heavy-charged particles have a high probability of absorption and are stopped in the surface region of semiconductor devices. Standard photodiodes can be used for their detection.

The absolute quantum efficiency includes the overall process of absorption and collection and is defined as the ratio of the amount of charge collected to the theoretical amount that would be created if the incident radiation emitted from a source were to be converted to charge carriers. An extensive theoretical analysis is carried out for the device in Fig. 10.[11] For practical purposes, if the device is fully depleted, the quantum efficiency will equal

$$\eta = \exp(-\alpha x_{pn}) - \exp(-\alpha x_n) \tag{36}$$

where α is the absorption coefficient and x_{pn} and x_n are the depths shown in Fig. 10. The absolute quantum efficiency for 15 keV X-rays of a fully-depleted basic device, such as that shown in Fig. 10, fabricated on 4000 Ω-cm n-type silicon material 380 μm thick is approximately 55%.[11] If the device were fabricated on silicon with a resistivity of 0.1 Ω-cm, the quantum efficiency would drop to approximately 20%.[11]

A number of parameters are important when comparing the usefulness of different semiconductor materials for the detection of high-energy electromagnetic and nuclear-particle radiation. Important material parameters include the net concentration of dopants, the bandgap, the ionization energy, E_i, the charge-carrier mobilities and lifetimes, the drift and diffusion lengths and the atomic number(s) of the semiconductor substrates.[2] Values for the net concentration of dopants in the purest materials available, the bandgaps, the ionization energies, the mean electron and hole mobilities, the mean electron and hole lifetimes and the atomic numbers of various semiconductor materials are found in Table 1.

Due to their well-developed processing technologies, silicon and germanium were the first semiconducting materials used to fabricate semiconductor-radiation detectors.[1,3] Silicon was used primarily to image light photons, while germanium, with its higher atomic number, was used in the detection of high-energy electromagnetic radiation.

The lithium-ion-compensation technique was introduced in the early 1960s to reduce the net dopant concentration of silicon and germanium substrates and to allow the production of thicker detectors that require lower reverse-bias voltages for full depletion. Lithium-drifted silicon detectors up to 10 mm in thickness can be fabricated and are known for their excellent energy resolutions.[1,3]

The lithium-drift process involves the initial diffusion of lithium on one side of a p-type silicon wafer forming an n^+ contact. A trench is cut into this side of the wafer with an ultrasonic cutter. The detector is then heated to 100–150°C at reverse-bias voltage of 100 to 1000 V. This procedure causes the lithium ions to drift from the n^+ contact into the p-type bulk. Lithium ions compensate the boron impurities, thus producing a nearly intrinsic device.[1,3]

The lithium-drift procedure is simple and requires very little equipment. However, lithium tends to form complexes with oxygen, which is generally present in silicon and germanium crystals. This effect and the presence of microdefects in substrates complicate the drift process. Moreover, lithium ions diffuse at relatively low temperatures. Ge(Li) detectors must be stored and operated at liquid-nitrogen temperature. Si(Li) detectors can be stored at room temperature, but should be operated at liquid-nitrogen temperature for the best results.[1,3]

High-purity germanium ingots with excellent material properties became available in the early 1970s, primarily for use in the fabrication of gamma-radiation detectors.[1,3] The purification procedures are based on float-zone refining techniques rather than on compensation techniques. High-purity

TABLE 1 Parameter Values for Various Materials Used in Fabricating Semiconductor-Radiation Sensors

Material	$\|N_D - N_A\|$ (cm^{-3})	E_g (eV)	E_i (eV)	μ_e (cm^2/V s)	μ_h (cm^2/V s)	τ_e (s)	τ_h (s)	Z (a.m.u.)
Si	3×10^{10}	1.12	3.61	1350	480	5×10^{-3}	5×10^{-3}	14
Ge	$>5 \times 10^{9}$	0.67	2.98	3900	1900	2×10^{-5}	2×10^{-5}	32
CdTe	10^{12}–10^{13}	1.56	4.43	1050	100	1×10^{-6}	1×10^{-6}	48, 52
HgI$_2$	Semi-insulator	2.13	4.20	100	—	1×10^{-6}	2×10^{-6}	53, 80
GaAs	10^{12}–10^{13}	1.42	4.70	8500	450	5×10^{-8}	5×10^{-8}	31, 33

germanium has a lower density of trapping centers than compensated material. It is important to collect all the charge carriers generated in a semiconductor sensor to obtain an accurate measurement of the energy of the incident radiation. It is, therefore, important that the material be damage free and contain a minimal amount of trapping centers. Room-temperature storage and handling of high-purity germanium is possible, although these detectors should be operated below 150 K to eliminate the effects of thermally-generated leakage-current noise.

More recently, high-purity silicon crystals with resistivities from 10,000 to 50,000 Ω-cm, lifetimes from 100 to 5000 μs and low defect concentrations have become available.[1,3] This material is fabricated either through the use of float-zone growth and refinement methods or by neutron transmutation. High-purity silicon is used in the fabrication of devices that must operate under a high voltage, for example power devices, thyristors, and high-energy electromagnetic and nuclear-particle radiation sensors. Due to its higher bandgap, high-purity silicon does not need to be cooled in most applications. However, cooling the device will lower its leakage current and noise. High-purity silicon will also have a lower density of trapping centers than compensated material.

Research on CdTe and HgI$_2$ began in the mid 1960s and in the early 1970s, respectively, in an effort to find semiconducting materials with high atomic numbers for use in the detection of high-energy photons.[49,50] Because of the higher atomic numbers of the elemental constituents of the compound semiconductors, their absorption coefficients are higher, Eqs. 11, 12 and 13, and therefore their efficiencies for X- and gamma-radiation detection are significantly higher than those values associated with Si and Ge. The bandgaps of CdTe and HgI$_2$ are large enough to allow the fabrication of detectors for use in room-temperature X- and gamma-radiation spectroscopy with excellent energy resolutions.[49,50] There are, however, disadvantages associated with the use of these semiconducting materials. As is true of all crystal-growing methods, especially those associated with the production of compound semiconductors, it is difficult to grow large-volume crystals that have acceptable material properties and that are uniform in purity, stoichiometry and charge-collection properties.

Three methods are currently available for the fabrication of CdTe crystals: the zone-melting method, the traveling-heater method and the Bridgman method. High-quality crystals can be produced using the zone-melting method, but the resistivities are low. High-resistivity crystals can be produced using the two other methods, but only at the expense of doping with elements such as chlorine. Chlorine doping has been associated with polarization, that is with time-dependent changes in the depleted thickness or in the charge-collection properties.[1,3,49] Use of a high-pressure Bridgeman method may eliminate the need for chloride doping.[51]

Single HgI$_2$ crystals can be grown from the vapor phase or by solution regrowth. However, the fabrication of HgI$_2$ detectors is complicated by the softness of the material, its interaction with many materials, for example aluminium and gold, its small thermal resistance, its solubility in most solvents and its toxicity. In addition, HgI$_2$ is unstable at temperatures above 70°C and it is nearly impossible to apply metal contacts to the semiconductor, because the mercury tends to amalgamate with metals. The low hole mobility of HgI$_2$ causes incomplete charge collection over distances greater than approximately 1 mm. This limits the sensitive thickness of the detector, even when high electric fields are applied.[50]

GaAs was also investigated for its usefulness in radiation detection in the early 1960s when high-resistivity, compensated material became available.[2] Like CdTe and HgI$_2$, its bandgap is sufficiently wide to permit room-temperature operation. In addition, the detection efficiency for high-energy photons resembles that of germanium, since the elemental constituents of the compound semiconducting material have atomic numbers (31 and 33) that are on either side of germanium (32). However, the performance of the prototype detectors were poor due to the strong trapping effects in the material. With the development of liquid- and vapor-phase epitaxial techniques to grow high-purity GaAs films on either GaAs or Si substrates, radiation detectors were fabricated that had excellent energy resolutions of gamma photons.[50] However, as high-quality epilayers cannot be grown in excess of 100 μm, the efficiencies of such devices are minimal and therefore they are not used in X- and gamma radiation detection.

Other compound semiconductors with large bandgaps have been investigated for possible use as room-temperature radiation detectors: AlSb, Bi$_2$S$_3$, CdSe, GaSe, PbI$_2$, ZnSe, ZnTe, and CdZnTe. However, the yields of all the compound-semiconductor materials are very low, their availabilities are restricted, and their prices are high. Thus, although CdTe and HgI$_2$ detectors have demonstrated good performances as room-temperature X- and gamma-radiation spectrometers, the crystal-growth methods and the long-term stabilities of these two and the other compound semiconductor materials listed have to be improved before they become widely available for commercial use. A promising alternative may be CdZnTe. Different stoichiometries of this material have recently been manufactured by a high-pressure Bridgman method with promising results.[51] Gamma-radiation detectors manufactured from this material are commercially available.[51]

The material parameters listed in Table 1 are interrelated. The thermal generation of carriers decreases as the bandgap increases. Therefore, to reduce the background current, one would like to maximize E_g. With respect to this demand, silicon is superior to germanium. Negligible values for the junction-leakage current of germanium detectors can only be obtained by operating the device at liquid-nitrogen temperature. However, as the bandgap increases, the ionization energy, E_i, also increases. As E_i increases, the total number of electron–hole pairs generated by a given radiation source decreases, $N = h\nu/E_i$, and the intrinsic energy resolution given by the full-width-at-half-maximum, FWHM, degrades.

$$\mathrm{FWHM} = 2.25(FEE_i)^{1/2} \tag{37}$$

where F is the Fano factor, which is defined as the ratio of the observed variance in the number of charge carriers produced to the variance predicted by pure Poisson statistics.[2]

In addition, to be collected, the charge-carrier drift and diffusion lengths must be larger than the distance the carriers have to migrate in the depletion layer or in the undepleted layers, respectively, before reaching a contact. If these lengths are long, the signal becomes independent of the incident photon's point of absorption within the detector's sensitive volume. The drift and diffusion lengths for electrons and holes are directly related to the charge-carrier mobilities and lifetimes and are given respectively by

$$L_{\mathrm{drift}} = \mu \tau \mathscr{E} \tag{38}$$

$$L_{\mathrm{diffusion}} = (\tau D)^{1/2} \tag{39}$$

where μ is the charge-carrier mobility, τ is the charge-carrier lifetime, $\mathscr{E}$ is the electric field and D is the carrier diffusivity. Relatively high charge-carrier mobilities and long charge-carrier lifetimes can be obtained in Si and Ge substrates due to the advanced state of their crystal growth and purification technologies.[52] However, these values in compound semiconductors such as CdTe, HgI_2 and GaAs are from one to one hundred thousand times lower due to the difficulty in fabricating high-purity material.

Lastly, the efficiency of the photoelectric absorption of incident radiation is proportional to the atomic number of the semiconductor material to the fifth power, Eq. 11. Consequently, compound semiconductors such as CdTe, HgI_2 and GaAs have a definite advantage over Si and Ge in the efficient detection of high-energy electromagnetic radiation. Compromises must therefore be made when deciding what semiconductor should be used to fabricate a particular radiation sensor.

6.6 SILICON DRIFT CHAMBER X-RAY SENSORS

The position of incident high-energy electromagnetic or nuclear-particle radiation cannot be determined from a basic semiconductor-radiation sensor such as the one shown in Fig. 10. A one-dimensional position-sensitive semiconductor-radiation sensor can be realized by dividing the top electrode of the basic device, that is the p^+ implantation covered with aluminum, into strips. This type of device is referred to as a strip detector. When the substrate is silicon, the device is given the abbreviation SSD.[52]

Strip detectors are typically placed under a reverse-bias that will ensure full depletion of the substrate. Under operating conditions, the Si/SiO_2 interface at the n^--substrate separating the p^+ strips will accumulate electrons. The electrons match the positive fixed-oxide charges normally associated with Si/SiO_2 interfaces. Each strip, therefore, acts as a separate detection element.[11] Electrons and holes created by absorbed radiation will be separated by the applied-electric field within the device. Holes will be collected at the nearest p^+ strip, while electrons will be collected at the n^+ contact. Conventional SSDs are capable of one-dimensional spatial resolutions on the order of 5 μm.[52]

In an effort to develop a two-dimensional position-sensitive radiation detector, double-sided silicon-strip detectors were designed and fabricated, Fig. 11a.[52] As in a conventional SSD, a double-sided SSD has p^+ strips on the radiation-incident top side, but it also has strips on the bottom of the device. These strips are n^+-doped and are orthogonal to the stips on the top. Isolation strips are fabricated between the n^+ strips to separate the elements, Fig. 11b.[52] Silicon-strip detectors were developed for and are primarily used in high-energy physics experiments and have been made in sizes up to 5×5 cm^2 in diameter.[52] However, the precision of these devices has been achieved at a huge expense. Thousands of channels need to be read out, each of which requires low-noise electronics, analog-to-digital converters, multichannel analyzers, and storage buffers. In an effort to minimize the number of read-out channels necessary to provide unambiguous two-dimensional spatial resolution of incident radiation, a sensor based on a novel transport scheme called a semiconductor drift chamber (SDC) was introduced by Dr. E. Gatti, from the University Politecnico di Milano in Italy, and Dr. P. Rehak, from Brookhaven National Laboratories in Long Island, N.Y., in 1982.[53] A schematic of the original device is shown in Fig. 12.[53] In the n substrate, p^+ rectifying strip junctions are fabricated on both surfaces, the ones on the top are parallel to those on the bottom. On one of these surfaces, an n^+ blocking strip contact is also formed.

When the device is reverse-biased, depletion regions will extend from the p^+–n junctions on the top and the bottom of the device into the bulk. With increasing bias, the depths of the depletion regions increase until they meet in the middle of the device. At their union, a potential-minimum (PM) channel for electrons results that is parallel to the surface. A second electric field, independent of the first, is superimposed to transport electrons collected in the channel to the small n^+ anode contact. This is accomplished by proper biasing of the two arrays of p^+ electrodes at both surfaces of the device. Electrons created by the absorption of radiation will be focused to the buried channel and will move by drift and diffusion to the collecting anode. The cloud of electrons produces an output pulse when it arrives close to the anode. The collected charge is a measure of the energy of the radiation.[53]

Similarly to the SSD, the holes are collected at the nearest p^+ contact. However, the hole pulse in this case serves as the trigger to begin a measurement of the time it takes the electron cloud to reach the anode. The electron transport time is linearly related to the carrier path, which ideally is straight through the center of the device. For a uniform electric field, the transit time is directly related to the straight-line distance of the incident radiation to the anode. If

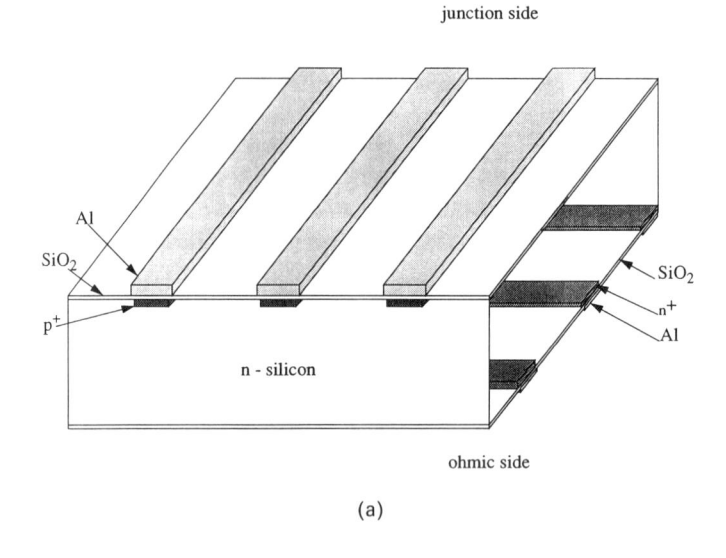

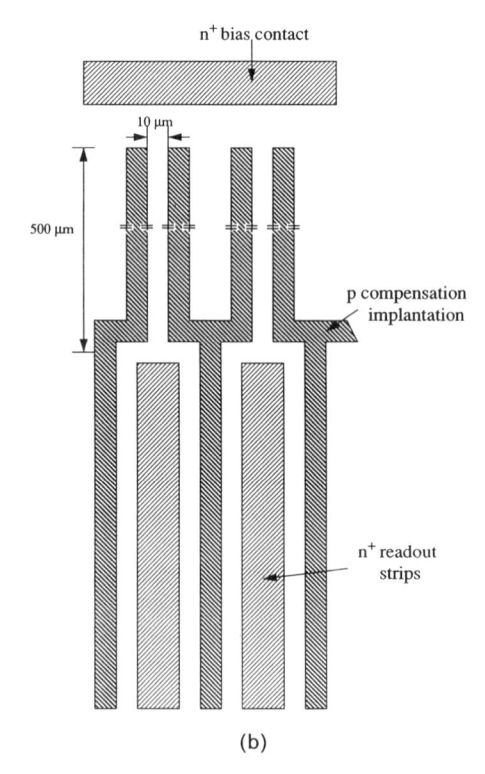

Fig. 11 (a) Schematic of a double-sided silicon-strip detector with capacitive-change coupling. No isolation structures are drawn on the n side. (b) Geometry of the n side surface with p implantations, defining the resistance between n^+ strips.

6.6 SILICON DRIFT CHAMBER X-RAY SENSORS 317

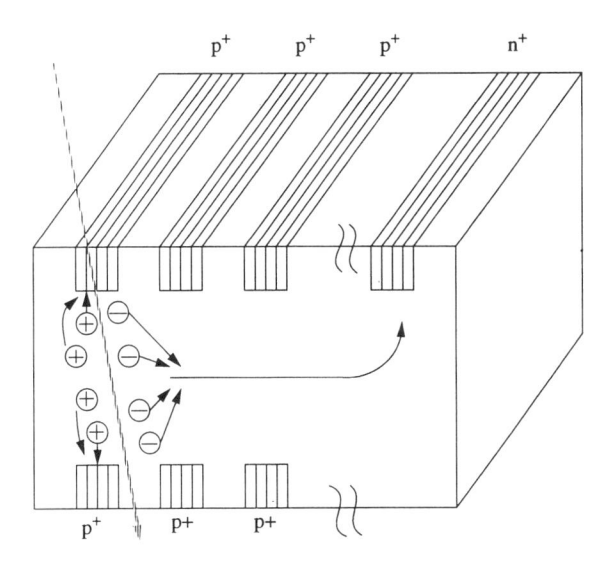

Fig. 12 Schematic of the original semiconductor drift chamber design.

the strip anode is divided into pads, two-dimensional position resolution of incident radiation can be obtained.

The number of outputs an SDC must incorporate to accurately read-out position and energy information is hundreds of times less than that required by a SSD of similar size.[54] In addition, the double-sided SSD provides only projections into two coordinates. If two particles pass through the detector, the SSD gives four possible points, two of which are real and the other two are phantom. In contrast, results of the position measurement of incident radiation obtained with the SDC are unambiguous. Since the SDC was introduced, a number of variations of the initial design have been conceived, simulated, fabricated, and tested.[54-58] Applications of this device are found primarily in semiconductor charged-particle and X-ray detection. The application as X-ray sensors is stressed below and techniques to optimize the sensors for the detection of soft X-rays are detailed.

An important advantage of the SDC over the basic semiconductor-radiation sensor and the SSD is that the capacitance of the collecting anode is very low and independent of the active area of the detector. A large area detector can therefore be built with a small output-detector capacitance. Reducing the effective detector capacitance reduces the effect of amplifier series noise. The noise performance of the SDCs is greatly improved over basic diodes and SSDs. The low effective capacitance of SDCs can be exploited for X-ray detection.

The design of the original SDC was modified by researchers from Brookhaven National Laboratories and Politecnico di Milano working together with groups from the Technischen Universität and the Max Planck Institut in Munich, Germany to optimize its performance for the detection of X-rays.[54-57] The n^+ anode of one such device is 200 μm in diameter and is in the center of the device.[54] The capacitance of the anode is 0.06 pF. Thirty-four circular p^+ electrodes surround the anode. The diameter of the active volume is 1 cm. A voltage is placed across the inner and outermost p^+ electrodes thereby self-biasing the rings by the chain effect.[54] The drift field created has a cylindrical symmetry with electrons drifting radially towards the center anode. External n^+ guard anodes prevent charges generated outside the active region from reaching the measuring anode. This device is termed a cylindrical drift chamber.[54,55]

In a system configuration, the output of a radiation detector, the anode, is connected to the gate of the first transistor of a preamplifier. The optimum system resolution of a detector, preamplifier and signal processor can be expressed in terms of equivalent-noise charge as[26]

$$\text{ENC}_{opt} = (2qkT/\pi)^{1/4}(I_1 C_d/f_t)^{1/4}[(C_d/C_i)^{1/2} + (C_i/C_d)^{1/2}]^{1/2} \qquad (40)$$

where q is the electronic charge, k Boltzmann's constant, C_d the capacitance of the detector, and C_i and f_t are the capacitance and the bandwidth of the input device of the preamplifier, respectively. The best resolution will be obtained when the gate capacitance of the first transistor of the preamplifier, C_i in Eq. 40, is matched to the output capacitance of the detector, C_d in Eq. 40, and all sources stray capacitance are minimized. The equivalent-noise charge originally reported at room temperature with the cylindrical drift chamber described above was 300 electrons, which corresponds to a FWHM at Mn Kα of 2.6 keV.[54] An improved measurement, FWHM equal to 930 eV at Mn Kα, was reported after an input field-effect transitor with a lower gate capacitance (2–3 pF) was used.[54] Cooling the device to liquid-nitrogen temperature provided additional optimization; a FWHM of 306 eV at Mn Kα was achieved.[54]

Cylindrical drift chambers with active areas of 2 mm^2 were recently designed in Milan, produced in Brookhaven, and tested at Metorex International, Oy in Espoo, Finland. Measurements of the energy resolution, FWHM at Mn Kα, obtained at several different temperatures were 542 eV, Fig. 13a, 474 eV, Fig. 13b, and 408 eV, Fig. 13c, at 20°C, -10°C, and -30°C, respectively. The gate capacitance of the input field-effect transistor used in these cases was also 2–3 pF. Further improvements in energy resolution can be achieved if a gate capacitance equivalent to the detector capacitance can be realized, as the detector capacitance is only 1% of the total input capacitance.

To detect soft X-rays, that is radiation with energies ranging from approximately 100 eV to 5 keV, the entrance dead-layer must be minimized. The detection of soft X-rays with a SDC device was improved by modifying the cylindrical drift chamber. Instead of p^+–n strip junctions on both sides of the device, a large area p^+–n junction was fabricated on the radiation-incident top side to eliminate oxide from this interface, thereby reducing the dead-layer. The p^+–n strips and the n^+ anode were placed on the bottom side.[56,57] Electrical contacts were only necessary at the rear. The concentric p^+–n rings of the cylindrical detector were replaced with a continuous spiral p^+–n electrode. In

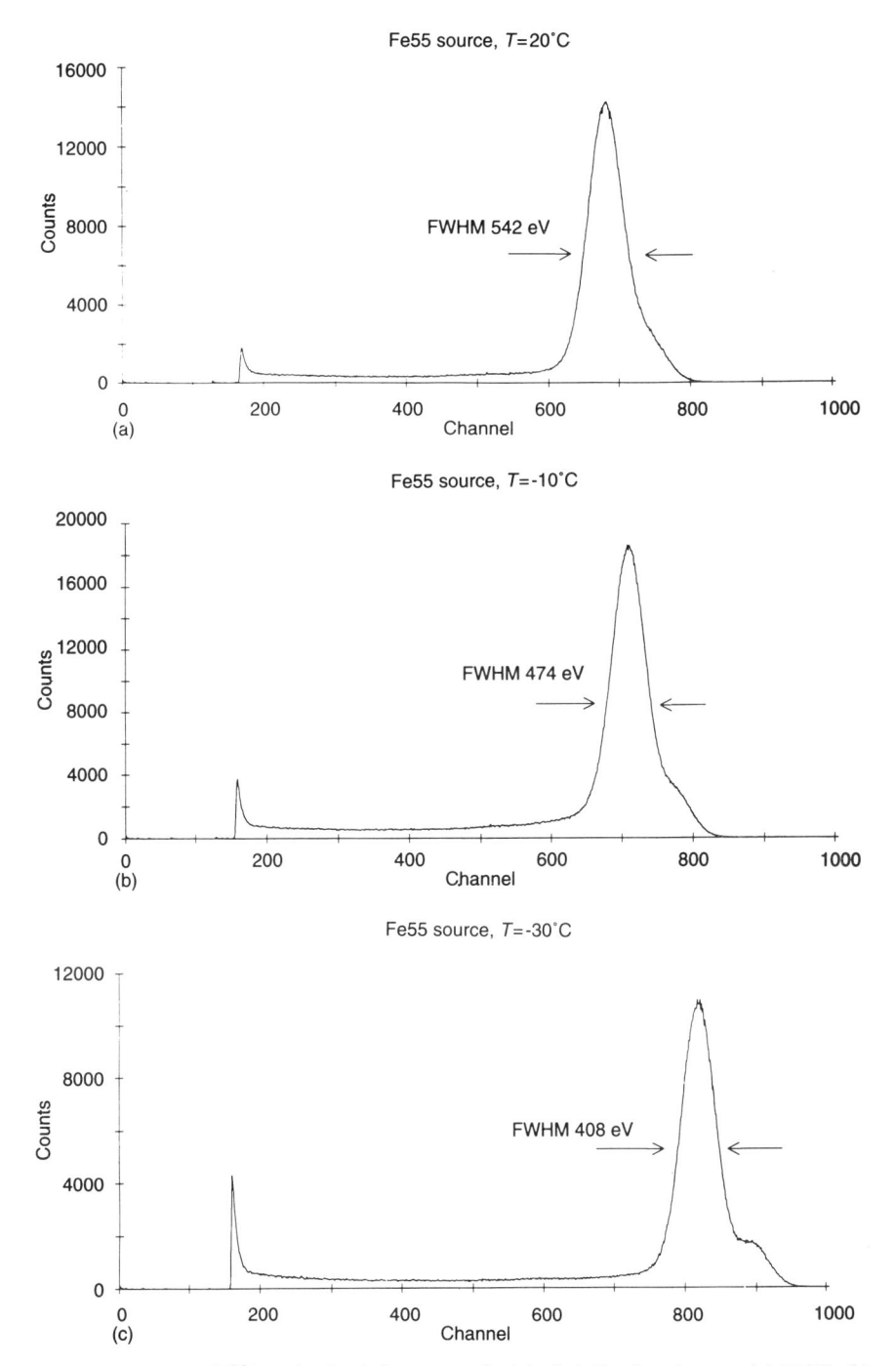

Fig. 13 Spectra of ^{55}Fe obtained from a cylindrical drift chamber at (a) 20°C, (b) −10°C, and (c) −30°C.

addition, all electrons generated at the silicon–silicon dioxide interface were collected on a guard sink to decrease the detector leakage current and to reduce the parallel noise of the detector. This device is termed a spiral drift chamber.[57] One problem with the spiral drift chamber is that because one surface of the device is a continuous electrode, the size of the device is limited. A solution to this problem has recently been published.[58] Essentially, the design involves several spiral drift chambers in series with a small oxide gap separating each chamber. The amount of oxide covering the top of the device is minimized and the size is now limited only to the diameter of the starting material.

To minimize the total input capacitance, the first-stage transistor should be integrated adjacent to the output of the detector on the same substrate. This would minimize the stray capacitance introduced by the connection. JFETs have been the preferred choice as the first-stage transistor due to their minimal gate currents and relatively high resistance to radiation damage.

Nonconventional JFETs have been designed and fabricated on totally depleted, high-resistivity silicon substrates.[26,59–61] In addition to the requirements that the transistors have to be operational on detector-grade silicon and have a gate capacitance matched to the output detector capacitance, the JFETs have to meet several other severe requirements: the gate current must be lower than the detector leakage current (1 nA/cm^2), the bandwidth must be wider than 200 MHz for high-energy resolution, and the processing involved in the transistor fabrication cannot degrade the operation of the detector. Devices with radiation detection, amplification, and charge-storage capabilities have been produced and tested.[60,61] Different designs include a MOSFET and a JFET fabricated on depleted substrates, termed depleted p-channel MOS (DEPMOS) and depleted JFET (DEPFET) respectively (see Fig. 14) and a pixel device that allows random access, nondestructive read-out.[60,61]

The DEPMOS and DEPFET are fully depleted by applying the proper bias voltages to the top and bottom p^+–n junctions. As in the SDC, a potential minimum for electrons results, in this case, beneath the transistor channel. Electrons generated by the absorption of incident radiation, or generated by thermal processes, will collect in the potential minimum. The conductance of the transistor channel is modulated by this build-up of electrons and therefore, the potential minimum has been termed an "internal gate". After measurement, a clearing electrode removes the charge stored in the potential minimum.[60] The integration of a matrix of DEPMOS transistors would produce a random-access pixel detector.[60] One further step involves adding buried grids into the bulk of the detector underneath the top read-out structure to realize a multievent-storage random-access pixel detector.[60]

The best results in the area of silicon drift chambers for use in the detection of soft X-rays have been achieved by the researchers at Bookhaven National Laboratories, Politecnico di Milano, Technicschen Universität and Max Planck Institut with fully-depleted junction charge-couple devices (FDJ CCDs).[26,61] These devices have been developed specifically for use in a German satellite project.[26,61] Fully-depleted junction charge-coupled device prototypes with

6.6 SILICON DRIFT CHAMBER X-RAY SENSORS

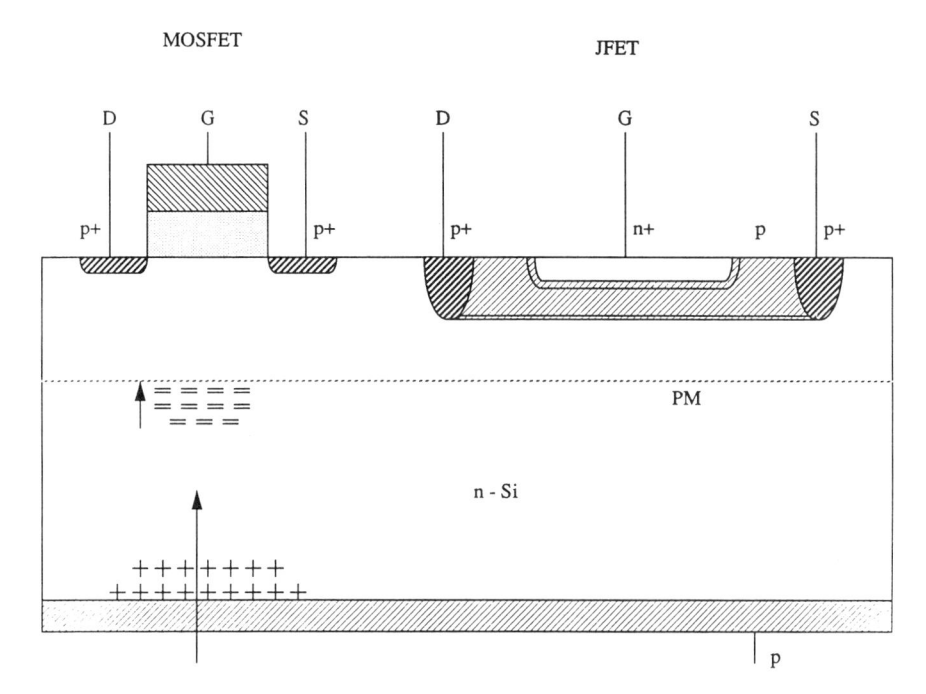

Fig. 14 DEPMOS and DEPFET detector-amplification structures.

integrated JFETs for first-stage amplification have been designed, fabricated and tested for X-ray detection.[61] An X-ray imager with a sensitive area of 36 cm^2 and a pixel size of 150×150 μm^2 is currently being fabricated as part of the European Space Agency's X-ray Multi-Mirror Mission (XMM). A simplified schematic of the basic configuration of a FDJ CCD topology derived from the silicon drift chamber principle is shown in Fig. 15.[61]

The position of the potential minimum for electrons in a semiconductor drift chamber can be controlled by varying the electrostatic field parallel to the surface. The electron potential minimum is shifted towards the top surface by allowing the bottom p^+ contact to be negatively biased with respect to the top shift electrodes. Local potential minima for electrons are created by appropriately biasing the transfer registers ϕ_1, ϕ_2 and ϕ_3. The voltages on the transfer registers can be changed with time to allow a shift of the signal charges toward the read-out anode.[61] The fact that a FDJ CCD has a sensitive volume equal to the thickness of the substrate allows a theoretical quantum efficiency of greater than 90% for X-rays with energies from 2–10 keV when the substrate is 500 μm thick.

The JFETs, which are not shown in Fig. 15, are integrated on the same high-resistivity silicon as the detector in FDJ CCD devices. The transistor is similar to a standard n-channel one, however, it has only one gate. The gate is implanted on the same surface of the FDJ CCD as the pumping structure of

322 RADIATION SENSORS

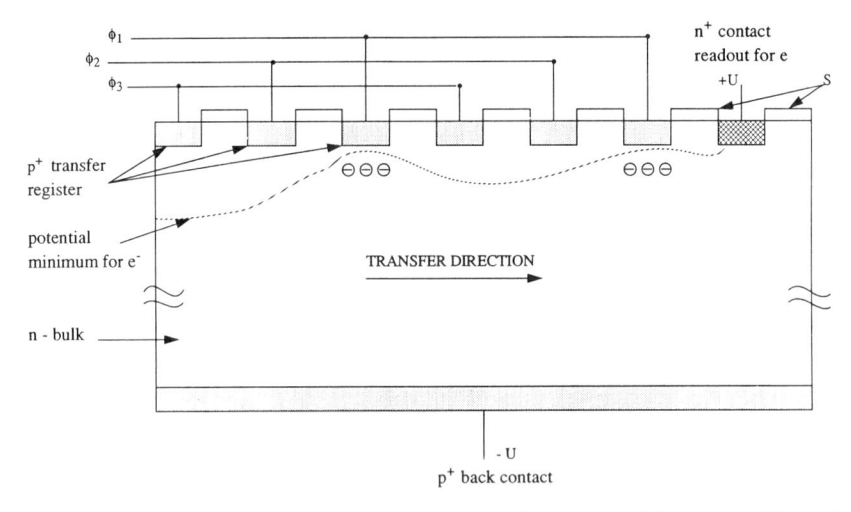

Fig. 15 Basic configuration of the FDJ CCD topology derived from the silicon drift chamber principle.

the CCD. Details of the design, the production and the tested characteristics have been published.[26]

The dependence of the square of the equivalent-noise charge (ENC)^2 of a FDJ CCD as a function of the shaping time at temperature 150 K is shown in Fig. 16. The minimum value of the electronic noise of two electrons is reached at the longest practical shaping time of 10 μs. Contributions from three different noise sources: series, parallel, and $1/f$, are also shown in Fig. 16. The best noise performance at room temperature is found for ENC equal to eight electrons at a shaping time of 1 μs.

A spectrum of ^{55}Fe X-rays obtained at a temperature of 150 K is shown in Fig. 17. The FWHM at MnKα is 119 eV. The pulse peak is also shown. The width of the ionization peaks is dominated by the statistics of the pair production in silicon and is practically unaffected by the electronic noise.

6.7 SUMMARY AND FUTURE TRENDS

A review of the characteristic properties of electromagnetic and nuclear-particle radiation and their interactions with semiconductors provides an essential basis to understanding the operation of radiation sensors. A solid foundation in semiconductor and radiation physics allows creativity in designing semiconductor sensors for a specific application. An example of this creativity is discussed in Section 6.4 on visible-light color sensors. A unique solution to the costly requirement of the standard three photodiodes per pixel was invented by using the wavelength dependence of the silicon absorption depth to resolve the energy of the incident radiation within the visible-light spectrum. It seems

6.7 SUMMARY AND FUTURE TRENDS   323

ENC² vs shaping time τ

T = 150K

Series 1/f noise
Series white noise

Shaping time (μm)

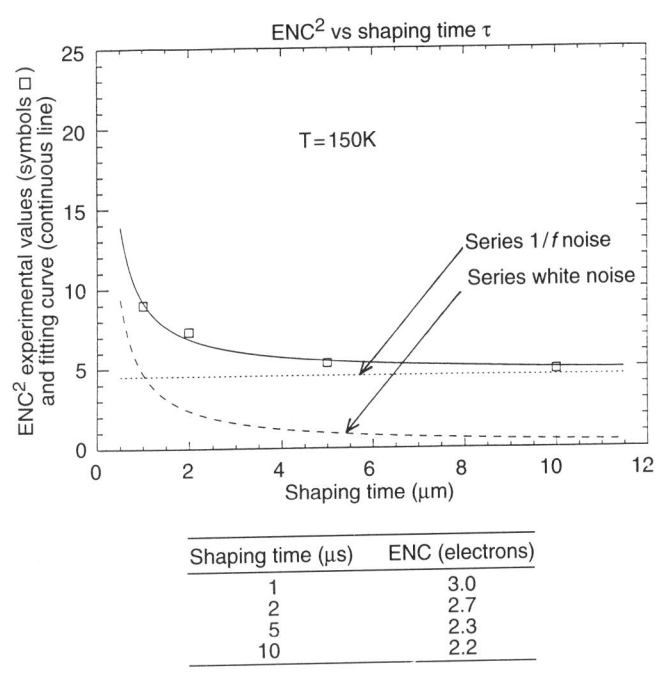

Shaping time (μs)	ENC (electrons)
1	3.0
2	2.7
5	2.3
10	2.2

Fig. 16 Plot of the equivalent-noise charge squared as a function of shaping time at 150 K on a FDJ CCD.

× 100

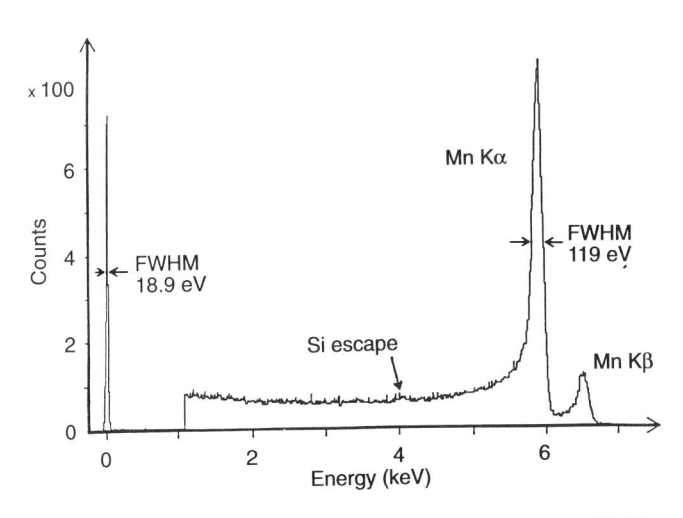

Mn Kα

FWHM 119 eV

FWHM 18.9 eV

Si escape

Mn Kβ

Energy (keV)

Fig. 17 Spectra of ^{55}Fe obtained from a FDJ CCD at 150 K.

inevitable that optical communications will provide the commercial volume required to mature the long-wavelength infrared technology. The advantages of HgCdTe indicate it will evolve as the dominant material in this growing market with superlattice-based devices providing future performance and manufacturing advantages. The development of HgCdTe epitaxial growth on alternative substrates even suggests the possibility of integrated control and read-out circuitry in Si or GaAs for high-speed application. A clear requirement for integration involves the consideration of on-chip optical communication as a means of reducing interconnect parasitics, which have become more of a performance limitation with each generation of integrated-circuit technology. In addition, variations of a basic p–i–n diode that began by fabricating the device on a lightly-doped substrate to achieve a thick, sensitive volume for the detection of X-rays has led to the creation of large area, low capacitance, two-dimensionally position-sensitive X-ray radiation sensors with optimized on-chip electronics for maximum sensitivity and performance. Strip detectors and drift chambers have been fabricated in high-purity germanium as well as in high-purity silicon. In addition, advancements are continuing in fabricating room-temperature X-ray and gamma-radiation sensors from compound semiconductors. However, the lack of commercially available one- and two-dimensional semiconductor radiation-sensor arrays is now holding up scientific experiments currently being conducted at international accelerator and synchrotron research institutes.

ACKNOWLEDGMENTS

We are grateful to Dr. George Michalowicz of National Semiconductor for thoroughly reviewing the chapter and for providing technical advice. We are indebted to Mr. Mark Steigerwald, Mr. Luke Steigerwald and Ms. Marie Kuszewski for their assistance in documentation.

PROBLEMS

1. Derive and plot expressions for the electric field, depletion-layer width and junction capacitance versus reverse bias for abrupt and linearly graded silicon p–n diodes. Section 2.3 of Ref. 4 is recommended for review.

2. Realizing that space-charge effects will limit the voltage drop across a photoconductor, use the simplyfying assumption of a parallel plate capacitor to derive an expression for the maximum photoconductor gain.

3. Derive the ideal diode equation in the presence of constant radiant illumination.

4. Using Eqs. 18 and 19, and $\langle i_{1/f}^2 \rangle = 10^{-23}$ A^2/s/f, plot the 1 Hz bandwidth noise current of spectra for an illuminated, reverse-biased diode with a 1 MΩ series resistance and a dark current, I_D of 100 µA $\times \exp(-0.05 \text{ eV}/kT)$

at 4, 77 and 300 K. Use $\tau = (N_t v_{th} \sigma_0)^{-1}$ for the carrier lifetime with trap concentrations, N_t, of 10^{12} and 10^{14} cm^{-3}, a capture cross-section, σ_0, of 10^{-15} cm^2 and $v_{th} = \sqrt{(3kT/m^*)}$ assuming the carrier effective mass equals the electron rest mass.

5. Plot $\sqrt{\alpha}$ versus hv curves at 4, 77 and 300 K for silicon and germanium using Eqs. 8 and 9, and dominant phonon energies of 0.063 eV for Si and 0.037 eV for Ge.

6. Using Eqs. 32 and 33, and assuming $N_c = N_v$ with the common anode rule, draw the HgCdTe heterojunction band diagrams for the following conditions where step junctions exist for both doping and composition.
   a) 5×10^{15} cm^{-3} p-type and 5×10^{16} cm^{-3} n-type, $x = 0.5$ both sides.
   b) 5×10^{15} cm^{-3} p-type and 5×10^{16} cm^{-3} n-type, $x = 0.5$ on the p side and 0.273 on the n side.
   c) 5×10^{15} cm^{-3} p-type and 5×10^{16} cm^{-3} n-type, $x = 0.273$ on the p side and 0.5 on the n side.

7. A collimated 10 keV X-ray bean is incident upon a fully depleted silicon sensor 0.05 cm thick with an active area of 30 mm^2. The sensor is cooled to a temperature of 77 K.
   a) What percentage of the incident radiation will be absorbed within the sensitive volume, assuming no radiation is absorbed in the space between the source and the sensor?
   b) Given that an incident X-ray interacts with a silicon atom, which type of interaction is most likely to occur?
   c) What is the most probable origin of the photoelectron ejected by the silicon atom following the interaction with the 10 keV X-ray?
   d) What is the energy of the ejected photoelectron?
   e) Given that the energy of an incident X-ray is fully absorbed within the sensitive volume of the silicon sensor, how many electron/hole pairs will be created?
   f) How many electron/hole pairs would be created if a 10 keV X-ray were fully absorbed within a fully depleted high-purity germanium sensor 0.05 cm thick with an active area of 30 mm^2 cooled to a temperature of 77 K?

8. a) A Fano factor of 0.123 was measured on a particular high-purity germanium sensor at 77 K. Calculate the statistical FWHM for the 122 keV gamma radiation emitted by a ^{57}Co source.
   b) The high-purity germanium sensor was replaced with a stable pulse generator. The measured response of the remainder of the system showed a fluctuation primarily due to electronic noise at 63 eV. Another independent source of fluctuation, primarily due to sensor drift, was

calculated to be 37 eV. Given that these three sources of noise predominate, calculate the overall FWHM using the answer to part (a).

9. a) What is the primary interaction of heavy charged particles and fast electrons with matter?
   b) Explain the difference between excitation and ionization of the absorber atom.
   c) What is the definition of the range and the mean range of a charged particle in matter?

10. a) Given a high-purity silicon sensor with an active area of 30 mm² 0.02 cm thick operating at room temperature, what substrate doping density is necessary if a voltage of 300 V is needed for full depletion?
    b) The structure of the sensor is that shown in Fig. 10. What is the output capacitance of the detector given that the geometry of the output anode is 0.05×0.05 cm²?
    c) What should the input capacitance of the preamplifier be to minimize the system noise and to obtain the best energy resolution?

REFERENCES

1. G. F. Knott, *Radiation Detection and Measurement*, 2nd ed., Wiley, New York, 1989.
2. W. H. Tait, *Radiation Detection*, Butterworths, London, 1980.
3. S. Middelhoek and S. A. Audet, *Silicon Sensors*, Academic Press, New York, 1989.
4. S. M. Sze, *Physics of Semiconductor Devices*, 2nd ed., Wiley, New York, 1981.
5. K. Krane, *Modern Physics*, Wiley, New York, 1983.
6. H. Haug, *Optical Nonlinearities and Instabilities in Semiconductors*, Academic Press, New York, 1988.
7. M. W. Scott, "Energy gap in HgCdTe by optical absorption," *J. Appl. Phys.* **40**, 4077 (1969).
8. R. Wolfenbuttel, *Integrated Silicon Color Sensors*, Ph.D. Thesis, Delft University of Technology, Delft, The Netherlands, 1988.
9. G. L. Hansen, J. L. Schmit, and T. N. Casselman, "Energy gap versus alloy composition and temperature in HgCdTe," *J. Appl. Phys.* **53**, 7099 (1982).
10. S. T. Pantelides, *et al.*, "Energy-gap reduction in heavily doped silicon: causes and consequences," *Solid State Electron.* **28**, 17 (1985).
11. S. A. Steigerwald Audet, *High-Purity Silicon Soft X-Ray Sensor Arrays*, Ph.D. Thesis, Delft University of Technology, Delft, The Netherlands, 1990.
12. M. J. Buckingham, *Noise in Electronic Devices and Systems*, Wiley, New York, 1985.
13. A. S. Grove, *Physics and Technology of Semiconductor Devices*, Wiley, New York, 1967.
14. E. H. Nicollian and J. R. Brews, *MOS Physics and Technology*, Wiley, New York, 1982.

15. T. G. M. Kleinpenning, "$1/f$ noise in p–n junction diodes," *J. Vac. Sci. Technol.* A, **3**, 116 (1985).
16. F. N. Hooge, "$1/f$ noise," *Physica* **83B**, 14 (1776).
17. W. A. Radford and C. E. Jones, "$1/f$ noise in ion-implanted and double layer epitaxial HgCdTe photo diodes," *J. Vac. Sci. Technol.* A, **3**, 183 (1985).
18. Y. Nemirovsky, "Surface passivation and $1/f$ noise phenomena in HgCdTe photodiodes," *J. Vac. Sci. Technol.* A, **8**, 1159 (1990).
19. A. van der Ziel and P. H. Handel, "$1/f$ in n^+–p diodes," *Trans. IEEE Electron Devices* **ED-22**, 1802 (1985).
20. G. Bohm, J. I. Kim, and J. Kemmer, "Qualification of the fabrication process for Si detectors by neutron activation analysis," *Nucl. Instrum. Methods* **A305**, 587–99 (1991).
21. X. Xiao, et al., "Silicide/Strained $Si_{1-x}Ge_x$ Schottky-Barrier Infrared Detectors," *Trans. IEEE Electron Device Lett.* (1993).
22. M. C. Teich et al., "Time and frequency response of conventional avalanche photodiodes," *Trans. IEEE Electron Devices* 1511 (1986).
23. K. Brennan, "Theoretical study of multi-quantum well avalanche photodiodes made from GaInAs/AlInAs material system," *Trans. IEEE Electron Devices* 1502 (1986).
24. H. Vydyanath, "Mechanisms of incorporation of donor and acceptor dopants in HgCdTe alloys," *J. Vac. Sci. Technol.* B, **9**, 1716 (1991).
25. Lemaire, "Reliability of optical fibers exposed to hydrogen: prediction of long term loss increases," *Opt. Eng.* 780 (1991).
26. V, Radeka, P. Rehak, S. Rescia, E. Gatti, A. Longoni, M. Sampietro, G. Bertuccio P. Holl, L. Struder, and J. Kemmer, "Implanted silicon JFET on completely depleted high-resistivity devices," *IEEE Electron Device Lett.* **10**, 91–3 (1989).
27. J. L. Schmit and E. L. Stelzer, "Temperature and alloy composition dependencies of the energy gap of HgCdTe," *J. Appl. Phys.* **40**, 4865 (1969).
28. Y. Nemirovsky and E. Finkman, "Intrinsic carrier concentration of Hg," *J. Appl. Phys.* **50**, 8107 (1979).
29. W. M. Higgens, et al., "Standard relationships in the properties of HgCdTe," *J. Vac. Sci. Technol.* B, **7**, 269 (1989).
30. Shih, et al., "Photoemission studies of core level shifts in HgCdTe, CdMnTe and HgZnTe," *J. Vac. Sci. Technol.* A **5**, 3031 (1987).
31. S.-H. Wei, et al., "Electronic structure and stability of II–VI semiconductors and their alloys: the role of metal d bands," *J. Vac. Sci. Technol.* A, **6**, 2597 (1988).
32. J. F. Wager, et al., "Surface characterization of HgCdTe native oxides," *J. Vac. Sci. Technol.* A, **3**, 212 (1985).
33. S. W. McClure, et al., "The magnetophonon effect: a tool for characterizing HgCdTe," *J. Vac. Sci. Technol.* A, **3**, 271 (1985).
34. J. Tregilgas, et al., "Type conversion of HgCdTe induced by redistribution of residual acceptor impurities," *J. Vac. Sci. Technol.* A, **3**, 150 (1985).
35. R. Tribuulet, "THM, a breakthrough method in HgCdTe bulk metallurgy," *J. Vac. Sci. Technol.* A, **3**, 95 (1985).
36. G. L. Destefanis, "Indium ion implantation in HgCdTe/CdTe," *J. Vac. Sci. Technol.* A, **3**, 168 (1985).

37. Y. L. Tyan, et al., "Analysis of excess carrier lifetime in p-type HgCdTe using a three level Shockley-Read model," *J. Vac. Sci. Technol.* B, **10**, 1560 (1992).
38. T. N. Casselmann, "Calculation of the Auger lifetime in p-type HgCdTe," *J. Appl. Phys.* **52**, 848 (1981).
39. O. L. Doyal, et al., "Photoabsorptance and electron lifetime measurement in HgCdTe," *J. Vac. Sci. Technol.* A, **3**, 259 (1985).
40. C. E. Jones, et al., "Status of point defects in HgCdTe," *J. Vac. Sci. Technol.* A, **3**, 131 (1985).
41. P. Capper, "A review of impurity behavior in bulk epitaxial HgCdTe," *J. Vac. Sci. Technol.* B, **9**, 1667 (1991).
42. R. Sporken, et al., "Current status of direct growth of CdTe and HgCdTe on silicon by molecular-beam epitaxy," *J. Vac. Sci. Technol.* B, **10**, 1405 (1992).
43. L. O. Bubulac, et al., "Ion implant junction formation in HgCdTe," *J. Vac. Sci. Technol.* A, **5**, 3166 (1987).
44. J. Wong, et al., "A diffusion model for indium in HgCdTe, *J. Vac. Sci. Technol.* A, **9**, 2258 (1991).
45. M. Yoshikawa, "The behavior of oxygen in HgCdTe," *J. Vac. Sci. Technol.* A, **3**, 153 (1985).
46. G. Bahir and E. Finkman, "Ion beam milling effect on electrical properties of HgCdTe," *J. Vac. Sci. Technol.* A, **7**, 348 (1989).
47. P. Leech and G. Reeves, "Specific contact resistance of indium contacts to n-type HgCdTe," *J. Vac. Sci. Technol.* A, **10** (1992).
48. V. Krishnamurthy, et al., "Oxide interfacial layers in Au ohmic contacts to p-type HgCdTe," *Appl. Phys. Lett.* **56**, 925 (1990).
49. G. Entine, P. Waer, T. Tiernan, and M. R. Squillante, "Survey of CdTe nuclear detector applications," *Nucl. Instrum. Methods* **A283**, 282–90 (1989).
50. J. S. Iwanczyk, Y. J. Wang, N. Dorrie, A. J. Dabrowski, T. E. Economou, and A. L. Turkevich, "Use of mercuric iodide X-ray detectors with alpha backscattering spectrometers for space applications," *IEEE Trans. Nucl. Sci.* **38**, 574–9 (1991).
51. J. F. Butler, F. P. Doty, and C. Lingren, "Recent developments in CdZnTe gamma ray detector technology," *SPIE 1992 Int. Symp., July 19–24, 1992, San Diego, CA.*, SPIE, Bellingham, WA, 1992.
52. H. Becker, T. Boulos, P. Cattaneo, H. Dietl, D. Hauff, E. Lange, G. Lutz, H. G. Meser, A. S. Schwartz, R. Settles, L. Struder, J. Kemmer, U. Prechtel, T. Ziemann, and W. Buttler, "New developments in double sided strip detectors," *IEEE Trans. Nucl. Sci.* **37**, 101–6 (1990).
53. E. Gatti and P. Rehak, "Semiconductor drift chamber—an application of a novel charge transport scheme," *IEEE Trans. Electron Devices*, **ED-29**, 1388 (1982); Proc. 2nd Pisa Meeting on Advanced Detectors, Grosseto, Italy, in *Nucl. Instrum. Methods* **225**, 608–14 (1984).
54. P. Rehak, E. Gatti, A. Longoni, J. Kemmer, P. Holl, R. Klanner, G. Lutz, and A. Wylie, "Semiconductor drift chambers for position and energy measurements," *Nucl. Instrum. Methods* **A235**, 224–34 (1985).
55. J. Kemmer and G. Lutz, "New detector concepts," *Nucl. Instrum. Methods* **A253**, 365–77 (1987).

56. J. Kemmer, G. Luz, E. Belau, U. Prechtel, and W. Wesley, "Low capacity drift diode," *Nucl. Instrum. Methods* **A253**, 378–81 (1987).
57. P. Rehak, E. Gatti, A. Longoni, M. Sampietro, P. Holl, G. Lutz, J. Kemmer, U. Prechtel, and T. Ziemann, "Spiral silicon drift detectors," *IEEE Trans. Nucl. Sci.* **36**, 203–9 (1989).
58. G. Bertuccio, A. Castoldi, A. Longoni, M. Sampietro, and C. Gauthier, "New electrode geometry and potential distribution for soft X-ray drift detectors," *Nucl. Instrum. Methods* **A312**, 613–16 (1992).
59. P. Rehak, S. Resica, V. Radeka, E. Gatti, A. Longoni, M. Sampietro, G. Bertuccio, P. Holl, L. Struder, J. Kemmer, U. Prechtel, and T. Ziemann, "Feedback charge amplifier integrated on the detector wafer," *Nucl. Instrum. Methods* **A288**, 168–75 (1990).
60. J. Kemmer, G. Lutz, U. Prechtel, K. Schuster, M. Sterzik, L. Struder, and T. Ziemann, "Experimental confirmation of a new semiconductor detector principle," *Nucl. Instrum. Methods* **A288**, 92–8 (1990).
61. L. Struder, *et al.*, "The MPI/AIT X-ray imager (MAXI)—high-speed *pn* CCDs for X-ray detection," *Nucl. Instrum. Methods* **A288**, 227–35 (1990).
62. A. Turner, "HgTe contacts to *p*-HgCdTe," *J. Vac. Sci. Technol.* B, **10**, 1534 (1992).

7 Thermal Sensors

A. W. VAN HERWAARDEN
Xensor Integration
Schoemakerstraat 97
2628 VK Delft, The Netherlands

G. C. M. MEIJER
Delft University of Technology
Department of Electrical Engineering
PO Box 5031
2600 GA Delft, The Netherlands

7.1 INTRODUCTION

This chapter describes the major groups of thermal sensors, sensors that measure physical quantities by transducing their signals into thermal quantities first and then transducing the thermal quantities into electrical quantities. A thermal sensor operates in three steps (with the exception of temperature sensors, which transduce from the thermal to the electrical signal domain only). The following steps are involved, which are treated in the next three sections.

1. First, a non-thermal signal is transduced into a heat flow: $A \rightarrow P$ (Section 7.2).
2. Second, the heat flow is converted, within the thermal signal domain, into a temperature difference: $P \rightarrow \Delta T$ (Section 7.3).
3. Third, the temperature difference is transduced into an electrical signal with a temperature (difference) sensor: $\Delta T \rightarrow U$ (Section 7.4).

A description of the various thermal sensors made with silicon or related technologies follows in Section 7.5, the main part of this chapter. Six major

---
Semiconductor Sensors, Edited by S. M. Sze.
ISBN 0-471-54609-7 © 1994 John Wiley & Sons, Inc.

332 THERMAL SENSORS

groups of thermal sensor are described, followed by a description of integrated temperature sensors.

The chapter finishes with summary and future trends, some instructive problems, and the references. A very general reference is "Thermal Sensors", by Meijer and van Herwaarden.[1] For those who wish to go into this subject a little more thorough than is possible in a chapter, this book is highly recommended.

7.2 HEAT TRANSFER

This section is a short introduction to the theory of heat and heat transfer. Our intuitive description of the nature of heat, and the theory of conductive and radiative heat transfer are fairly straightforward. The theory of convective heat transfer is not that simple, and we will only go into it superficially. If you have to deal with convective heat transfer, you may want to refer to specialized books for a better insight.

7.2.1 Heat

The Nature of Heat. Heat, also called thermal energy, can in a simple, intuitive way be viewed as the internal kinetic energy of a collection of molecules (or atoms). For gases, the heat is closely related to the average velocity of the molecules, and in the case of multi-atom molecules, the rotations and vibrations of the atoms within the molecules have to be taken into account as well. The temperature is directly linked to the concept of thermal energy, because systems with equal thermal energy (density) have equal temperatures. For liquids, the situation is similar to that of gases. For solids, the situation is different, in that the molecules cannot move about freely. Here the internal kinetic energy is stored as so-called phonons, which are the coordinated movements (vibrations) of the atoms about their fixed lattice position. In some solids, the electrons are also free to move and they can also store thermal energy. Simple diffusion will ensure that if ever there is a surplus of heat (i.e., of fast molecules or a higher density of phonons) in some area, this heat will flow towards areas with less heat until thermal equilibrium has been established.

Specific Heat and Thermal Capacitance. The internal energy of a system is composed of thermal (kinetic) energy and of potential energy, the energy related to the average distance between the molecules. In the case of a system that has constant pressure, the specific heat, the heat required to increase the temperature of a material by 1 K, is defined as:

$$c_p = \left(\frac{dH}{dT}\right)_p \qquad (1)$$

In Eq. 1 T is the absolute temperature (in K). The enthalpy H is defined as the sum of the internal energy E and the product of pressure p and volume V, and takes the external work into account. For solids, the specific heat approaches a constant value at high temperatures of $3R_0$ (given in J/K-mol), where R_0 is the universal gas constant and $3R_0$ is approximately 25 J/K-mol. Because the densities of many solids in terms of atoms per volume lie in a narrow band, the values for the specific heat expressed in J/K-m^3 are much closer than when expressed in J/K-mol. This is of interest when estimating the thermal capacitance, C_{th}, of structures. The thermal capacitance can be defined as the heat necessary to increase the temperature of a given system by 1 K and equals the mass of the system m times the specific heat, c_p, $C_{th} = c_p m$. When analyzing thermal problems using electrical analogs, the thermal capacitance is analogous to the electrical capacitance.

7.2.2 Conduction

Conductivity. When there is a temperature gradient present in a substance, heat will flow from the hotter to the colder region, and the heat flow P'' (in W/m^2)† will be proportional to the temperature gradient (taken in the x direction here):

$$P'' = -\kappa \frac{dT}{dx}. \qquad (2)$$

In Eq. 2 κ is the thermal conductivity (in W/Km). The heat transport is carried out by diffusion of phonons, molecules and/or electrons.

Conduction in Gases. The thermal conductivity of a gas by diffusion of molecules is fairly independent of pressure, although it varies with temperature. Most gases have a conductivity of the order of 15–30 mW/K-m, radon and xenon have the lowest at 4–6 mW/K-m, and hydrogen has the highest at 185 mW/K-m. For conduction between two parallel surfaces at distance d the heat transfer coefficient G'' (in W/K-m^2) becomes:

$$G''_{\text{dense gas}} = \frac{\kappa}{d}. \qquad (3)$$

The situation is different, when the heat transfer occurs in a system where the mean free path between intermolecular collisions is much larger than the distance between the surfaces (as in micromachined structures at low pressures). Then, the heat transfer is carried out by individual molecules instead of by the

†The single, double and triple indices refer to values that are specified per unit of length, area and volume, respectively. For example, P' is in W/m.

gas as a whole, and the heat transfer is directly proportional to the density of the molecules ρ (in kg/m^3), that is to the absolute pressure, p. At low pressures, the heat-transfer coefficient can be approximated by:[2]

$$G''_{\text{low pressure}} = G''_0 \frac{p}{p_0} \approx \rho c_v u_{\text{mol}} \tag{4}$$

where G''_0 is the heat-transfer coefficient at the reference pressure p_0, c_v is the specific heat at constant volume, and u_{mol} is the average molecular velocity. With $p_0 = 1$ Pa (100 kPa = 1 atmosphere), G''_0 is typically on the order of 1 W/K-m^2 for gases such as air.

A generalized formula, valid over the full pressure range, is the combination of Eqs. 3 and 4:[3]

$$G''_{\text{intermediate pressure}} = \frac{G''_0(p/p_0)(\kappa/d)}{G''_0(p/p_0) + \kappa/d}. \tag{5}$$

In typical microstructures with a plate distance d of about 300 μm, the pressure at which $G''_0 p/p_0$ and κ/d are equal is on the order of 100 Pa, if κ is 0.03 W/K-m and $G''_0/p_0 = 1$ W/K-m^2-Pa. For surface-micromachined structures where gaps of 1 μm are feasible, the thermal conduction by the gas in the gaps will be pressure-dependent up to almost atmospheric pressure.

Conduction in Solids. In solids, the thermal conductivity is carried out by two mechanisms. The first is that of electron conduction, which resembles conduction in gases, and occurs in metals and highly degnerated semiconductors. In a pure metal crystal at normal temperatures the thermal conductivity is independent of the temperature. Real metals show only a very small (usually negative) variation of κ with temperature. The absolute value of κ is high, typically on the order of 100–400 W/K-m for pure crystalline metals and 20–100 W/K-m for most alloys.

The second mechanism, occurring in all solids, is that of heat transport by phonons. In pure crystals the conductivity by phonons can be very high because they have a very long path before being scattered in the material. Diamond is the best known thermal conductor at room temperature and silicon is also a very good conductor, with values of κ of 660 and 150 W/K-m, respectively. Just as with the electrical conductivity in silicon, the thermal conductivity in pure diamond and nondegenerated silicon exhibits a peak at low temperatures (see Fig. 1). In amorphous materials, the electrons and phonons cannot travel very far and, therefore, their thermal conductivity is very low (down to 0.1–1 W/K-m).

Conduction in Liquids. For liquids, no simple models exist. Their thermal conductivity usually lies between the values for gases and those for solids. Figure 1 shows how the thermal conductivity varies with temperature. As an example,

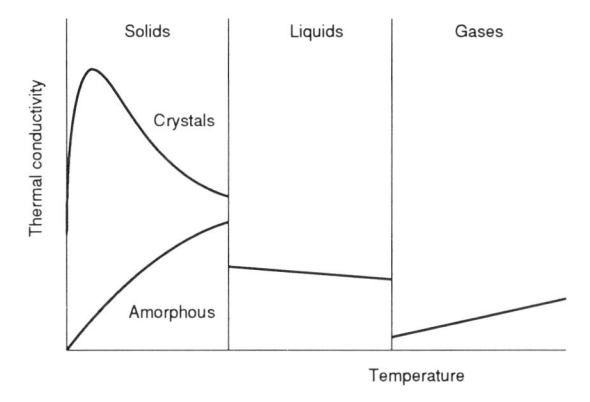

Fig. 1 The thermal conductivity as a function of temperature (not to scale). (After Ref. 1)

the values for the various phases of water are:

ice (272 K) = 2.22 W/K-m,
water (373 K) = 0.68 W/K-m,
water vapour (380 K) = 0.025 W/K-m.

7.2.3 Convection

Heat Transfer in Flows and Boundary Layers. Heat transfer to flowing fluids is about the most difficult subject in thermal sensing theory. We will, therefore, not go into too much detail here, since this would take up too much space. We will briefly introduce some of the main parameters and numbers used to characterize the heat transfer in forced convection.

In the case of a fluid (liquid or gas), heat can be transferred by the movement of lumps or elements in the fluid. Because a mass with a certain temperature represents an amount of heat, the transport of matter at the same time constitutes a transport of heat. If the fluid motion is forced externally, we speak of forced convection. If buoyancy forces due to temperature differences cause the fluid motions, we speak of free or natural convection. In our small structures, free convection is usually not of importance.

The flow can further be distinguished into laminar and turbulent flow. In laminar flow, which predominates at low flow speeds, the fluid motion takes place along regular streamlines. In turbulent flow, a strongly fluctuating flow component is superimposed on the mean flow pattern, which is the result of instabilities in the flow occurring at higher flow speeds. As this intensifies the exchange of heat in the fluid, turbulent flow results in an increased heat transfer. The conditions at which a flow is laminar or turbulent are governed by many parameters, like the flow speed, the flow geometry (e.g., the shape and size of a channel), and the material properties of the fluid.[4] For most of the applications which we will consider, flow speeds and sensor sizes are quite small and laminar flow can be assumed.

For a given flow configuration, with a characteristic length L (e.g., the inner diameter of a pipe or the length of a plate in a free flow), the flow of an incompressible viscous fluid is characterized by the value of the (dimensionless) product UL/v, with U as the flow velocity and v as the kinetic viscosity of the fluid. This product is called the Reynolds number, Re. For a specific flow situation, a critical value of Re exists. Below this value, the flow is stable, small disturbances in the flow are suppressed, and a laminar flow results. Above the critical value, the flow is unstable and turbulent flow results.

When considering the operation of thermal structures, it is generally not necessary to study the heat transfer throughout the entire flow, but only the heat transfer from the solid surface, which is exposed to the flow, into the boundary layer. Because of molecular interaction, the fluid at the surface adheres to the surface and adopts its temperature. In the boundary layer near the surface, the velocity and temperature change gradually to their values in the outer stream. Figure 2 shows this effect, here the fluid velocity in the boundary layer varies between zero at the surface to the free-stream fluid velocity outside the boundary layer, and the temperature of the fluid in the boundary layer varies between the wall temperature at the surface and the free-stream fluid temperature outside the boundary layer. Because the visocisity and thermal conductivity of most common fluids are very low, the boundary layer is usually very thin and the main stream remains unaffected. Figure 3a shows what happens when a free stream of fluid encounters a heated plate, the extent of the boundary layer is indicated by the solid line. This shows that the thicknesses of the thermal and mechanical boundary layers increase with the distance from the leading edge. When an initial part of the plate is not heated, as shown in Fig. 3b, the thermal boundary layer develops only after this unheated region, and will in general remain thinner than the mechanical boundary layer.

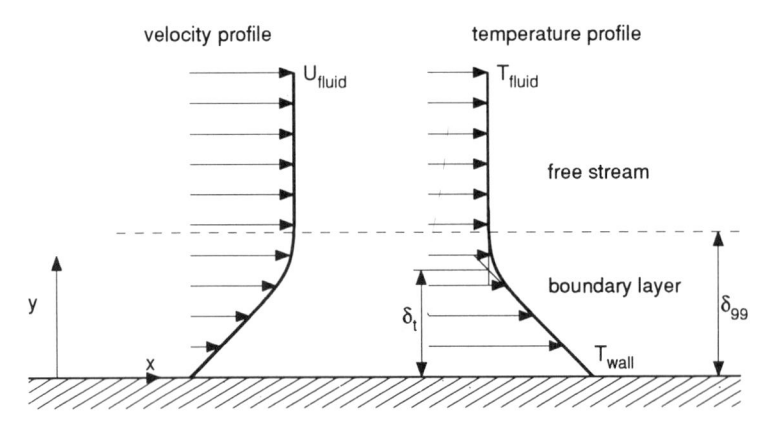

Fig. 2 Velocity and temperature profiles in a boundary layer. (After Ref. 1)

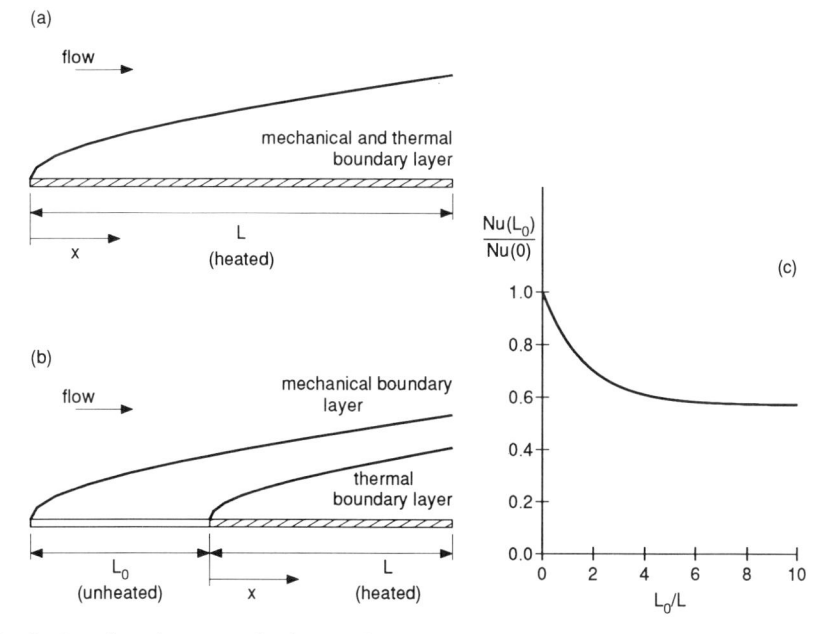

Fig. 3 Laminar heat transfer from a flat plate with and without an initial cold length. (After Ref. 1)

The dimensionless ratio of the kinematic viscosity v (in m^2/s) and the thermal diffusivity α ($=\kappa/c_p\rho$, also in m^2/s) is called the Prandtl number $Pr=v/\alpha$, which describes the relation between the thermal and mechanical boundary layer. For $Pr=1$, they coincide. For real gases, Pr is about 0.7 and the thermal boundary layer increases a little more rapidly with the distance from the leading edge (that is, at any place the thermal boundary layer will be thicker than the mechanical boundary layer if they start at the same place).

The heat-transfer coefficient G'' is defined as the ratio of the heat flux P'' leaving the surface and the temperature difference between the wall and the fluid in the free stream:

$$G'' = \frac{P''_{\text{wall}}}{T_{\text{wall}} - T_{\text{free stream}}} = \frac{\kappa}{\delta_t} \tag{6}$$

where δ_t is the length as defined in Fig. 2. In convection problems, the dimensionless Nusselt number (Nu) is often used. Nu is the dimensionless form of the heat transfer coefficient and is defined as:

$$Nu = G''L/\kappa. \tag{7}$$

In this case, L is a characteristic length, such as the length over which a boundary layer has developed (see Fig. 3a).

Forced-Convection Heat Transfer. As mentioned before, we will not go into convection problems at length, and we will, therefore, give attention only to the case of boundary-layer flows over flat plates, which are the most relevant for silicon sensors. Moreover, they give relatively-easy-to-interpret results. An important example of a boundary-layer flow is that along a flat, hot plate parallel to the oncoming flow (Fig. 3a). The local flow regime at distance x downstream from where the plate begins is determined by the value of the local Reynolds number $Re_x = Ux/v$, where U is the constant flow velocity in the free stream. For $Re_x < 5 \times 10^5$ we have laminar flow, for $5 \times 10^5 < Re_x < 3 \times 10^6$ transitional flow, and for $Re_x > 3 \times 10^6$ turbulent flow.

The local Nusselt number Nu (at any place x) and average Nusselt number Nu_{av} (averaged over the entire surface) for a plate of length L are given by the following expressions (valid for Prandtl numbers $Pr > 0.6$):[5]

$$Nu = G''(x)\frac{L}{\kappa} = 0.664 Pr^{1/3} Re^{1/2} \frac{1}{2}\left(\frac{x}{L}\right)^{-1/2} \quad \text{(laminar flow)} \qquad (8)$$

$$Nu_{av} = G''_{av}\frac{L}{\kappa} = 0.664 Pr^{1/3} Re^{1/2} \qquad (9)$$

where x is measured from the leading edge. For turbulent flow, we find approximately:

$$Nu_{av} = 0.037 Pr^{1/3} Re^{0.8} \quad \text{(turbulent flow)}. \qquad (10)$$

A different situation occurs when the flat plate has an initial, unheated part of length L_0 and a heated part of length L (see Fig. 3b). The average laminar heat transfer has been calculated numerically and plotted in Fig. 3c as a function of L_0/L. For L_0/L larger than 4 the following approximation can be used (with deviations of less than 2%):[6]

$$\frac{Nu_{av}(L_0)}{Nu_{av}(L_0=0)} = 0.825\left(\frac{L}{L_0}\right)^{1/6}. \qquad (11)$$

7.2.4 Radiation

Black-Body Radiation. The third way in which bodies may exchange heat is by emitting or absorbing thermal radiation. This radiation is inherently electromagnetic. Kirchhoff's law states that if radiation is incident on a surface, the sum of the absorbed, reflected and transmitted radiation is equal to the total incident radiation, for each wavelength. The fraction of radiation that is absorbed is called the absorptivity α, which is a function of the wavelength of the radiation. A body that absorbs all of the radiation ($\alpha = 1$) is called a black body. According to Kirchhoff, for each wavelength, it holds that the emissiviity $\epsilon(v)$ is equal to the absorptivity $\alpha(v)$ (the fraction of radiation that is emitted, compared to that emitted by a black body is the emissivitity). The Stefan–Boltzmann law states that the power emitted by a gray or black body is equal to:[7]

$$P''_{rad}(T) = \epsilon \sigma T^4 \qquad (12)$$

in which $P''_{rad}(T)$ is the total heat flux emitted by a body, ϵ is the emissivity of the surface (for a black body $\epsilon = 1$), and σ is the Stefan–Boltzmann constant: $\sigma = 56.7 \times 10^{-9}$ W/m²-K⁴.

For a black body, Eq. 12 is the integral (multiplied by $c/4$) over all wavelengths of the famous formula of Max Planck, that describes the energy density of black-body radiation in a cavity at temperature T. Figure 4 shows the spectral radiancy, which is the power emitted by a black body at the indicated temperature, of 1 m² over a solid angle of 1 sr (there are 4π sr on a full sphere) per unit of wavelength (here taken as 1 μm). The spectral radiancy is found from Planck's formula by multiplying by $c/4\pi$, to convert from radiation energy density to emitted radiation power per solid angle. The diagonal line shows

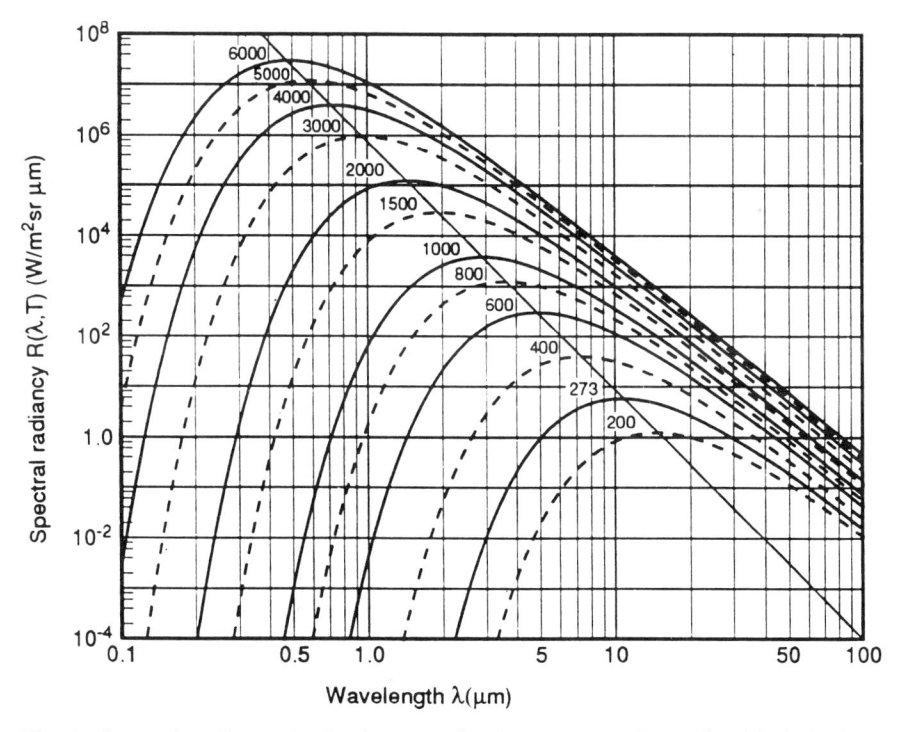

Fig. 4 Spectral radiancy (emitted power density per sr) $R(\lambda, T)$ of a black body at temperatures given in Kelvin. The diagonal line shows Wien's displacement law. (After Ref. 1)

Wien's displacement law: the wavelength at the maximum of the curve is inversely proportional to the absolute temperature.

Heat Transfer by Infrared Radiation. With Eq. 12 we can calculate the energy exchange between two parallel plates, one at room temperature T ($=300$ K), and the other at a slightly higher temperature $T+\Delta T$, where ΔT is small with respect to T. Series expansion of Eq. 12 shows that the neat heat transfer from the hot to the cold plate is given by:

$$G''_{rad} = 4\epsilon\sigma T^3 = \epsilon \times 6 \text{ W/K-m}^2. \tag{13}$$

Silicon, however, is somewhat transparent to infrared radiation at wavelengths above 1.1 μm, and for $T=300$ K the infrared radiation has a maximum at approximately 10 μm, as can be seen in Fig. 4. Therefore, silicon microstructures have an effective emissivity and absorptivity to infrared radiation that is significantly lower than unity, that is more like 0.3–0.1. This makes the heat transfer by thermal radiation negligible in many silicon sensors. However, if the devices are coated with highly absorbent layers, significant heat transfer may occur.

In infrared sensors, the sensor looks through a window at objects whose infrared radiation is to be measured. Assume the sensitive area is A_{sen}, and we observe an object with area A_{ob} at distance d, where d is much larger than the length of either surface. The sensor and object surfaces are in parallel and completely visible to each other. According to Lambert's law[7] the net heat transfer from the object to the sensor is then given by:

$$P = \epsilon\sigma A_{sen} A_{ob} \frac{(T^4_{ob} - T^4_{sen})}{\pi d^2} \tag{14}$$

in which it is assumed that the sensor is black and at a near-ambient temperature and that the object is gray with an emissivity $\epsilon \leqslant 1$.

For a small temperature difference ΔT between the object and the sensor, we can approximate the right-hand side of Eq. 14 by using the first terms of the Taylor series and define a thermal 'conductance' G_{ob}, expressing the radiative heat exchange between the object and the sensor as:

$$G_{ob} = \frac{P}{\Delta T} = \frac{4\epsilon\sigma A_{sen} A_{ob} T^3_{sen}}{\pi d^2}. \tag{15}$$

7.3 THERMAL STRUCTURES

7.3.1 Modeling Practical Structures

Systematical Design. In practical sensor structures, thermal effects are induced in the sensor by physical effects that interact with the sensor. The sensitivity and accuracy have to be as large as possible, while the influence of other physical effects (such as heat 'leakage' along connections and suspensions, and self-heating effects) has to be minimized.

It is usually possible to design structures in such a way that only a few well-known parameters dominate their behavior. When certain parameters are not well known, their influence ought to be negligible. Therefore, a good design of a thermal structure is simple and can be described by simple models. The validity of the approximations and assumptions in the model are checked by comparing the results of simulations with experimental results. When it is found that the approximations and assumptions are not correct, it does not automatically mean that a more complex model has to be used. Often, it is better to change the thermal structure to avoid the undesirable influence of additional parameters!

In temperature sensors, the thermal signal is the temperature induced in the sensor material by the body in which the sensor is mounted. Because most temperature sensors measure their own temperature, the main objective when measuring the temperature of a certain body is to bring the sensor in good thermal contact with that body.

In other thermal sensors, the thermal signal is the temperature difference induced in the sensor by a physical effect. Here, optimizing the structure is aimed at optimizing the conversion of the power P to the temperature difference ΔT, where the thermal resistance $R_{th} = \Delta T/P$. This has led to a widespread use of very thin membranes in which the thermal resistance between the physical interaction area and the ambient is maximized.

Choice and Optimization of Structure. When choosing and optimizing a thermal structure, the thermal-sensing element is an important factor, because its presence influences the thermal resistances and capacitances. For example, with transistors, diodes, or resistors in micromachined structures, connection leads can be made as thin and long as desired because they have little influence on the thermal resistance of the connection between ambient and sensitive areas. On the other hand, a thermopile often forms both the sensing element and the thermal connection between the hot and cold areas of the sensor, and its design directly influences the thermal resistance between the ambient and the sensitive area.

An important design aspect of a sensor structure is the physical transduction process that is the basis of the sensor. Some sensors require large interaction areas, for example infrared sensors, while others need hardly any interaction area at all, for example true rms converters (rms denotes the root of the mean of the square of a signal, and this is proportional to the root of the power of an AC voltage or current).

Another crucial aspect is the packaging of the sensor chip, that is whether the sensor will be exposed to harsh conditions (flow sensors) or will be hermetically sealed (infrared sensors, true rms converters, temperature sensors).

Sensor vulnerability is, therefore, critical during both use and production because it influences other important design criteria—the yield and the resulting cost.

Another choice involved in selecting a structure is that of technology—thin-film membranes or silicon membranes. In general, thin-film structures are more sensitive, while silicon structures are less vulnerable and somewhat easier to model.

We can choose from three basic structures:
1. the closed membrane (this includes wafer-thick devices),
2. the cantilever beam (and the bridge, which can be considered as two beams with their ends connected),
3 the floating membrane (consisting of a beam with an enlarged end).

The closed membrane has a smaller effective sensitive area and a lower vulnerability. The cantilever beam and the bridge form the intermediate case, while the floating membrane has the opposite characteristics—large effective sensitive area and high vulnerability.

Electrical–Thermal Analogies. The behavior of the heat flow and temperature in thermal systems is described mathematically by the same equations as those used in electrical currents and voltages in electrical systems. This enables thermal systems to be described by means of electrical equivalents, which is convenient because of the many excellent tools available for electrical circuit analysis, and the familiarity of solving electrical-network problems.

The electrical equivalents of the thermal parameters are listed in Table 1, together with the SI-units in which they are expressed. Note that prescribed temperature differences are equivalent to a voltage source, and prescribed heat flows to a current source. Some examples of thermal resistances for different geometries are given below. Note that the thermal capacitances are simply the mass of the body multiplied by the specific heat. When we multiply the thermal capacitance of an element and the thermal resistance between this element and a heat source, we obtain the thermal time constant, characteristic of the time

TABLE 1 Electrical Equivalents of Thermal Parameters

Thermal Parameter	Electrical Parameter
Temperature: T (K)	Voltage: V (V)
Heat flow, Power: P (W)	Current: I (A)
Heat: Q (J = W s)	Charge: Q (C = A-s)
Resistance: R (K/W)	Resistance: R (Ω = V/A)
Conductance: G (W/K)	Conductance: G (S = Ω^{-1})
Capacity: C (J/K)	Capacitance: C (F = A-s/V)
Thermal resistivity: ρ_{th} (K-m/W)	Electrical resistivity: ρ_{el} (Ω-m)
Thermal conductivity: κ (W/K-m)	Electrical conductivity: σ (S/m)
Specific heat: c_p (J/kg-K)	Permittivity: ε (F/m)

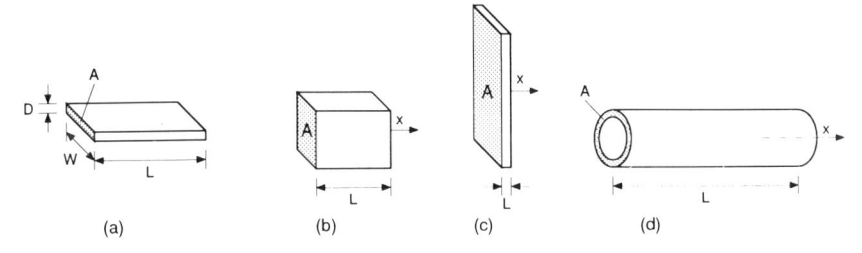

Fig. 5 One-dimensional heat flow along the axis of rods, wires, rectangular plates and tubes. The symbol L denotes the linear dimension in the direction of the heat flow. (After Ref. 1)

needed to heat the element, similar to the electrical situation of capacitances and resistances.

Some Examples of Thermal Resistances

Axial heat flow in bodies having an arbitrary but uniform cross-section. Let us consider the stationary one-dimensional heat flow along the axis of a body of length L and uniform cross-sectional area A (see Fig. 5), which is exposed to a temperature difference $\Delta T = T_1 - T_2$ between its ends.

In the steady state, the temperature distribution is given by a constant temperature gradient equal to $\Delta T/L$, and the total flow through the rod is $\kappa A \, \Delta T/L$. Hence, for this configuration we find a thermal resistance R_{th}:

$$R_{th} = \frac{L}{\kappa A} \quad \text{(structure with length } L \text{ and uniform cross-section } A\text{).} \tag{16}$$

This expression can be used for the one-dimensional heat flow in all kinds of structure with a uniform cross-section, such as rods, plates or wires as shown in Fig. 5.

For example, the thermal resistance of a flat plate with a rectangular cross-section on length L, width W and thickness D (Fig. 5a) is equal to:

$$R_{th} = \frac{L}{W} \frac{1}{\kappa D}. \tag{17}$$

For a square plate, where the length L is equal to the width W, the thermal resistance is equal to $(\kappa D)^{-1}$, which is called the thermal sheet resistance R_{st} of the plate.

Radial heat flow in circular bodies. Multiple thermal resistances can be combined in series or in parallel, and the resulting resistance is calculated just as in the electrical case. This method can be used to calculate complex geometries.

344 THERMAL SENSORS

If the geometry is such that the heat flow is one-dimensional or if a cylindrical or spherical symmetry makes it possible to assign surfaces of equal temperatures, the effective thermal resistance follows from the integration between the beginning and the end of the structure.

For instance, the thermal resistance for radial heat flow in circular bodies with height D can easily be found by calculating the thermal resistance of a tube-shaped cylindrical element with height D, thickness dr and radius r, which is $dR = dr/(\kappa 2\pi r D)$. The thermal resistance of a structure with inner radius r_{inn} and outer radius r_{out} can then be obtained by means of integration:

$$R_{th,rad} = \frac{1}{2\pi\kappa D} \ln \frac{r_{out}}{r_{inn}} = \frac{1}{2\pi\kappa D} \ln \frac{d_{out}}{d_{inn}} \quad \text{(radial heat flow in tubes)} \quad (18)$$

where d_{out} and d_{inn} are the corresponding diameters of the tube and $1/\kappa D$ is the thermal sheet resistance of the body (tube or membrane).

Radial heat flow in spheres. The thermal resistance for a radial heat flow of a spherical-shell element with thickness dr and radius r is $dR = dr/(4\pi\kappa r^2)$. Integration between the limits r_1 and r_2 yields the thermal resistance between concentric spherical surfaces:

$$R_{th,sphere} = \frac{1}{4\pi\kappa}\left(\frac{1}{r_1} - \frac{1}{r_2}\right) \quad \text{(radial heat flow in spheres).} \quad (19)$$

Example: Self-Heating of a Cylindrical Temperature Sensor in a Cylindrical Hole. A cylindrical, resistive, temperature sensor with a temperature-dependent resistance R of about 100 Ω, a length l_s, and a diameter d_s, is mounted in a hole of a thermal conducting body with a diameter d_h to measure the body temperature T_B (Fig. 6). Due to tolerances the hole diameter must be slightly larger than that of the sensor. The measurement current for the sensor, which is 10 mA, creates a self-heating effect in the sensor.

To measure this self-heating effect, we will consider the thermal conductance of the body and the sensor to be infinite compared to the low conductance of the air in the gap between the sensor and the body. Owing to the measurement current, the power dissipation in the sensor equals $I^2R = 10$ mW. To get an impression of the self-heating effect, we simplify the problem by neglecting the heat flow in the axial direction. Assuming radial symmetry we can use Eq. 18 to find the thermal resistance R_{th}. If we use $\kappa_{air} = 0.026$ W/K-m and for example, the dimensions $d_h = 3$ mm, $d_s = 2.8$ mm and l_s ($=D$) $= 25$ mm, we find that $R_{th} = 17$ K/W. The temperature rise due to self-heating then amounts to $PR_{th} = 0.17$ K, which is unacceptably high in many applications. The self-heating effect may be reduced by filling the air gap with a thermally conducting compound or by lowering the measurement current.

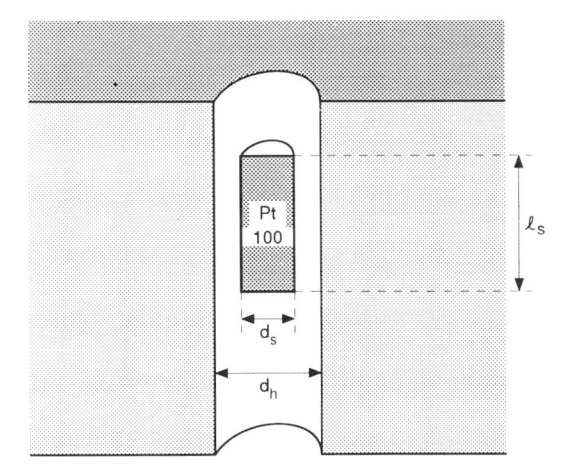

Fig. 6 A cylindrical temperature sensor mounted in the hole of a body to measure its temperature. (After Ref. 1)

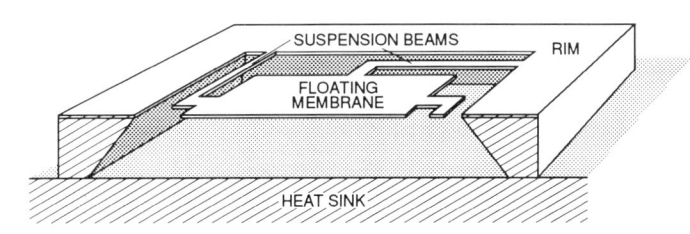

Fig. 7 Floating-membrane structure with a large floating membrane suspended from the wafer-thick rim by four long and narrow cantilever beams (two and half beam visible). (After Ref. 1)

Numerical Modeling. In many cases a numerical solution of a model is required, since no (easy) analytical solutions are at hand. In these cases we use numerical modeling of the physical situation. By using electrical analogies, electrical-circuit analysis tools such as SPICE can be used for time-dependent solutions. Alternatively, software packages such as ANSYS or other finite-element modeling packages can be used. For relatively simple problems, one can do numerical calculations on a PC using simple software routines based on elementary models.[8]

7.3.2 Floating Membranes

The simplest membrane structure in terms of a thermal model is the floating membrane. In silicon, a floating membrane is made by etching a (closed) membrane in the silicon wafer. Next, a large piece of the membrane is etched free from the rim, leaving it suspended by only a few suspension beams, as shown in Fig. 7.

The large floating membrane then becomes an interaction area in which the non-thermal physical signal can influence the thermal signals (transduction step 1 of the sensor action, see Section 1). The suspension beams, together with any heat transfer from the interaction area directly to the ambient, define the thermal resistance (conversion step 2 of the sensor action). The leads of the temperature-difference transducer are incorporated in the suspension beams (transduction step 3 of the sensor action).

Thermopiles may be used in the suspension beams to measure the temperature difference between the 'floating' membrane and the rim at an ambient temperature. This structure produces a very large hot area that has a very high thermal resistance because the suspension beams are usually quite long and narrow. This leads to a fairly well-defined thermal structure having an interaction area (i.e., the floating membrane) that may be considered as an isothermal plane, connected to the ambient by some well-defined thermal resistances (i.e., the suspension beams). The thickness of the rim to which the suspension beams are connected ensures that the beams will have a perfect thermal ground on the other side.

The Thermal Model. The floating membrane can be represented by a simple, discrete-element model of the thermal structure (see Fig. 8a). The thermal conductance of the suspension beams is represented by $1/R_{beam}$, the floating membrane constitutes the thermal capacitance of the structure C_{flm}, and a parasitic conductance G_{flm} describes the undesired heat loss in the floating membrane caused by convection, radiation, and conduction through the gas. The variable conductance G_{sen} represents the desired conductance created by the physical signal.

We can use this model to calculate the transfer function and the time constant of the floating-membrane structure. In the steady-state situation the temperature rise in the sensor above the ambient that results from the power dissipation P is equal to $\Delta T_{flm} = T - T_{amb} = P/(1/R_{beam} + G_{flm} + G_{sen})$. The time response of the system when the heating power is abruptly changed from zero to a constant

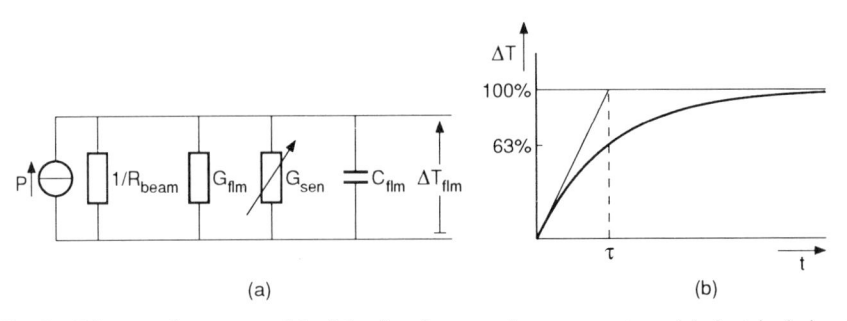

Fig. 8 Discrete-element model of the floating-membrane structure: (a) electrical circuit and (b) step response curve for the first-order circuit. (After Ref. 1)

value P_0 at time $t=0$ is given by:

$$\Delta T_{\text{flm}}(t) = \frac{P}{1/R_{\text{beam}} + G_{\text{flm}} + G_{\text{sen}}} (1 - \exp(-t/\tau_{\text{flm}})) \tag{20}$$

(see Fig. 8b). A similar result is found for an abrupt change in T_{amb} or in/$(1/R_{\text{beam}} + G_{\text{flm}} + G_{\text{sen}})$. The time constant τ_{flm} follows from the total thermal conduction and thermal capacitance:

$$\tau_{\text{flm}} = \frac{C_{\text{flm}}}{G_{\text{flm}} + G_{\text{sen}} + 1/R_{\text{beam}}}. \tag{21}$$

For this simple RC time constant, the step response shows the time constant in several ways: the derivative of the curve at the step time crosses the final value after one time constant, and the curve reaches 63% of its final value after one time constant, as shown in Fig. 8b.

In general, the factor R_{beam} can be designed independently of the factors C_{flm}, G_{flm} and G_{sen}, which are closely related. For example, enlarging the interaction area will usually enlarge G_{sen}, G_{flm} and C_{flm} by approximately the same factor.

7.3.3 Cantilever Beams and Bridges

The integrated cantilever beam (Fig. 9) is very similar in mathematical description to the uniform transmission line. Below, some simple results from this theory will be given, interested readers are referred to the literature.[9]

The cantilever beam is a rectangular beam etched out of a thin membrane, attached on one side to the rim of a silicon chip and free on the other planes. This structure is characterized by a large thermal resistance between the tip of the beam and the base where it is attached to the rim. Heat dissipated at the tip of the beam will flow through the silicon to the rim, creating a temperature difference in the beam. In addition, heat may be lost to the ambient through

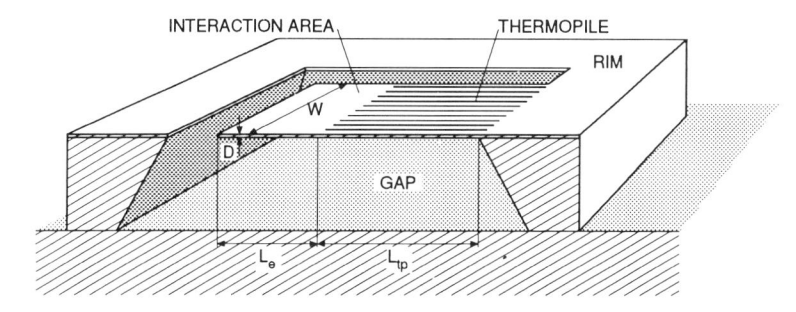

Fig. 9 Cantilever-beam structure with thermopile length L_{tp} and width W and a hot region of length L_e beyond the thermopile. (After Ref. 1)

the emission of infrared radiation and through conduction and convection if gases are present.

In vacuum, with no heat loss from the cantilever beam through gas conduction or infrared radiation, the thermal resistance of the cantilever beam is given by the thermal sheet resistance R_{st} ($=1/(\kappa D)$) times the length–width ratio L/W of that part of the cantilever beam across which the temperature difference is being measured, for instance, by a thermopile of length L_{tp}:

$$R_{th} = R_{st} \frac{L_{tp}}{W}. \qquad (22)$$

Just as with the floating membrane, a separate interaction region can be created by extending the beam beyond the thermopile, to enhance the interaction with the physical signal to be measured. This region can be made rather large. The total length L of the cantilever beam can be much larger than the length L_{tp} of the thermopile. A total interaction region of area $L_e W$ is available, provided the length extending beyond the thermopile is $L - L_{tp} = L_e$.

Let us examine the thermal characteristics of the cantilever beam more closely. In the case of heat loss by gas conduction or infrared radiation, we obtain a thermal transmission line as mentioned above. Using transmssion-line theory we obtain the transfer function of the cantilever beam:

$$R_{th} = R_{st} \frac{L}{W} \frac{\tanh(\gamma L)}{\gamma L} \qquad (23)$$

where R_{st} is the thermal sheet resistance of the beam and W the width of the beam, where $\gamma^2 = G_p'' R_{st}$ (G_p'' is the shunt conductance from the beam to the ambient per unit of area, expressed in W/K-m^2). For small values of γL, Eq. 23 can be approximated by the first term of a power series:

$$R_{th} = R_{st} \frac{L}{W} \left(1 - \frac{R_{st}(G_p'' + j\omega C_{th})L^2}{3}\right) \qquad (24)$$

where C_{th} is the thermal capacitance of the beam per square meter.

When we study the influence of parallel conduction on the thermal resistance of the beam, the term $R_{st} G_p'' L^2 / 3$ denotes the sensitivity of the transfer function of Eq. 24 to parallel conductance. Thus, it is an important parameter for the sensitivity of the beam as applied to flow sensors and vacuum sensors. The thermal time constant τ_{cant} of the cantilever beam is found from Eq. 24 by solving the corresponding homogeneous differential equation, which yields, for small values of γL:

$$\tau_{cant} \simeq R_{st} \frac{L^2}{3} C_{th}. \qquad (25)$$

7.3.4 Closed Membranes

In closed membranes, we encounter radially directed heat transfer in a (very flat) cylindrical structure in which there is heat loss at the bottom and top of the cylinder. In the radial direction, the thermal series resistance and the thermal parallel conductance and capacitance depend on the distance from the center, where r is present in all quantities. This can be seen in Fig. 10, where a circular membrane is assumed (practical micromachined structures are mostly rectangular). The dependence on the radius r means that there is a nonuniform transmission line for which the (relatively simple) equations of the previous section are not valid. Instead, more complicated differential equations have to be solved. We will refrain from giving the full derivation here, and refer the interested reader to standard mathematics textbooks.[10] We assume that heat is lost to the ambient by means of parallel conduction G_p'' (in W/m²-K) and that heat P (in W/m or in W/m²) is generated by means of dissipation in a resistor or by interaction with the ambient (infrared radiation, heat of a catalytically promoted chemical reaction). Then, if $\gamma^2 = G_p'' R_{st}$, where R_{st} is the thermal sheet resistance of the cylindrical membrane, the general solution for the temperature is given by the sum of the modified Bessel functions:

$$T(r) = A_1 I_0(\gamma r) + A_2 K_0(\gamma r) + P''/G_p'' \tag{26}$$

where A_1 and A_2 are constants dependent on the boundary conditions. K_0 and I_0 are modified Bessel functions of the zeroth order. There are two main cases to be discussed: homogeneous heating and resistor heating.

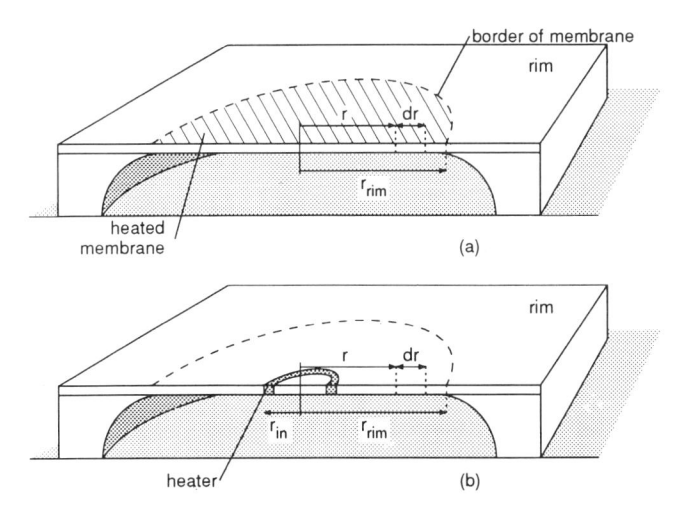

Fig. 10 Geometry of the nonuniform transmission line (a) homogeneous heating and (b) resistor heating. (After Ref. 1)

Homogeneous Heating. Consider a closed circular membrane which is homogeneously heated with a power density P'' (in W/m^2), see Fig. 10a. As the boundary conditions we use the temperature derivative in the center (should be zero because of symmetry) and the temperature elevation $T(r_{rim})$ at the edge of the rim (is by definition zero). If G_p'' approaches zero, and the values of γr are small, the solution approaches:

$$T(r) = P'' R_{st}(r_{rim}^2 - r^2)/4. \tag{27}$$

The result is plausible since the total heat generation is proportional to $P'' r_{rim}^2$, while the thermal resistance of the membrane is proportional to R_{st}.

Resistor Heating. For a closed membrane in which heating with power P occurs by means of a concentric resistor (see Fig. 10b) which is circular around the center and has a radius of r_{in} and where heat loss also occurs by parallel conduction, we obtain the following boundary conditions:

$$\left.\frac{dT}{dr}\right|_{r=r_{in}} = \frac{PR_{st}}{2\pi r_{in}} \qquad T(r_{rim}) = 0. \tag{28}$$

In this equation we ignore any heat loss within the center inside the heating resistor. If necessary, a set of two equations can be defined, one of which takes account of the heat loss in the center inside the resistor. But even without this factor, the solution based on these boundary conditions is very complicated. We will, therefore, give the result for a design in which the heat loss due to parallel conduction (G_p'') is maximized. Extensive calculations show that this is achieved for $(r_{rim}/r_{in}) = 5$, which then yields:

$$T(r_{in}) \approx \frac{PR_{st}}{4}\left(1 - \frac{G_p'' R_{st} r_{rim}^2}{10}\right). \tag{29}$$

If no significant heat transfer is present through ambient gas, Eq. 29 simplifies to Eq. 18 (where $\ln(5)/2\pi = 1/4$).

7.4 THERMAL-SENSING ELEMENTS

7.4.1 Introduction

Temperature-sensing elements constitute an important part of thermal sensors. Here we will discuss the main elements suitable for silicon sensors, and some non-IC elements that are of importance for thermal-sensor work.

For integrated sensors, thermocouples and transistors are the main elements, although diodes and integrated resistors are also sometimes used for convenience. Thermocouples and piles of them (thermopiles) have the advantage of measuring temperature differences without any offset (a self-generating measurement of the temperature difference), but they cannot measure the absolute temperature. For this, transistors are very suitable, within the operating range of transistors (-50 to $+180°C$). For sensors operating outside this range diodes, or resistors, made of polysilicon or thin film if necessary, can be used.

As discrete elements, platinum resistors are well known for their range, accuracy and stability, serving as a temperature standard over a wide range, while for high temperatures, thermocouples in connection with an absolute reference, are also used for this purpose.

7.4.2 Resistors

Integrated and Thin-Film Resistors. As mentioned above, in most cases thermopiles or transistors (or diodes) are the optimum choices for temperature and temperature-difference measurement. Stress dependence, voltage dependence and their wide tolerances make resistors less attractive. Outside the temperature range of these devices, resistors can sometimes be used to advantage. In silicon technology, the absolute resistance values of integrated resistors are usually not very accurate ($\pm 20\%$ typically), but matching of the ratio of two resistances is very good ($\pm 0.1\%$). A bridge-type of measurement is, therefore, appropriate with resistors.

Thin-film sensors have the advantage of a wider operating range than integrated silicon sensors, due to the absence of p–n junctions that give bad electrical isolation at high temperatures. Polysilicon, which is standard in many foundries but stress dependent, and platinum (not standard) are often used for the temperature-sensitive elements in thin-film sensors.

Pt100 and Other Platinum Resistors. The discrete platinum resistor with a resistance value of $100\,\Omega$ at $0°C$, the Pt100, is a widely used reference temperature sensor. It is available in many qualities, the basic quality being the DIN-IEC 751 standard, with a temperature dependence of approximately $0.38\%/K$ and a tolerance of $\pm 0.15\,K$ at $0°C$. Better qualities ($\frac{1}{2}$DIN, etc.) are available at higher prices. Other resistance values are also widely used, for instance Pt1000, for lower measurement currents.

7.4.3 Thermocouples

Thermocouples are two-lead elements that measure the temperature difference between the ends of the wires (see Fig. 11). They are based on the thermoelectric Seebeck effect, that a temperature difference ΔT in a (semi)conductor also creates an electrical voltage ΔV:

$$\Delta V = \alpha_S \Delta T \tag{30}$$

where α_S is the Seebeck coefficient expressed in V/K. The Seebeck coefficient

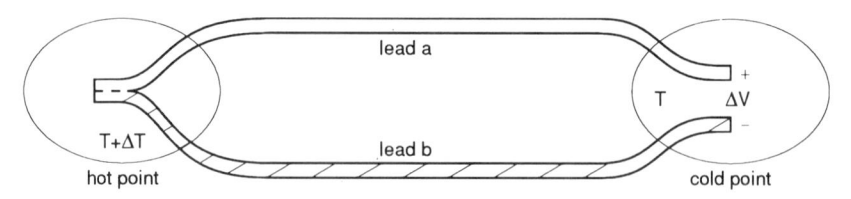

Fig. 11 The Seebeck effect: an electrical voltage ΔV is generated due to a temperature difference ΔT. (After Ref. 1)

α_S is a material constant. By taking two wires of materials with different α_S, there will be different electrical voltages across the wires, even when the wires experience the same temperature gradients. With a junction of the wires at the hot point, the voltages are subtracted and an effective Seebeck coefficient will remain.

The Seebeck effect results from the temperature dependence of the Fermi energy E_F, and the total Seebeck coefficient for nondegenerate n-type silicon is given by:

$$\alpha_S = -\frac{k}{q}\left[\ln\left(\frac{N_c}{n}\right) + 2.5 + s_n + \varphi_n\right] \tag{31}$$

in which N_c is the conduction-band density of states (of the order of $10^{25}/m^3$), n the electron density (determined by the doping concentration), and k the Boltzmann constant. The factor s_n (of the order of -2 to $+2$) is the exponent describing the relation between the relaxation time (mean free time between collisions) and the charge-carrier energy. The factor φ_n denotes the phonon drag effect (phonons dragging charge carriers towards the cold part of the crystal), and is negligible for highly-doped silicon. For low-doped silicon, φ_n is approximately 5 at room temperature and can rise up to about 100 at low temperatures (100 K) (see Fig. 12).[11] The same formula applies to p-type silicon, except that in this case the sign is positive instead of negative.

Seebeck Coefficients. In practice, the Seebeck coefficient for silicon may be approximated as a function of electrical resistivity for the range of interest for use in sensors at room temperature:

$$\alpha_S = \frac{mk}{q}\ln(\rho/\rho_0) \tag{32}$$

with $\rho_0 \approx 5 \times 10^{-6}$ Ω-m and $m \approx 2.5$ as constants.[12] The doping concentrations used in practice lead to Seebeck coefficients on the order of 0.3–0.6 mV/K.

The absolute Seebeck coefficients of a few selected metals and some typical values of silicon are shown in Table 2 at two different temperatures.[12–14] This table shows that the Seebeck coefficients for metals are much smaller than for

7.4 THERMAL-SENSING ELEMENTS

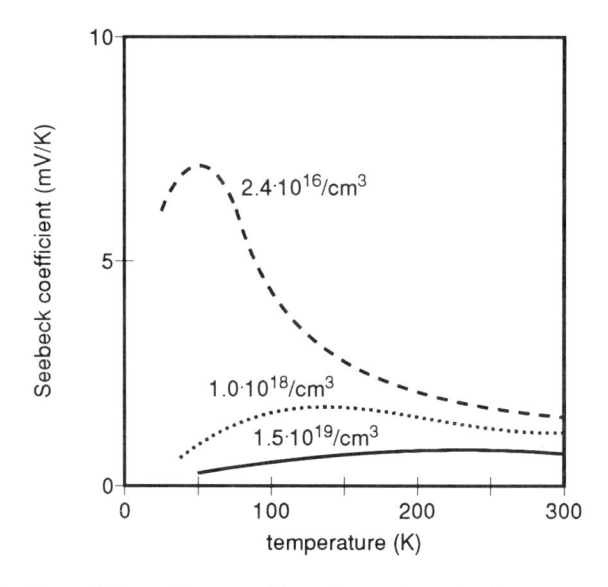

Fig. 12 Seebeck coefficient of *p*-type silicon for various doping concentrations. (After Ref. 1 and Ref. 11)

TABLE 2 Seebeck Coefficient for Some Metals and Mono- and Poly-Silicon (in μV/K)

Material	273 K	300 K
p-type mono silicon (Si)		300 to 1000
Antimony (Sb)	43[a]	
Chrome (Cr)	18.8	17.3
Gold (Au)	1.79	1.94
Copper (Cu)	1.70	1.83
Aluminum (Al)		−1.7
Platinum (Pt)	−4.45	−5.28
Nickel (Ni)	−18.0	
Bismuth (Bi)	−79[a]	
n-type polysilicon (Si)		−200 to −500

[a]Averaged over 0 to 100°C.

silicon, and that the influence of aluminum interconnections on chips is negligible compared to the Seebeck coefficient for silicon.

Thermopiles consisting of series-connected Si and Al couples have been used to measure the Seebeck coefficient of many different types of doped silicon made in planar IC technology. The resulting values are in good agreement with the values measured in bulk silicon samples with the same carrier concentration. Also, IC-compatible thermocouples of, for instance, polysilicon–aluminum, polysilicon–gold and bismuth–antimony have been used to make many thermal sensing devices. Although the bismuth–antimony (and other, more complicated materials of typically bismuth–telluride compositions) yield the more sensitive thermopiles, polysilicon thermopiles are attractive for their compatibility with standard IC-processes. Polysilicon thermostrips have been used to measure the thermo-electric power of polysilicon films. For phosphorous- and boron-doped polysilicon, an α_S on the order of ± 200 to ± 500 μV/K has been obtained for a 0.4 μm thick, 250 to 2500 $\Omega/\square$ film.[14]

Designing the Thermopile. In a thermal sensor, the geometry of the thermopile is usually determined by structural considerations to have a specific length L_x, a width W and a beam thickness D. The only degree of freedom left for the thermopile design is the number of strips, or, in other words, the thermopile electrical resistance R_{tp}. If we neglect the area required for electrical isolation between individual strips, for a rectangular thermopile having N thermostrips we find

$$N \simeq \sqrt{\frac{R_{st}W}{R_{se}L_x}} \qquad (33)$$

where R_{se} is the (technology-dependent) electrical sheet resistance. The thermopile sensitivity is simply $N\alpha_S$, the number of strips times the Seebeck coefficient.

There is an optimum value for the thermopile resistance, with the rule-of-thumb that, for efficiency, the thermopile strips should not be narrower than the required separation between them. In practice thermopile resistances are chosen in the 5 to 200 kΩ range, not too high—to minimize interference, not too low—to minimize the influence of the offset of the electronics.

7.4.4 Transistors

The temperature-sensing capabilities of the transistor depend upon the behavior of a p–n junction in the silicon. In a first-order approach, there is no difference between diodes and transistors in a two-terminal configuration (with short-circuited base–collector terminals). However, the accuracy in transistors is better due to a number of physical nonidealities in the diodes. Therefore, in the following, we restrict ourselves to the transistor behavior. Diodes can, however, be applied in those cases where transistors cannot be fabricated.

When the transistor is operated at a collector–current I_c, which is constant or proportional to the absolute temperature (PTAT), the base–emitter voltage V_{BE} decreases almost linearly with the temperature:

$$V_{BE} = V_{BE0} - \lambda T \qquad (34)$$

where λ is a constant that depends on the bias-current density and the process parameters, and T denotes the absolute temperature. In practical circumstances, the sensitivity of the transistor λ is on the order of 2 mV/K, while the $V_{BE}(T)$ curves intersect the vertical axis at the value $V_{BE0} = 1.27 V$. This value is independent of the process parameters, bias-current level, and transistor geometry. This is an important property in view of the calibration of transistor temperature sensors for the following reasons:

1. With a single measurement of $V_{BE}(T_{ref})$ at an arbitrary reference temperature T_{ref}, the complete $V_{BE}(T)$ curve can be known over a wide temperature range.
2. Any spread that may occur in the value of V_{BE} because of process tolerances is compensated for by adjusting the bias current for the desired value of $V_{BE}(T)$. This adjustment results in identical $V_{BE}(T)$ curves.

We will next discuss the details of the $V_{BE}(T)$ behavior. Let us assume that the collector–base voltage of the sensor transistors is biased at zero volts by short-circuiting these terminals. This is desirable because changes in the collector–base voltage effect the base width (base-width modulation or Early effect) and therefore the base–emitter voltage. In this case, it holds that

$$I_C = A_e J_s \exp \frac{qV_{BE}}{kT} \tag{35}$$

where A_e is the emitter area, J_s is the saturation-current density which depends on the doping profile, T is the absolute temperature, q is the electron charge and k is Boltzmann's constant ($k/q = 86.17 \mu V/K$).

Calibration and Accuracy of Transistor Temperature Sensors. Because of production tolerances, transistor temperature sensors initially deviate from their nominal temperature characteristics. They can be calibrated by adjusting the biasing current or the emitter area. This can be illustrated by rewriting Eq. 35 as:

$$V_{BE} = \frac{kT}{q} \ln \frac{I_C}{A_e J_s}. \tag{36}$$

At 300 K, changing I_C/A_e by a factor of 2 changes V_{BE} by $(kT/q) \ln 2 = 18$ mV. This kind of change does not affect V_{BE0}, which is equal for all types of transistors and current levels. Because of this very important property, it is sufficient to trim the transistors at a single temperature to make the transistor characteristic equal the nominal one.

Some manufacturers supply transistors with a V_{BE} tolerance of approximately 3 mV.[15] The required range of the collector current needed to adjust the base–emitter voltage amounts to approximately $\Delta V_{BE}/(kT/q) \approx \pm 12\%$.

Integrated transistors, such as those used in smart sensors, show a process spread in V_{BE} of about 6 mV (standard deviation). Once they have been calibrated, transistors operate accurately as temperature sensors, with a good long-term stability in the 10 to 40°C range. During thermal cycling over a larger range some drift is to be expected. For instance, for the temperature range at 20 to 110°C, empirical drift values of the order of 20 to 150 mK (depending on the manufacturer) have been reported.[16]

7.4.5 Other Elements

Acoustic-Wave Sensors. Temperature can also be measured by detecting the delay time of acoustic waves (AW) propagating in piezoelectric substrates (see also Chapter 3). The acoustic waves that are the most suitable for this purpose are surface acoustic waves (SAW) and plate waves (PW),[17] which can be generated and detected by means of an interdigital transducer (IDT) (see Fig. 13). The material properties and dimensions of the substrate depend on the temperature, resulting in a dependence of the delay time on temperature. The main part of an AW temperature sensor is composed of an acoustic sensing element and dedicated electronic circuits, forming a feedback loop which oscillates with a temperature-dependent frequency.[17,18]

Several experimental results have been published so far. Hauden et al.[19] used special crystal cuts of piezoelectric quartz with linear oscillator-frequency versus temperature characteristics to realize SAW temperature sensors with a high sensitivity and good linearity. The temperature range from −100 to 200°C. The sensitivity was 2.8 kHz/°C at an oscillation frequency of 97.3 MHz (180 ppm/°C).

Neumeister et al.[20] realised a highly linear temperature sensor with a large resolution using a YZ–LiNbO$_3$ SAW delay-line oscillator, with a temperature coefficient of +94 ppm/K. The temperature ranged from −40 to 160°C. The oscillator frequency was 43 MHz with a sensor sensitivity of 4 kHz/°C. The short-term stability (⩽1 second) was approximately 2 Hz, while the long-term stability over 10 days was about 200 Hz at 30°C. An overall accuracy of 0.2°C could be achieved.

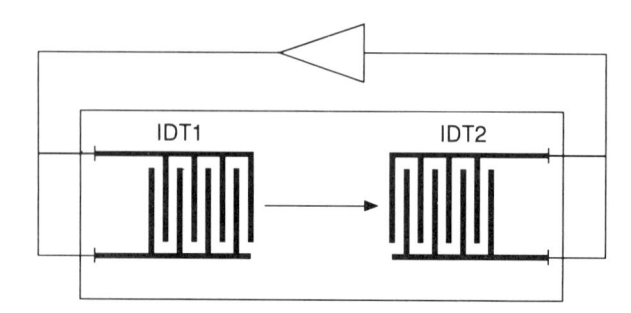

Fig. 13 Physical electronic system for an AW sensor. (After Ref. 1)

Integrated Bimetal Actuator. An interesting effect for implementation in micromachined structures is the bimetal effect. Micro-actuators have been made of silicon beams with gold layers on top of them, making bimetal layers.[21] By using a heating sensor at the tip of the beams, the beams could be bent by controling the heating power. Thus, a temperature-dependent bending of the beam is achieved. This is, of course, the same effect that is undesiredly obtained when stress in multilayer beams (silicon–silicon dioxide, for instance) causes bending. For the Si–Au cantilever beams of 500 μm in length, a bending of 0.1 μm/K has been observed.[21]

7.5 THERMAL AND TEMPERATURE SENSORS

7.5.1 Transduction Mechanisms and Operation Modes

Self-Generating and Modulating Sensors. We will review some thermal sensors that are the result of recent research work. We can distinguish two groups of thermal sensors: self-generating and modulating sensors.

In self-generating sensors the input signal supplies (generates) the power P for the output signal ΔT, and the sensor transfer is determined through its known thermal conductance G. In these sensors, the input signal follows from the product of the measured temperature difference ΔT and the known thermal conductance G. Examples are the infrared radiation sensor, the true rms converter and the micro-calorimeter. If we use thermopiles for the temperature-difference sensing (thermocouples are self-generating elements), we have a tandem transduction (from non-thermal to thermal and then from thermal to electrical), in which both transductions are self-generating. So, we do not need any power to operate these sensors and they operate without any offset. For these sensors, only one (passive) operation mode exists.

In modulating sensors, the input signal modulates the thermal conductance G that determines the sensor transfer. In these sensors, the heating power P that generates the output signal ΔT is supplied by the sensor, and it is, therefore, known. In these sensors, the input signal follows from the ratio of the known supplied heating power P and the measured temperature difference ΔT. Examples are the flow sensor and the vacuum sensor, which are modulating in that the first transduction step is modulating (the non-thermal signal influences a heat flow). Even if the second step is self-generating, we still speak of a modulating sensor.

Constant-Power versus Constant-Temperature Operation. Modulating sensors can be operated in two different ways: maintaining a constant power dissipation or a constant temperature. In both modes, the ratio of P and ΔT gives the input-signal-dependent thermal conductance G.

With a constant power dissipation the power P is kept constant and ΔT varies with the input signal (e.g., flow velocity), see Fig. 14a. This operation is

358 THERMAL SENSORS

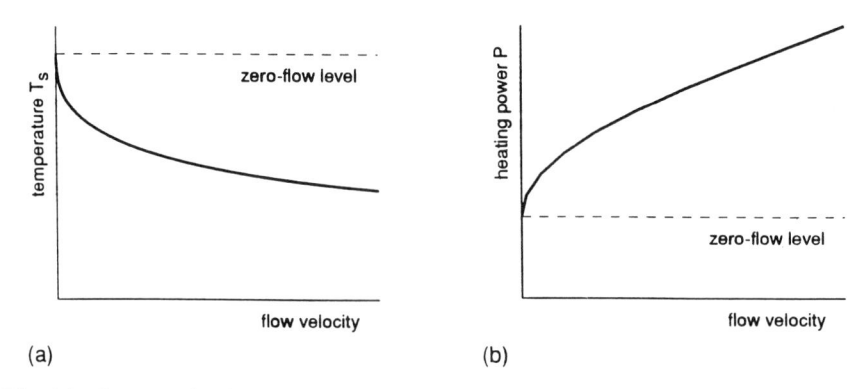

Fig. 14 Output of a flow sensor when used in: (a) constant power mode; (b) constant temperature mode. (After Ref. 1)

very simple, it can be simplified even further when the supply voltage or the current is kept constant. In some cases this constant voltage or current operation even offers an inherent temperature compensation due to the temperature dependence of the heater. The response time τ is determined by the thermal time constant of the sensor, $\tau = C/G$.

With a constant temperature the output signal ΔT is kept constant by applying feedback in the sensor's thermal control, so that now the heating power P varies with the input signal, see Fig. 14b. Because of the feedback, a considerably faster response is obtained, with an effective time constant:

$$\tau_{fb} = \frac{C}{G + G_{fb}} = \frac{G}{G + G_{fb}} \tau \qquad (37)$$

where G_{fb} is the (thermal) feedback transconductance of the circuit (expressed in W/K), and with $\tau = C/G$ being the "structural time constant" of the sensor encountered in the constant-power operation. In practice, the maximum allowable amount of feedback must be limited to avoid instability in the feedback loop, caused by higher-order and parasitic effects, for example the thermal delay in the sensor due to the spatial separation of the heater and the temperature-sensing element. Consequently, the maximum increase in sensor speed is limited.

7.5.2 Flow Sensors

In many domestic, industrial, and scientific situations we require information about flow of gases and liquids. There exists a variety of flow measurement instruments, one group being the thermal flow sensors. We can distinguish four types of thermal flow measurement:

1. The cooling of a hot object by the flow (boundary-layer-flow measurement, anemometers).

2. Determining the heat capacity of the flow (calorimetric method, mass-flow measurement).
3. Thermal tracing (time-of-flight measurement), in which a heat pulse is injected in the flow, and is traced for its time-of-flight from the heater to a detector downstream.
4. Thermal detection of flow oscillation caused by fluidic oscillators or vortex shedding.

The first three sensors lend themselves to integration in silicon. We will discuss some examples of the first type of sensor, because research on this type has progressed the most.

Boundary-Layer-Flow Sensors Cooled by the Flow. Microelectronic thermal anemometers were first introduced by simply replacing discrete self-heating components by their integrated counterparts. The early microelectronic thermal anemometers consisted of devices that were integrally heated. The thermal isolation of such bulk devices depends on the way they are mounted to their substrate, which is often difficult to reproduce accurately in a production environment. Moreover, their large thermal mass results in a slow time response, typically on the order of several seconds. Micromachining has successfully been used to create devices in which a small, thermally-isolated area acts as the actual sensing part.[22,23] This improves the time response considerably, and a time constant of only a few milliseconds can be obtained. Furthermore, the thermal-isolation structure allows the heated sensor and the temperature reference to be combined on the same device.

A number of silicon sensors have been reported in which all the silicon was etched to increase the thermal isolation between heater and ambient even further, using a dielectric material (porous silicon, silicon dioxide or silicon nitride) as the mechancal carrier of the anemometer components. In this case, thin-film or polysilicon resistors were used.

An example is the double-bridge flow sensor manufactured by Honeywell,[24] which is illustrated in Fig. 15. Thermal anemometers, typically, respond to the absolute value of the flow velocity or velocity component. Additional measurements are needed for the temperature reference and to separate the flow-dependent output component. To obtain direction sensitivity, the Honeywell sensor has two 500 μm long silicon nitride bridges suspended over an etched pit. Each bridge has a separate heater and temperature sensor, in the form of thin-film metal resistors. When the heaters are subject to equal heating, the temperature difference between the two bridges is a measure of the flow rate, and also reveals the sense of direction of the flow. The time response of the sensor is less than 5 ms, and flow rates down to 1 cm^3/min can be measured. The sensor is commercially available in a package with a flow channel, including a dust filter to protect the structure against damage. The sensor can also be used to measure pressure differences down to 0.2 Pa.

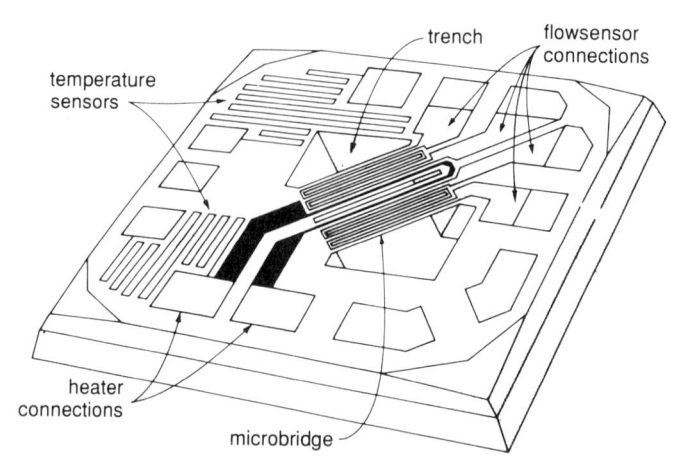

Fig. 15 Double-bridge thermal mass-flow sensor made by Honeywell (After Ref. 24)

Boundary-Layer Wind-Flow Sensors. The double-bridge sensor is an example of how directional flow measurements can be carried out by employing the difference in heat transfer between the upstream and downstream part of a sensor. As is indicated by the local Nusselt number (see Section 7.2.3), the heat transfer is higher upstream, since downstream the flow is already heated and will not cool the sensor chip as vigorously.

This effect can be implemented straightforwardly in a single silicon chip by heating the sensor symmetrically and measuring the resulting flow-induced temperature difference in the sensor.[25] By combining measurements in two directions perpendicular to each other a two-dimensional sensitivity can be obtained, allowing both the velocity and the direction of wind over the surface to be determined.

Two silicon-integrated two-dimensional thermal flow sensors containing thermopiles to measure the flow-induced temperature differences in two directions are depicted in Fig. 16. The first sensor (Fig. 16a) is a wafer-thick device, and four thermopiles are integrated. The two at the top (indicated in Fig. 16a) and bottom measure the flow in the x direction, the thermopiles on the left and the right measure the flow in the y direction. The heater consists of four parts (one of which is indicated in Fig. 16a). In the second (Fig. 16b) the sensivitity of the structure was greatly increased by using micromachining techniques and placing the sensor components on a thermal isolation area in the form of a floating membrane (see Fig. 7).[26] Figure 17 shows the typical output of the wafer-thick sensor as a function of the flow velocity and angle. In Fig. 17 the combined output of the thermopile at the top and bottom is V_{12}, that of the thermopiles at the left and right is V_{34}. For laminar flow this sensor measures with an accuracy of about one degree in the flow direction and about two percent in the flow velocity.

7.5 THERMAL AND TEMPERATURE SENSORS **361**

(a)

(b)

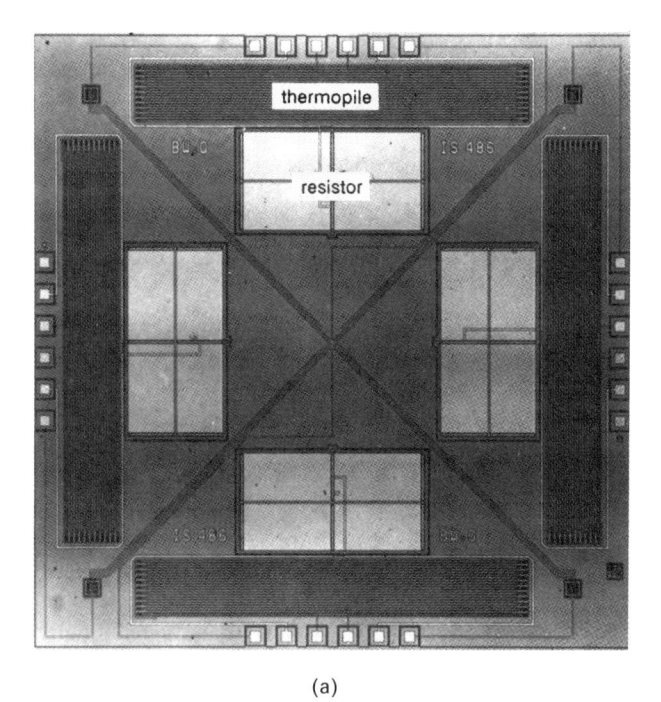

Fig. 16 Integrated flow sensors with thermopiles measuring the flow-induced temperature differences in two directions; total dimensions are 6 mm by 6 mm: (a) wafer-thick sensor; (b) floating-membrane sensor. (After Ref. 1)

362   THERMAL SENSORS

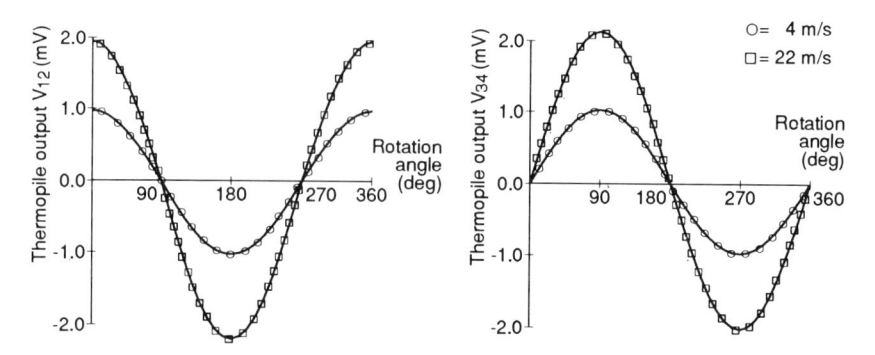

Fig. 17 Measured thermopile output voltages of the flow sensor in Fig. 16a as a function of the flow angle, for two different flow velocities. (After Ref. 1)

7.5.3 Infrared Radiation Sensors

There are two main types of sensors to detect infrared radiation. In phonon detectors the radiation induces electron–hole pairs in the sensor material. These sensors are very fast and have a high sensitivity. They suffer from so-called dark currents (often necessitating cooling), and have a wavelength-dependent response, which is cut off above a certain wavelength. In thermal sensors, the (thermal) energy of the radiation is transduced into heat. These sensors are orders of magntiude slower and less sensitive. Nevertheless, their easy operation and flat response over the entire spectrum makes them very popular for certain applications, such as noncontact temperature measurement and burglar alarms.

Of the various types of thermal infrared-radiation sensors, the most interesting, when looking at silicon devices, are the thermopile sensor, which is self-generating and therefore without offset and without biasing, and the bolometer, which has a resistive temperature-sensing element that needs biasing and has offset, but is potentially more sensitive than the thermopile. A rising star is the pyroelectric sensor, using PVDF (polyvinylidene fluoride) polymers applied to fully processed wafers, allowing sensor matrices.[27] It is not discussed here any further.

The performance of infrared sensors is often described with a few figures of merit. One of them is the time constant τ, another the responsivity R, which is the output voltage of the sensor relative to the power of the radiation incident on the sensor (in V/W). A third figure of merit is the relative detectivity D^*, incorporating the sensitive area A_D, the bandwidth B, the noise $(4kTBR_{tp})^{1/2}$ and the responsivity R:

$$D^* = \left(\frac{A_D B}{4kTR_{tp}}\right)^{1/2} R. \tag{38}$$

For thermal detectors, D^* can have values up to 10^9 cm(Hz)$^{1/2}$W.

7.5 THERMAL AND TEMPERATURE SENSORS

Thermopile IR Sensors. Thermopile sensors are the oldest radiation sensors. Currently, these sensors are made in thin-film technology, using a closed-membrane structure with silicon-nitride and/or silicon-dioxide layers as the membrane and silicon as the rim. A thermopile measures the temperature increase of the middle of the membrane with respect to the rim, and various black coatings are applied to the membrane to absorb the infrared radiation.

It is the coating that performs the first transduction step, from the radiative signal domain to the thermal signal domain. A black coating can have a nearly flat absorptivity of 99–100% over the 0 to 14 μm wavelength spectrum. Often, a compromise has to be made between blackness, thickness (which influences the thermal capacitance and time constant), and ease-of-handling. For instance, a thin layer of gold-black, gold deposited in a porous layer, is extremely black and very thin, but is also extremely vulnerable, and cannot be touched. Thin, tough, gray polymer layers are sometimes preferable.

Various materials are used for thermopiles. Some sensors are based upon monocrystalline silicon/aluminum thermopiles embedded in a silicon membrane. The most sensitive structures are the floating membrane structures. The performance of a commercially available device has a responsivity of 7.5 V/W, and a D^* value of 4×10^7 cm(Hz)$^{1/2}$/W for a 1.8 mm^2 floating membrane.[28] More sensitive are the sensors with closed nitride/oxide membranes with thin-film thermopiles. Basically three groups exist. The first group uses bismuth/antimony thermocouples, made in a relatively simple process giving thermopiles with high sensitivity. One such device that is commercially available has an R of 40 V/W and a D^* of 3×10^8 cm(Hz)$^{1/2}$/W, at 1 mm^2 sensitive area.[29] The second group uses telluride compounds with different doping, leading to a fairly complex processing. These sensors are still under development. The third group is IC-compatible, using polysilicon/aluninum thermocouples. Although these sensors are less sensitive, they are interesting because they use standard technology and have the possibility of making smart sensors with electronics on the sensor chip. The performance of a typical sensor of this group, which is commercially available, shows an R of 22 V/W, and a D^* of 1×10^8 cm(Hz)$^{1/2}$/W, at 1.4 mm^2 sensitive area.[30]

Bolometers. In the bolometer, the heat absorbed by a black coating is measured using a temperature-sensitive resistor. Because of the required biasing to measure the resistance value, often two identical resistors are used in a Wheatstone-bridge configuration, one exposed to the radiation, the other shielded. Because the sensor is not offsetless, chopping of the radiation by means of a butterfly is often applied. Resistor materials such as platinum, bismuth, or telluride are used, or nickel on silicon nitride membrane as fabricated by Lang et al.[31] As can be seen in Fig. 18 the sensor is mounted on a ceramic carrier, which in its turn is glued inside a TO-8 housing. One resistor of the sensor is shielded, the other exposed. A third sensor on the rim of the sensor chip allows monitoring of the sensor base temperature.

Because of the better thermal isolation possible with resistance measurement

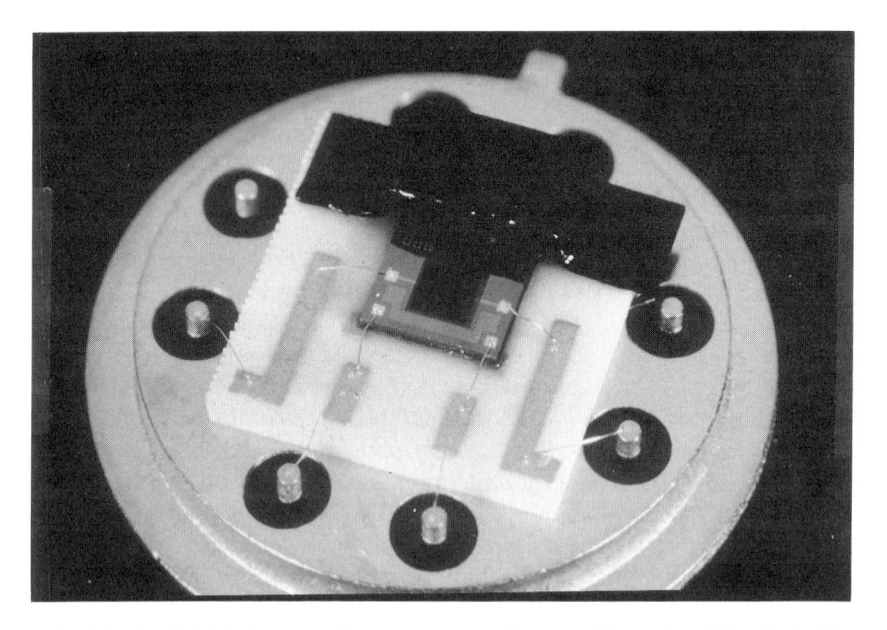

Fig. 18 A nickel bolometer detector mounted on an IC header. (After Ref. 31)

(only two very thin connection leads are present on the dielectric membrane), somewhat higher sensitivities are obtained for the bolometers, with values of 100 V/W or more. The detectivities are comparable to those of the thermopile sensors.

7.5.4 Thermal-Conductivity Sensors

Just like flow, there are several other physical quantities that can be measured by the influence they exert on the thermal characteristics of a thermal structure, and in particular their influence on the thermal resistance between a heated area of a thermal sensor and the ambient.

In the vacuum sensor, the (pressure-dependent) thermal conductance of a low-pressure gas enveloping the sensor structure influences the thermal resistance to the ambient. Vacuum is defined as that range of pressures between zero pressure and atmospheric pressure.

In gas-type sensors and in liquid-type sensors, the thermal conductivity of the surrounding fluid, whether it is a gas or a liquid, influences the overall thermal resistance between the interaction area and the ambient. Thus, the type of gas can be determined from its thermal conductivity, which can differ significantly for different kinds of gas. Because the physics of the vacuum sensor and the gas-type sensor are similar, they are treated together as thermal-conductivity sensors.

Below, we will describe the performance of floating-membrane vacuum

7.5 THERMAL AND TEMPERATURE SENSORS

sensors, which provide the highest thermal isolation between interaction area and ambient.

Thermal Model of the Floating-Membrane Structure. Let us assume that the floating membrane has an area of A_{flm} and that the suspension beam has a length L and a width W. Let us further assume that the suspension beam is made of a thin silicon sheet with thermal sheet resistance R_{st}. The temperature increase in the interaction area of the floating-membrane structure ΔT_{flm} can then be calculated by using the following equation, which is closely related to the model shown in Fig. 8 and Eq. 20:

$$P = \Delta T_{flm}\left(A_{flm}G'' + \frac{W}{LR_{st}}\right) \qquad (39)$$

where A_{flm} is the sensor's interaction area, and the heat transfer coefficient G'' (in W/Km^2) is given by Eq. 5.

The heat-transfer coefficient G'' can then be deduced from the sensor measurement by measuring ΔT_{flm} with a thermopile and measuring the heating power P electrically, where A_{flm} and W/LR_{st} (the offset) are both known. The pressure is deduced from Eq. 5. In practice, the power is measured as a voltage (V_{heat}) across the heating resistor used to heat the interaction area. When the temperature increase in the floating membrane is measured by an integrated thermopile, the pressure measurement reduces to measuring two voltages and calculating the pressure.

Thermal conductivity sensors based on the integrated-thermopile floating-membrane structure have been fabricated in bipolar silicon technology, using etching from the back-side. For a device with a suspension beam curled all around the floating membrane, with $A_{flm} = 1.4$ mm^2 and an offset $W/LR_{st} = 50$ kK/W, we find the sensitivity of the output signal for a pressure of 13%/Pa,[32] assuming $G''/p_0 = 1$ W/K m^2 Pa. The typical output of such a sensor is shown in Fig. 19, which also illustrates the functioning as thermal conductivity sensor, distinguishing helium, nitrogen, and argon.

7.5.5 True-rms Converters

Although true-rms converters are not truly sensors, in their construction and physics they very much resemble thermal sensors, therefore, they fit nicely into the thermal-sensor family.

True-rms converters are used to transfer the effective value or, as it is most commonly called, the root mean-square (rms) value of an AC voltage or current to its equivalent DC value. This is achieved by measuring the temperature increase of a heater resistance caused by the AC voltage or current, and then inducing the identical temperature increase with a DC voltage or current. This DC value represents the true-rms value of the AC signal.

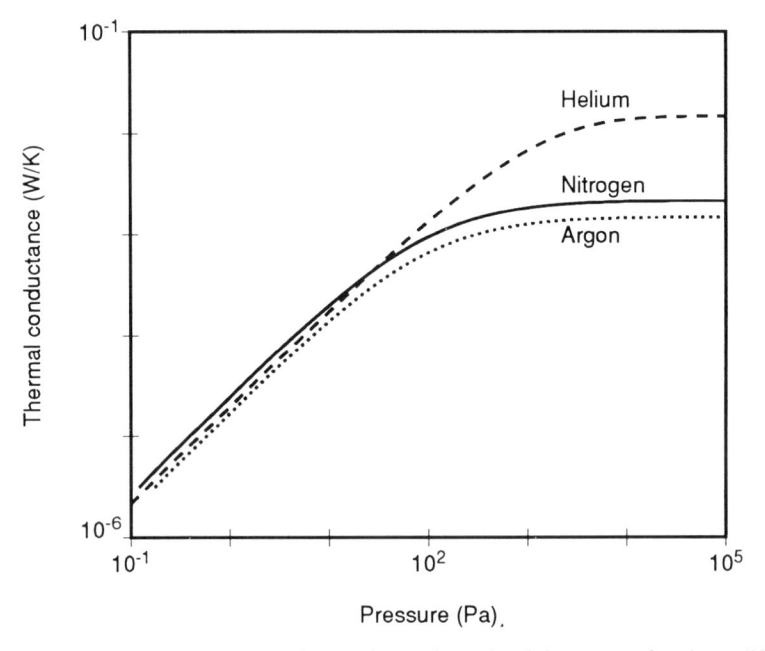

Fig. 19 Output of a floating-membrane thermal conductivity sensor for three different gases. (After Ref. 1)

Solid-state devices can be used very effectively in commercial instruments, whereas standard thermal converters and resistor-mounts are used as the most accurate AC–DC transfer standards for the calibration of commercial devices. At high frequencies in the range of 1 MHz to 1 GHz. the bolometer (Section 7.5.3) is commonly used. At lower frequencies, closed-membrane devices with thermopiles or floating-membrane devices with transistors are being developed for practical use.

Thermal Converters with Transistor Sensors. Thanks to integrated-circuit technology, it has been possible to develop solid-state thermal converters using transistors for temperature sensing. A commercially used version of a transistor-sensed thermal converter is shown in Fig. 20.[33] The design uses thin-film resistors on thermally-isolated floating-membrane structures that are supported only by leads extending across the air gaps between the islands and the surrounding frame. The frame and the islands are formed by anisotropic etching which leaves, underneath the islands, some silicon into which the transistors are diffused.

As the transistor is equally sensitive to changes in the ambient temperature and the heater's temperature, it always uses twin converters in a balanced configuration to reduce the common-mode sensor drift. The DC voltage is approximately 2 V and can easily be measured by a digital voltmeter. By using a sophisticated measurement and error-correction scheme, the sensor, used in

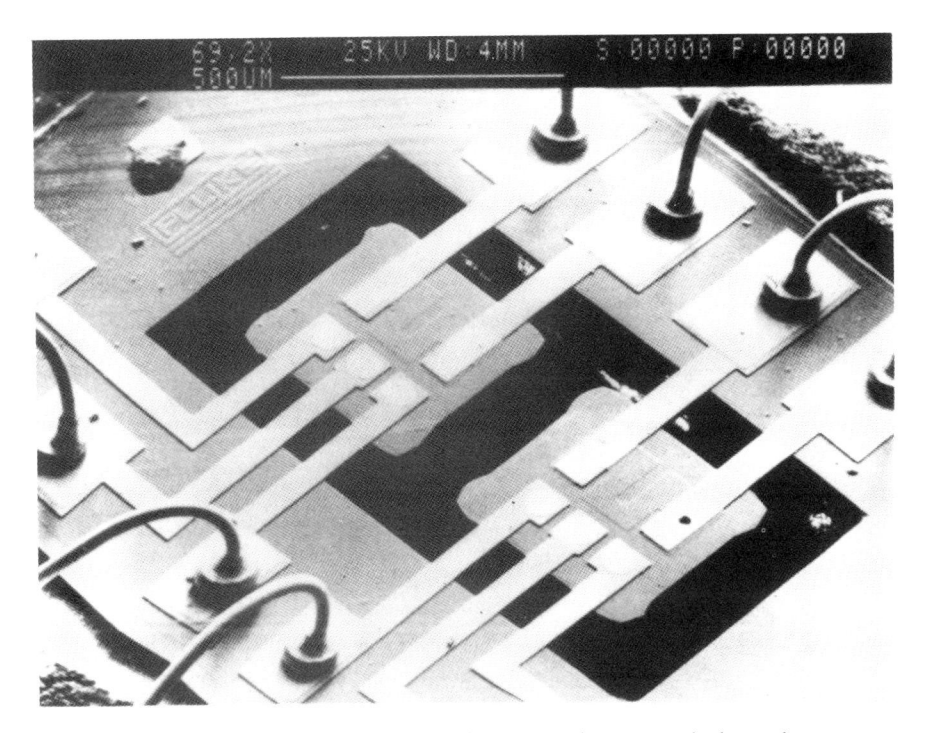

Fig. 20 Scanning electron micrograph of the transistor-sensed thermal converter, showing the two islands. (Courtesy of John Fluke Mfg. Co. Inc., Everett, WA 38206, USA)

a commercially available multimeter, gives an accuracy on the order of 120 ppm over a 40 Hz–20 kHz frequency range.

7.5.6 Humidity Sensors

Humidity is defined by the concentration of water molecules in a space filled by a humid gas. Common measures for the humidity are the partial pressure of water vapor, the absolute humidity (the weight of water vapor per unit of volume), the relative humidity (the ratio between the actual vapor pressure and the saturation pressure at the same temperature) and the dew-point temperature, usually shortened to the dew point.[34]

Because of the particular thermal properties of water, many of the methods of measuring humidity are based on thermal principles. In two of these methods, humidity is even expressed in terms of temperature only: the dew point and the psychrometric temperature difference. The dew point is the temperature to which a humid gas must be lowered isobarically (i.e., at a constant pressure) in order to just saturate the water vapor. The psychrometric temperature difference is measured by a dry and a wet temperature sensor, simultaneously exposed to a stream of humid gas. A remarkable fact is that both principles are the basis for secondary humidity standards.

368 THERMAL SENSORS

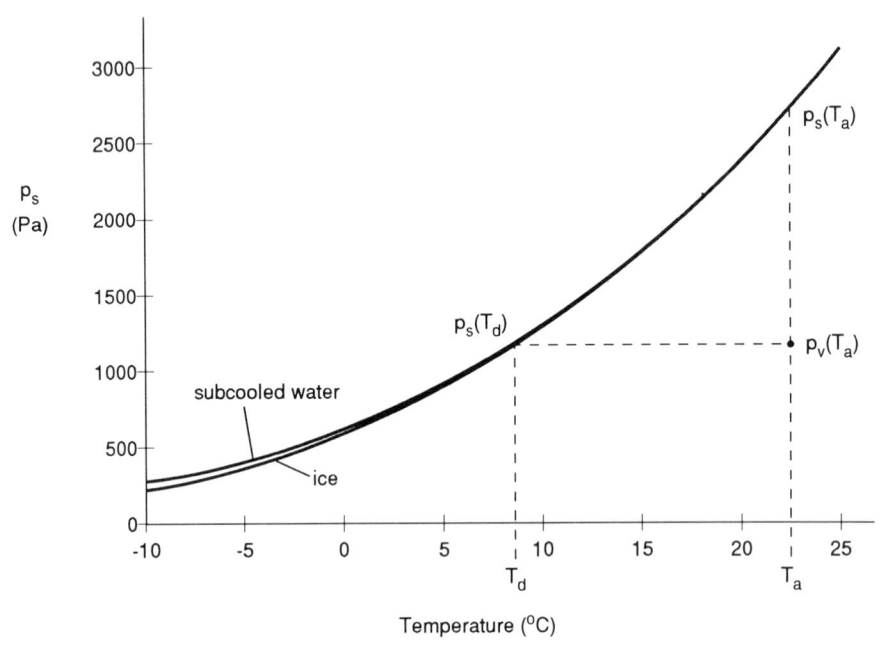

Fig. 21 Saturation vapor pressure of water plotted versus temperature. T_d is the dew-point temperature; the relative humidity is p_v/p_s. (After Ref. 1)

Below we will show an example of an integrated dew-point sensor (no integrated psychrometers have been made so far).

Dew-Point Sensors. The dew-point method employs the accurately known relationship between the saturation vapor pressure p_s of water and the temperature (Fig. 21). Irrespective of the ambient temperature T_a, the vapor pressure p_v is expressed by the dew-point temperature T_d, for which $p_s(T_d) = p_v(T_a)$. The method consists of cooling the gas, detecting the dew and measuring the temperature of the cooled gas.

The cooling of the gas is done with a Peltier element. The detection of dew on the cold side of the Peltier element can be done optically (by observing the dew drops or frost on a mirror mounted on the Peltier element). An interesting alternative is using a capacitive element on the Peltier cooler and detecting the change in capacitance between two electrode pairs (with interdigitated structure). When condensation occurs, the impedance between the electrodes drops sharply, due to an increase in the dielectric constant, which ranges from 1 for humid air to 80 for liquid water. Two advantages of electric dew detection over optical detection are the smaller overall dimensions due to the flat structure and a simpler construction because components do not require alignment. These properties facilitate the construction of probe-shaped transducers that can be inserted into a pipe or vessel containing the humid gas. Probe transducers do

Fig. 22 Experimental integrated dew-point sensor, On the top, the capacitor (with two connection leads), on the bottom, the transistor (with three leads). (After Ref. 1)

not require an air pump, as in most other optical instruments. Moreover, planar capacitive structures can be fabricated in silicon technology, enabling the integration of a temperature sensor on the same chip, giving high accuracy of the dew-point temperature measurement. Figure 22 shows an experimental example of such a sensor in which a large area is reserved for the interdigitated capacitance and a bipolar transistor is used for temperature measurement.[35]

7.5.7 Micro-Calorimeters

Micro-calorimeters are self-generating sensors in which the heat of a chemical reaction is detected. Generally, an output voltage is obtained as a function of the concentration of the substance to be measured. Micro-calorimeters can be realized for use in both gaseous and liquid media. In both cases, the substance to be measured is converted into another substance by means of a catalytic or enzymatic reaction that produces heat. The catalyst or enzyme is deposited on the sensitive interaction area of the thermal sensor. Below, we discuss a typical example of liquid micro-calorimeters based on thermopiles.

An Integrated Thermopile Liquid Micro-Calorimeter. Many types of microcalorimeters have been realized, based on thermistors, on Mylar films with

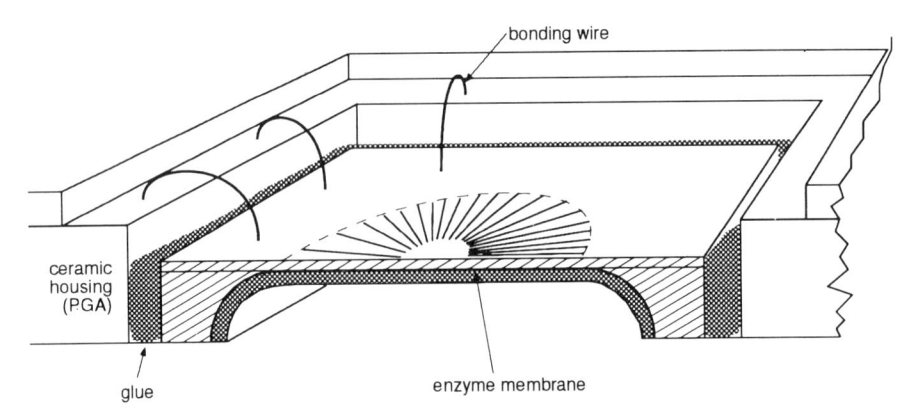

Fig. 23 Cross-section of the liquid micro-calorimeter LCM-2506 with an enzyme membrane deposited on the back of the closed silicon-membrane structure. (After Ref. 1)

bismuth–antimony thermopiles,[36] but there are also integrated silicon thermopile micro-calorimeters and versions with nitride membranes having polysilicon thermopiles. The integrated devices consist of a silicon closed-membrane structure, where the enzyme membrane has been deposited on the back-side of the silicon membrane, thus separating the liquid and the electronics side of the sensor.

Bataillard et al.[37] performed experiments on the integrated silicon closed-membrane sensor shown in Fig. 23. This sensor is based on a silicon-aluminum thermopile with a sensitivity of the order of 70 mV/K and a transfer of 1.5 V/W in flowing water. The silicon membrane is 4 μm thick, 3.5 mm wide by 3.5 mm long. The devices were tested by measuring the concentration of glucose in water with an enzyme membrane containing glucose-oxidase and catalase in a flow-injection analysis system. The sensitivity to glucose appeared to be 5–10 μV/mmol/l, which corresponds very well to the theoretical estimates.

7.5.8 Temperature ICs

PTAT Temperature Sensors. In integrated-circuit (IC) temperature sensors, electronic circuits for amplification, biasing, linearization and analog-to-digital (A/D) conversion can be integrated with the sensor transistors on the same chip. The best-known IC temperature sensor is the so-called PTAT sensor, which generates an output current or a voltage proportional to the absolute temperature (PTAT).

The basic signal in PTAT sensors is the difference ΔV_{BE} between the base–emitter voltages of two transistors operated at a constant ratio of their emitter-current densities. When both transistors are at the same temperature

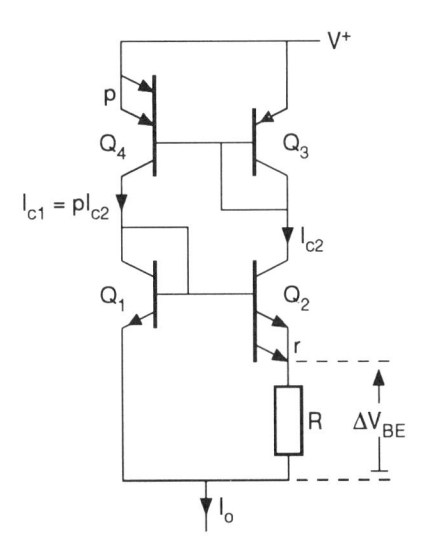

Fig. 24 Principle of an all-*npn* PTAT current source.

T, it is found from Eq. 36 that

$$\Delta V_{BE} = V_{BE1} - V_{BE2} = \left(\frac{kT}{q}\right) \ln\left(\frac{I_{C1} J_{s2} A_{e2}}{I_{C2} J_{s1} A_{e1}}\right). \tag{40}$$

For identical transistors fabricated on the same chip, it holds that $J_{s2} = J_{s1}$. When the emitter area ratio is denoted by $r = A_{e2}/A_{e1}$ and when the collector-current ratio $p = I_{C1}/I_{C2}$ is kept constant it holds that:

$$\Delta V_{BE} = \left(\frac{kT}{q}\right) \ln(pr). \tag{41}$$

This voltage is proportional to the absolute temperature (PTAT). Figure 24 shows the implementation of the basic PTAT current source. Transistors Q_1 and Q_2 implement the emitter ratio r, and transistors Q_3 and Q_4 form a current mirror with current ratio p.

The PTAT current sources[38] provide a calibrated output current of, for instance, 1 μA/K which is proportional to the absolute temperature (PTAT) and stabilized against changes in the supply voltage. The basic PTAT voltage is amplified and buffered. The amplifier gain is adjusted at the wafer so that interchangeable, calibrated devices are obtained.

Temperature Sensors with Offset Reduction. A drawback of both PTAT sensors and single-transistor sensors is that at ordinary temperatures there is a large initial "offset" signal. When the temperature range of interest is small, it is

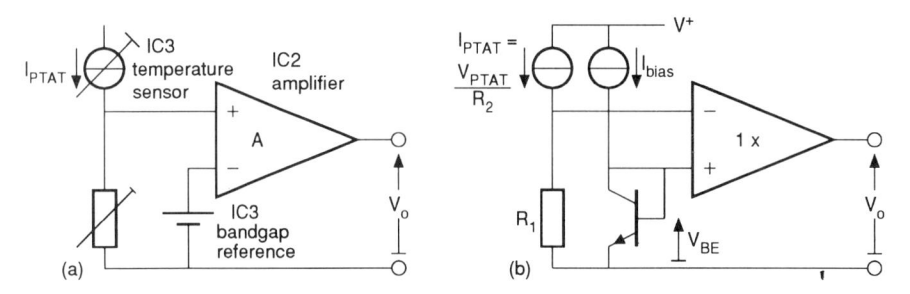

Fig. 25 A temperature-measurement system with an output signal on a °C, °F, or another scale: (a) with bandgap reference; (b) with intrinsic reference.

advantageous to have a temperature transducer with its "zero" at a temperature in or close to the range of interest. A signal on a °C, °F, or another scale can be obtained with the system shown in Fig. 25a, which has been implemented with a PTAT temperature sensor, a voltage reference, and a differential amplifier.

The circuit can also be implemented by using an additional transistor temperature sensor instead of the band-gap reference (Fig. 25b). When the transistor and PTAT current source are at the same temperature we find using Eq. 34 that:

$$V_0 = V_{BE} - I_{PTAT}R = V_{BE0} - \lambda T - I_{PTAT}R. \tag{42}$$

From this equation it can be concluded that the temperature dependence of V_{BE} adds to the overall sensitivity. Moreover, extrapolating Eq. 42 to $T = 0$ K gives $V_0 = V_{BE0} \approx 1.27$ V. This value is determined mainly by the bandgap voltage of silicon, depending little on process or biasing parameters. The linear temperature characteristic of $V_0(T)$ is completely determined by the reference points, and since nature has provided us with a well-known reference point at $T = 0$ K, only a single second reference point has to be fixed in order to fix $V_0(T)$ for all temperatures. This is done, for instance, at a temperature of 300 K, by changing a resistor value R for the desired value of V_0.[39]

A three-terminal integrated temperature sensor with a microcontroller interface has been designed, in which these principles have been applied. Two linear combinations of a PTAT voltage and a base–emitter voltage, both converted into currents, are used to produce a duty-cycle output block wave, which has a duty cycle that depends upon the temperature. The sensor has a 1 mW power consumption, a range of -45 to $130°C$, and a basic absolute inaccuracy of 0.5°C. The output signal frequency is of the order of 2 kHz, and the duty cycle is $0.320 + 0.00470t$, where t denotes the temperature in degrees Celsius. The interested reader is referred to the data sheet and the literature.[40,41]

7.6 SUMMARY AND FUTURE TRENDS

7.6.1 Summary

This chapter has given a description of thermal sensors made with silicon technology (or related technologies like thin-film technology). The operation of thermal sensors generally can be described in three steps, as given in Sections 7.2–7.4. In the first step the non-thermal quantity is transduced into a thermal quantity by either transducing the power of the non-thermal quantity directly into a heat flow (the self-generating sensors), or by exerting influence by the non-thermal signal on a heat-flow generated by the sensor itself (the modulating sensors). In the second step, the heat flow in the sensor is converted into a temperature difference by means of a thermal resistance. In silicon sensors, micromachining has proved to be a powerful tool for obtaining optimized thermal structures. Closed membranes, cantilever beams and bridges, and floating membranes are often encountered structures in which thermal resistances and parallel conductances can be defined very accurately in a simple way. In the third step, the temperature difference is transduced into an electrical signal. The main elements used for this step are transistors or resistors that measure the absolute temperature and are suited for smart sensors, and thermocouples which are interesting for measuring temperature differences, as they can do this without offset and will not spoil the offsetless character of self-generating sensors.

After these introductory sections, Section 7.5 describes the various thermal sensors. Flow sensors can measure gas and liquid flow by the heat loss of a heated interaction area to the flowing fluid. In silicon technology a complete wind meter that measures wind velocity, direction, and air temperature can be made on a single chip. Infrared-radiation sensors are less sensitive than photon detectors, but their ease of operation and flat response over the entire spectrum makes them attractive for many every-day applications. In particular, the thermopile-based sensors, which are completely self-generating and need no chopping of the radiation. Thermal-conductivity sensors can be used to measure the absolute gas pressure (in a range from 10^{-4} to 100 mbar) based on the pressure dependence of the thermal conductivity of gases. The sensors can also distinguish gases by their different thermal conductivities. True rms converter developments are based on special thin-film thermocouples on dielectric membranes using silicon technology. Versions based on transistors and micromachining are already used in commercially available multimeters. Humidity sensors based on absolute temperature measurement (dew-point sensors) can be integrated in silicon to some extent, the Peltier cooling element is not yet suited for integration. Micro-calorimeters are those very rare chemical sensors that are both without offset (they are self-generating sensors) and are very selective due to the application of enzymatic catalysts. Integrated versions show good results for measurement of glucose, which has important medical applications, ureum, and penicillin. Temperature ICs exploit the possibility of making smart sensors, by having signal-processing circuitry on the sensor chip to calibrate, amplify, and buffer the temperature-dependent transistor voltage. Some novel devices incorporate computer-interfacing circuitry.

7.6.2 Future Developments

Future progress will stem from three major sources, developments in silicon-sensor technology, advances in integrated electronics, and demands from the marketplace.

Technology. New technologies such as micromachining form the basis for many of the thermal sensors presented here. Because of new etching and wafer-bonding techniques, complex three-dimensional silicon structures can be fabricated and mass produced. Moreover, wafer-bonding techniques offer new possibilities for the mechanical and chemical protection of the sensitive and vulnerable parts of the thermal sensors. This may solve some of the packaging problems for thermal sensors, which form the bottleneck in the development of commercial products based on flow sensors, micro-calorimeters, and humidity sensors.

The temperature-sensing elements also for an interesting subject for today's research. It seems to be possible to produce transistor temperature sensors in which the performance is comparable to that of platinum resistors. Silicon is also an interesting material for higher temperatures up to 300°C, as is illustrated by the introduction of spreading-resistance temperature sensors.[42,43]

Electronics. A very important development will be integration of the sensors on the same chip as the electronic-processing circuitry, thus forming smart sensors. Compared to conventional sensor systems, smart-sensor systems can be implemented at relatively low costs because they eliminate or reduce the external circuitry, wiring and printed-circuit boards, and because they reduce the calibration and assembling effort. Integrated sensors will be considered in Chapter 10.

Market. Sensor development will inevitably be stimulated by demands from the marketplace. Sensors are urgently needed in environmental control. Many temperature, flow and chemical sensors will be needed to optimize energy-saving processes and limit the waste of materials, and to monitor and safeguard the quality of our environment. Other demands come from the consumer and high-volume market. Smart sensors, thermal sensors, and microcontrollers will enable our smart houses and buildings to be automated. They will be applied, for instance, to control, measure and monitor the temperature, flow and infrared radiation in air-conditioning systems, ovens, refrigerators, heating systems, vacuum cleaners, washing machines, burglar and fire-protection systems, etc. The same types of sensors are needed for the automotive market. However, higher demands will be placed on reliability, resistance against corrosion and mechanical damage, and protection against electrical pulses.

ACKNOWLEDGMENT

The authors would like to thank Rob Janse who made all the drawings, Paul P. L. Regtien for proofreading the manuscript, and Adam Hilger (Publishers, Bristol, UK) for their permission to use the figures and parts of the text of the book *Thermal Sensors*.

PROBLEMS

1. **Convective Heat-Transfer.** Consider a flow sensor with a surface $A = 3$ mm $\times$ 3 mm, which at one side is exposed to laminar flow of 1 m/s. The sensor is uniformly heated at a constant 5 K temperature increase with respect to the flow, see Fig. P1. Find the answer to the following two questions:

   a) What is the heat transfer to the flow in this case?
   b) How is this heat transfer divided over the upstream half and the downstream half of the sensor?

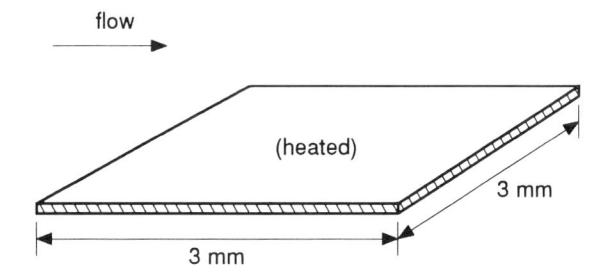

   Fig. P1 Convective heat transfer from a heated surface. (After Ref. 1)

   Note that the viscosity of air $v = 15 \times 10^{-6}$ m^2/s and that the Prandtl number of air is $Pr = 0.7$.

2. **Rectangular Thermopile Design.** A thermopile is 1.5 mm in length and width. It is made of integrated silicon-aluminum thermocouples having a sheet resistance of 100 $\Omega/\square$ and a Seebeck coefficient of 0.6 mV/K, while the thermopile has an overall resistance of 10 kΩ. Neglect the required separation between the silicon strips (the aluminum requires no extra room, gives no extra resistance, and yields no extra output signal). What is the sensitivity of the thermopile?

3. **Direction-Sensitive Flow Measurement.** Consider again the flow sensor of problem 1, and assume that the difference between the heat transferred to the flow upstream and downstream flows through the sensor chip, from halfway along the downstream end to halfway along the upstream end (i.e.,

over a distance of 1.5 mm in total). The chip is 500 μm thick, and the thermal conductivity of silicon is 150 W/K-m. A thermopile with sensitivity $N\alpha_S$ (with N the number of strips and α_S the Seebeck coefficient), for example that of Problem 2, is used to measure the temperature difference that this heat flow causes across the chip.

a) Calculate the output voltage of the thermopile.

b) How is the sign of the output voltage related to the flow?

4. **Micromachining or Not?** Consider again a flow sensor chip as in Problem 1, but now treated with micromachining such that a floating membrane is 3 mm × 3 mm and 4 μm thick, inside a rim of 0.5 mm width on all sides (which may be ignored for the calculations). Again, the thermopile of Problem 2 is present in the floating membrane to measure the temperature difference between upstream and downstream, caused by the difference in the heat transfer coefficient. What is the first estimate for the output voltage of the thermopile now, if the differential heat flow is the same as in Problem 1?

5. **Infrared Radiation Heat Transfer.** Consider an infrared radiation sensor with a floating membrane sensitive area with 100% absorptivity and an area of 1 mm × 2 mm. It views an object at 56 cm distance, with an area of 1 cm × 1 cm, also perfectly black. The sensor is at 300 K and the object is 10 K hotter, see Fig. P2. Answer the following questions:

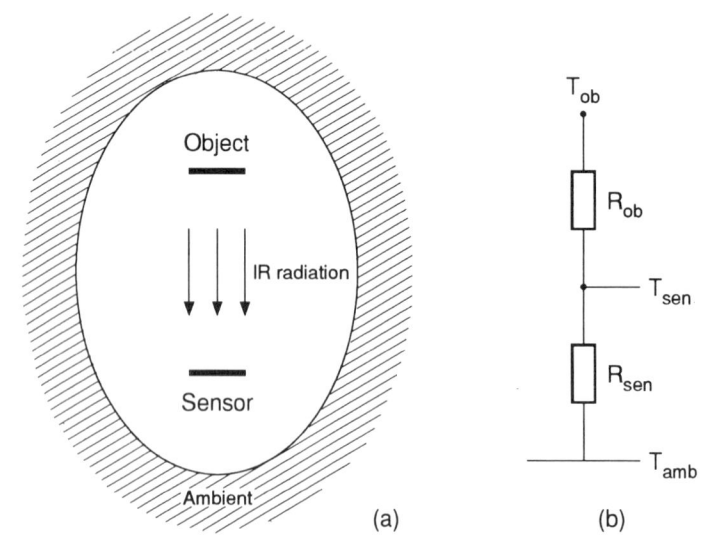

Fig. P2 Infrared sensor viewing an object: (a) physical model; (b) electrical model. (After Ref. 1)

a) What is the net power exchange between sensor and object?
b) What is the temperature increase of the floating membrane of the sensor, if its thermal conductance to the ambient is 330 μW/K?
c) Make an estimate of the three major components making up the thermal conductance from the floating membrane to the ambient if the sensor is encapsulated in nitrogen, $\kappa_N = 26$ mW/K-m.
d) Estimate the improvement if the sensor were encapsulated in xenon ($\kappa_{Xe} = 6$ mW/K-m) instead of in nitrogen?

6. **Thermal Model of Floating-Membrane Sensor.** A floating-membrane sensor has a suspension beam 2 mm in length, 200 μm in width and 5 μm in thickness ($\kappa_{Si} = 150$ W/K-m). At its end a floating membrane is suspended with an area of 2 mm^2, between heat sinks at 0.5 mm distance under and above the membrane. The sensor is encapsulated in argon ($\kappa_{Ar} = 18$ mW/K-m).

a) What is the beam's thermal resistance compared to that of the floating membrane?
b) How do the thermal time constants of the beam and the overall system (using a simple model) compare? Note, that the specific heat of silicon is $c_p = 1.6$ MJ/m^3-K.

7. **Transitions Pressures in Gas Conduction.** A surface micromachined polysilicon resistor is used for thermal pressure measurement. It is suspended above a silicon surface at 1 μm distance. Heat dissipated in the resistor is conducted to the silicon via the gas present between the resistor and the silicon substrate. Calculate the transition pressure, the pressure at which the thermal conductance as given by low-pressure conductance and dense-gas conduction are equal. Do this for:

a) Nitrogen ($\kappa = 26$ mW/K-m, and an experimental value for $G_0''/p_0 = 1.0$ W/K-m^2-Pa).
b) Helium ($\kappa = 151$ mW/K-m, and an experimental value for $G_0''/p_0 = 0.8$ W/K-m^2-Pa).
c) If we used bulk micromachining, leading to a gap of 525 μm, what would these pressures be then?

8. **Circular Thermopile Design.** A thermopile is integrated in a closed silicon membrane. The membrane is monocrystalline silicon with a diameter of 3.6 mm and a thickness of 5 μm ($\kappa_{Si} = 150$ W/K-m). The integrated silicon-aluninum thermopile has the hot junctions at a diameter of 1.8 mm, the cold junctions are at the edge of the rim at 3.6 mm diameter, as shown in Fig. P3. The Seebeck coefficient of the silicon is 0.6 mV/K (that of the aluminum is negligible at approximately -1.7 μV/K). The sheet resistance of the silicon is 50 $\Omega/\square$ (again the aluminum can be ignored). What is the

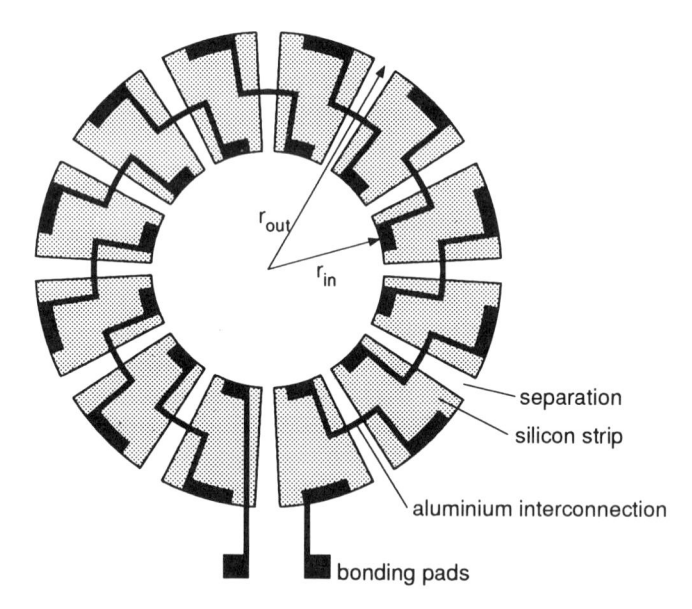

Fig. P3 Circular silicon thermopile integrated in a closed membrane, consisting of 12 Si–Al thermocouples.

sensitivity of the thermopile, if its electrical resistance is 80 kΩ (if no separation between the silicon strips is required, and the aluninum does not require extra space).

9. **Sensivitity of a Micro-Calorimeter.** A silicon closed-membrane micro-calorimeter is used to measure glucose concentrations in water. For this an enzymatic layer (glucose–oxidase) is deposited on the all-silicon membrane, which has a diameter of 3.6 mm and a thickness of 5 μm ($\kappa_{Si} = 150$ W/K-m) (see Fig. 23). The sensor measures the heat generated by the enzymatically promoted conversion of glucose with an integrated silicon–aluminum thermopile, for instance the one of Problem 8. Assume a heat generation by the glucose-oxidase enzyme of 1 W/m² for a 1 mmol/l glucose solution.

   a) What is the sensitivity of this micro-calorimeter for glucose?
   b) What is the resolution in a 1 Hz band?

10. **Self-Heating in a Transistor (or Resistor).** A transistor, used as a temperature-sensing element, is located at the surface of a silicon chip of 0.5 mm thickness (κ_{Si}), as shown in Fig. P4. Its junction has an area of 25×25 μm. The transistor is biased with a current of 10 μA, at a collector–emitter voltage of 0.6 V. What is the error in the temperature measurement caused by the self-heating of the transistor? Simplify the problem by assuming radial heat flow in a half-sphere.

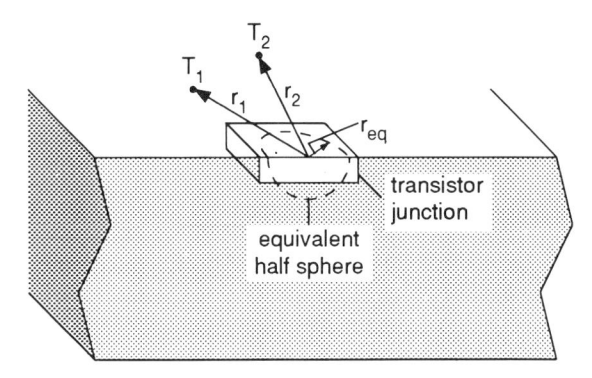

Fig. P4 Self-heating and thermal gradients caused by a flat transistor at the surface of a silicon chip. (After Ref. 1)

11. **Biasing Sensitivities of Temperature Sensors.** Two temperature sensors, a resistive Pt100 element and a transistor (see Fig. P5a and b) are biased with a current source I_{bias}. The output voltages V_0 represent the temperature-dependent output signals. An undesired change in I_{bias} causes an error in V_0, which is equivalent to a temperature error of the sensing elements. Calculate the equivalent temperature errors for a 1% change in I_{bias} for both sensors at 0°C. Note that $V_{BE}(273\text{ K}) = 650$ mV.

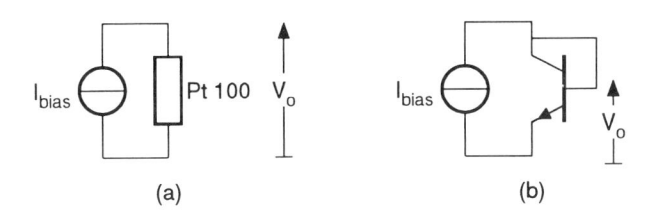

Fig. P5 Biasing of temperature sensors: (a) resistive element; (b) transistor element.

12. **Temperature Sensor with Intrinsic Reference.** In a temperature sensor with an intrinsic reference, such as depicted in Fig. 25b, the resistor R_1 is trimmed to calibrate the sensor to zero output at 0°C: $V_0(273\text{ K}) = 0$ V. The PTAT current amounts to 1 μA/K, while $V_{BE} = 650$ mV at 0°C.

a) Calculate the required value of R_1.
b) What is the extrapolated value of V_0 at 0 K?
c) What is the temperature sensitivity of the output, the derivative dV_0/dT?

REFERENCES

1. G. C. M. Meijer and A. W. van Herwaarden, *Thermal Sensors*, Adam Hilger, Bristol, 1994.
2. A. Roth, *Vacuum Technology*, 2nd ed., North-Holland, Amsterdam, 1982, Ch. 2.
3. A. W. van Herwaarden and P. M. Sarro, "Performance of integrated thermopile vacuum sensors," *J. Phys. E: Sci. Instrum.* **21**, 1162–7 (1988).
4. H. Schlichting, *Boundary Layer Theory*, 7th ed., McGraw–Hill, New York, 1979, pp. 449–544.
5. A. J. Chapman, *Heat Transfer*, 4th ed., MacMillan, New York, 1984, pp. 268–70.
6. E. R. G. Eckert and R. M. Drake, *Analysis of Heat and Mass Transfer*, McGraw–Hill, New York, 1972, pp. 315–19.
7. A. J. Chapman, *Heat Transfer*, 4th ed., MacMillan, New York, 1984, pp. 264–7.
8. D. A. Anderson, J. C. Tannehill, and R. H. Pletcher, *Computational Fluid Mechanics and Heat Transfer*, Hemisphere, New York, 1984; G. Dahlquist and A. Bjork, *Numerical Methods*, Prentice-Hall, Englewood Cliffs, NJ, 1974.
9. G. C. M. Meijer and A. W. van Herwaarden, *Thermal Sensors*, Adam Hilger, Bristol, 1994, Sect. 2.3.4.
10. M. Abramowitz and I. A. Stegun, *Handbook of Mathematical Functions*, Dover, New York, 1965.
11. R. R. Heikes and R. W. Ure, *Thermoelectricity: Science and Engineering*, Interscience, New York, 1961. T. H. Geballe and G. W. Hull, "Seebeck effect in silicon," *Phys. Rev.*, **98**, 940–947 (1955).
12. A. W. van Herwaarden and P. M. Sarro, "Thermal sensors based on the Seebeck effect," *Sens. Actuators* **10**, 321–46 (1986).
13. F. J. Blatt, P. A. Schroeder, C. L. Foiles, and D. Greig, *Thermoelectric Power of Metals*, Plenum, New York, 1976; R. D. Barnard, *Thermoelectricity in Metals and Alloys*, Taylor & Francis, London, 1972; T. J. Quinn, *Temperature*, Academic Press, London, 1983.
14. F. Völklein and H. Baltes, "Thermoelectric properties of polysilicon films doped with phosphorus and boron," *Sens. Mater.* **6**, 325–34 (1992).
15. *Motorola Silicon Temperature Sensors MTS 102 and 105* Databook DS 2536, 1978.
16. G. C. M. Meijer and K. Vingerling, "Measurement of the temperature dependence of the $I_c(V_{be})$ characteristics of integrated bipolar transistors," *IEEE J. Solid-State Circuits* **SC-15**, 237–40 (1980).
17. A. Venema, J. C. Haartsen, M. J. Vellekoop, G. W. Lubking, and A. J. van Rhijn, "Acoustic wave physical-electronic systems for sensors," *Fortschritte der Akustik der 16. Deutsche Arbeitsgemeinschaft für Akustik* 1155–8 (1990).
18. M. J. Vellekoop, A. J. van Rhijn, G. W. Lubking, and A. Venema, "All-silicon plate wave oscillator system for sensor applications," *Proc. IEEE Ultrasonic Symp., New York*, 1990.
19. D. Hauden, G. Jaillet, and R. Coquerel, "Temperature sensor using SAW delay line," *Proc. IEEE Ultrasonic Symp.* 148–51 (1981).
20. J. Neumeister, R. Thum, and E. Lueder, "A SAW delay line oscillator as a high resolution temperature sensor," *The 5th Int. Conf. Solid-State Sensors and Actuators*, 1989, 670–2.

21. W. Riethmüller and W. Benecke, "Thermally excited silicon microactuators," *IEEE Trans. Electron Devices* **ED-35**, 758–63 (1988).

22. O. Tabata, "Fast-response silicon flow sensor with an on-chip fluid temperature sensing element," *IEEE Trans. Electron Devices* **ED-33**, 361–5 (1986).

23. G. N. Stemme, "A monolithic gas flow sensor with polyimide as thermal insulator," *IEEE Trans. Electron Devices* **ED-33**, 1470–4 (1986).

24. R. G. Johnson and R. E. Higashi, "A highly sensitive silicon chip microtransducer for air flow and differential pressure sensing applications," *Sens. Actuators* **11**, 63–72 (1987).

25. B. W. van Oudheusden, "Silicon flow sensors," *IEE Proc. D* **135**, 373–80 (1988).

26. B. W. van Oudheusden and A. W. van Herwaarden, "High-sensitivity 2-D flow sensor with an etched thermal isolation structure," *Sens. Actuators* **A21–A23**, 423–30 (1990).

27. C. Lucas, "Infrared detection: some recent developments and future trends," *Sens. Actuators* **A25–A27**, 167–72 (1991).

28. Xensor Integration, Delft, the Netherlands, *Data sheet XI-IR3704 Infrared Sensor*, 91.053704.

29. Th. Elbel, J. E. Muller, and F. Volklein, "Miniaturisierte thermische Strahlungssensoren: Die neue Thermosaule TS-50.1," *Feingeratetechnik (Berlin)* **34**, 113–15 (1985).

30. Xensor Integration, Delft, the Netherlands, *Data sheet XI-IR3774 Infrared Sensor*, 91.083774.

31. W. Lang, K. Kuhl, and E. Obermeier, "A thin-film bolometer for radiation thermometry at ambient temperature," *Sens. Actuators* **A21–A23**, 473–7 (1990).

32. A. W. van Herwaarden, D. C. van Duyn, and J. Groeneweg, "Small-size vacuum sensors based on silicon thermopiles," *Sens. Actuators* **A25–A27**, 565–69 (1991).

33. R. Goyal and B. T. Brodie, "Recent advances in precision ac measurements," *IEEE Trans. Instrum. Meas.* **IM-33**, 164–7 (1984).

34. Many fundamental articles on hygrometry can be found in the four-volume standard work *Humidity and Moisture*, A. Wexler, Ed. More recent developments are described by several authors in the *Proc. 1985 Int. Symp. on Moisture and Humidity, Washington, DC, 1985*, published by the ISA. A concise overview of current humidity measurements systems is given in *Moisture Sensors in Process Control* by K. Carr-Brion (1986).

35. P. P. L. Regtien, "Silicon dew-point sensor for accurate humidity measurement systems," *Ph.D. Thesis* Delft University Press, Delft, 1981.

36. M. J. Muehlbauer, E. J. Guilbeau, and B. C. Towe, "Applications and stability of a thermoelectric enzyme sensor," *Sens. Actuators* **B2**, 223–32 (1990).

37. P. Bataillard, E. Steffgen, S. Haemmerli, A. Manz, and H. M. Widmer, "An integrated silicon thermopile as biosensor for the thermal monitoring of glucose, urea and penicillin," *Biosens. Bioelectron.* **8**, no. 2, 89–98 (1993).

38. M. P. Timko, "A two-terminal IC temperature transducer," *IEEE J. Solid-State Curciuts* **SC-11**, 784–8 (1976).

39. G. C. M. Meijer, "An IC temperature transducer with an intrinsic reference," *IEEE J. Solid-State Circuits* **15**, 370–3 (1980).

40. G. C. M. Meijer, "A three-terminal integrated temperature transducer with microcomputer interfacing," *Sens. Actuators* **18**, 195–206 (1989).
41. Smartec, *Specification sheet SMT160-30*, Breda, The Netherlands, 1991.
42. G. Raabe, "Silizium Temperatur Sensoren von −50°C zu +350°C," *NTG-Fachberichte* **79**, 248 (1982).
43. Philips, *Silicon Temperature Sensors Databook*, Philips, Eindhoven, 1990.

8 Chemical Sensors

S. R. MORRISON
Simon Fraser University
Burnaby, British Colombia
Canada

8.1 INTRODUCTION

In this chapter we discuss the detection of gaseous species, with the primary emphasis on detecting combustible gases such as CO, H_2, alcohols, propane, and other hydrocarbons. The detection of such gases in air with semiconductor sensors is done primarily with semiconducting metal oxides although studies have been made using FETs (field-effect transistors), primarily for hydrogen sensing. The two approaches to chemical sensing are quite different. In the case of a metal-oxide semiconductor, such as SnO_2, a chemical reaction between oxygen and the combustible gas occurs at the surface of the solid, changing the resistance of the solid. To make the resistance sensitive to such chemical activity, we must select metal oxides with special forms or properties, and with special additives. In the case of the FET, the chemical action of the gas to be detected occurs where the gate would normally be for a MOSFET, and the chemical activity changes the potential at the "gate". Because the sensor is a modified MOSFET, there is a built-in amplification which leads to high sensivitity.

FET-based sensors have also been used for the detection of ions and for biosensors. The detection of ions will be discussed in Section 8.7. Biosensors will be discussed in the next chapter.

Metal-oxide sensors have been commercially available for many years. The dominant manufacturer is Figaro Engineering in Japan. The type of metal-oxide sensor sold by Figaro is termed the "Taguchi sensor", after the developer of the SnO_2 sensor.[1] The sensors have several problems, but for many applications

---

Semiconductor Sensors, Edited by S. M. Sze.
ISBN 0-471-54609-7 © 1994 John Wiley & Sons, Inc.

these problems are well compensated for by the low cost of the sensor and by its sensitivity in the detection of combustible gases (such as hydrogen or hydrocarbons in air. The operation of metal-oxide sensors is based on the decrease of resistance of a layer of powdered SnO_2 if a combustible gas is present in the ambient atmosphere.

The most quoted model to explain the resistance change in a metal-oxide semiconductor sensor is that, in air, oxygen adsorbs on the surface, dissociates to form O^-, where the electron is extracted from the semiconductor. This electron extraction tends to increase the resistance (assuming an n-type semiconductor). In the presence of a combustible gas, say hydrogen, the hydrogen reacts with the adsorbed O^- to form water and the electron is re-injected into the semiconductor, tending to decrease the resistance. A competition results between oxygen removing electrons and the combustible gas restoring these electrons. So, the steady-state value of the resistance depends on the concentration of the combustible gas. To illustrate we could have the competing reactions

$$O_2 + 2e^- \rightarrow 2O^- \qquad (1)$$

$$H_2 + O^- \rightarrow H_2O + e^- \qquad (2)$$

and the more H_2 present the lower the density of O^-, the higher the electron density in the semiconductor, and thus the lower the resistance.

Another model which could exist or coexist is that the combustible gas, if chemically active, extracts a lattice oxygen from the metal oxide, leaving vacancies that acts as donors. The oxygen from the air tends to re-oxidize the metal oxide, removing the donor vacancies. Thus, there is a competition between the oxygen removing donor vacancies and the combustible gas producing donor vacancies. The density of donor vacancies (and hence the resistance) then depends only on the concentration of combustible gas because the oxygen pressure is constant (as when operating in air).

A catalyst is usually provided to accelerate the reaction rate so that the response of the sensor is rapid. Its action is described in Section 8.3.

From this brief introduction, it is clear that to understand the operation of chemical sensors based on semiconductors we must examine in more detail how adsorbed gases affect the semiconductor resistance, how the extraction of lattice oxygen affects the semiconductor resistance, how the combustible gases interact with the adsorbed oxygen or with the lattice, how catalysts improve the combustible gas/oxygen reaction, how adsorbing or absorbing gases affect the gate voltage of a FET, and finally, how all these chemical and physical processes can be controlled to yield a useful gas sensor.

8.2 INTERACTION OF GASEOUS SPECIES AT SEMICONDUCTOR SURFACES

8.2.1 Surface States and Double Layers

Because most metal-oxide sensors are based on n-type material, we will emphasize n-type semiconductors and reactions occurring on their surface. The

p-type semiconductors are usually more unstable than the n-type semiconductors, in the sense that they interact with the ambient gases to give (often irreversible) changes in the bulk properties. Although this does not necessarily remove p-type semiconductors from consideration, it will be clear from symmetry what changes are needed for the models to apply to p-type material.

When a solid is terminated by a surface, the surface atoms are incompletely coordinated; one or two nearest neighbors are missing, and there are "dangling bonds" that are unsatisfied, that is, unshared with neighbors.

The most important case to us is the metal oxide, an ionic crystal, such as SnO_2. Here both the cations and anions have poor coordination. The positively-charged Sn ions at the surface have an incomplete shell of negative oxide ions around them. With too few negative ion neighbors, the positively charged ions are more attractive to electrons. So their conduction-band-like orbitals can be at a lower energy than the conduction-band edge and can capture electrons from the bulk. They also can bond well (sharing two electrons) to a "basic" molecule such as OH^-, which has an electron pair to give to the bond. The surface anions, on the other hand, do not have their quota of positive ions around them so their (occupied) anionic-like orbitals can be at an energy level higher than the valence band edge. They can capture holes or give up electrons to the bulk. They also can bond well (again sharing two electrons) to an acid molecule such as H^+ which has a pair of unoccupied orbitals. The broad features of this behavior are described by the Madelung model.[2] Actually, at low temperature we expect ionic solids to be normally covered by chemisorbed water, the OH^- bonded to the surface at cationic and the H^+ at anionic sites. The influence of such water in gas sensing has not been clarified, although we note that many metal-oxide sensors change their response at high humidity. At high temperature the adsorbed water can be driven off leaving the active sites available for interaction with bases or acid gaseous species or with electron-donor or electron-acceptor gaseous molecules.

The next important case for gas sensing is that of bonding sites on a noble metal catalyst. In this case the orbitals of interest are d orbitals, bonding or antibonding, directed out from the surface, that can share electrons in bonds to adsorbing gases.

For completeness, we mention homopolar materials such as silicon where an electron in an orbital oriented toward the surface is not shared and is, therefore, at a higher level than the corresponding electrons in the valence band. This electron in silicon is localized in an sp^3 orbital at the surface and the orbital can: (a) act as a donor with the dangling electron easily excited into the conduction band in the case of a semiconductor; (b) act as an acceptor, lowering the system energy by capturing (pairing its dangling electron with) a conduction-band electron; (c) act as an adsorption site, bonding to an adsorbing gas. Such materials are of little interest here because, normally, they will have an oxide layer at the surface when exposed to air, and with the oxide layer they belong in the "ionic crystal" description.

In all cases described in the above paragraph the electronic energy levels in

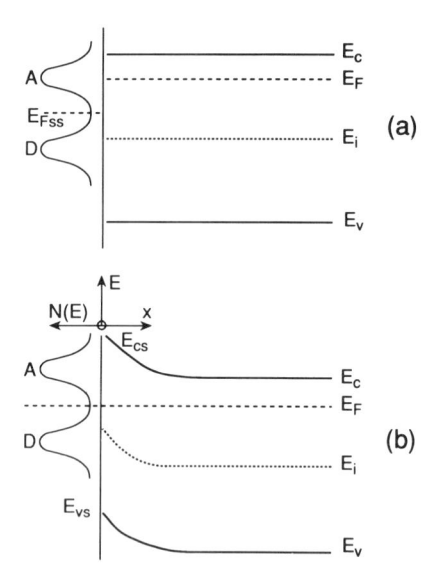

Fig. 1 Band model of a semiconductor surface show charging of surface species. Donor, D and acceptor, A, surface states are indicated as bands of energy levels of dentisy $N(E)$: (a) no charge exchange between the semiconductor and the surface states; (b) the band bending where electrons from the surface region of the semiconductor have moved to the surface states to reach equilibrium.

the bandgap are termed "surface states" and (for semiconductors) there are both donor and acceptor surface states present on the solid. Figure 1 shows the band model for the surface. E_c is the conduction-band edge, E_v the valence-band edge, E_i the mid-gap energy, and E_F the Fermi energy. The surface states are drawn as a band of energy levels of density $N(E)$, a function of E. There are several reasons for band formation. For example, with surface steps and various crystal faces exposed, all the surface states are not at the same energy. We show two bands of energy levels. There are unoccupied acceptor states, here termed A, associated with the anions as described above. Figure 1a is the energy-band model if there is no charge exchange between the surface states and the semiconductor ("flat bands"). We note, however, that in this case the Fermi energy in the semiconductor does not necessarily have any relation to the "Fermi energy" of the surface states. (We use the term $E_{F_{ss}}$ to describe the effective Fermi function.) In Fig. 1b we show the equilibrium case. Electrons have moved from the region of high E_F, the near surface region of the semiconductor, to the region of low E_F, the surface states. This separation of charge leads to a double-layer voltage that, in this example, raises the energy levels at the surface (including E_{cs} and E_{vs}, the band edges at the surface). When the double-layer voltage is sufficient to make E_F constant throughout the system, we have equilibrium. (The Fermi function describes the occupancy of all energy levels.) This movement of the bands near the surface is called "band bending". Note that because of the form of the Fermi distribution function with its exponential dependence of occupancy versus E_F, the Fermi energy will be approximately half way between a set of donors and acceptors as indicated in Fig. 1. If there is only one energy level, by the same argument, the Fermi energy will be close to that energy level for equilibrium. This provides a rule-of-thumb estimate for the amount of "band bending" required to reach equilibrium.

Of greatest interest to us is the analysis of species that can be adsorbed and provide surface states that will inject electrons into the semiconductor (reducing agents) or accept electrons from the semiconductor (oxidizing agents). For example, hydrogen when adsorbed may inject an electron into the semiconductor and adsorb as a proton; oxygen when adsorbed may capture an electron from the semiconductor and adsorb as an O^-. The mathematics describing such cases is similar to the case of surface states and, except where indicated, we will use the term surface state to describe the surface energy levels on adsorbed species. A third type of surface state must be considered, that is the one where a reducing agent such as CO undergoes a surface reaction with a metal oxide such as in Eq. 3, with the lattice oxygen O_L going to the gas phase:

$$CO + M^{2+} + O_L^{2-} \rightarrow CO_2 + M^{2+} + 2e^-. \qquad (3)$$

Here the metal cation M is left behind, and if it is not volatile or does not diffuse into the solid it may donate its excess charge to the semiconductor and again behave as a surface state. This reduction process is discussed further in Section 8.2.5.

The electrical double layer formed as shown in Fig. 1b can be of three types. We will discuss these cases with a simplified-model adsorption-based surface state, assuming only one energy level. The energy level can be a donor that is neutral when unoccupied (such as $H \rightarrow H^+ + e^-$), or an acceptor that is also neutral when unoccupied (such as $O_2 + e^- \rightarrow O_2^-$).

If extra electrons are injected into an n-type semiconductor an "accumulation layer" develops as indicated in Fig. 2a. A double layer develops between the positively charged donor surface states and the donated (injected) electrons. Because the injected electrons are mobile, they can move close to the surface and so the resulting electrical double layer is not deep, extending only a few tens of angstroms into the semiconductor. The Fermi energy at the surface shifts in energy by $d\psi$, where ψ is the double-layer potential in Fig. 2a. The distance between the charged layers, d, can be related to the double-layer capacity (farads per unit area) by

$$C = qdN/d\psi = \varepsilon\varepsilon_0/d \qquad (4)$$

with N the surface-state charge per unit area, ε the dielectric constant, and ε_0 the permittivity of free space. For the accumulation layer, d is relatively small, so to attain equilibrium N can be large. Charge transfer into the semiconductor will continue until equilibrium where the Fermi energy of the semiconductor

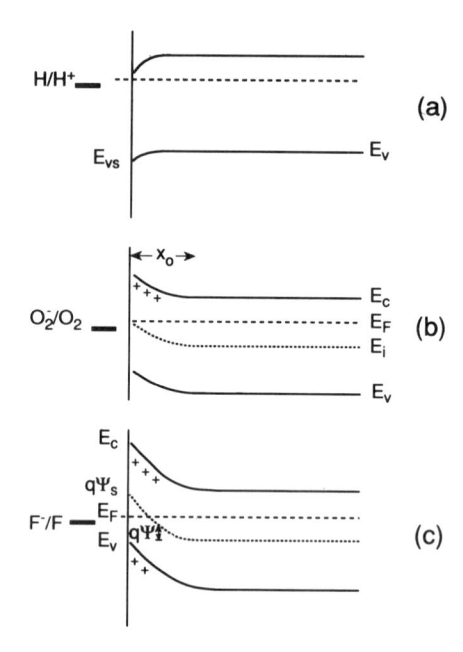

Fig. 2 Types of surface double layers. (a) The accumulation layer where a strongly electropositive surface species injects electrons to the conduction band, leaving a positively charged surface and negatively charged semiconductor. (b) The depletion layer, where conduction-band electrons are trapped at the surface, compensated by the positively charged donors near the semiconductor surface. (c) The inversion layer, where the electron trap is so deep that it extracts electrons from both the conduction band and the valence band. Ψ respresents the potential (as a function of x) relative to the potential deep in the bulk. The surface states are represented, for simplicity, by a single level; if the level is occupied, the species is the reducing agent (e.g. O_2^-), if it is unoccupied, the species is the oxidizing agent (e.g. O_2).

describes the occupancy of the surface states. An accumulation layer is formed on a p-type semiconductor when holes are injected into the valence band (electrons are extracted from the valence band by a strong acceptor). Here an accumulation layer is formed between the negatively charged surface states and the extra holes in the valence band.

If electrons are extracted from the conduction band of the n-type semiconductor by a moderately strong acceptor surface state, a "depletion" or "exhaustion" or "space-charge" layer develops at the surface, as indicated in Fig. 2b. Here the double layer arises between the negatively charged surface states and the positive donor ions (immobile) in the n-type semiconductor. Again, electron extraction will proceed until the Fermi energy in the semiconductor describes the occupancy of the surface states. There is normally a significant charge (providing several tenths of a volt) in the double layer. However, in the semiconductor, the charge must be provided by the donor ions, and these are presented in limited quantity. To provide the needed charge, the double layer extends a significant distance into the semiconductor. In other words d (Eq. 4) will be large. We will show below that, in the depletion layer case, the density of charged surface states is low, of the order of 0.001 monolayers area at most. Again, a depletion region arises for a p-type semiconductor when holes are extracted from the valence band by a donor surface state, leaving a double layer between the negatively charged bulk acceptor ions near the surface and the positively charged surface states.

A third type of surface layer, which is of little interest in gas sensing except in the case of FET-based devides, is the "inversion layer". With an n-type semiconductor, an inversion layer develops when a very strong oxidizing agent (acceptor surface state) adsorbs on the surface. (In Fig. 2c we suggest fluorine as the strong oxidizing agent.) If the surface-state energy level is near the valence-band edge, then to bring the Fermi energy near to the surface-state energy for equilibrium the Fermi energy at the surface must be close to the valence band. If the Fermi energy is closer to the valence band than the conduction band, a substantial hole concentration is present. In other words the acceptor surface state is so low in the band diagram that it extracts valence band electrons. This is shown in Fig. 2c. The surface is "inverted", the n-type material has become p-type at the surface. In the case of FETs the inversion layer is induced by an applied voltage to the gate. A high electric field is applied to induce positive charges at the surface of the n-type material. If the field is strong enough, some of the positive charges will be valence-band holes.

Undoubtedly the most important type of surface layer in gas sensors is the depletion layer. Fortunately, it is also the most amenable to analysis. We will analyze the case using a "simple" surface state, for example that due to the oxygen molecule as an acceptor on an n-type semiconductor. (We note that this choice avoids the complication of dissociation of the oxygen molecule {chemical energy required} that would arise analyzing Eq. 1.) We will assume again only one energy level for the surface state and assume the dopant concentration to be independent of distance. We are particularly interested of course in the effect of such adsorption on the resistance of the semiconductor.

8.2.2 Analysis of Oxygen Adsorption

Figure 2b shows a band model for the surface of an n-type semiconductor with O_2^- adsorbed, showing the case at equilibrium, where a double layer has formed that shifts the Fermi energy at the surface to a position where the O_2^-/O_2 ratio is described by the Fermi energy. Because of the nature of the Fermi distribution function, the Fermi energy at equilibrium will not be far from the energy level of the surface state. As oxygen is absorbed, the electrical double layer is formed in the n-type semiconductor, with the negative layer being the adsorbed oxygen, and the positive charge being the bulk donors that have given up their electrons. The total charge in the space-charge layer must equal the charge on the adsorbed oxygen, so we have

$$N_s = N_D x_0 \tag{5}$$

where N_s is the density of electronic charges on the surface in (in m^{-2}), N_D is the donor density in the sample (we assume here immobile donors), and x_0 is the thickness of the space-charge layer, that is the thickness of the region that is depleted of electrons. We solve Poisson's equation in one dimension:

$$d^2\Psi/dx^2 = -qN_D/\varepsilon\varepsilon_0 \tag{6}$$

where Ψ is the potential relative to the bulk potential. As indicated in Fig. 2, for the case of a depletion layer, we define $\Psi = 0$ at $x = x_0$ (in the bulk), decreasing to Ψ_s at the surface. From integration we find:

$$q\Psi = -(qN_D/2\varepsilon\varepsilon_0)(x_0 - x)^2 \tag{7}$$

using the boundary conditions $\Psi = 0$ and $d\Psi/dx = 0$ at $x = x_0$. Then, at the surface ($x = 0$) we have

$$q\Psi_s = -qN_D x_0^2/2\varepsilon\varepsilon_0 \tag{8}$$

where Ψ_s is the potential at the surface relative to the bulk and $-\Psi_s + \mu$ represents an activation energy (barrier) for the movement of electrons to the surface. Here μ is the energy difference between the conduction-band edge and the Fermi energy. From Eqs. 5 and 8 we have

$$q\Psi_s = -qN_s^2/2\varepsilon\varepsilon_0 N_D \tag{9}$$

giving the barrier as a function of the density of adsorbed oxygen. Equations 8 and 9 are forms of the Schottky equation. The density of carriers at the surface n_s is given by the Boltzmann factor as

$$n_s = N_D \exp(q\Psi_s) \tag{10}$$

As described above, the oxygen will adsorb to an equilibrium value such that the energy of the oxygen level is near the Fermi energy, that is the Fermi energy describes the occupancy of electrons on the O_2/O_2^- level at an energy E_{O_2}. Because the concentration of O_2 is the concentration of physically adsorbed oxygen, it is, in principle, a constant if the temperature and oxygen pressure are constant. From Fermi statistics

$$[O_2^-]/[O_2] = \exp\{-(E_F - E_{O_2})/kT\} \tag{11}$$

where $[O_2^-] = N_s$, the charge concentration at the surface, and E_{O_2} is the energy level associated with the O_2^-/O_2 redox couple (Fig. 2b).

From Eqs. 4 and 11 we have an expression for x_0, the thickness of the depletion layer at the surface, and can evaluate the conductance change occurring because the layer is depleted of current carriers. The conductance of thin-film sample with conductivity σ, and t, L and W the thickness, length and width respectively, is given by

$$G = \sigma(Wt/L)(1 - x_0/t) \quad (12)$$

assuming only one side is exposed to the atmosphere. If x_0 is close to t the conductance will be sensitive to x_0. From Eq. 12 we find

$$dG/G = -dx_0/(t - x_0). \quad (13)$$

With the reducing agent present, the value of N_s varies with the partial pressure of the reducing agent and x_0 varies with N_s.

8.2.3 Adsorption onto Surface States

A final possibility for describing oxygen presence on the semiconductor surface is the case of adsorption onto surface states. Examples would be gallium arsenide (GaAs) and chromium oxide (Cr_2O_3). In the case of GaAs, the adsorbed oxygen interacts with the As on the surface.[3] One can view it as forming a form of arsenate, but the structure and the stoichiometry is wrong. Effectively, the chemisorbed oxygen shares electrons in a bond to the surface arsenic atoms, allowing up to a 1/1 ratio of As/O. This oxygen activation, although of basic interest, is probably not of interest for a sensor, certainly not in air, because in air the GaAs will be oxidized slowly to Ga_2O_3 and arsenic to arsenic oxide.

A more interesting example is chromium oxide. The surface chromium is poorly coordinated, and can form bonds to adsorbing oxygen,[4-6] resulting in $Cr^{4+}-O^-$ or $Cr^{4+}-O_2^-$ groups. Again, one can have a full monolayer of oxygen or CO bound to these active Cr surface states changing their energy level and, thus, the surface barrier. The material is of interest because the indications are that, in this case, lattice oxygen is not involved in the combustion of combustible gases, and, therefore, as a combustible gas sensor it can be expected to be more stable than a sensor that can be reduced.

8.2.4 Interaction of Combustible Gases with Adsorbed Oxygen

The chemical steps involved in the oxidation of combustible gases such as CO or H_2 by adsorbed oxygen can be very complex. In our discussion we will assume the active adsorbed species is O^-, based on many observations. Particularly strong support for this simplification is provided by electron-spin resonance studies, where O_2^- gives a triplet resonance signal, O^- a doublet. It is found that H_2, CH_4, C_2H_6, and CO quench the O^- signal rapidly but not the O_2^-. Lunsford[7] reviews evidence that O^- is much more active than O_2^-.

However, O_2^- is not entirely inactive, for example, O_2^- is the active form in oxidizing ethylene oxide on a silver catalyst.[8-10].

As described in more detail elsewhere[11] the ratio of $[O^-]/[O_2^-]$ expected for an ideal case with no chemical bonding shifts is

$$[O^-]/[O_2^-] = [O_2]^{-1/2} \exp[(-E_{O_2} - E_O - \Delta G/2)/kT] \tag{14}$$

where E_O is the energy required to move an electron from an O atom into the conduction band, and ΔG is the dissociation energy of the oxygen molecule. It is found experimentally[12,13] on SnO_2, ZnO, and TiO_2 that at high temperature the O^- (the catalytically active) form is dominant. For ZnO at moderate oxygen pressure the O^- form becomes dominant above 180°C, for TiO_2 it becomes dominant above 400°C, and for SnO_2 it becomes dominant above about 150°C.[14,15] For these materials, a gas sensor will probably be possible only for temperatures above those listed. For these three oxides, no low-temperature (about room temperature) sensors have been reported.

A simple model for a catalyst-free surface has been described.[2] The assumption is that three reactions occur:

$$e^- + O_2 \underset{k_{-1}}{\overset{k_1}{\rightleftharpoons}} O_2^- \tag{15}$$

$$e^- + O_2^- \overset{k_2}{\rightarrow} 2O^- \tag{16}$$

$$R + O^- \overset{k_3}{\rightarrow} RO + e^- \tag{17}$$

where the k's as rate constants. R is the reducing agent. Letting n_s be the concentration of electrons e^- at the surface, we have at steady state:

$$d[O_2^-]/dt = k_1 n_s [O_2] - k_{-1}[O_2^-] - k_2 n_s [O_2^-] \tag{18}$$

$$d[O^-]/dt = 2k_2 n_s [O_2^-] - k_3[R][O^-] = 0 \tag{19}$$

and from Eqs. 18 and 19 we find the surface charge, N_s a $= [O^-] + [O_2^-])$ is given by

$$N_s = \{k_1 n_s [O_2]/(k_{-1} + k_2 n_s)\}\{1 + 2k_2 n_s/k_3[R]\}. \tag{20}$$

As [R] varies, N_s and n_s vary. Since (from putting Eq. 19 in Eq. 10) n_s varies exponentially with N_s, we can consider that the left side of Eq. 18 is relatively insensitive to N_s compared to the right side with its dominating parameter, n_s. We solve Eq. 20 as a quadratic to determine n_s as a function of [R]. As will be discussed, it was found to be proportional to n_s, and with Eq. 20, we can relate G to [R], the concentration of combustible gas.

8.2.5 Reduction of Metal Oxide

In the introduction we mentioned an alternative way by which the metal oxide could gain or lose oxygen, which is by the reduction of the metal oxide. The combustible gas could extract lattice oxygen from the surface of the solid, leaving oxygen vacancies that are donors. A reaction such as Eqs. 1 or 21

$$R + O_L^{2-} = RO + V_O^+ + e^- \tag{21}$$

may occur. Here V_O is an oxygen ion vacancy.

For stable sensitivity with an oxide used as a gas sensor, normally one prefers an oxide that is not reduced in this way by the combustible gases that are to be detected. The suitability for gas detection using an oxide that depends on such a reaction depends on the diffusion constant (mobility) of the defects created. For example, if we assume that reduction of the oxide leads to oxygen-ion vacancies (as in Eq. 21), two cases lead to a material suitable for sensing, namely a diffusion constant approaching zero and a very high diffusion constant. If the diffusion constant is near zero, then the vacancies remain at the surface and the re-oxidation can be rapid so the resistivity can come quickly to steady state. The removal of oxygen can be equated to the appearance of a donor surface state due to the metal atom (or poorly coordinated metal ion) left behind as the O_L^{2-} is extracted, If, on the other hand, the diffusion constant is very high, the vacancies will diffuse into the bulk rapidly and a steady state between reduction by R and re-oxidation by air occurs rapidly. Then, the stoichiometry of the bulk (the density of bulk donors) is single-valued as a function of the combustible gas pressure.

Difficulties arise if the diffusion constant is intermediate. Oxygen vacancies will slowly diffuse into the sample as long as the sample is exposed to the combustible gas. Long transient changes in donor density (and hence conductance) will result. The time required for re-oxidation to a steady-state value then depends on how long the sample was in the combustible gas.

Aso et al.[16] studied the removal of lattice oxygen by propylene for several oxides of interest in combustible-gas sensing. Between 400 and 550°C few bulk vacancies appeared with TiO_2, SnO_2, In_2O_3, and WO_3, although a monolayer or so of surface vacancies could be formed. Oxygen is extracted from ZnO to leave a few monolayers. These, then, are possible materials for gas sensors based on "adsorbed" oxygen, where, if the lattice is reduced, a restoration to a fully oxidized state is straightforward.

8.3 CATALYSIS, THE ACCELERATION OF CHEMICAL REACTIONS

8.3.1 General

Metal-oxide gas sensors need a catalyst deposited on the surface of the semiconductor to accelerate the reaction and increase the sensitivity.

A catalyst is a material that increases the rate of chemical reactions wothout itself changing. It does not change the free energy of the reaction but lowers the activation energy. For a simple example, we can consider the catalyst Pt in the oxidation of hydrogen. Without the catalyst we would have the reactions:

$$H_2 \rightarrow 2H \tag{22}$$

$$O_2 \rightarrow 2O \tag{23}$$

$$2H + O \rightarrow H_2O. \tag{24}$$

Equations 22 and 23 require a huge input of energy and will not happen at moderate temperature. True, the energy is regained in Eq. 24, but the reaction is stopped kinetically by the "activation energy", the energy to induce the reactions in Eqs. 22 and 23. On the other hand with Pt available as a catalyst we have the possibilities

$$H_2 + 2Pt \rightarrow 2Pt-H \tag{25}$$

$$O_2 + 2Pt \rightarrow 2Pt-O \tag{26}$$

$$2Pt-H + Pt-O \rightarrow 3Pt + H_2O. \tag{27}$$

The reactions in Eqs. 25 and 26 require very little energy input. A hydrogen molecule adsorbed on the Pt surface readjusts its bonds easily to form Pt–H groups as in Eq. 25. The Pt effectively dissociates the hydrogen molecule and presents the hydrogen in an active form. Thus, the overall reaction has a very low activation energy and can proceed at relatively low temperature. The overall free energy change with Eqs. 22–24 and with Eqs. 25–27 is the same, but the kinetic-controlling activation energy is much higher for the former. So with the catalyst, the reaction occurs much faster. If the sensor depends on such a reaction, the catalyst will lower the response time to an acceptable number of seconds.

A promoter is a second additive, one that improves the catalyst's performance. The obvious mechanisms for promoting action are: (a) stabilizing the surface so the reaction cannot irreversibly change the surface; (b) inducing phase changes of the catalyst to provide a more active phase; (c) stabilizing the surface area; or (d) stabilizing a favorable valence state (such as Cu^+ rather than $Cu + Cu^{2+}$).

The catalyst/promoter chosen influences the selectivity of the sensor. Ideally, if one wants to detect a particular gas in a mixture of gases, one would like a catalyst/promoter combination that catalyzes the oxidation of the gas of interest and does not catalyze the oxidation of any other combustible gas. Unfortunately, such ideal combinations are not easily found. A catalyst that catalyzes propane oxidation, for example, will catalyze all other combustible gases to some extent. However, there is some selectivity. One catalyst/promoter combination may be more active in propane oxidation, another more active for other gases. We will discuss arrays of sensors that have a spectrum of catalyst/promoter additive that lead to preferential sensing of a spectrum of gases, and the possibility that, by analyzing the response of such an array, it can be determined which gases are present in the air.

8.3.2 The Role of Catalysts in Gas Sensing Semiconductors

In the preceding subsection we have discussed oxidation and its acceleration by catalysts. We have not discussed how a catalyst improves the response time and sensitivity of semiconductor gas sensors. In fact, at first inspection one feels it should either not affect the gas sensing or should lower the sensitivity. If we simply assume that only Eqs. 25 to 27 result from adding the catalyst, we find there is no reference to the metal oxide and the only action of the catalyst is to remove the combustible gas that would otherwise affect the conductivity of the metal oxide.

There are two possible actions of the catalyst that will strongly affect the conductivity of the metal oxide. The first is "spillover" and the second is "Fermi energy control".

It is well known from the catalytic literature[17,18] that species from the catalyst surface "spill over" onto the surface of the support. (The term "support" is used in the catalytic literature to mean an ionic solid such as Al_2O_3 on which the catalyst is deposited. In our case the "support" is the semiconductor sensor.) Thus, if Eq. 25 occurs on the catalyst, the resulting hydrogen can spill over (in its active state) onto the semiconductor. If Eq. 26 occurs on the catalyst, the active oxygen can spill over onto the semiconducting oxide support. Of course, there is no bond on the support for the neutral hydrogen or oxygen such as the Pt–H or Pt–O bonds, so, to form a bond the species must become ionized as they spill over. Thus the catalyst particles dispersed on the semiconductor support provide a source of the O^- ions discussed in Section 8.2.1, and the same catalyst particles can provide a source of active protons to remove these O^- ions. This means that the reaction can go to completion (high sensitivity) and it can go to completion rapidly (fast response time).

The other possible action of the catalyst particles is Fermi energy control.[19] If the particles of catalyst are small (e.g., 5 nm diameter) and well dispersed across the surface of the semiconductor support, then a metal/semiconductor interface is obtained. At equilibrium, the Fermi energy of the semiconductor is isoenergetic with the Fermi energy of the metal. Actually, this case is directly analogous to that of Fig. 1, where we now have the Fermi energy of the metal catalyst to deal with instead of the "Fermi energy" of surface states. An equivalent depletion region to that shown in Fig. 1b results in the semiconductor at the metal/semiconductor contact. It is possible to develop an accumulation or inversion region with highly reactive metals such as Li, but we will consider only a depletion region.

With the small metal particles there is a high surface/volume ratio. Thus, it is possible that sufficient oxygen is adsorbed on the surface of the metal to raise the Fermi energy of the catalyst particles and hence the barrier $q\psi_s$ in the depletion layer. As the barrier changes so does the resistance of the metal-oxide powder. The presence of a reducing agent then affects the density of adsorbed oxygen, and indirectly the conductance of the semiconductor (Eq. 28 below). A metal catalyst was considered as the example in the above discussion, but the same effect would be obtained with a metal-oxide oxidation catalyst—in this case, too, the adsorption of oxygen could affect the "bulk" Fermi energy of a small particle.

8.4 THE ELECTRICAL PROPERTIES OF COMPRESSED POWDERS

The quantitative description of conductance in a compressed powder is difficult, and we must be content with a semi-quantitative one. The description becomes even more complex if we allow the particles to be "sintered", that is fused together at high temperature. Such sintering is done primarily to provide mechanical strength. The best description of the sintered powder might be with grains of powder connected by "necks" where the grains have fused.[20] Something close to Eq. 12 may be appropriate in an analysis but with the various values of t, the thickness of the necks, some sort of average must be used.

A semi-quantitative model for powder conduction with a pressed pellet can be derived based on Fig. 3. For conductance, electrons must move from grain to grain, but each grain has a surface barrier (depletion region) at the surface that must be overcome for electron passage. The activation energy is $-q\psi_s$. We assume that the surface barrier $-q\psi_s$ is the same for all particles, and that variations in the contact area and number of intergranular contacts appear in

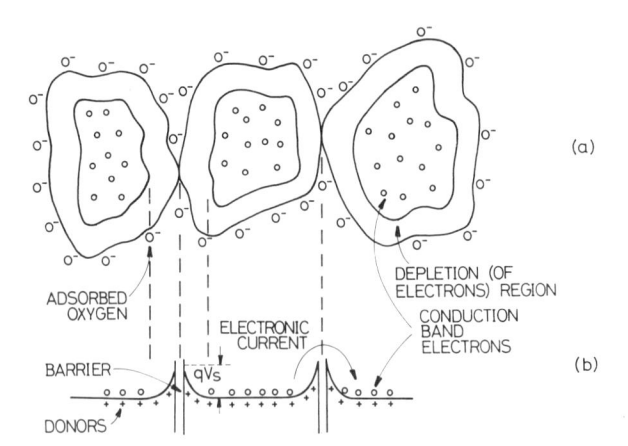

Fig. 3 Barriers at intergranular contacts on a pressed pellet. (a) Three gains with adsorbed oxygen providing surface depletion layers. The depleted layers cause a high contact resistance. (b) The corresponding band model for a more quantitative analysis where, for conductance, electrons must cross over the surface barriers.

the less sensitive linear term G_0. So, with Eq. 10

$$G = 1/R = G_0 \exp(q\psi_s/kT). \tag{28}$$

In this case, the sensitivity arises because electrons must overcome the barrier $q\psi_s$, therefore, the conductance varies exponentially with the barrier height. This can be compared to the "thin film" case of Eq. 12 where the conductance varies linearly with x_0, and, from Eq. 8, x_0 varies as the square root of $-q\psi_s$. In the powder case, the conductance is much more sensitive to ψ_s.

To find a relation between the concentration of combustible gas [R] and the conductance of the pressed pellet, we simply introduce Eqs. 15 to 17. Even with the simple reactions of Eqs. 15 to 17, the resulting expression is cumbersome. A simpler expression arises if we determine (as is typically reported for the commercial Taguchi sensors) the log of the resistance R as a function of the log of the gas pressure where $R = -\alpha/n_s$ with α a constant. Letting s be the slope:

$$s = d(\log R)/d(\log P_R) \tag{29}$$

where P_R, the pressure, is assumed proportional to [R], the concentration of combustible gas. We find, taking the derivative

$$s = -\{1 + (1 - 2aR/b)^{-1}\}/2 \tag{30}$$

where

$$a \equiv k_{-1} N_s \qquad b \equiv \alpha(k_1[O_2] - k_2 N_s) \tag{31}$$

So, if the term in parentheses in Eq. 30 is small the resistance varies inversely with the square root of the partial pressure of combustible gas. Such a behavior is common with Taguchi-type sensors.

8.5 THIN-FILM SENSORS

8.5.1 Oxide with a High Defect Mobility

As was discussed in Section 8.2.5, if the concentration of bulk defects (e.g. oxygen-ion vacancies) is controlled by the ambient atmosphere, and responds rapidly to changes in the ambient atmosphere, the application of the device as a thin, evaporated-film sensor becomes particularly desirable. The only example in the literature of a low temperature (350°C) sensor based on this principle is that of bismuth molybdate.[21] This material has been studied extensively as a partial oxidation catalyst and it has been shown[22,23] that oxygen-ion vacancies move rapidly through the material at temperatures above about 300°C. Thus, the stoichiometry, determined by the surface reaction, penetrates throughout a film. Because the oxygen vacancies are donors, the film resistance responds rapidly and reproducably to the amount of oxide ion extraction.

If the defect mobility is high, it is possible that grain boundaries do not influence the resistance substantially. The effect of grain boundaries is described in Eq. 28 for low mobility materials in terms of electrons being activated over a potential barrier Ψ_s. For high defect-mobility materials, any charges (e.g., O^-) on the grain bounary have as their counter-charge the mobile ionized defects, for example V_O^+, as in Eq. 21. These counter-charges can move very close to the grain boundary, or even be localized on the grain boundary. Then, tunneling through the thin barrier, if such a barrier exists, will occur and will not contribute to the resistance of the film. Thus, an evaporated film is expected to behave as a thin single crystal where the conductance changes arise by changes in the donor concentration. The lack of grain boundary problems results in excellent reproducibility in the characteristics. As long as the level of oxide-ion removal is not so large that the lattice collapses irreversibly to another phase, one expects good stability.

Studies of bismuth molybdate as a sensor show the expected stability.[21] It was found that this sensor is particularly sensitive to alcohols and aldehydes and insensitive to alkenes and alkanes.

Titanium dioxide at high temperature also shows a high lattice-oxygen mobility, and has been used for air/fuel (A/F) ratio measurements of an automobile exhaust.[24] Other cases where mobile defects probably dominate are such high-temperature A/F sensors. There is, of course, substantial interest in using metal-oxide sensors for this application.[25-30].

8.5.2 Thin Films with Low Defect Mobility

Equation 12 represents the expected behavior of ideal (single-crystal) thin films where adsorption/desorption of oxygen is the dominant reaction or where only the surface monolayer of the metal oxide is reduced. This is the case where the resulting defect, say an oxygen vacancy, has a low mobility and cannot diffuse into the crystal. The sensitivity arises by the penetration of the space-charge layer into the sample with more oxygen adsorbed, changing the effective sample thickness. For high sensitivity, the thickness of the depletion region, x_0, must be close to the sample thickness, t (Eq. 13). The depletion region is typically 100 to 400 nm thick (Eq. 8), which means that the single-crystal thickness should approach this value for good sensitivity. Reproducibility in crystal thickness in this range is very difficult. Also, if the thickness t is less than x_0, so that, in principle, all the electrons from the sample are exhausted, the theory becomes very complicated because the adsorption of oxygen becomes limited by the lack of bulk electrons. Thus, in the region of interest the sensitivity depends strongly on the exact value of the thickness t, and reproducible, sensitive, samples are hard to make.

Because of these problems, the use of single crystals in gas sensing is not simple. With evaporated films, where it is easier to reproduce the sample thickness, there is a problem controlling the almost-inevitable grain boundaries. A barrier similar to the surface barrier arises because of majority-carrier trapping at the grain boundaries so most of the resistance of the film will be due to the intergranular contact resistance. In addition, at low temperature, diffusion of defects along grain boundaries is much more rapid than defect diffusion in the bulk. So with a sample temperature selected (a) high enough so that catalysis will be possible and (b) low enough so that bulk diffusion will not occur, grain boundary diffusion still could occur. With a thin enough sample, so that sensitivity according to Eq. 12 will be satisfactory, say 100 nm, such diffusion along the grain boundaries can penetrate into or through the sample and dominate the resistance changes. For example, if oxygen diffuses along the grain boundaries, it will behave as a surface state in a way analogous to Fig. 3 and in n-type material the grain-to-grain contact resistance will dominate the sample resistance. Then, the steady-state resistance depends on oxygen ions moving in and out of the grain boundary and the response time depends on the defect mobility along the grain boundaries. Dibbern et al.[31] and Yamasaki et al.[32] made more direct measurements showing that grain-boundary effects dominated such thin oxide films. Advani et al.[33] showed improvement in the sensing behavior of thin films of TiO_2 by diffusing gold along the grain boundaries, shorting them out.

With these grain boundary problems[34] the thin-film sensor has not been successful except when based on a high bulk mobility of defects.

8.6 THICK-FILM AND PRESSED-PELLET SENSORS

8.6.1 General

Sensors prepared from semiconducting oxide powder follow a conductance/pressure similar to Eq. 30, with the high sensitivity arising because the contact resistance dominates the conductance. Unfortunately, contact resistances are by nature unstable and nonreproducible, leading to the problems to be discussed in this section.

The semiconducting oxide most studied is SnO_2, although good results from Fe_2O_3 have been reported.[35] As discussed in Section 8.5.1, TiO_2 has been used commercially as a "thin-film" sensor, but in a temperature range where defect mobility is high.

The most popular catalyst is Pd, possibly because Pd is somewhat unstable in oxygen at the temperature of interest (350–400°C). Many other additives have been examined for use as co-catalysts or promoters in attempts to improve the selectivity of the sensor. The operating temperature is chosen empirically to provide the highest sensitivity to the combustible gases. Presumably, the sensitivity is low when the temperature is low because the catalytic activity is low. The sensitivity is low at high temperature because the combustible gas is burned at the surface of the thick film or pressed pellet and its effect is not felt

400 CHEMICAL SENSORS

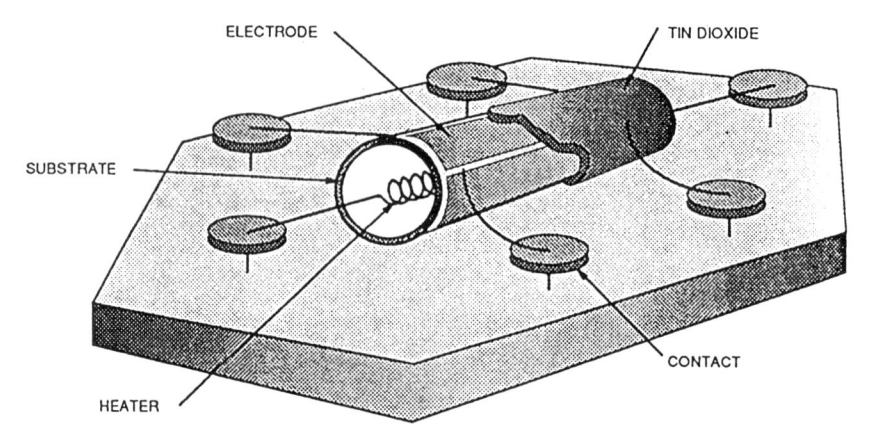

Structure of a Taguchi gas sensor

Fig. 4 The Taguchi sensor, a ceramic tube with a $SnO_2 + Pd$ paste deposited on the outside, a heater element on the inside.

in the main body of the oxide sensor. For each sensor/gas combination, an optimum temperature between these limit is used.

One form of the sensor is that of the commercial "Taguchi" sensor. Figure 4 shows the structure. Gold electrodes are deposited on a small ceramic tube. A paste of semiconducting powder is applied on the outside of the tube. The paste is prepared in a series of steps as illustrated in Fig. 5. It illustrates the complexity of these powder-based sensors. With such a series of preparation steps, exact reproducibility of sensor characteristics cannot be expected. The structure is calcined. A heater element in the center of the tube brings the oxide coating to the desired operating temperature.

Another form, of interest because it promises to provide a simpler and more automated preparation technique, is the "thick film", where the powder is deposited on a flat substrate by a technique such as silk screening.[36] The deposition of the heating element is more difficult with this form. Silk-screened RuO_2 or evaporated Pt are possible heater materials.

The metal-oxide semiconductors that have been tested for use have included WO_3, Fe_2O_3, $LaCrO_3$, CoO, TiO_2, NiO, Nb_2O_5, and In_2O_3. The additives for catalysts/promoters include many oxides, for example transition metal oxides, oxides of Group II, and Li. Madou and Morrison[37] list many of the combinations tested and include references. A few are noted in Table 1 to illustrate the possible combinations. For use in air, semiconductors such as Si, Ge, or sulfides are not expected to be useful because they oxidize at the temperature normally used. Metals such as Pd and Pt can be catalysts because their oxides tend to dissociate at such temperatures. For a contact, gold is preferred because it does not form an oxide.

Obviously the catalyst material must have the same stability when heated in air that the sensor semiconductor has. Thus, we expect catalysts and

8.6 THICK-FILM AND PRESSED-PELLET SENSORS 401

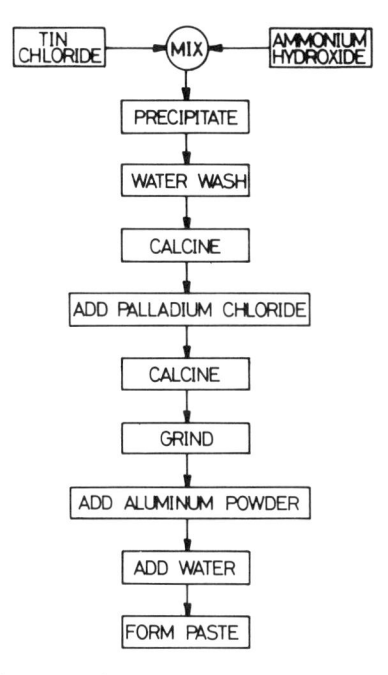

Fig. 5 Typical preparation steps for the SnO_2 paste used as a thick film. The steps include the formation of the hydroxide and calcining to the oxide, then the addition of the catalyst (Pd), followed by a binder to make the coating physically strong.

TABLE 1 Example Combinations of Semiconductor/Additive/Gas Combinations for Metal-Oxide Sensors

Semiconductor	Suggested Additives	Gas to be Detected
SnO_2	Pt + Sb	CO
SnO_2	Pt	alcohols
SnO_2	Sb_2O_3 + Au	H_2, O_2, H_2S
ZnO	V, Mo	halogenated hydrocarbons
WO_3	Pt	NH_3
Fe_2O_3	Ti-doped + Au	CO

It has been suggested in the literature that these catalyst/promoter combinations may provide some degree of selectivity for the gas indicated.

promotors are those from the group of oxides and noble metals, especially those that have been found in the catalytic literature to be catalytically active in oxidation catalysis. (We note, as discussed in Section 8.6.2, that a good catalyst for selectively forming a particular product, as needed for normal catalysis, may not be a good catalyst for sensors.)

Binders, where needed to provide a mechanically strong layer, can be tetraethylsilicate, silica sol, $SnCl_2$, or an organic binder.

8.6.2 Selectivity in Metal-Oxide Sensors

The ultimate objective for metal-oxide sensors would be a series of sensors each of which would respond to only one gas, that is, it would be highly selective. Unfortunately, as is clear from the proceding discussions, the nature of the sensor leads to sensitivity to all combustible gases. A catalyst such as Pd will catalyze the oxidation of CO, H_2, and all organic molecules, and the reaction will be reflected in conductivity changes in the semiconducting support.

To induce some selectivity, we have several parameters available: the catalyst, the promotor, the temperature, and, perhaps, "filters" that restrict the presence of some gases on the sensor. Even the particle size of the metal-oxide sensor can affect the selectivity.

The temperature is important because some gases (e.g., alcohols or CO) are easier to oxidize than others (e.g., CH_4 and alkanes). Thus, a low temperature would induce selectivity towards alcohols or CO, and a high temperature selectivity towards CH_4. In the latter case, as mentioned in Section 8.6.1, at the high temperature alcohols or CO will be oxidized fast at the outermost periphery of the sensor, so only this outermost layer will change conductivity and the overall response will be low. Because of this temperature selectivity, a time-varying temperature is sometimes used, to attempt to show a spectrum of gases. The temperature "spectrum" can be emphasized with a sensor design such that there is slow exchange of ambient gas in the sensor compartment with the external atmosphere. The easily oxidizable gases can be removed first from the compartment at a low temperature then the temperature can be increased to make the measurement of gases such as alkanes more accurate.

Another approach to selectivity is that of filters. Their use is, to a great extent, empirical. For example, Fukui and Komatsu[38] claim H_2 is kept from their sensor by depositing SiO_2 on their SnO_2 sensor. Ogawa et al.[39] claim that ultrafine SnO_2 rejects methanol. Carbon cloth or low porosity pellets have been used to prevent highly reactive or large molecules from reaching the sensor.[40,41]

Different combinations of catalysts and promoters can change the selectivity significantly. The theory behind their use is not entirely clear at present. The extensive literature on catalysis is only of limited help in providing theory or even identifying useful combinations. The reason is that in normal catalysis the objective has always been selectivity in the product gas, while for sensors we want selectivity in the reactant gas. That is, in normal catalysis we have a pure feed stream of reactants flowing to the catalyst and we want the reaction to yield a particular product—the catalyst should show product selectivity. In sensor catalysis, the gas flow into the catalyst is an unknown combination of gases and we want the catalyst to oxidize one selectively. We do not care what the product is, we want "reaction selectivity". Research on catalysts for sensors is needed.

An approach to selective gas sensing is to provide an array of sensors, each one with a significantly different response spectrum for various gases. For example, one sensor could be chosen to change resistance by a factor of 2 in 10 ppm hydrogen, by a factor of 4 in 100 ppm carbon monoxide. Another sensor in this small "array" could be chosen such that it changes resistance by a factor of 4 in 100 ppm hydrogen and 2 in 100 ppm carbon monoxide. Measurement of a H_2/CO mixture by the two sensors would, in principle, provide the complete analysis. Such an approach to selectivity should be highly effective with these metal-oxide sensors, but only if the stability problem described in the next subsection is overcome. Such an array of Taguchi sensors has been used[42] to distinguish types of coffee and to monitor air pollution.[43]

8.6.3 Stability: Drift in Metal-Oxide Sensors

A gas sensor should be stable in a variable ambient atmosphere. It must also be reversibly unstable in the presence of the gases to be detected. These requirements are difficult to meet simultaneously. Slow changes in the properties of the bulk material or slow changes in the properties of the near-surface region are almost inevitable. Such slow changes that follow a change in the ambient atmosphere or changes in the temperature are called "drift". SnO_2 has been the material of choice because of its stability and its convenient resistivity. Reduction by propylene does not affect[44] the bulk characteristics of SnO_2 (among other oxides, namely TiO_2, In_2O_3, and WO_3). SnO_2 is remarkably resistant to acids. So, SnO_2 interacts in general only at the surface. However, even with tin oxide, the high temperature of an initial sintering step may produce bulk vacancies and hence bulk conductivity in excess of equilibrium at the operating temperature (about 350°C). These may require years to diffuse out at 350°C. The surface of tin oxide or other oxides can be unstable due to "poisons". Sulfur (as H_2S for example) is a potential poison that can block the catalytic activity of Pd at the surface. Wagner et al.[45] found instability due to the presence of H_2S in a commercial SnO_2-based catalyst. Other poisons include many reactive species, for example chlorine gas.

A long-term drift problem present with SnO_2-based sensors, and probably present with other types, provides a good example of drift. After a SnO_2/Pd-based sensor has been held at room temperature for a time and then heated to operating temperature, it indicates the presence of gases. A high conductivity spike is observed. This temporary high conductivity occurs whether or not a combustible gas is in the ambient atmosphere. The conductivity spike, lasting about a minute or two, can be useful, because when it disappears (a) one can be certain the sensor is operating, and (b) the disappearance indicates a low level of combustible gas is present. However, although this high conductivity spike *almost* vanishes after a few minutes, a residual excess conductivity is retained for days or weeks after storage at room temperature. If one wants a reasonably accurate measure of a low combustible-gas partial-pressure, one must "burn-in" the sensor for several days until the initial spike has decayed to a value low enough to be considered a negligible background. Manufacturers burn-in the sensors for 3 or 4 days before they calibrate it and recommend a similar time for the user before they use the resulting calibration. Even then, there is significant drift of SnO_2 sensors.[46]

The probable reason for this spike is the adsorption of combustible gases from the atmosphere while the sensor is stored at room temperature. Room temperature is too low to support catalytic oxidation of the contaminating gases. When the sensor is heated to operating temperature the pre-adsorbed gases react with the surface oxygen ions, injecting electrons, and giving a false signal. The bismuth molybdate sensor, not depending on a noble metal, does not show a significant spike, suggesting most of the "room-temperature adsorption" is adsorption on the Pd.

8.6.4 Reproducibility in Preparation of Metal-Oxide Sensors

It is clear that the electrical properties of a pressed (and possibly sintered) powder cannot be expected to be closely reproducible. Intergranular contacts will vary in area; favored high-conductance paths for the current will be present and vary from sensor to sensor. Such variations lead to the need to calibrate each sensor individually if one requires accuracy.

Another nonreproducibility, which is difficult to overcome and difficult to understand completely, is the relative sensitivity to a spectrum of gases. For example, if one calibrates four sensors using 100 ppm propane, and finds sensor A is most sensitive, B next, and so on, each can now be used as a propane sensor, but, if these sensors are used to measure ethyl alcohol, one may find C the most sensitive of the group, followed by B, then A. Thus, it is desirable that the sensor be calibrated for the gas that is to be detected. The manufacturer may not have the facilities to calibrate for highly noxious gases, so the calibration quoted for a given sensor may be inaccurate if used for noxious gases. If the sensor is to be used in gas mixture, satisfactory calibration is very difficult.

8.7 FET DEVICES FOR GAS OR ION SENSING

8.7.1 The ChemFET

Field-effect transistors (FETs) can be sensitive to some gases or ions if the gate is exposed. The most studied case for gases is a FET with a Pd gate used for H_2 detection. The H_2 dissolves in the Pd and affects the gate voltage. Such a FET, together with other chemical sensing FETs, can be termed a ChemFET.

A FET is a silicon-based structure as in Fig. 6. In the example of Fig. 6, an inversion layer (Fig. 2c) is created at the surface of the *p*-type silicon by the application of a positive voltage to the gate contact. The application of the voltage to the gate induces a charge at the silicon surface in accordance with a capacitance law: $Q = CV$. If the voltage on the gate is high and positive, some of the charge appears as electrons in the inversion layer. The current from the drain to the source through this inverted region is measured.

8.7 FET DEVICES FOR GAS OR ION SENSING

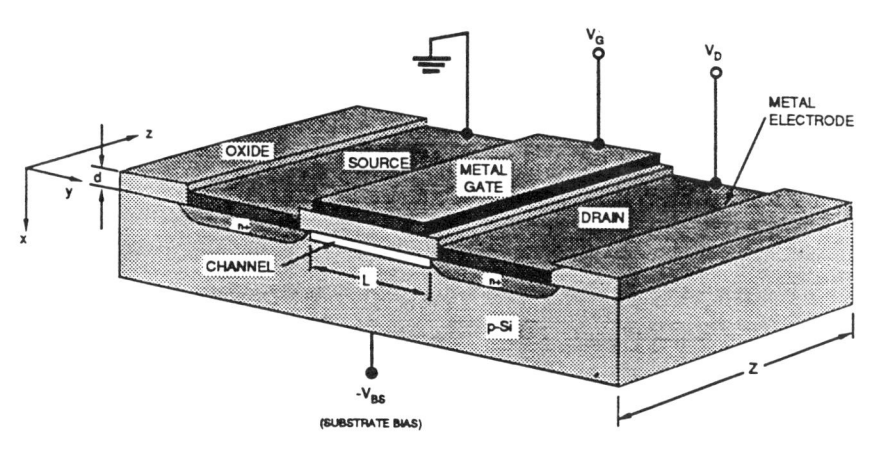

Fig. 6 The essentials of a FET. An inversion layer (channel) is induced by the voltage V_G, and the conductance of the channel is measured by the current between the drain and the grounded source.

The insulator shown in Fig. 6 must be SiO_2 near the silicon (although, as discussed below, other insulating oxides are normally covering the silica). The reason is that the formation of an inversion layer requires that there are few "interface states" at the Si/insulator interface. Interface states are localized energy levels due to lattice mismatch, impurities, or other defects at the interface. If there are many interface states, the charge induced by V_G, that should be in the inversion layer, is trapped in the interface states and there is no conductivity change when a gate voltage is applied. Fortunately, we can have an excellent oxide thermally grown on silicon, SiO_2. With this as the insulator we can reduce interface state densities to less than 10^{10} cm^{-2}. Other insulators are not as accommodating.

With the inversion layer (termed the "channel" when discussing FETs) current can flow between the source and the drain, as the n^+ regions (highly doped n-silicon) make good contact to the electrons in the induced channel. A positive voltage V_D is applied to the drain and the current is measured. As V_D increased the current saturates and, normally, the FET is operated with V_D high enough to be in this current saturation region.

Now a small change in the potential V_G at the gate leads to a significant change in the saturation current. For a ChemFET, the external voltage V_G is held constant and adsorption or absorption of the gas or ion to be measured causes a small electric field that changes the effective gate voltage and the resulting drain–source current.

8.7.2 The GasFET

The principle example of a ChemFET for gas sensing is the Pd-gate H_2 sensor, studied mostly by Lundstrom and his coworkers.[47] They suggest that hydrogen dissolved in the Pd (at about 150°C), moves to the Pd/SiO$_2$ interface, and forms a dipole layer. The dipole changes the work-function difference between the metal and the SiO$_2$ (or, equivalently, the Si). As the threshold voltage of the FET (the gate voltage needed to induce the inversion layer) depends on this work-function difference, this dipole translates into an effective change in the gate voltage. The change in the threshold voltage is

$$\Delta V_T = -\mu N \theta \varepsilon_0 \tag{32}$$

where μ is the dipole moment of the interfacial hydrogen, N is the density of sites, θ is the fraction of sites covered, and ε_0 is the permittivity of free space.

Another model described by Mariucci et al.,[48] which is not greatly different, suggests that the work-function change is due to a palladium hydride phase at the interface. They suggest an alternative form, where the FET material is amorphous silicon. The sensitivity is much lower than that for the silicon-crystal FET, which is commonly 10 ppm of H$_2$ in air.

Other gases, specifically H$_2$S and NH$_3$, can be detected using the Pd-gated MOSFET, where the molecule to be detected can dissociate to produce hydrogen. Other gases can be detected with a porous gate, where pores are provided so the gas can reach the Pd/SiO$_2$ interface. Dobos and Zimmer[49] have prepared CO-sensitive MOSFETs with a Pd/PdO gate where pores are available for CO penetration to the interface. The CO interacts more strongly with the PdO than it does with the metal Pd. Therefore, the Pd is slightly oxidized to PdO. Lundstrum and Sodeberg[50] report sensitivity to CCl$_4$ with a porous gate.

Another form of FET-based gas sensor is the adsorption-FET (ADFET) and similar devices where the insulating oxide is made very thin. Adsorption of polar molecules from the gas phase onto an ultra-thin oxide coating (5 nm) leads to an electric field sufficient to affect the charge in the channel. H$_2$O, NH$_3$, HCl, CO, NO, NO$_2$, and SO$_2$ were shown[51] to cause changes equivalent to a change in potential at a gate. The key is the very thin oxide so the weak field associated with a layer of dipoles can penetrate.

8.7.3 The ISFET

The original ion-sensitive, field-effect transistor (ISFET) to measure pH or to measure ions in solution was reported by Bergveld.[52] A MOSFET, the essentials of which are given in Fig. 6, is prepared but without the metal gate, as in the example of Fig. 7. In this section we give a simplified discussion of the origin of the sensitivity to pH or ion concentration—for a rigorous analysis see Ref. 37.

As with all FETs a channel is induced. If it is necessary to produce the channel, a gate voltage can be applied to the reference electrode in the ion-conducting solution shown in Fig. 7. A reference electrode is an electrode that accurately reflects the potential in the solution independent of changes in the dissolved species or in the pH of the solution.

8.7 FET DEVICES FOR GAS OR ION SENSING

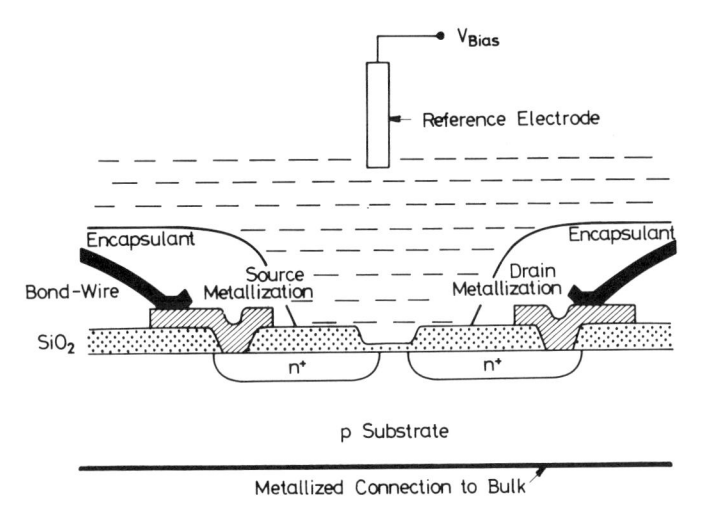

Fig. 7 A simple ISFET. The metal gate is removed from the MOSFET of Fig. 6, a voltage to induce a channel is applied, if necessary, to the reference electrode, and the adsorbing species to be measured induces a double-layer voltage across the oxide that coats the channel.

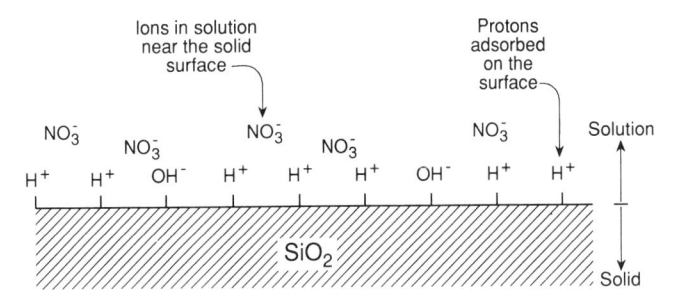

Fig. 8 The origin of a double-layer voltage at the SiO$_2$ surface in a strong nitric acid solution. Protons are adsorbed at the surface forming one side of the double layer, and negative ions in the solution are attracted to the near-surface region forming the other side of the double layer.

Now at the oxide/solution interface ions are adsorbed. For example, with pH very low protons are the dominant adsorbate, with pH high OH$^-$ ions are the dominant adsorbate. The adsorbed charges attract and are neutralized by counter-ions in the solution. This double layer is termed a Helmholtz double layer. A double-layer potential difference ψ_0 develops at the interface. For example, if we have the acid HNO$_3$ present at low enough pH, a charge distribution as in Fig. 8 would arise. Effectively the potential at the surface of the SiO$_2$, the potential that produces the channel, is given by $V_G + \psi_0 +$ constant, where the constant is associated with the reference electrode. Because ψ_0 depends on pH, the channel current varies with pH at constant V_G. The Helmholtz double-layer potential can arise by adsorption of ions other than H^+ and OH^-. It can be shown[37] that for simple cases with strong ionic bonds between the adsorbed species and the SiO_2

$$q\psi_0 = 2.303kT(\text{pH}_{pzc} - \text{pH}) \tag{33}$$

where pH_{pzc} is the pH for zero net adsorbed charge, the "point of zero charge".

The main difficulties with these devices are twofold, lack of stability and difficulty in providing a reliable reference electrode. Much of the stability problem is associated with the insulating oxide. SiO_2 is the worst case, consequently it is not used by itself. Processes such as hydration of the oxide occur (especially when immersed in aqueous solutions), changing the effective thickness or equivalently the bulk dielectric constant of the insulator.

The hydration problem is overcome to a great extent by covering the SiO_2 layer with Si_3N_4, IrO_x, Ta_2O_5, or Al_2O_3. Si_3N_4 suffers from degradation in part due to surface oxidation[53] and from the requirement that the surface be etched with HF to provide sensitivity.[54] In general, all devices show drift, although in some cases the drift can be anticipated and accounted for. It has been suggested[55] that intermittent sampling (flow injection analysis) can lessen the exposure of the ISFET to contaminating species and decrease the drift.

Selectivity in such devices can, in some cases, be provided by a semi-permeable membrane over the ISFET. The use of these for biosensors is discussed in Chapter 9.

The requirement of a reference electrode is difficult to meet while utilizing the advantage of planar technology built onto a silicon chip. A reference electrode utilizes a chemical reaction to move ions into solution from an electrode. For example, with the silver/silver chloride reference electrode the movement of chloride ions from or to the solid Ag/AgCl electrode carries the current:

$$e^- + AgCl \leftrightarrow Ag + Cl^-. \tag{34}$$

The Nernst equation (an expression for the electrical potential due to chemical species, arising in thermodynamics) for the potential with such a reaction is

$$V = V_0 - (RT/F)\ln[Cl^-] \tag{35}$$

where V is the potential of the silver electrode, V_0 a constant, R the gas constant, F the Faraday constant, T the temperature and $[Cl^-]$ the concentration of chloride ions. If there is a constant concentration of chloride ions, the potential V is stable as desired. Such a stable concentration is simple to obtain with a typical reference electrode where the silver/silver chloride is isolated from the solution to be tested by a frit (or a salt bridge) and KCl is provided on the silver/silver chloride side of the frit. No such simple answer is easily available if the "reference electrode" is, for example, a silver film deposited somewhere on a silicon chip for compatability with FET technology.

8.8 SUMMARY AND FUTURE TRENDS

All the forms of semiconductor chemical sensors have one major problem. In order to detect the chemical species of interest, the sensors must be exposed, unprotected, to the ambient solution or gas. It is difficult to make them reversibly reactive to the gases of interest and nonreactive with respect to all other possible chemical species that may appear in the atmosphere or liquid. Fortunately, in most cases, the form of interference is known and an ideal sensor is not required. For example, the degrading effect of H_2S or Cl_2 on some sensors is no problem if the user is sure these particular species will not be present.

Sensors from semiconducting metal oxides have the desired feature of low cost, good sensitivity, and convenient form of response (a simple change in resistance). These features have made, and undoubtedly will continue to make, these sensors popular. However, the sensors have problems in reproducibility, stability and selectivity. Every improvement in these aspects will undoubtedly increase the usage of the devices.

Unfortunately, all of the sources of poor reproducibility, stability, and selectivity are not known. A large part of the problem is associated with grain boundaries. Because for high sensitivity the resistance must be associated with current passing through grain boundaries or intergranular contacts, great improvement in reproducibility and stability will be difficult. It is hoped that there will be a breakthrough in producing regular arrays of identical grain boundaries, but even small improvements in this area can improve the characteristics of the sensor substantially.

The selectivity problem (the sensitivity of these sensors to all combustible gases) is being studied by many workers. New metal-oxide semiconductors, temperature programming, filters and members, and catalyst/promoter combinattions are being investigated to improve selectivity.

GasFET sensors have fewer problems than ISFET sensors. Because they are not used in a chemically very-reactive environment and have fewer problems with hydration, their drift is less. However, because they must be operated at a much lower temperature than the metal-oxide sensors to retain good device characteristics, they are limited in the catalytic reactions possible. As discussed in Section 8.7.2, the Pd-gated sensor for detecting hydrogen and the non-catalyst sensing of polar molecules using the ADFET or a similar structure are the applications of interest.

The ISFET, on the other hand, can become a much more useful detector. Sensors are produced commercially, based, in general, on alumina or silicon nitride as the insulator. Their application in biosensing shows great promise as will be discussed in the next chapter.

PROBLEMS

1. Show that Eq. 11 derives from statistics.

410 CHEMICAL SENSORS

2. Consider Eq. 13. Give an equation for the sensitivity in terms of N_s (instead of x_0).

3. Consider Fig. 8.2c. Derive an expression showing what electric field must be present at $x=0$ to make the valence band closer to the Fermi energy than the conduction band.

4. What other reasons than those discussed in Section 8.2.1 can be the cause of band formation in surface states?

5. Discuss why one expects adsorbed O^- to be much more reactive than O_2^-. Why does one expect a p-type semiconductor to show catalytic activity with lattice oxygen as the oxide (Eq. 21)?

6. Under the assumption that there are enough empty surface states so that Richardson's equation

$$J = AT^2 \exp\{-(q\psi_s + E_c - E_f)/kt\}$$

with $A = 1.2 \times 10^6$ A/m²-K² is valid for electron transfer to the surface, determine the rate of capture of electrons when a depletion layer (Fig. 8.2b) is present. As ψ_s becomes more negative, the rate of electron capture as described by the Richardson equation becomes lower and lower. If equilibrium must be reached in less than 10 s for a practical sensor, estimate the allowable limit of band bending. For simplicity, assume as the criterion that the rate of electron capture at equilibrium as given by Richardson's equation must suffice to transfer the surface state charge N_s in 10 seconds. Assume the temperature is 300 K, $E_c - E_F = 0.15$ eV, the donor density is 10^{23} m^{-3} and $\varepsilon\varepsilon_0$ is 10^{-10} F/m. Discuss this result in respect to the need for a Pd catalyst.

7. Derive Eq. 30 from Eq. 20, with the resistance R inversely proportional to $n_s(a = n_s R)$.

8. Derive an expression for the resistance of an n-type sample of length L, area W^2 as a function of N_t, where N_t is the density of charge trapped at a grain boundry. Assume there is only one grain boundary extending across the sample, located at L/W^2. Use Richardson's equation from Problem 6. Assume the applied voltage is small.

REFERENCES

1. N. Taguchi, Gas detecting device, *U.S. Patent* 3,631,436 (1971).
2. J. D. Levine and P. Mark, "Theory and observations of intrinsic surface states on ionic crystals," *Phys. Rev.* **144**, 751 (1966). See also S. R. Morrison, *The Chemical Physics of Surfaces*, Plenum, New York, 1977, second edition 1990.
3. R. Dorn, H. Luth, and G. J. Russell, "Adsorption of oxygen on cleaved (110) GaAs surface," *Phys. Rev.* B **10**, 5049 (1974).

4. M. P. McDaniel and R. L. Burwell, "Excess oxygen of chromia," *J. Catal.* **36**, 394 (1975).
5. D. DeCogan and G. A. Lonergain, "The influence of some impurities on the conduction properties of Cr2O3 and Fe2O3," *Solid State Commun.* **15**, 1519 (1974).
6. S. R. Morrison, "Surface states on a chromia catalyst," *J. Catal.* **47**, 69 (1977).
7. J. H. Lundsord, "ESR of adsorbed oxygen species," *Catal. Rev.* **8**, 135 (1973).
8. W. M. H. Sachtler, "The mechanism of the catalystic oxidation of some organic molecules," *Catal. Rev.* **4**, 27 (1970).
9. A. W. Czandra, "The adsorption of oxygen on silver," *J. Phys. Chem.* **68**, 2765 (1964).
10. P. A. Kilty and W. M. H. Sachtler, "The mechanism of the selective oxidation of ethylene to ethylene oxide," *Catal. Rev.* **10**, 1 (1974).
11. S. R. Morrison, *The Chemical Physics of Surfaces*, 2nd ed., Plenum, New York, 1990.
12. H. Chon and J. Pajares, "Hall effect studies of oxygen chemisorption on ZnO," *J. Catal.* **14**, 257 (1969).
13. V. Lantto and P. Romppainen, "Electrical studies on the reactions of CO with different oxygen species on SnO_2 surfaces," *Surf. Sci.* **192**, 243 (1987).
14. N. Yamazoe, J. Fuchigama, M. Kishikawa, and T. Seiyama, "Interactions of tin oxide surface with oxygen, water and hydrogen," *Surf. Sci.* **86**, 335 (1979).
15. S. C. Chang, "Oxygen adsorption on SnO_2," *Proc. 2nd Int. Meeting on Chem. Sensors, Bordeaux, July, 1986*, p. 78.
16. I. Aso, M. Nakao, N. Yamazoe, and T. Seiyama, "Study of metal oxide catalysts in the olefin oxidation from their reduction behavior," *J. Catal.* **57**, 287 (1979).
17. P. A. Sermon and G. C. Bond, "Hydrogen spillover," *Catal. Rev.* **8**, 211 (1973).
18. W. C. Conner, Jr., G. M. Pajonk, and S. J. Teichner, "Spillover of adsorbed species," *Adv. Catal.* **34**, 1 (1986).
19. S. R. Morrison, "Selectivity in semiconductor gas sensors," *Sens. Actuators* **12**, 425 (1987).
20. S. R. Morrison, "ZnO, semiconductor and adsorbent," *Adv. Catal.* **7**, 259 (1955).
21. N. Hykaway, W. Sears, R. F. Frindt, and S. R. Morrison, "The gas sensing properties of bismuth molybdate thin films," *Sens. Actuators* **15**, 105 (1988).
22. G. W. Keulks, "The mechanism of oxygen atom incorporation into the products of propylene oxidation over bismuth molybdate," *J. Catal.* **19**, 232 (1970).
23. R. D. Wragg, P. G. Ashmore and J. A. Hockey, "Selective oxidation of propense over bismuth molybdate catalysts," *J. Catal.* **22**, 49 (1971).
24. M. J. Esper, E. M. Logothesis, and J. C. Chu, "Titania exhaust gas sensor for automotive applications," *S.A.E. Technical Papers*, Series 790140 (1979).
25. T. A. Jones, J. G. Firth, and B. Mann, "The effect of oxygen on electrical conductivity of some metal oxides in inert and reducing atmospheres at high temperature," *Sens. Actuators* **8**, 281 (1985).
26. H. Arai, C. Yu, Y. Fukuyama, Y. Shimizu, and T. Seiyama, "Application of *p*-type semiconducting perovskite-type oxides to a lean-burn oxygen sensor," *Proc. 2nd Int. Meeting on Chem. Sensors, Bordeaux, July, 1986*, p. 142.
27. K. Park and E. M. Logothetis, "Oxygen sensing with $Co_{1-x}Mg_xO$ ceramics," *J. Electrochem. Soc.* **124**, 1443 (1977).

28. T. A. Jones, J. G. Firth, and B. Mann, "The effect of oxygen on the electrical conductivity of some metal oxides at high temperatures," *Sens. Actuators* **8**, 281 (1985).
29. J. Gerblinger and H. Meixner, "Influence of dopants on the response time and signals of lambda sensors based on thin films of strontium titanate," *Sens. Actuators* **B6**, 231 (1992).
30. P. T. Moseley, "Materials selection for semiconductor gas sensors," *Sens. Actuators* **B6**, 149 (1992).
31. U. Dibbern, G. Kuersten, and P. Willich, "Gas sensitivity, sputter conditions and stoichiometry of pure tin oxide layers," *Proc. 2nd Int. Mtg. on Chemical Sensors, Bordeaux, France, 1986*, p. 127.
32. T. Yamasaki, U. Mizutani, and Y. Iwama, "Electrical properties of SnO_2 polycrystalline thin films and single crystals exposed to oxygen and hydrogen gases," *Jpn. J. Appl. Phys.* **22**, 454 (1983).
33. G. N. Advani, Y. Komem, J. Hasenkopf, and A. G. Jordan, "Improved performance SnO_2 thin film gas sensors due to gold diffusion," *Sens. Actuators* **2**, 139 (1981/1982).
34. G. Sberveglieri, "Classical and novel techniques for preparation of SnO_2 thin film gas sensors," *Sens. Actuators* **B6**, 239 (1992).
35. Y. Nakatini, M. Matsuoka, and Y. Iida, "Gamma hematite ceramic gas sensors," *IEEE Trans. Comp. Hybrides and Mfg. Technol.* **CHMT-5**, 522 (1982).
36. P. Dutronc, B. Carbonna, F. Menil, and C. Luact, "Influence of the nature of the screen-printed electrode metal on the transport and detection properties of thick film semiconductor gas sensors," *Sens. Actuators* **B6**, 279 a1992).
37. M. J. Madou and S. R. Morrison, *Chemical Sensing with Solid-State Devices*. Academic Press, Boston, 1989.
38. K. Fukui and K. Komatsu, *Proc. Int. Meeting on Chem. Sensors, Fukuoka, Japan, 1983*, p. 51.
39. H. Ogawa, A. Abe, M. Nishikawa, and S. Hayakawa, "Electrical properties of tin oxide ultrafine particle films," *J. Electrochem. Sco.* **128**, 2020 (1981).
40. C. E. Allman, *Adv. Instrum.* **38**, 399 (1988).
41. S. J. Gentry and P. T. Walsh, "Poison-resistant catalytic flammable-gas sensors," *Sens. Actuators* **5**, 239 (1984).
42. J. W. Gardner, H. V. Shurman, and T. T. Tan, "Application of an electronic nose to the discrimination of coffees," *Sens. Actuators* **B6**, 71 (1992).
43. J. Mizsei and V. Lantto, "Air pollution monitoring with a semiconductor gas sensor array system," *Sens. Actuators* **B6**, 223 (1992).
44. I. Aso, M. Nakao, N. Yamazoe, and T. Seiyama, "Study of metal oxide catalysts in the olefin oxidation from their reduction behavior," *J. Catal.* **57**, 287 (1989).
45. J. P. Wagner, A. Forkson, and M. May, "Comparative performance of ionization vs. photoelectric fire detectors," *J. Fire Flammability* **7**, 71 (1976).
46. Y. Matsuura, K. Takahate, and S. Matsuura, "Mechanism of the long term resistance change of SnO_2 gas sensors," *Proc. Third Int. Meeting Chem. Sensors, Cleveland, September 1990*, p. 44 and p. 73.
47. I. Lundstrom, M. S. Shivaraman, and C. M. Svensson, "A hydrogen sensitive Pd-gate MOS transistor," *J. Appl. Phys.* **46**, 3876 (1975).
48. L. Mariucci, G. Fortunato, A. Pecora, A. Bearzotti, P. Carelli, and R. Leone,

"Hydrogenated amorphous silicon technology for chemical sensing thin film transistors," *Sens. Actuators* **B6**, 29 (1992).

49. K. Dobos and G. Zimmer, "Performance of CO-sensitive MOSFETs with metal oxide semiconductor gates," *IEEE Trans. Electron Devides* **ED-32**, 1165 (1985).
50. I. Lundstrom and D. Sodeberg, "Hydrogen sensitive MOS structures, part 2: characterization," *Sens. Actuators* **2**, 105 (1981/1982).
51. P. Cox, "Environmental monitoring device and system," *U.S. Patent* 3,831,432 (1974).
52. P. Bergveld, "Development of an ion-sensitive solid state device for neurophysiological measurements," *IEEE Trans. Biomed. Eng.* **BME-17**, 70 (1990).
53. H. Abe, M. Esashi, and T. Matsuo, "ISFET's using inorganic gate thin films," *IEEE Trans. Electron Devices* **ED-26**, 1939 (1979).
54. A. Sibbald, P. Whalley, and A. Covington, "A miniature flow-through cell with a four-function chemFET integrated circuit," *Anal. Chem. Acta* **159**, 47 (1984).
55. S. Alegret, J. Bartroli, C. Jimenez-Jorquera, M. del Valle, C. Dominguez, J. Esteve, and J. Bansells, "Flow through pH ISFET + reference-ISE as integrated detector in automated FIA determinations," *Sens. Actuators* **B7**, 555 (1992).

9 Biosensors

A. S. DEWA
Micropump Corporation
Vancouver, WA 98668, USA

W. H. KO
Department of Electrical Engineering and Applied Physics
and Electronic Design Center
Case Western Reserve University
Cleveland OH 44106, USA

9.1 INTRODUCTION

9.1.1 The Concept of a Biosensor

Biosensors are a special class of chemical sensors that take advantage of the high selectivity and sensitivity of biologically active materials. This high selectivity and sensitivity of the biological material is a result of millions of years of evolution of life on earth, since much of the communication among biological organisms is based on chemical signals, whether the senses of smell and taste, or immunological reactions, or pheromones, or "hunting" of single-celled organisms. Even the senses of vision, hearing, and touch are transmitted by chemical communication through the nervous system. These communication processes can be considered to be "bio-recognition" processes. Thus, the potential to use these bio-recognition processes as inputs to a sensor is apparent. The diversity of life is reflected in the large variety of biosensors, since there are biological chemicals, organelles, cells, tissues, and organisms that react to everything from small inorganic molecules, such as oxygen, to large, complicated proteins and carbohydrates.

A good operational definition of a biosensor (and also a well-quoted one) is that a biosensor is a device incorporating a biological sensing element with a

Semiconductor Sensors, Edited by S. M. Sze.
ISBN 0-471-54609-7 © 1994 John Wiley & Sons, Inc.

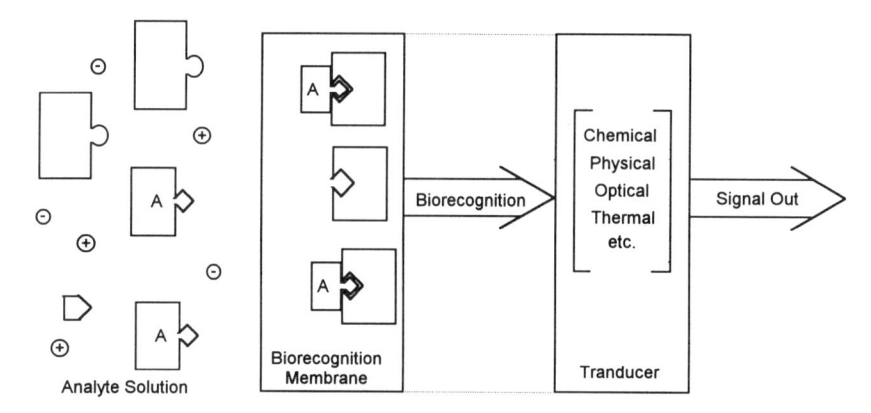

Fig. 1 Schematic drawing of the general biosensor. It consists of a traditional transducer with a biological-recognition membrane intimately in contact with the transducer. The biologically active material recognizes the analyte molecule "A" through a shape-specific recognition. In affinity-based biosensors, the binding of the analyte and bioactive molecule is the chemical signal that is detected by the transducer. In metabolic biosensors, the biologically active material converts the analyte, and any co-reactants, into product molecules. The transducer converts the result of that reaction into the output signal. The transduction can be through measurement of concentration changes in a product or co-reactant concentration, or the heat liberated in the reaction.

traditional transducer, such as physical or chemical transducers.[1] The biological sensing element selectively recognizes a particular biological molecule through a reaction, specific adsorption, or other physical or chemical process, and the transducer converts the result of this recognition into a usable signal, usually electrical or optical. Figure 1 illustrates this definition.

There are two classes of bio-recognition processes, bio-affinity recognition and bio-metabolic recognition, which are differentiated by the general method of detection. Both bio-recognition processes involve the binding of a chemical species with another which has a complementary structure. This is referred to as "shape-specific binding". In bio-affinity recognition, the binding is very strong, and the transducer must detect the presence of the bound receptor–analyte pair. The most common bio-affinity recognition processes are receptor–ligand binding, and antibody–antigen binding. In bio-metabolic recognition, after binding, the analyte and any other co-reactants are chemically altered to form the product molecules. The transducer must detect the change in the concentration of either the products or the co-reactants, or the heat liberated during the reaction. Bio-metabolic processes include enzyme–substrate reactions and the metabolism of specific molecules by organelles, tissues, and cells.

As can be seen from the definition of a biosensor and Fig. 1, the biological recognition element is usually immobilized on the surface of the traditional transducer in some type of membrane. Thus, the bio-recognition element is actually a bioreactor on top of the traditional transducer, so the response of the biosensor will be determined by the diffusion of the anlyte, reaction products, co-reactants or interfering species, and the kinetics of the recognition process.

The modern biosensor was essentially defined by the work of Clark and Lyons.[2] They developed the first enzyme electrode where the enzyme, glucose oxidase, was immobilized on an electrochemical oxygen electrode. Since then, the field has grown to include biosensors based on chemically sensitive semiconductor devices, fiber optics, thermistors, chemically mediated electrodes, surface-acoustic-wave devices, piezoelectric micro-balances, and many other transducers.

9.1.2 Selectivity and Sensitivity

The major difference between chemical sensors and biosensors is that the biosensor has a much greater sensitivity and selectivity. Over the millions of years of natural selection, biological molecules have evolved that have very sensitive responses to very specific molecules or groups of molecules. The concept of "shape-specific recognition" is commonly used to explain the high sensitivity and selectivity of biological molecules, especially enzyme–substrate and antigen–antibody systems. Here, the analyte molecule has a complementary structure to the enzyme or antibody, and upon binding the bound pair is in a lower energy state than the two separate molecules. In the case of affinity recognition processes, such as antibodies, this binding is very difficult to break. For enzymes, binding is the basis for the catalytic reaction, and once the analyte is converted to the product, it is freed from the enzyme.

9.1.3 Recognizable Biomaterials

The biomaterials that can be recognized by the bio-recognition elements are as varied as the different reactions that occur in biological systems. The major classes of the reaction are given in Table 1. As can be seen in Table 1, almost all types of biological reactions, chemical or affinity, can be exploited for biosensors.

TABLE 1 Classes of Recognizable Biological Chemicals and Some Examples

Analyte	Examples
Metabolic chemicals	Oxygen, methane, ethanol, other nutrients
Enzyme substrates	Glucose, penicillin, urea
Ligands	Neurotransmitters, hormones, pheromones, toxins
Antigens and antibodies	Human Ig, anti-human Ig
Nucleic acids	DNA, RNA

TABLE 2 Biosensor Components

Biological Elements	Transducer Type	Transcucer Example
Organisms	Electrochemical:	
Tissues	a. Potentiometric	Ion selective field-effect transistors
Cells		and micro-electrodes
Organelles	b. Amperometric	Micro-electrodes
Membranes	c. Impedometric	Micro-electrodes
Enzymes	Optical	Fiber optodes and luminescence
Receptors	Calorimetry (thermal)	Thermistors and thermocouples
Antibodies	Acoustic (mass)	Surface acoustic wave delay-lines and
Nucleic Acids		bulk acoustic wave microbalances

A biosensor consists of a biological sensing element (column one) plus a transducer (column two). Examples of specific transducers are given in column three next to the transduction principle.

9.1.4 Recognition Elements for Biosensors

The possible biological sensing elements are as diverse as life itself, ranging in size from whole organisms down to specific molecules. Table 2 gives a partial list of biological sensing elements and a list of the possible transduction principles, with examples of actual devices next to their respective transduction principle. To design a biosensor, a biological sensing element from column one is chosen, then the appropriate transduction principle is selected from column two. Next to that principle, in columm three, is a list of actual transducers that can be used.

Those biological elements in Table 2 that are larger than a molecule are used because they contain the biologically active material of interest, and provide an easy way to immobilize the active material in its natural environemnt (i.e., inside the cell). These cellular and multicellular structures are good for co-enzyme systems or coupled reactions, where it takes multiple biologically active materials to convert the analyte into a molecule that can be detected by the transducer. Nature has designed these coupled reaction systems very well, so immobilizing whole cells, organelles, or tissues is the easiest, and, in some cases, the only way to use these systems. However, keeping these living cellular and multicellular structures alive can be difficult, and thus, they are mainly used in bio-assay systems where the instrument is operated in the controlled environment of a laboratory and the cells and tissues can be replenished from a controlled stock by trained technicians.

To get an understanding of the reaction kinetics that can take place in the biologically active materials, simple examples of an affinity reaction, the binding of an antibody with its antigen, and a metabolic (or biocatalytic) reaction, the enzyme–substrate reaction, will be considered.

The simple binding of an antibody to its antigen is described in the following chemical reaction and equilibrium constant of association

$$Ab + Ag \underset{k_d}{\overset{k_a}{\leftrightarrow}} Ab \cdot Ag \qquad K_a = \frac{k_a}{k_d} = \frac{[Ab \cdot Ag]}{[Ab][Ag]} \qquad (1)$$

where k_a is the binding reaction rate, k_d is the un-binding reaction rate, K_a is the equilibrium constant of the binding reaction, $[Ab]$ is the concentration of the antibody, $[Ag]$ is the concentration of the antigen, and $[Ab \cdot Ag]$ is the concentration of the bound complex antibody and antigen. In a biosensor, one of these concentrations is fixed by immobilizing either the antibody or antigen in or on a suitable membrane, which is coupled to a transducer. If the antibody concentration is fixed, $[Ab]_{tot} = [Ab]_{eq} + [Ab \cdot Ag]_e$, where the subscript, tot, signifies the total amount of antibody and the subscript, eq, signifies the equilibrium concentration species. The bound antibody–antigen complex concentration as a function of the antigen concentration (i.e., the analyte) is given by

$$[Ab \cdot Ag]_{eq} = \left(\frac{[Ag]_{eq}}{(1/K_a) + [Ag]_{eq}} \right) [Ab]_{tot}. \qquad (2)$$

This relationship is plotted in Fig. 2. Of course, this is an ideal case; antibody–antigen binding can be a very complicated process when multiple binding sites are involved.

For the enzyme–substrate system, the reaction kinetics are of interest since they determine how fast the substrate is converted into the product. For the simplest case, a single enzyme acting on a substrate molecule, where any other reactants are assumed to be in excess so as not to limit the reaction, the reaction

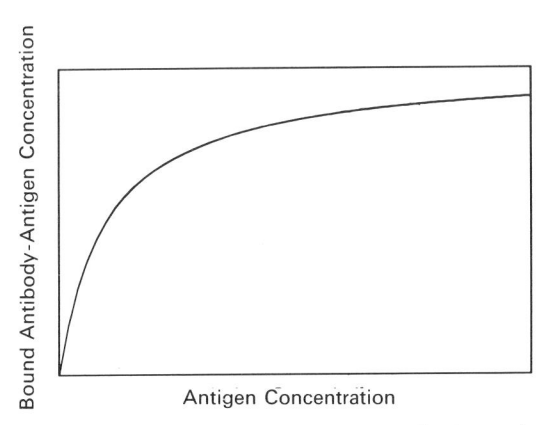

Fig. 2 The equilibrium concentration of the bound antibody–antigen complex as a function of antigen concentration in a system with a fixed antibody concentration as plotted from Eq. 2.

follows the Michaelis–Menton kinetics[3]

$$S + E \underset{k_{-1}}{\overset{k_1}{\leftrightarrow}} ES \overset{k_2}{\to} P + E \qquad (3)$$

where E is the enzyme, S is the substrate, ES is the bound enzyme–substrate complex, P is the product and k_1, k_{-1}, k_2 are the reaction rate constants for the enzyme–substrate binding reaction, the un-binding of the enzyme–substrate complex and the formation of the product, respectively. It is assumed that the formation of the enzyme–substrate complex is a reversible reaction but the conversion to product is not. In an enzyme biosensor, the enzyme is immobilized in a membrane on the transducer's surface. Thus, the total amount of enzyme is fixed at $[E]_{tot} = [E] + [ES]$. (Again, the brackets, [ ], denote concentrations.) The rate of product formation is given by

$$v = \frac{\partial [P]}{\partial t} = k_2 [ES]. \qquad (4)$$

where the rate, v, has units of mol/l-s and t is time. To relate the rate of product formation with the substrate concentration, it is assumed that the reaction is in steady state so that

$$\frac{\partial [ES]}{\partial t} = k_1 [E][S] - k_{-1}[ES] - k_2 [ES] = 0. \qquad (5)$$

Solving Eq. 5 for $[ES]$ and substituting into Eq. 4 yields

$$v = \frac{k_2 [E]_{tot}[S]}{([S] + K_M)} \qquad (6)$$

where the Michaelis constant is defined as

$$K_M \equiv \frac{(k_{-1} + k_2)}{k_1}. \qquad (7)$$

Equation 6 has the same shape as Eq. 2, and thus can also be represented by Fig. 2.

Few enzymes follow the Michaelis–Menton kinetics over a wide range of experimental conditions. Most enzyme-substrate interactions have more complex kinetics. However, for these cases with complex kinetics, K_M is still a useful measure of the enzyme–substrate interaction, if it is defined as the concentration where half the enzymes are bound to the substrate, or alternatively, as the substrate concentration where the reaction at half its maximum rate. Thus, it can be used to compare enzyme–substrate systems under different conditions.

Most enzyme molecules are proteins. Thus, they are amphoteric molecules which have many acidic and basic groups on the surface of the molecule.[3] These groups are subject to solution equilibrium, and the degree of ionization of the enzyme depends on the pH of the solution. If these ionizable groups are near the catalytic site on the enzyme molecule, the change in charge density can reduce or eliminate the catalytic activity of the enzyme. Therefore, the kinetics of enzymes are strongly dependent on environmental conditions (see Problem 2 as an example of this).

As can be seen from these two simple examples, the response of the biological-recognition processes to analyte concentration is highly nonlinear because of the kinetics of the recognition interactions. To fabricate a viable biosensor with a linear response to the analyte concentration, it must be ensured that the device operates under mass flow controlled conditions, not reaction kinetic control. This is discussed further in Section 9.3.

9.1.5 Biomimetic Structures

Biomimetic structures are artificial structures that are built up to mimic the processes that occur in cell membranes. The cell membrane is one of the most sophisticated biosensor/actuator systems known. Some receptors in the cell membrane recognize different molecular species and cause a change in membrane permeability to certain other chemical species (receptor-controlled ion channels). These ion channels can be either directly controlled by the receptor, or indirectly controlled by using a messenger protein to open the ion channel.[4] There are also receptors that bind proteins and other molecules and bring them into the cell for use by the cell.[5] Many of these natural transduction mechanisms also include "chemical amplification" of the signal, with gains of up to 1000.[6] Using these "natural" sensing systems and coupling them to artificial transducers is a difficult challenge due to the very fragile nature of cell membranes, but the potential rewards are great.

Cell membranes are made up of proteins that float in a fluid bilayer of phospholipid molecules. The phospholipids are organized so that the hydrophobic tails of the molecules are inside the membrane and the hydrophilic head groups are on the outside of the bilayer, as shown in Fig. 3. Thus, the membrane has a very low trans-membrane conductance to aqueous ions. In general, the membrane is not made up of a single type of phospholipid molecule, but is a mixture of different phospholipids, depending on the type of cell. The distribution of phospholipids is also different on the interior and exterior of the cell membrane.[7]

The membrane proteins can either span or not span the membrane.[5] These proteins form the receptors and ion channels. In general, there is a 40 to 90 mV potential across the membrane, with the interior of the cell negative with respect to the outside environment.

Extra-cellular Medium

Trans-Membrane Protein
Membrane Protein
Phospholipid Molecules

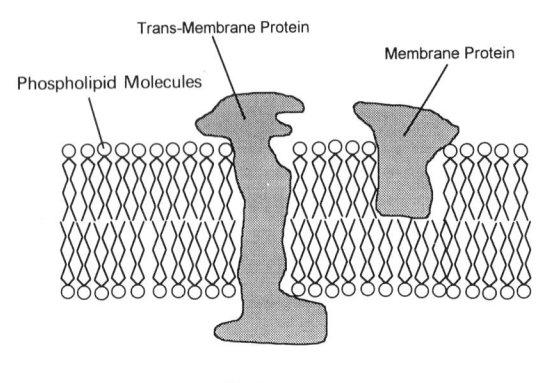

Cytoplasm

Fig. 3 The structure of a cell membrane. The membrane consists of a fluid bilayer of phospholipid molecules, with proteins embedded in it, that separates the interior of the cell from the extra-cellular medium. The proteins can either be trans-membrane or just exist on one side of the membrane. The trans-membrane proteins can be ion channels or receptors.

Artificial bilayer lipid membranes (BLM) are very fragile and are usually supported on artificial (hydrophilic) supports. They may be either cast from solution[8] or deposited by the Langmuir–Blodgett (LB) technique.[9] The major biosensor application of biomimetic structures has been ion-channel sensors.[10,11] Studies show the potential of ion-channel sensors, but no stable sensor structures have been made, due to the fragile natural of the BLM. There has also been much work on the formation of more stable artificial lipid membranes by adding polymerizable groups to lipid tails.[12]

In summary, the biomimetic concept offers many potential advantages for biosensors. However, the fragility of the BLMs has limited their use to laboratory tests. As more is learned about the physical chemistry of the BLMs and the biochemistry of the receptor proteins and cell functions, the impact on biosensors will be great.

9.1.6 Scope of the Chapter

This chapter will discuss the fundamentals of biosensor fabrication and the basic theory that governs the response characteristics. As noted in Section 9.1.1, a biosensor is a bioreactor coupled to a transducer, and thus, the response of the biosensor is a function of both the biological reaction that occurs in the recognition membrane and the transducer's response to the result of that reaction. The details of transducer fabrication and transduction principles have been discussed in previous chapters (specifically, Chapters 3, 6, 7, and 8).

This chapter will begin by discussing the basics of the immobilization of biologically active materials on the transducer. The different classes of immobilization will be covered separately. Then, the basics of mass transport and loading of the immobilization membrane will be discussed, as well as their effects on the biosensor's response. Next, a summary of the different types of biosensors will be discussed with respect to their transduction principle. The basics of the operation of each type of device will be presented. Finally, one of the most important, and often neglected, parts of the design of a successful biosensor, the packaging, will be covered.

The biosensor field is much too large to be covered adequantely by a single chapter, and many excellent review books have already been published.[13-18]. The topic of biosensors covers the fields of biology, molecular biology, chemistry, physics, biophysics, materials science, chemical engineering, mechanical engineering, and electrical engineering. Obviously, in-depth knowledge of all of these topics, as they relate to all the different biosensors, is beyond any one person. Therefore, this chapter will present a basic discussion of what governs the functioning of the different biosensors using simple analytical models, basic chemistry, and fundamental concepts. For an in-depth study of a particular biosensor, the reader should consult the pertinent literature.

9.2 IMMOBILIZATION OF BIOLOGICAL ELEMENTS

The immobilization of the biological element on the physical transducer is one of the keys to a high-sensitivity, long-lived biosensor. The immobilization must: (1) confine the biologically active material on the transducer and keep it from leaking out over the lifetime of the biosensor; (2) allow contact to the analyte solution; (3) allow any products to diffuse out of the immobilization layer; and, most importantly, (4) not denature the biologically active material. The last requirement is critical, since enzymes, antigens, organelles, cells, and tissues are all fragile biological materials that can easily be rendered inactive by mechanical damage, heat or freezing, chemical toxins, lack of certain chemicals, chemical modification of the wrong part of the biological material, or even changes in the conformation of the molecules.

Most of the biologically active materials used in biosensors are proteins or contain proteins in their chemical structures. The fundamental unit in protein structure, the α-amino acid, is shown in Fig. 4. The twenty naturally occurring protein units are differentiated by the functional group, R. The functional groups include hydrogen, methyl, isopropyl, and isobutyl groups, acid groups, alcohol

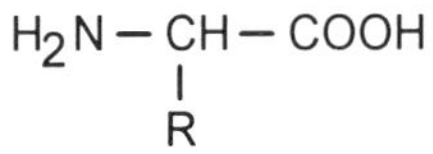

$$H_2N - CH - COOH$$
$$|$$
$$R$$

Fig. 4 The general structure of the protein moiety, for all twenty naturally occurring protein units. The group "R" distinguishes the different protein units.

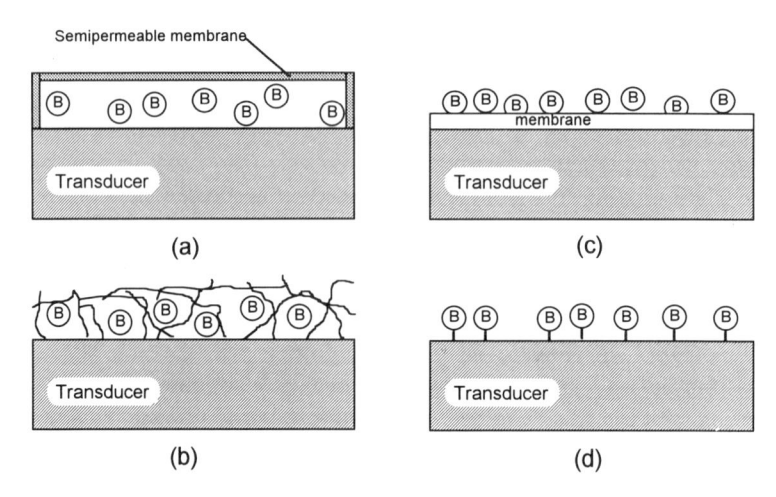

Fig. 5 The four different immobilization schemes: (a) membrane entrapment, (b) matrix entrapment, (c) physical adsorption and (d) covalent bonding. "B" stands for the biologically active material, which can include antibodies, enzymes, receptors, organelles, cells, tissues, and organisms.

groups, amine groups, aromatic rings, and thiols. The polypeptide chains are formed by linking the acid group of one base to the amine group on another and splitting off a water molecule ($-COOH + H_2N- \rightarrow -CO-NH- + H_2O$).[19] The many different functional groups that exist on the protein chain allow for many different coupling reactions.

There are two basic types of immobilization techniques that depend on the basic mechanics of the immobilization: binding or physical retention. The binding techniques involve attaching the biologically active material directly to the surface of the transducer, or the surface of a base membrane on the transducer. Adsorption and covalent binding are the two types of binding techniques. The retention techniques involve separating the biologically active material from analyte solution with a layer on the surface of the transducer, which is permeable to the analyte and any products of the recognition reaction, but not to the biologically active material. The two types of retention techniques are membrane confinement and matrix entrapment. The classifictation of the immobilization techniques is schematically summarized in Fig. 5. These four techniques will be discussed individually in the following subsections.

9.2.1 Membrane Confinement

The most straightforward physical retention immobilization technique is membrane confinement. Conceptually, the idea is simple: entrap a solution containing the biologically active material on the surface of the transducer using a semipermeable membrane.[19,20] The membrane must be chosen such that the pores are large enough to let the analyte, the products, and the solution through, but small enough to retain the biologically active material. Ultrafiler membranes, usually based on polymers of polyamide or polyether sulfon, and dialysis membranes are good candidates. The membrane must be matched to the biosensor so that it does not appreciably effect response of the transducer.

9.2.2 Matrix Entrapment

Matrix entrapment involves the formation of the porous encapsulation matrix around the biologically active material, usually enzymes, antigens, or whole cells.[19-21] This is accomplished by either crosslinking a mixture of the matrix-forming material and the biologically active material or through the formation of a gel containing the biologically active material. The entrapment matrix must have pores large enough to let the analyte solution and the products diffuse through it, while stopping the biologically active material. Typically, natural materials are the easiest to use as the matrix, since they are not toxic to the biological materials. The polymerization reactions occur under conditions that are compatible with the biological materials, and they do not produce toxic by-products. The number of synthetic polymers that can be used as the entrapment matrix is limited by the toxicity of the monomers, cross-linking agents, and by-products of the polymerization reaction, as well as the polymerization conditions.

The most widely used cross-linking agent is the bifunctional reagent, glutardialdehyde, which is usually called glutaraldehyde [$HOC-(CH_2)_3-COH$]. It can be used to cross-link biological materials, such as bovine serum albumin (BSA), collagen, and gelatin, as well as synthetic polymers, such as cellulose acetate and the methacrylates. Two popular classes of synthetic matrix polymers are the polyacrylamides and the polymethacrylates.

9.2.3 Adsorption

Adsorption is the simplest method of immobilization.[19-21] The transducer surface is exposed to a solution of the biologically active material for a period of time, then the surface is washed to remove the loosely bound material and the biosensor is ready to be used. The biologically active material is held on the surface by a combination of van der Waals forces, hydrophobic forces, hydrogen bonds, and ionic forces. This wide variety of molecular forces is because biological materials are complex, large molecules, or structures of molecules, which can have proteins, carbohydrates, phospholipids, etc., as their constituents. Therefore, different parts of the molecule or structure are attracted to the surface by different forces.

The advantage of adsorption, besides its ease of use, is that the forces that bind the biologically active material are relatively "gentle" and, thus, are not likely to distort the "active" conformation of the molecules, which can denature the material. However, some unfolding of the biologically active molecules will always occur. Co-factors and co-enzymes can be simultaneously adsorbed in their natural state of complexation.

The one disadvantage to adsorption is that the molecules are weakly bound to the transducer. A change in temperature, analyte concentration, pH, or ion concentration can desorb the biologically active material. Another operational problem is that it is difficult to quantify the amount of biologically active material that has been adsorbed on the transducer.

9.2.4 Covalent Bonding

Covalent bonding provides a more permanent binding of the biological material to the transducer surface.[19-21] The transducer's surface, or the surface of the membrane material on the transducer's surface must be treated to have reactive groups to which the bioactive material can bind. The surface treatment must have two terminal functional groups: one which binds covalently to the transducer or membrane surface and one that binds to the biologically active material. Thus, these surface treatments can be highly asymmetric molecules. For example, alkyl ethoxysilanes are widely used to bind molecules to glass, or silicon dioxide and silicon nitride gate insulators. These molecules can be prepared with a terminal functional group that can bind with the proteins, such as amine ($-NH_2$) or cyano ($-CN$) groups, as shown in Fig. 6. Table 3 gives a partial list of surface treatments for binding biological materials directly on the sensor surface.

The advantages of covalent binding are: (1) the biologically active material is directly on the surface of the transducer, reducing the response time of the biosensor since the diffusion time of the products of the reaction to the transducer is reduced, and (2) the bond to the transducer is much stronger than physical adsorption, so the sensor lifetime is longer. Major disadvantages are that covalent binding may chemically modify the important binding sites on the biological material and may also partially or totally denature the molecules. In addition, all the reagents shown in Table 3 react with amino groups, so that there is little control of the site of attachment on the protein chain, since most protein chains have many amino groups.

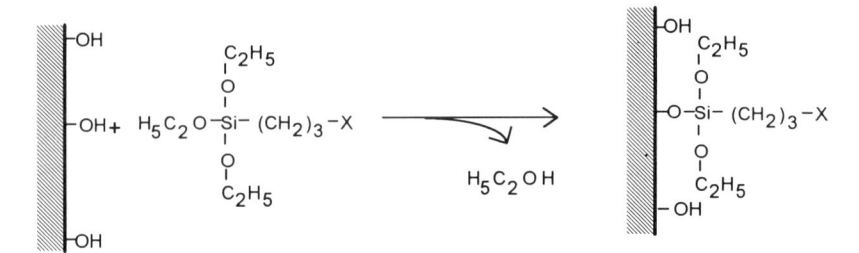

Fig. 6 Surface treatment of an hydroxide-terminated surface, such as SiO_2, for covalent protein immobilization using alkyl ethoxysilanes. The terminal group "X" can be a variety of different reactive groups, such as amino, cyano, etc.

TABLE 3 Some Examples of Surface Treatments for Covalent Binding of Biologically Active Materials on the Surrace of Transducers

Surface	Surface Treatment	Intermediate	Attachment Point	Bound Structure

The first column shows the active surface groups. The second lists the different surface treatments for the particular surface. Column three shows the intermediate structure formed before binding and the fourth column shows the specific attachment point on the biologically active molecules. Finally, the last column shows the bound structure. This data is compiled from selected references.[6,19,20]

9.3 MASS TRANSPORT IN BIOSENSORS

As seen in Fig. 1, in most biosensors the recognition reaction takes place in a membrane or molecular layer on the surface of the transducer. In fact, a biosensor is actually a bioreactor coupled to a transducer. Therefore, the transport of the analyte into the membrane, the transport of reaction products to the transducer, and the transport of the products out of the membrane all greatly affect the response characteristics of the biosensor. Even a potentiometric biosensor depends on the flow of analyte into the membrane to generate the change in local concentration that the potentiometric transducer detects. For example, in a theoretical model of an enzyme field-effect transistor (ENFET), the assumption used concerning the response characteristics of the ion selective field-effect transistor (ISFET) was that pH reaction on the ISFET's surface was infinitely fast, and, thus, the whole response curve was due to the diffusion of the substrate, products, and buffer through the enzyme immobilization membrane.[22]

This section will discuss the basics of mass transport as it applies to biosensors. A very detailed discussion is beyond the scope of this chapter, because solutions to the transport equations for specific sensor structures require a detailed knowledge of the diffusion coefficients for all species involved in the reactions and the geometry of the sensor structure. In addition, the partial differential equations usually require numerical solutions.

9.3.1 Mass Transport in Sensor Systems

The basic mass transport processes that are possible in a biosensor system are diffusion, convection, and migration. Of these three, migration of the analyte molecules under the influence of an electrical potential can be neglected because, in a typical biological sample, there are many small, very mobile counter-ions which carry the necessary charge to screen the analyte from the applied potential (this applies to an electrochemical biosensor). Thus, the motion of the analyte molecules is not significantly affected by the applied potential. Therefore, this discussion will be limited to the transport of the analyte and products to and from the transducer part of a biosensor by the diffusion through the sensor membrane and by convection in the analyte solution.

In the biosensor membrane, the flow of analyte and products is governed by diffusion, and thus can be described by Fick's laws of diffusion. The first law states that the amount of material diffusing across a surface in a given time is proportional to the concentration gradient at that surface. In one dimension, this can be represented by the following differential equation

$$j(x, t) = -D \frac{\partial C(x, t)}{\partial x} \qquad (8)$$

where j is the flux across the surface ($m^{-2} s^{-1}$). $C(x, t)$ is the concentration (m^{-3}), which is a function of both position and time, and the constant of proportionality, D, is called the diffusion coefficient (m^2/s). It is assumed that the diffusion coefficient is a constant for this example, not a function of the concentration or charge state of the molecules. This flux of material results in a change in the concentration as time progresses, and is described by Fick's second law of diffusion

$$\frac{\partial C(x, t)}{\partial t} = D \frac{\partial^2 C(x, t)}{\partial x^2}. \tag{9}$$

Since these equations are partial differential equations, they must be solved with respect to the initial and boundary conditions. The solutions are usually numerical, except for a few ideal, highly symmetric cases. Thus, it can be expected that there is a different solution for each specific sensor geometry and experimental conditions.

The effects of diffusion on the sensor response can be seen by considering the simplest case: an instantaneous, complete recognition reaction (i.e., the reaction consumes all the analyte reaching the sensor's surface) takes place at the sensor's surface in a stagnant solution. Therefore, the sensor output is proportional to the flux of anlyte arriving at the surface. Under these boundary conditions, $C(0, t) = 0$ and $C(\infty, t) = C_{bulk}$, the solution to Eq. 9, as a function of time and distance from the surface, is the error function

$$C(x, t) = C_{bulk} \, \text{erf}\left(\frac{x}{2\sqrt{Dt}}\right). \tag{10}$$

Figure 7 shows the concentration profiles as time increases. Since the sensor output is proportional to the flux of analyte arriving at the sensor's surface (the slope of the concentration profiles at $x = 0$, from Eq. 8), the response of the sensor decreases with time. This is obviously not a desirable situation.

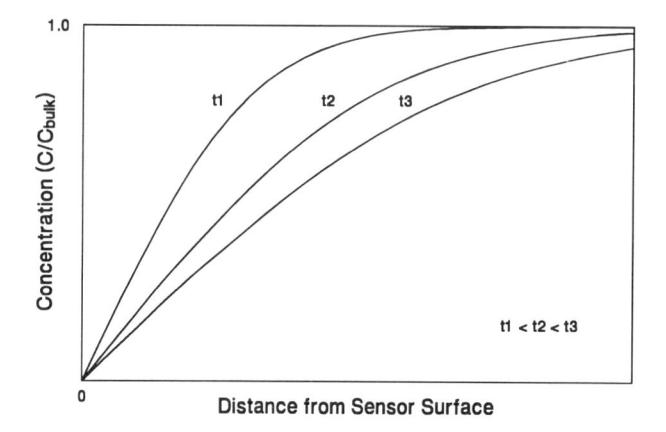

Fig. 7 Concentration profiles in a stagnant solution for a sensor that instantaneously consumes all the analyte that arrives at its surface, calculated from Eq. 10. As time progresses, the concentration in solution decreases, as does the flux of analyte at the sensor's surface, thus decreasing the sensor output.

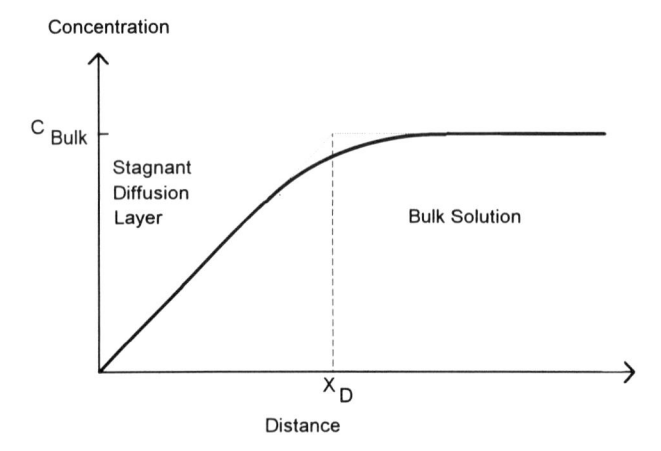

Fig. 8 The concentration profile in a well-stirred analyte solution. In the first-order approximation, the concentration can be treated as a linear concentration gradient from $x=0$ to $x=X_D$. Beyond X_D, the concentration is constant at the bulk value. This is described in Eqs. 11 and 12.

In practical biosensor systems, either the solution is well stirred, or the sensor is part of a flow injection analysis (FIA) system. The mixing action of the flow sets up a quasi-equilibrium condition by replenishing the analyte, which diffuses into the sensor's membrane to be consumed by the recognition reaction, thereby maintaining the solution buffer concentration in the membrane. A simple model of convective mass transport is to divide the fluid near the sensor's surface into layers of constant velocity. At the surface of the sensor membrane, the velocity of the fluid must be zero. The first layer of fluid out from the surface layer moves with a slight velocity, and the next layer with a slightly larger velocity and so on until the bulk solution velocity is reached. The viscosity of the solution determines the distance into the solution from the surface where the bulk velocity is reached. An exact solution to the flow problem depends on the flow rate, the surface smoothness, the viscosity of the analyte fluid, etc.[23]

The general characteristics of the concentration gradient due to the mixing are shown in Fig. 8. The first-order model of this concentration profile assumes that there is a linear concentration profile up the stagnant diffusion-layer thickness, X_D, and beyond that the concentration is at its bulk value. Thus, the flux at the sensor surface is proportional to the bulk concentration as given by

$$j = k_D(C_{\text{bulk}} - C_{\text{surface}}) \tag{11}$$

where the mass transport constant k_D is related to the diffusion coefficient and the diffusion layer thickness

$$k_D = \frac{D}{X_D}. \tag{12}$$

Thus, a linear concentration dependence can be achieved in this ideal case.[23]

9.3.2 Trans-Membrane Transport

The time response of a biosensor is mostly determined by the time response of the membrane containing the biological sensing elements. In the usual case, the analyte solution is well stirred and thus, there should be a linear dependence of the concentration at the surface of the sensor membrane and the bulk concentration. When the biosensor is placed in the analyte solution, the initial concentration of analyte in the sensor membrane is zero. The time for the sensor to go from zero to full response is determined by the diffusion of the analyte into the sensor membrane. This section will consider the effects of this diffusional transport on the time response of the biosensor.

In the membrane, only diffusion can occur, so the mass flow is described by Fick's second law, Eq. 9. Since it is a second-order partial differential equation, the solution for specific sensor structures, and their particular boundary conditions, will require numerical solutions. However, to get a qualitative feel for the effects of membrane transport on the time response of a biosensors, the problem can be simplified to one that has an analytical solution.[24] There are two types of biorecognition reactions, affinity and metabolic. This section will consider the metabolic case.

The simplest case of a metabolic sensor (e.g., an amperometric or potentiometric enzyme biosensor) is to consider the solution in one dimension under the following assumptions: (1) response of the transducer is instantaneous; (2) the transport from the bulk to the outer surface of the membrane is instantaneous, and (3) all the analyte will be consumed before it can diffuse from the outer surface of the membrane to the transducer's surface (i.e., the inner surface of the membrane). This leads to the following boundary conditions: (1) at the inner surface of the membrane the concentration of analyte is always zero; (2) the concentration at the membrane's outer surface is always the bulk concentration, and (3) the initial condition is that the concentration of analyte in the membrane is initially zero. This problem can be solved using the Fourier series method[24] which yields the following result for the concentration as a function of time and distance

$$C(x, t) = C_b \left[\frac{x}{l} + \sum_{n=1}^{\infty} \frac{2(-1)^n}{n\pi} \sin\left(n\pi \frac{x}{l}\right) \exp\left(-\frac{(n\pi)^2 Dt}{l^2}\right) \right] \quad (13)$$

where l is the membrane thickness, D is the diffusion coefficient for the analyte, and C_b is the analyte concentration in the bulk solution. It is again assumed that the transducer response is proportional to the flux that arrives at the transducer surface, $x=0$. The response can be found by applying Eq. 8 to Eq. 13, and substituting in $x=0$. If the response is normalized to the steady-state response, $j_{ss} = DC_b/l$, the result is given by

$$\frac{j(t)}{j_{ss}} = 1 + \sum_{n=1}^{\infty} 2(-1)^n \exp\left[\frac{-(n\pi)^2 Dt}{l^2}\right]. \quad (14)$$

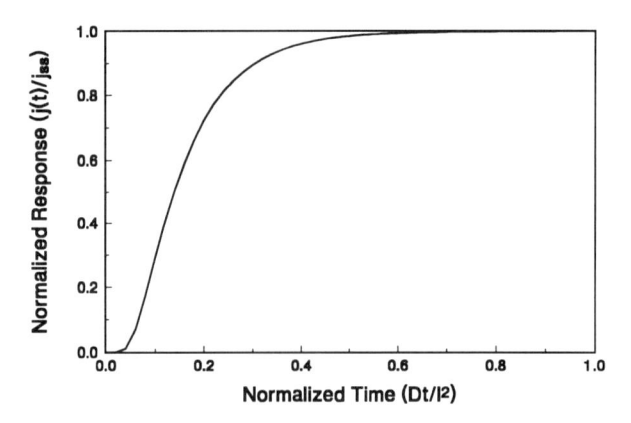

Fig. 9 Normalized biosensor time response as plotted from Eq. 14. The important features of this response curve are that there is a time lag as the analyte initially diffuses into the biological recognition membrane, and, then, the response rises to its steady-state value.

This is plotted in Fig. 9 as $j(t)/j_{ss}$ versus the dimensionless time parameter, Dt/l^2. This simple model shows some important features of the biosensor response curve. There is an initial delay with no response while the analyte diffuses through the membrane, and then the response begins to rise to its steady-state value. The normalized time is scaled by the diffusion coefficient and the membrane thickness squared. Obviously, a thinner membrane decreases the diffusion time; however, in most cases, the amount of biologically active material immobilized on the sensor is proportional to the membrane thickness. This is especially true for membrane and matrix immobilization. Thus, there is a trade-off between response time and sensitivity. This model is only valid for the diffusion-limited case and does not apply for membranes that are thinner than the diffusion-limited thickness.[24]

In general, the biosensor sensitivity curve (analyte concentration versus sensor output) will be "S" shaped. This is because, at very low analyte concentration, the response is limited by the transducer's detection or signal-to-noise limit, and at high concentrations, the biologically active material will be saturated. In between, there should be a linear region while the sensor is operating in the diffusion-limited case.

A similar simplified treatment of an affiinity sensor can also be made. The interested reader is referred to Ref. 24.

9.3.3 Membrane Loading Effects

The amount of biologically active material in the biosensor's membrane will greatly affect the sensititivy, saturation level, and time response. This section will consider the effects of the amount of bioactive material in the sensor membrane and its effect on the steady-state response. To show these effects, an enzyme-based, potentiometric biosensor will be considered. In the case of potentiometric detection, it is assumed that the output of the transducer is proportional to the concentration of the product molecules at the surface of the transducer.

The enzyme reaction rate follows the Michaelis–Menton kinetics, Eq. 3, and the concentration of the substrate and the product in the enzyme membrane are described by the diffusion–reaction equation, which is just Fick's second law with a term added to account for the consumption or production of a species[22]

$$\frac{\partial C}{\partial t} = D \frac{\partial^2 C}{\partial x^2} \pm R(C) \tag{15}$$

where $R(C)$ is the reaction term and the sign is dependent on whether the species is produced or consumed. The purpose of this model is to show the effects on the steady-state response, so the time derivative will be set to zero. Using the Michaelis–Menton kinetics, Eq. 6, as the reaction term, Eq. 15 becomes, for the substrate,

$$D_s \frac{\partial^2 [S]}{\partial x^2} - \frac{k_2[E][S]}{K_M + [S]} = 0 \tag{16}$$

and for the product,

$$D_P \frac{\partial^2 [P]}{\partial x^2} + \frac{k_2[E][S]}{K_M + [S]} = 0. \tag{17}$$

An exact solution to these two coupled partial differential equations requires numerical techniques. However, analytical solutions can be found for the two limiting cases, $[S] \ll K_M$ and $[S] \gg K_M$. The first case represents enzyme kinetics that are much faster than the transport through the membrane so the substrate concentration is the limiting factor. The second case is for very high substrate concentration that saturates the enzymes.

In the first case, $[S] \ll K_M$, Eq. 16 reduces to

$$D_s \frac{\partial^2 [S]}{\partial x^2} - \alpha [S] = 0 \tag{18}$$

where

$$\alpha = \frac{k_2[E]}{K_M D_s}. \tag{19}$$

α is the enzyme loading factor, which is the "effective concentration" of the enzyme, taking into account the concentration of enzyme in the membrane, the kinetics of the enzyme reaction, and the mass transport of the substrate molecules through the membrane. The boundary conditions for Eq. 18 are

$$x=0 \quad \frac{\partial [S]}{\partial x}=0$$
$$x=l \quad [S]=[S]_l. \tag{20}$$

The physical meaning of these boundary conditions is that at the transducer's surface, $x=0$, there is no transport flux and at the outer edge of the membrane, $x=l$, the substrate concentration is fixed at the value of the analyte solution at the membrane's outer surface. The solution for the substrate concentration is

$$[S]=\left(\frac{\cosh(x\sqrt{\alpha})}{\cosh(l\sqrt{\alpha})}\right)[S]_l \tag{21}$$

where the term, $l\sqrt{\alpha}$, includes the effects of the kinetics of the enzyme reaction, the diffusion through the membrane, and the total amount of the enzyme contained in the membrane. The term of interest is the product concentration at the transducer's surface, $x=0$. To find a relationship of product concentration from the substrate concentration, the first step is to add Eqs. 16 and 17

$$D_s \frac{\partial^2 [S]}{\partial x^2} + D_P \frac{\partial^2 [P]}{\partial x^2} = 0. \tag{22}$$

This equation can be integrated once to yield

$$D_s \frac{\partial [S]}{\partial x} + D_P \frac{\partial [P]}{\partial x} = \text{constant}. \tag{23}$$

This equation represents the diffusion fluxes in the enzyme layer. At the membrane's outer surface, $x=l$, the product and substrate fluxes must balance in the steady state, since no material is being created or destroyed. Thus, the value of the constant must be zero throughout the membrane. Integrating again results in the mass balance equation for the membrane

$$D_S[S] + D_P[P] = \text{constant}. \tag{24}$$

Substituting in Eq. 21, the product concentration becomes

$$[P] = \frac{D_S}{D_P}\left(1 - \frac{\cosh(x\sqrt{\alpha})}{\cosh(l\sqrt{\alpha})}\right)[S]_l + [P]_l. \tag{25}$$

9.3 MASS TRANSPORT IN BIOSENSORS

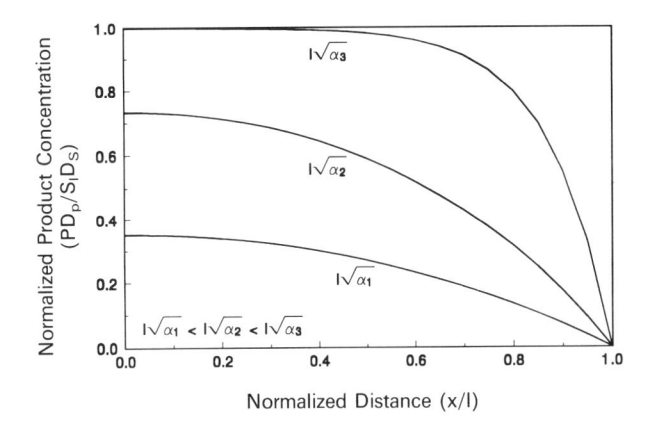

Fig. 10 The effects of the membrane loading, $l\sqrt{\alpha}$, on the product concentration at the transducer's surface, plotted from Eq. 26, with $[P]_l = 0$. ($x/l = 0$ is the transducer's surface and $x/l = 1$ is the membrane's surface.) For higher membrane loading, all of the substrate is consumed and the normalized product concentration approaches unity. As noted in the text, the membrane parameter contains terms for the thickness of the membrane, the concentration of the enzyme, the reaction kinetics, and the diffusion constants.

Finally, the surface concentration, which is proportional to the transducer's output signal in a potentiometric biosensor, is given by

$$[P]_0 = \frac{D_S}{D_P}\left(1 - \frac{1}{\cosh(l\sqrt{\alpha})}\right)[S]_l + [P]_l. \tag{26}$$

Equation 25, normalized to $D_p/(D_s[S]_l)$, is plotted in Fig. 10 for the case of zero product concentration at the membrane surface (i.e., the diffusion of product away from the surface is very rapid). As can be seen in Fig. 10, when the amount of enzyme and the kinetic rate are high enough to consume all the substrate, the transducer output will be maximal (assuming that the output of the transducer is proportional to the concentration of product at the surface of the transducer). Therefore, the "effective amount" of the enzyme, $l\sqrt{\alpha}$, should be above a certain threshold level to get the maximum amount of "chemical signal" to the chemical transducer's surface. It should be noted that this threshold "effective amount" depends not only on the enzyme concentration in the membrane, and the thickness of the membrane, but also on the kinetics of the recognition process and the transport properties of the membrane.

One thing to note from the expression for the product concentration, Eq. 26, is that there is a term for the concentration of product just outside the membrane. This term can arise from the slow diffusion of the product into the bulk solution or because the product is part of the solution. A very important example of this is of a pH electrode or a pH sensitive ISFET as the transducer for an enzyme–substrate system, such as penicillinase–penicillin. Since the hydrogen ions exist in every aqueous solution, and the pH can be influenced by parameters other than the enzyme–substrate reaction, such as the CO_2 partial pressure in the measurement environment, the use of differential measurements between two identical transducers, one with the biologically active material and one without, can be very advantageous.

In the second case, $[S] \gg K_M$, Eqs. 16 and 17 reduce to

$$D_s \frac{\partial^2 [S]}{\partial x^2} - k_2[E] = 0 \qquad (27)$$

and

$$D_P \frac{\partial^2 [P]}{\partial x^2} + k_2[E] = 0. \qquad (28)$$

These equations can be readily integrated, and both are subject to the boundary conditions of Eq. 20. The solution for the substrate concentration is

$$[S] = [S]_l + \frac{k_2[E]}{2D_S}(x^2 - l^2) \qquad (29)$$

and for the product concentration

$$[P] = [P]_l + \frac{k_2[E]}{2D_P}(l^2 - x^2). \qquad (30)$$

From these equations, it is obvious that the product concentration at the surface of the transducer, $x=0$, is a constant that depends on the immobilized enzyme concentration, the reaction kinetics, and the diffusional mass transport. Therefore, the output of the biosensor is constant. The physical meaning of this case is that the substrate concentration is so large that it has saturated the enzyme. In this case, there is an insufficient amount of enzyme in the membrane to detect changes in such a large substrate concentration.

From the analysis in this section, for a given amount of enzyme (or biologically active material) in the membrane, the output of the sensor will go from being approximately linearly dependent on the analyte concentration to a saturation value as the analyte concentration increases. In the linear region, there is a minimum amount of enzyme needed in the recognition membrane to achieve the maximum output for a given substrate concentration.

9.4 TRANSDUCTION PRINCIPLES

As shown in Table 2, the types of detection methods for biosensors include almost every type of traditional sensor. (See Chapters 3, 6, 7, and 8 for a discussion of the actual transduction principles.) The variety of detection techniques is just a reflection of the variety of biological processes. To achieve a high sensitivity and stable biosensor, the biological-recognition process must be matched to a proper sensor-detection mechanism. For example, if the biological-recognition process is a chemical reaction, such as an enzyme–substrate system, then the detection method must measure the concentration of a product or a reactant, or measure the heat reaction that has taken place; not, for example, the amount of adsorbed mass on the surface of the sensor, which would be much less selective and sensitive.

There is no single detection technique that is ideal for all the possible biological-recognition processes. For example the ISFET, a potentiometric sensor, is a good transducer for the enzyme–substrate reaction but not for the antibody–antigen adsorption recognition.[25] In the ENFET, the ISFET detects the potentiometric response of the ion-sensitive membrane to either the concentration change in one of the products or reactants in the enzyme catalyzed reaction. Since ion-sensitive membranes usually produce a large response, the ENFET is a successful biosensor. In the immunochemical field-effect transistor (IMFET), the adsorption of an antigen on the antibody immobilized on the gate of the FET is supposed to be detected by the change in the double-layer charge at the interface. However, there are finite leakage paths through this interface and small inorganic counter-ions in the sample solutions that mask the measurement of the antibody–antigen charge layer.[26,27] Therefore, the IMFET is not practical as a steady-state measurement device, but it is possible to use the IMFET to measure the transient change in charge upon antibody–antigen binding.

This section will discuss some of the major types of biosensor detection methods and give some examples.

9.4.1 Electrochemical

Electrochemical detection techniques are well suited for biosensor systems that depend on chemical reactions for biological recognition. (See Chapter 8 for a discussion of electrochemical transduction.) Since the electrochemical techniques measure the result of the biological recognition in terms of electrical parameters, they can be directly interfaced to an electronic computational system for further signal processing and possible use for the control of biological systems. The three basic electrochemical schemes are potentiometric, amperometric and impedometric detection, corresponding to the three basic electrical output variables. These will be individually discussed in the following three subsections.

Potentiometric. Potentiometric detection measures the potential across an electrochemical cell containing the biological sensing element. Usually, either the activity of a product or a reactant in the recognition reaction is monitored. The potential at the half cell containing the working electrode and the biological sensing element (i.e., the biosensor) is measured against a standard reference electrode. The reference electrode is chosen to maintain a standard fixed potential with respect to the solution over the operating range of the sensor.

In an ideal potentiometric measurement, the current through the electrochemical cell is zero and the electrodes are in thermodynamic equilibrium with the solution. Thus, the potential difference between the electrodes is a thermodynamic variable in the equations of state of the system. In practice, the measurement current through the cell is made small enough so that, within the time scale of the experiment, the system can be considered in thermodynamic equilibrium.

To understand the potentiometric response of an electrochemical cell, consider an electrochemical system which is represented by the following chemical equation

$$\text{Ox} + ne \leftrightarrow \text{Re} \tag{31}$$

where Ox is the oxidized species and Re the reduces species, e is the charge on an electron, and n the number of electrons in the reaction. The potential across the cell is given by the Nernst equation[28]

$$V = V^0 + \frac{RT}{nF} \ln\left(\frac{a_{\text{Ox}}}{a_{\text{Re}}}\right) \tag{32}$$

where V^0 is the potential at the standard state of the system, R is the gas constant, T is the absolute temperature, F is the Faraday constant, which is the charge on a mole of electrons (9.647×10^4 C), and a_{Ox} and a_{Re} are the activities of the oxidized and reduces species, respectively. The activities are usually assumed to be the concentrations in dilute solutions, given by [Ox] and [Re]. The formal potential is the equilibrium potential of the system (when $a_{\text{Ox}} = a_{\text{Re}}$). For a pH electrode or pH ISFET, the ideal, Nernstian response is 29.2 mV/pH at 25°C.

The potential across the cell is proportional to the Gibbs free energy of the system

$$\Delta G = -nFV_{rxn} \tag{33}$$

where V_{rxn} is the potential across the cell from right to left in Eq. 31. The Gibbs free energy is the maximum work attainable from the electrochemical cell.

For the cell to be in equilibrium, the electrochemical potentials of the charge carriers in the two phases, electrode and solution, must be equal. This means that the electrochemical potential (Fermi level) of the electrons in the electrode is the same as the electrochemical potential of the ions that carry the charges in the electrolyte solution.

For an ion-selective electrode (i.e., an electrochemical electrode with an ion-selective membrane on it) the potential response is given by the Nickolsky equation,[29] which is a modified version of the Nernst equation, Eq. 32, taking into account any intefering ions

$$V = \text{constant} + \frac{RT}{nF} \ln[a_i + k_{ij}(a_j)^{n/m}] \qquad (34)$$

where the subscript *i* refers to the analyte and the subscript *j* to the interfering species; *m* is the number of electrons involved in the interfering reaction; k_{ij} is the selectivity coefficient; *n* is the number of electrons in the recognition reaction and the *a*'s are the activities of the different species. Again, the activites can be replaced by the concentrations. The constant includes the standard potential of the electrode plus any membrane potential.

One consequence of zero measurement current, or as small as is practical, is that the impedance of the system is very high. Usually the electrodes are small with very small leakage currents. Therefore, the input amplifier must have a very high input impedance so that it will not draw any significant current from the cell. These high impedances cause a high noise susceptibility for these systems. Hence, care must be exercised in constructing the electrodes, designing the measuring circuit, and performing the measurements. The use of guarded and shielded cables on the cables to the working electrode might be necessary to increase the signal-to-noise ratio of the measurement.

Typical potentiometric biosensors use ion-selective electrodes (ISEs) or ISFETs. One advantage of an ISFET over an ISE is that, being an insulated gate field-effect transistor (IGFET), it performs an impedance transformation on the sensor chip. A FET is a transconductance amplifier, that is, it transforms an input voltage (high input impedance) to an output current (low impedance). Therefore, the signal leaving ISFET is much less susceptible to noise.

For ENFET measurements, the local pH of the membrane is the variable of interest. If a single ENFET is used, the solution pH, temperature, and ionic strength must be kept constant for the output of the ENFET to be a function of only the analyte concentration. To lessen these environmental interferences, a differential measurement between the ENFET and an identical device without the enzyme can be used (which is just an ISFET). Ideally, this ISFET should have the same response to the environment as the ENFET, except for the enzyme–substrate interaction.

Potentiometric biosensors have been fabricated using both ISEs and potentiometric gas sensors. ISE biosensor fabrication is simple, since high-quality ISEs are available commercially and conventional membrane immobilization techniques exist which allow for easy fabrication, as shown in

440 BIOSENSORS

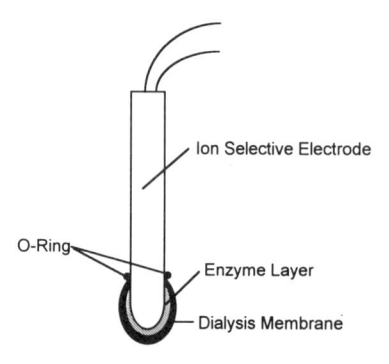

Fig. 11 A schematic of an enzyme electrode made from an ion-selective electrode. The enzyme layer, which can be adsorbed, covalently bonded, or entrapped in a matrix, is contained by a dialysis membrane on the end of the electrode. The membrane is held in place with a rubber O-ring.

Fig. 11. Gas-sensitive potentiometric electrodes can also be used if a hydrophobic gas-permeable membrane is inserted between the biologically active layer and the gas sensor surface. Some examples of potentiometric biosensors include: ammonia gas and ammonium ISE have been used for amino acids and urea, glucose sensors have been fabricated using both pH- and iodine-selective electrodes, and pencillin electrodes have been made using pH electrodes.[30,31] In general, the detection range of potentiometric biosensors is on the order of 10^{-5} M to about 10^{-1} M, with stability on the order of weeks to a few months. (The unit, M, is the abbreviation for Molar, which is defined as moles per liter.)

ENFETs based on pH ISFETs have been fabricated for glucose, penicillin, and urea.[32,33] Enzyme immobilization is much more difficult for ENFETs than for ISEs. Membrane entrapment is difficult with ENFETs, because the ISFET chips are planar structures. Therefore, matrix entrapment, covalent bonding to a membrane, or physical adsorption on a membrane are usually used. To make differential measurements against an identical ISFET on the same chip, structures on the chip must be formed so that both the ISFET and the ENFET have the same immobilization membrane, but only the ENFET's membrane contains the active enzyme. This can be accomplished either by depositing the enzyme membrane on the whole sensor chip and then denaturing the enzyme by irradiation or by selectively depositing the enzyme only on the ENFET's membrane. To accomplish selective deposition, wells have to be formed on the chip's surface over the active gate areas of the FETs, using photolithography or by structures created during the packaging of the sensor.

ENFET sensitivities and lifetimes are equivalent to the ISE-based enzyme biosensors, since good ISFETs have near Nernstian responses and the lifetime is dependent on the enzyme and immobilization procedure. In ENFET measurements, the pH buffer concentration of the analyte solution affects the dynamic range, sensitivity, and linearity of ENFETs.[22,34] Physically, what occurs is that, if the buffer concentration is too high, it will buffer the hydrogen ions that are produced by the enzyme-catalyzed reaction, thus reducing the output signal. The buffer is necessary because, as mentioned above, the catalytic action of the enzyme is maximum only in a small region of pH values. Therefore, a trade-off must be made.

Amperometric. Amperometric biosensors measure the concentration-dependent current through an electrochemical electrode, which is coated with the biologically active material. Amperometric transduction is based on oxidation or reduction of an electroactive species at an electrode surface. (See Chapter 8 for details of electrode reactions.) The way in which amperometric biosensors operate is that the biologically active material, usually an enzyme, oxidizes or reduces the analyte (the enzyme's substrate). This reaction is electrochemically connected to the electrode either directly, through mediator molecules, or by the oxidation or reduction of co-reactant molecules or product molecules.

The electrochemical electrodes are usually made of platinum, though gold and different forms of carbon are also used.[35] Recently, conducting organic salts have been studied because of their enhanced electron transfer to enzymes.[36]† All four types of immobilization are used with amperometric biosensors. However, in most cases, there is a protective membrane on the surface of the immobilization layer to keep other proteins and other large biological molecules out of the biologically active layer and as a permselective layer to help eliminate interfering species. At an electrochemical electrode, any species for which the applied potential is favorable for oxidation or reduction will do so and interfere with the analyte signal. This membrane can be permselective or even be negatively charged to exclude anions in the analyte solution.

The amperometric biosensor can be classed into three generations.[37] In the first generation, the transduction of the biological reaction is by the oxidation or reduction, at the electrode surface, of an electroactive product or reactant of the biological-recognition reaction. This is shown in the reaction schematic in Fig. 12a. First-generation sensors limit the usable biological-recognition reactions to those that produce or consume electroactive molecules. In general, biological molecules are not electroactive, so the transduction is limited to the small molecular species in biological fluids, such as oxygen, hydrogen and hydrogen peroxide. An example of a first-generation amperometric biosensor is Clark's use of an oxygen electrode to measure the oxygen consumed when glucose oxidase catalyzed the reaction of glucose and oxygen to form gluconolactone and hydrogen perioxide.[2]

The second-generation amperometric biosensors use mediator molecules to transfer the electrons from the enzyme, after it reduces or oxidizes the substrate, to the electrode. Figure 12b schematically illustrates the reaction kinetics. These

---

†In general, enzymes are not electroactive molecules; however, some enzymes contain a flavin redox center, such as glucose oxidase, or require a coenzyme which has the redox center.

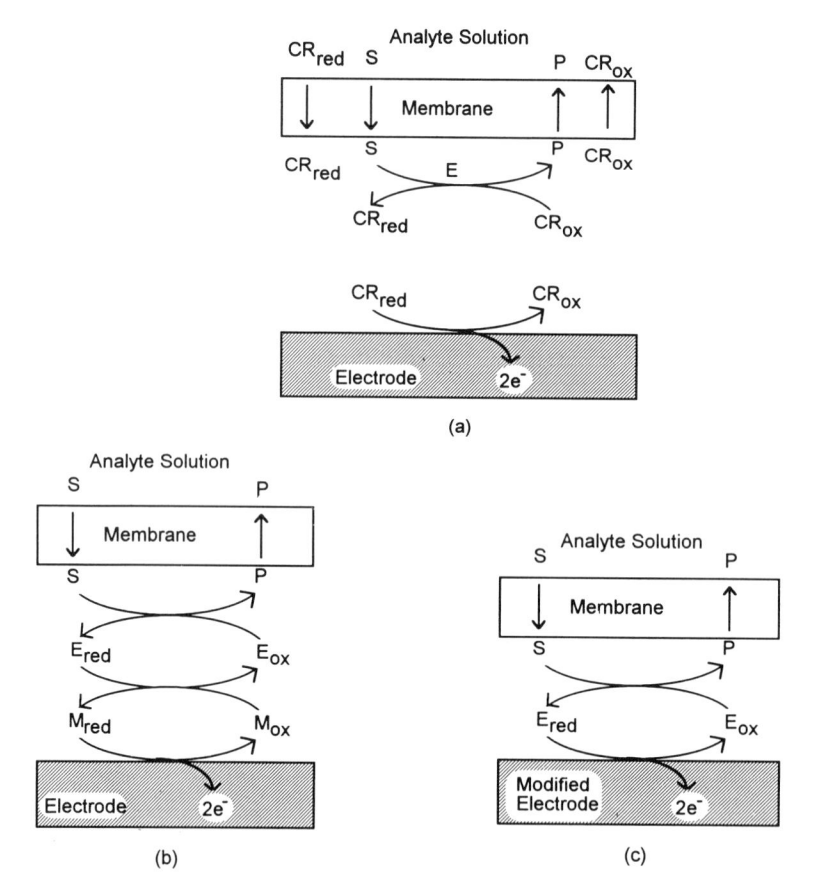

Fig. 12 Schematic of the three generations of enzyme electrodes. E is the enzyme, S is the substrate, P is the product, M is the mediator and CR is the co-reactant. The subscripts "ox" and "red" signify the oxidized and reduced state of the particular molecule. (a) shows the first-generation electrode; (b) shows the second-generation, or mediate-enzyme electrode; and (c) shows the third-generation, or modified electrode. In the first generation, the enzyme-catalyzed reaction is detected by the amperometric measurement of the concentration of the co-reactants. In the second generation, the mediator molecule links the enzyme reaction to the electrode, electrochemically. In the third generation, the modified electrode directly regenerates the enzymes.

mediator molecules should have the following properties:[38]

1. rapidly react with the reduced or oxidized enzyme,
2. have reversible heterogeneous kinetics,
3. the mediator should regenerate at the electrode at low overpotential, which is pH independent,
4. be stable in both the oxidized and reduced forms,

5. be non-toxic,
6. not react with oxygen.

The most widely used mediator is ferrocene (η^5-bis-cyclopentadienyl iron), which has been used with many enzymes, including glucose oxidase, lactate oxidase, pyruvate oxidase.[38] Some other mediators are hexacyanoferrate, N-methylphenazinium (NMP), quinones, and organic dyes.[38,39]

The third generation is the modified-electrode amperometric biosensors. In these sensors, the surface of the electrode is modified by the addition of molecules which allow the direct oxidation or reduction of the enzyme at the electrode. This is illustrated in Fig. 12c. The conducting organic salt, NMP$^+$TCNQ$^-$ (7,7,8,8-tetracyano-p-quinodimethane) has proven to oxidize glucose oxidase directly.[40] Third-generation electrodes are easier to fabricate and use than the second generation, because the mediator is eliminated.

A pseudo-second-generation class of amperometric biosensors are the β-nicotinamide adenine dinucleotide (NAD) co-enzyme biosensors. Over 250 dehydrogenase enzymes use NADH or the related β-nicotinamide adenine dinucleotide phosphate (NADP) as co-enzymes.[40] Thus, there is the potential to design many different biosensors based on the same structure and chemistry by only changing the dehydrogenase enzyme.[40] When in its oxidized state, NAD$^+$, it binds with the dehydrogenase enzyme to form a holoenzyme. The holoenzyme binds to the substrate molecule and catalyzes the reaction to form the product. (In the case of a dehydrogenase-NAD$^+$ holoenzyme, the reaction is the removal of a hydride ion from the substrate.) During the catalysis, the NAD$^+$ is reduced to NADH. After the catalysis the NADH is freed from the enzyme. The NADH must be regenerated to NAD$^+$ at the electrode for continued sensor operation. Since it is difficult to regenerate the NAD$^+$ at a bare metal electrode (1.1 V versus a standard calomel electrode for a platinum electrode), modified electrodes should be used.[39]

The models of the performance of amperometric biosensors is based on: (1) the reaction kinetics of all the electrochemical reactions for an electron to go from the substrate molecule to the electrode, and (2) the mass transport of all the species involved. A complete treatment requires detailed knowledge of the electrochemistry of all the reactions and the mass transport parameters. In keeping with the philosophy of this chapter, a simplified treatment of one case will be given as an example. Complete solutions can be found in Ref. 37.

Consider the case of the mediated enzyme electrode. The Michaelis–Menton kinetics, Eqs. 3 and 6, can be modified to the following form

$$S + E_O \underset{k_{-1}}{\overset{k_1}{\leftrightarrow}} E_O S \tag{35a}$$

and

$$E_O S \overset{k_2}{\to} P + E_R \tag{35b}$$

where E_O and E_R are the oxidized and reduced forms of the enzyme. The mediator reaction for the oxidation of the enzyme is

$$E_R + M_O \xrightarrow{k_e} E_O + M_R \qquad (36a)$$

and the reaction for the oxidation of the mediator at the electrode is

$$M_R \to M_O + e^- \qquad (36b)$$

where, again, M_O and M_R are the oxidized and reduced forms of the mediator. In this derivation, it is assumed that (1) the mediator oxidation at the electrode surface is very rapid, thus $[M_O] = [M]$, the total amount of added mediator, and (2) the enzyme layer is thin, so that the concentrations are uniform throughout the layer. The fluxes of the different species are: (i) for the membrane transport

$$j = k_D([S]_{\text{bulk}} - [S]) \qquad (37a)$$

(ii) for the formation of the enzyme–substrate complex

$$j = l(k_1[E_O] - k_{-1}[E_O S]) \qquad (37b)$$

(iii) for decomposition of the complex into the product and reduced enzyme

$$j = lk_2[E_O S] \qquad (37c)$$

and (iv) for the regeneration of the enzyme by the mediator

$$j = lk_e[E_R][M] \qquad (37d)$$

where k_D is the membrane mass transport rate constant and l is the enzyme layer thickness. Under steady-state conditions, all of the fluxes are equal and the current is proportional to the flux, $i = nFj$, where n is the number of electrons in the mediator electrode reaction and F is the Faraday constant. If the enzyme concentration is fixed, $[E_{\text{tot}}]$, the following expression for the flux can be found[24]

$$\frac{1}{j} = \frac{K_M}{lk_2[E_{\text{tot}}]\left([S]_{\text{bluk}} - \dfrac{j}{k_D}\right)} + \frac{1}{lk_2[E_{\text{tot}}]} + \frac{1}{lk_e[M][E_{\text{tot}}]}. \qquad (38)$$

The first term in Eq. 38 includes the effects of the Michaelis–Menton kinetics and the mass transport across the membrane. The second term is for the decomposition of the enzyme into the product and reduced enzyme. The last term describes the regeneration of the enzyme by the mediator. Any of these three terms can be the rate-limiting step.

As usual, this flux can be considered in two limiting cases. In the first case, assume that the decomposition of the enzyme–substrate complex and the oxidation of the reduced enzyme by the mediator are very rapid, and thus, the last two terms can be dropped. After rearrangement, Eq. 38 becomes

$$\frac{1}{j} = \frac{K_M}{lk_2[E_{tot}][S]_{bulk}} + \frac{1}{k_D[S]_{bulk}}. \tag{39}$$

In this case, the flux is proportional to the substrate concentration. The first term is due to the enzyme kinetics and the second is due to mass transport through the membrane. To make repeatable sensors that do not depend on the enzyme concentration or kinetics, the sensor should be used in the membrane diffusion limited case, where the first term can be neglected and the flux is proportional to the bulk substrate concentration.

In the second case, the diffusional flux is greater than the conversion of enzyme to product or the enzyme regeneration by the mediator. There will be a significant concentration of substrate in the enzyme layer and, in the limit, it can be assumed that $[S]_{bulk} \gg j/k_D$ and the concentration of substrate in the enzyme layer will be the bulk value. Equation 38 then reduces to

$$\frac{1}{j} = \frac{K_M}{lk_2[E_{tot}][S]_{bulk}} + \frac{1}{lk_s[E_{tot}]} + \frac{1}{lk_e[M][E_{tot}]}. \tag{40}$$

Here, the flux is very dependent on the concentration of the enzymes and mediator in the enzyme layer. In the reaction-limited operation, Eq. 40 can be used to determine rates of each step by varying one variable at a time, enzyme concentration, mediator concentration and the bulk substrate concentration.

The number of different enzymes used in amperometric biosensor is very large.[39] Some of the common substrates include glucose, urea, cholesterol, fructose, sucrose, and ethanol. Amperometric detection has also been used with tissues, and whole organisms.[41] In general, the detection limits of amperometric biosensors is about 10^{-9} to 10^{-8} M, with a linear range from 10^{-7} to about 10^{-3} M. The lifetimes of the amperometric biosensor depend on the biologically active material and the immobilization.

Impedometric. Impedometric detection measures the frequency response of the small-signal impedance of the electrochemical electrode and biological sensing element. Synchronous detection allows very high signal-to-noise ratios and thus, very small signals can be measured. However, this method is usually not used for working sensors, due to the large amount of data analysis necessary to yield results, but as a tool to investigate the structure and processes occurring at the electrode and in the membrane covering it.

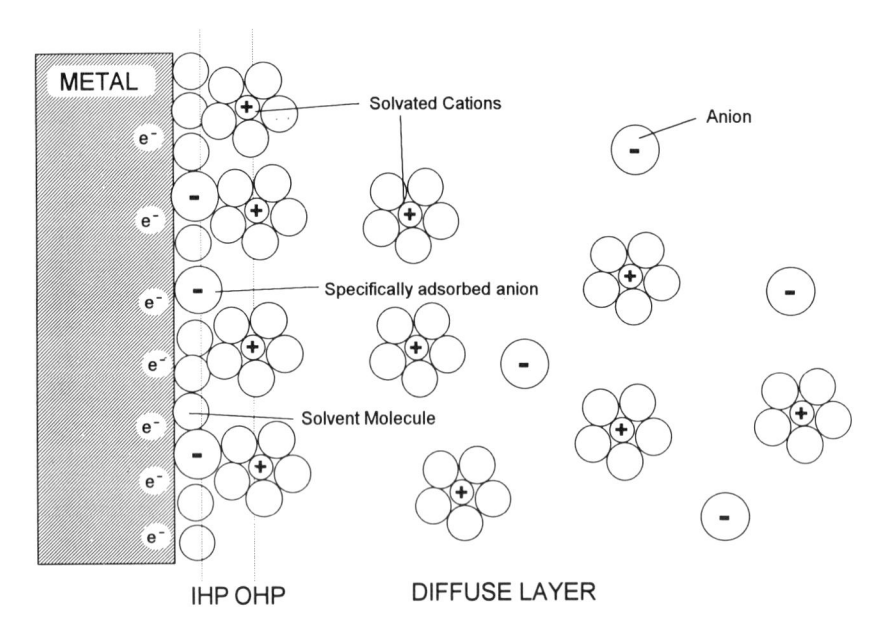

Fig. 13 Model of the double-layer region at the electrode-electrolyte interface, showing the inner Helmholtz plane (IHP), the outer Helmholtz plane (OHP) and the diffuse layer.

A bare electrochemical electrode can be modeled with passive circuit elements which can be related to the physical processes occurring at the electrode. There is a charge transfer resistance, R_{ct}, associated with the current flow from the electrode to the solution. Since there is a potential difference across this interface, there exists a charged double layer (i.e., an interfacial capacitance). On the metal side, there exists a layer of charge on the surface of the metal since the electric field inside a conductor must be zero. Because the system is in the steady state, an equal and opposite amount of charge must exist in the solution near the interface. This charge in the solution can be divided into two regions. There are specifically adsorbed ions on the surface of the electrode, the locus of whose centers form the inner Helmholtz plane (IHP), as shown in Fig. 13. There are also solvated ions nonspecifically adsorbed at the interface. Due to the size of the solvent molecules, they are held near the electrode by the long-range electrostatic force. These ions are distributed near the interface because of thermal agitation of the solution. This layer is called the diffuse layer. The locus of the centers of the nearest solvated ion is called the outer Hemholtz plane (OHP). The total capacitance between the metal and the solution (IHP and the diffuse layer) is called the double-layer capacitance. Figure 13 gives a schematic picture of the double-layer region.

The usual equivalent circuit of an electrochemical electrode is the double-layer capacitance, C_d, in parallel with the series combination of the charge transfer resistance, R_{ct}, and the Warburg impedance, all in series with

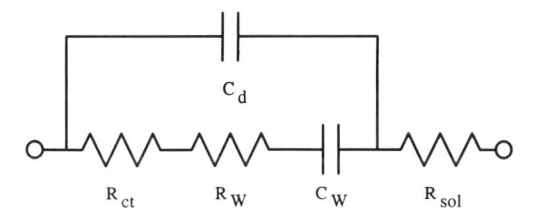

Fig. 14 Equivalent circuit for an electrochemical electrode. C_d is the double-layer capacitance, R_{ct} is the charge-transfer resistance, R_W and C_W are the Warburg impedance, and R_{sol} is the solution resistance.

a solution resistance, R_{sol}, as shown in Fig. 14. The Warburg impedance[28] is usually represented by a resistor, R_W, and a capacitor, C_W, in series. It represents the resistance to mass transfer. In a biosensor, circuit elements can be added for membrane capacitance and resistance for mass transfer through the membrane, or these can be lumped into the existing circuit elements.

Some enzyme reactions are linked by conductivity and capacitance changes in the solution and can be detected impedometrically, for example, the urea–urease reaction.[42] Changes in the cell concentration in a culture broth can also be related to a change in conductivity of the broth.[42] Although impedometric detection shows promise, it has yet to be exploited.

9.4.2 Optical

Optical detection involves measuring the change in optical absorption or emission of a molecule or molecules at a particular region of the spectrum. Infrared (IR) and ultraviolet–visible (UV–VIS) spectroscopy are classical examples of absorption detection, and X-ray fluorescence spectroscopy is an example of emission. However, these instruments are much too large to be considered practical sensors and require complicated data analysis. The basic operating principles of optical transducers are discussed in Chapter 6.

There are two schemes for optical detection, the direct method and the indirect (or competitive-binding) method.[43] In the direct method, when the analyte binds to the receptor, a change in optical absorption or fluorescence occurs. The direct method follows binding reaction kinetics similar to the antibody–antigen binding discussed in Section 9.1.4. For an indicating reagent, R, and an analyte, A, the following reaction applies

$$A + R \leftrightarrow A \cdot R \qquad K_R = \frac{[A \cdot R]}{[A][R]}. \tag{41}$$

The same analysis as in the antibody–antigen applies. Given that the total amount of reagent in the sensors is fixed, $[R]_{tot} = [R]_{eq} + [A \cdot R]_{eq}$, the equilibrium concentration of the bound analyte–reagent is given by

$$[A \cdot R]_{eq} = \left(\frac{[A]_{eq}}{(1/K_R) + [A]_{eq}} \right)[R]_{tot} \qquad (42a)$$

and the equilibrium free-reagent concentration is given by

$$[R]_{eq} = \left(\frac{1}{(1/K_R) + [A]_{eq}} \right)[R]_{tot}. \qquad (42b)$$

Depending on the chemical properties of the indicator, either the free reagent or the bound analyte–reagent complex can be detected by the optical sensor. The optical signal is proportional to the number of molecules, not the concentration, so the geometry of the sensing region is important.

In the indirect, or competitive-binding, method there are two competing reactions occurring, the analyte–receptor reaction and an analog–receptor reaction.[43] The analog molecules are fluorescently-labeled molecules that are similar in structure to the analyte molecules. Thus, the analog molecules will bind to the receptors, though not as strongly as the analyte molecules. The optical change, which is measured, is in the analog–receptor reaction. Figure 15 shows the competitive-binding scheme. Since there are fixed concentrations of receptor and analog immobilized in the chamber on the tip of the fiber, as

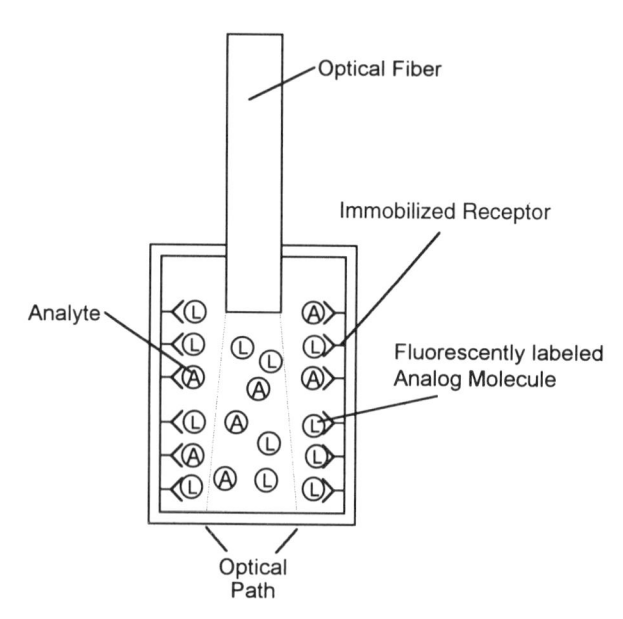

Fig. 15 The competitive-binding scheme for optical biosensors. The analyte molecules are indicated by A and the fluorescently labeled analog molecules are indicated by L.

the concentration of the analyte increases, the amount of analog bound to the receptors decreases, and thus the optical signal will change. In the example of Fig. 15, the antibody is immobilized out of the optical path, and as the analyte molecules, A, displace the labeled analog molecules, L, the number of analog molecules in the optical path increases, changing the optical signal.

Competitive-binding sensors have two competing reactions occurring in the sensing region: (1) the analyte–receptor reaction described by Eq. 41, and (2) the analog–receptor reaction, which is described by

$$L+R \leftrightarrow L \cdot R \qquad K_L = \frac{[L \cdot R]}{[L][R]} \tag{43}$$

where the analog molecules and the receptor molecules are represented by L, and R, respectively. As before, the total number of receptors is fixed, $[R]_{tot} = [A \cdot R]_{eq} + [L \cdot R]_{eq} + [R]_{eq}$, and the total amount of analog is also fixed, $[L]_{tot} = [L \cdot R]_{eq} + [L]_{eq}$. Solving these equations for $[A]$ yields

$$[A] = \frac{[A \cdot R][L]}{[L \cdot R]} \frac{K_L}{K_R}. \tag{44}$$

If the equilibrium constants are large, $[R] \ll [L \cdot R], [A \cdot R]$, then the $[R]$ can be dropped from the equation for the total receptor and the substitution for $[A \cdot R]$ can be made in Eq. 44 to yield

$$[A] = \frac{([R]_{tot} - [L \cdot R])[L]}{[L \cdot R]} \frac{K_L}{K_R}. \tag{45}$$

Thus, the analyte concentration can be determined from the optical detection of either the analog, L, or the bound analog–receptor complex, $L \cdot R$. This is a nonlinear relationship; however, it does allow the detection of analyte molecules that do not produce optical changes directly.

An example of the competitive-binding sensor is a glucose sensor using concanavalin-A immobilized on a dialysis tube covering the end of the optical fiber as the receptor, and the high molecular weight, fluorescently labeled sugar, FTIC dextrin, as the analog.[44] The concanavalin-A is immobilized on the wall of the dialysis tube, out of the line of sight of the optical fiber. As glucose diffuses through the dialysis tube, it displaces some of the FTIC dextrin, which was bound to the concanavalin-A, and, thus, the number of fluorescent FTIC molecules in the chamber increases, increasing the optical signal.

Optical sensors offer some advantages over electrochemical methods. First, no reference electrode is required; however, a reference intensity is necessary to minimize environmental effects on the system. Second, fiber-optic sensors are immune to electrical noise, but ambient light can be a problem. Finally, multicomponent measurements can be made by including different dyes and measuring the optical signal at different wavelengths in the same fiber. However, the linearity is usually limited to a very narrow range, usually two orders of magnitude in concentration, as compared to electrochemical or thermal detection.[45]

Fiber Optics. Fiber optics has revolutionized the field of optical sensors. Fiber sensors are called "optodes" in a manner analogous to "electrodes".[44] The fibers are a very convenient way to bring the light to and from the analyte. There are two methods of fabrication, the sensing dye or fluorophore can be immobilized on the tip of the fiber in a membrane, or the cladding can be stripped from the length of the fiber and the sensing dye or flurophore can be immobilized on the surface of the core where it is probed by the evanescent field.[44] This is shown schematically in Fig. 16.

The optical properties of the fibers limit the usable wavelengths. Plastic fibers, which are the least expensive and safest, must be used above 450 nm, glass fibers can go down to 380 nm, and fused silica, the most expensive, can extend the range down to the UV. The optical source also limits the usable wavelengths. Incandescent source can be used down to about 380 nm, but to go lower it is necessary to use arc lamps, which further increase the cost of the sensor.[44] Lasers can also be used as the optical source allowing only a specific transition in the molecule to be excited, and, hence, only specific wavelength emissions. These wavelength limitations also limit which dyes and fluorophores can be utilized.[44]

Semiconductor photonic devices provide optical sources and detectors that

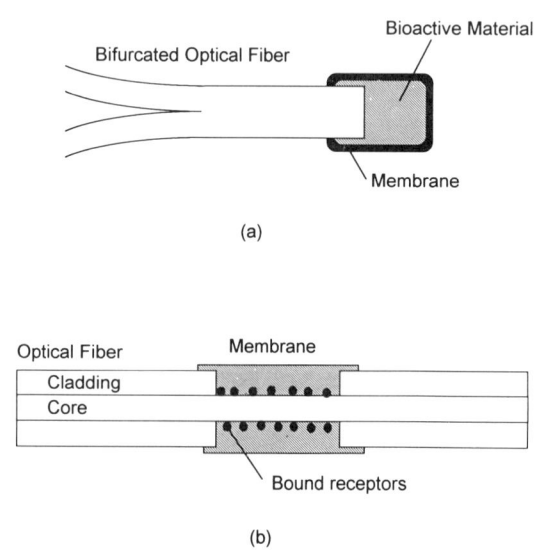

Fig. 16 The two modes of optical-fiber-based biosensors: (a) direct measurement and (b) evanescent wave measurement.

are inexpensive, small, and require low power. They allow the fabrication of inexpensive, portable equipment for clinical and field use. Light-emitting diodes (LEDs) offer illumination sources in the infrared, red, amber, green, and blue spectral ranges. For narrower spectral ranges, laser diodes can give monochromatic light. Photoconductors, photodiodes and phototransistors are small, compact, low-power detectors; although, they do not have the sensitivity of photomultiplier tubes, which can detect individual photons.

Fluorophore- and Chromophore-Based Biosensors. The two most widely used detection methods for optical biosensors are based on either a change in optical absorbance or a change in fluoroscence caused by a change in the concentration of a particular molecular species in the sensing region of the optical sensor. The chromophore and fluorophore molecules show a change in the absorbance or fluorescence, respectively, at a particular wavelength due to a change in the oxidation state of the molecule. Thus, chemical sensors can be constructed for the usual chemical parameters, such as pH, oxygen, and ammonia, which then can be coupled to enzyme reactions.[46] Additionally, some biologically active molecules, such as β-nicotinamide adenine dinucleotide (NADH) and glucose oxidase show changes in their fluorescence upon oxidation (NADH) or binding to glucose (glucose oxidase).[47]

Bioluminescence- and Chemiluminescence-Based Biosensors. The use of bioluminescent and chemiluminescent molecules in biosensors is a new exciting area. These reactions are the most sensitive "chemical" probes known, with detection limits in the picomolar ranges.[48] The simplest accepted mechanism for the production of light from a biological reaction is that of the crustacean, Cypridina hilgendorfi

$$\text{Luciferin} + \text{Luciferase} \xrightarrow[H_2O]{O_2} \text{Oxyluciferin}^*$$

$$\text{Oxyluciferin}^* \longrightarrow \text{Oxyluciferin} + \text{light}$$

(46)

where the luciferin is the molecule that emits light and the luciferase is the enzyme that oxidizes the luciferin to its excited state, which is denoted by the asterisk (*). The typical bioluminescence reaction is much more complicated, involving multiple enzyme-catalyzed reactions and enzyme co-factor, such as NADH.[48] The biochemistry is beyond the scope of the chapter, but the potential of picomolar detection makes the topic worth mentioning.

9.4.3 Thermal

Thermal detection measures the enthalpy of the detected reaction. (See Chapter 7 for details of the operation of the thermal transducers.) The transducers are usually thermistors or thermocouples in very carefully insulated and thermostated calorimeters. The enzymes or other biological-recognition elements are immobilized on supports, such as glass beads, in a column. The analyte solution flows through a heat exchanger to set its initial temperature, and the flows through the column. The temperature sensor measures the temperature of the solution flowing out of the column. The peak signal is proportional to the concentration of the analyte. The change in temperature that can be detected in a well-designed system is about 0.0001°C, but friction and turbulence in the column limit it to about 0.01°C.[49] Enzyme thermistors, where the enzyme was immobilized directly on a thermistor, were tried, but too much reaction heat was lost to the solution for adequate detection.

The total heat evolved in a chemical reaction is proportional to the molar enthalpy of the reaction

$$Q = -n\Delta H \qquad (47)$$

where Q is the total heat evolved in the reaction, n is the number of moles of the product formed, and ΔH is the molar enthalpy change.[50] The temperature change of the system, including the solution, is given by

$$\Delta T = \frac{Q}{C_s} = -\frac{n\Delta H}{C_s} \qquad (48)$$

where C_s is the heat capacity of system.

Enzyme reactions usually evolve a relatively large amount of heat and, thus, are well suited for calorimetry. For example, the urea–urease substrate–enzyme pair yields 61 kJ/mol, glucose–glucose oxidase yields 80 kJ/mol and penicillin-G–penicillinase yields 67 kJ/mol.[50] Typical detection ranges are 5×10^{-5} to 5×10^{-1} M for enzyme catalyzed reactions.

Some of the problems associated with enzyme calorimetry are: (1) the size of the sensor calorimeter; (2) the susceptibility of the transducer to environmental temperature variations, and (3) interfering chemical reactions and processes, such as mixing and adsorption, which give off or absorb heat. The one advantage of this detection method is the wide range of linear response, about five orders of magnitude.

9.4.4 Mass

Mass detection of analytes depends on the change in resonant frequency of a mechanical resonator when the analyte molecules are adsorbed on the surface of the transducer's biologically active membrane. (Chapter 3 covers the theory of piezoelectricity and the piezoelectric transducers in depth.) The typical mechanical resonators used are piezoelectric crystals for bulk-wave (BAW) oscillators (also known as quartz-crystal microbalances) or surface-acoustic-wave (SAW) delay-line oscillators. Since frequency change can be measured to very high precision ($\Delta f/f < 10^{-5}$ to 10^{-6}), very small mass changes can be measured, which lead to high-sensitivy biosensors.

The typical BAW uses an AT-cut quartz crystal, because of its small temperature coefficient (1 ppm/°C between 10–50°C).[51] The crystal is usually a thin disk, square, or rectangle with metal electrodes deposited on the opposite faces. The resonant frequency of an AT-cut quartz crystal is given by

$$f_0 = \frac{N}{d} = \frac{N\rho A}{m} \quad (49)$$

where N is the frequency constant, 0.167 MHz/cm for AT-cut quartz at room temperature, and d is the thickness of the crystal.[51] Substituting for the thickness of the crystal in terms of its area, A, density, ρ, and total mass, m, yields the second half of Eq. 49. In the pure shear-mode oscillations, there is no stress at the surface of the crystal due to the oscillations, and thus if material adsorbs on the surface of the crystal, the mechanical properties of the material will not affect oscillations of the crystal except for the addition of mass. The change in frequency due to the addition of mass can be calculated from Eq. 49 assuming that the change in mass is small with respect to the mass of the crystal ($\Delta m \ll m$)

$$\Delta f = -f_0 \frac{\Delta m}{m} = -2.3 \times 10^{-6} f_0^2 \frac{\Delta m}{A} \quad (50)$$

where Δm is in grams, A is in cm^2 and f_0 and Δf are in Hz, and the density of quartz, 2.65 g/cm^3, is used to calculate the constant. The typical commercially available AT-cut quartz oscillators are 10–16 mm disks, squares, or rectangles having f_0 of 5, 9, or 10 MHz.[51] For a 10 MHz, 1.0 cm diameter AT-cut quartz crystal, a sensitivity of about 0.3 Hz per nanogram can be expected from Eq. 50.

SAW transducers use Rayleigh waves that are excited from one pair of interdigitated electrodes and are picked up on another pair of electrodes spatially separated from the first. The transit time of the wave packet from one set of electrodes to the other will be lengthened by adsorption of mass on the delay-line's surface. As with the BAW transducer, SAWs are used in oscillator circuits and the change in resonant frequency is related to the analyte adsorbed. The advantage of SAW devices is that they can be used in a different mode with a reference device having an inactive membrane so that any nonspecific adsorption can be compensated.

The change in resonant frequency with adsorbed mass, assuming that there is no change in the mechanical properties of the biologically active layer, has the same form as for the BAW

$$\Delta f = (k_1 + k_2) f_0^2 \frac{\Delta m}{A} \quad (51)$$

where k_1 and k_2 are material constants of the SAW substrate.[52]

BAW and SAW sensors have been used extensively for the detection of inorganic and organic vapors, and used very sparingly in solution.[53] The main problem with using these devices in solution is that the energy loss in solution is large (except in the shear mode). The oscillation frequency depends on the viscosity, density, and conductivity of the solution. Thus, the small changes in solution parameters may alter the response and mask the sensor response, or even stop the oscillations. Packaging the acoustic transducer for solution is also difficult, since any mechanical contact to the piezoelectric surface can alter the response. Nonspecific adsorption of material is also a problem in biological solutions.

For biosensors, piezoelectric transducers have been used with affinity reactions, antibody–antigen reactions, and DNA recognition.[51] The measurement procedures for piezoelectric biosensors include "end-point" detection, competitive binding, the "sandwich" method, and direct liquid-phase exposure.

The most commonly used method is the "end-point" method. In this technique, the initial resonant frequency of the crystal coated with the biologically active material is measured. Then, the crystal is immersed in the analyte solution for incubation for a specified time. Once it is removed from the analyte solution, the crystal is washed with a high ionic strength solution to remove the nonspecifically adsorbed material. After the crystal is dry, the resonant frequency is measured again. The difference in frequency is related to the mass that is bound to the biologically active membrane. In this method, all resonant frequency measurements are made in air. Some examples of piezoelectric biosensors are a human transferrin sensor (detection range 10^{-4} to 10^{-1} mg/cm^3), a salmonella typhimurium sensor (detection range 10^5–10^9 cells/cm^3), a salmonella typhimurium DNA sensor and, of course, a human Ig sensor (detection range 10^{-6}–10^{-1} g/l).[51,53]

Some of the disadvantages of the method are: the sensors usually can be used only once, since removing the bound analyte without removing or denaturing the biologically active material is difficult; it is not a continuous measurement technique; and non-specific adsorption can interfere with the analyte signal. In addition, temperature must be held constant to achieve maximum accuracy, and the humidity must be tightly controlled, since most biological materials have hydrophilic regions in the molecules or are hygroscopic, thus the uptake of water in different humidity conditions will be different.

An indirect method for determining an amount of antigen in a solution is the competitive-binding approach.[54] The competitive-binding approach uses a crystal that has a well-controlled amount of the target antigen adsorbed on the surface of the crystal. The unknown sample of antigen has a known amount of antibody added to it, the crystal is exposed to this solution and, then, the resonant frequency is measured again. The antigen on the crystal competes with any free antigen in the analyte solution for antibodies. This shift in fequency from the original measurement can be correlated to the free antigen concentration in the analyte solution by using calibration curves.

A method to overcome the nonspecific binding is the "sandwich" method.[55] In this method, the piezoelectric crystal is coated with a controlled amount of antigen that is specific for the antibody being measured. Then, the crystal is exposed to the analyte solution, and its resonant frequency is measured. Next, the crystal is exposed to a solution containing an excess amount of another substance that specifically adsorbs on the antibody that is being measured, and the frequency shift is measured again. From this frequency shift, the concentration of the antibody can be determined from calibration curves.

Direct liquid-phase measurements with piezoelectric biosensors have met with some success for antibody–antigen binding. However, there is still much debate on the literature as to whether the change in frequency is due to mass adsorption or other effects, such as the changes in the viscoelastic properties of the biologically active membrane on the crystal's surface.[51]

Gas-phase detection has also been used for piezoelectric biosensors for odorants and pesticides with some success.[51] The parameters that must be controlled in gas-phase detection include temperature and humidity, since the biologically active layer requires a certain amount of humidity to function as if it is in solution. Differential measurements against a reference crystal help to reduce the effects of temperature, flow rate, and pressure fluctuations on the sensor's output.

9.5 PACKAGING OF BIOSENSORS

9.5.1 Functions of Packaging

No sensor can be used in measurements without proper packaging. In fact, the packaging should be an integral part of the sensor design process, not an afterthought. This is true for all sensors, including biosensors. The function of biosensor packaging are:

1. To protect the sensor from the environment it is working in, so that the sensor can perform the sensing function properly within the designated lifetime.
2. To protect the environment from the sensor material and operation so that it is biocompatible and there are no toxic reaction products that get into the environment that can affect it in any undesirable manner. For biosensors, the materials interfacing with the environment should be inert to the chemical and biological environment of the measurement and should not release toxic or undesirable products during the sensing operations.

The protection of sensors from degradation or breakdown in their environment may include:

1. Electrical isolation or passivation of leads and electronics from ions and moisture that causes conductive paths for leakage currents.

2. Mechanical protection to ensure the structural integrity and dimensional stability.
3. Optical and thermal protection, to prevent undesired effects of ambient light and heat that may alter the signal and sensor operation.
4. Chemical isolation of the sensor from the harsh chemical environment.

The protection of the environment from the sensor may include:

1. Sensor material selection to eliminate or reduce body reaction.
2. Sensor operation and packaging selection to avoid toxic products.
3. The sterilization of the sensors.

Packaging of sensors is a specialized field; it requires the integration of information from materials (metals, glasses and polymers), packaging techniques, sensor characterization, and evaluation. Sensor packaging encompasses an understanding of physics, chemistry, biology, materials science, and electrical and mechanical engineering. This section will discuss the requirements and design considerations in biosensor packaging and the materials commonly used. A few selected examples will be presented to illustrate the principles in packaging techniques. For additional information, sensor packaging references should be consulted.[56-61]

9.5.2 Packaging Requirements and Considerations

There is no unique and generally applicable packaging method for all biosensors. Each device works in a special environment and will have unique operational specifications. The packaging, therefore, will have to be designed to satisfy these conditions. However, there are basic requirements and common considerations that are useful in selecting a sensor package. An outline of these considerations is given below.

For electrical protection or passivation, the main considerations are: electrostatic shielding, moisture penetration, interface adhesion, interface stress, and corrosion of substrate materials. The shielding of sensors, particularly those with high impedance parts like potentiometric biosensors, should be considered in the design and application of sensors to reduce the external interferences and internal crosstalk. Moisture penetration is the major failure mechanism of sensors (especially biosensors). When moisture infiltrates the package and condenses onto the components and substrates, leakage currents flow, which generate noise, interference signals and finally result in catastrophic failure (electrical and mechanical breakdown). The most desirable approach is to use hermetic packages made of impermeable glass, ceramic, and metal that can provide an effective barrier against the influx of moisture and ions for many years. However, not all biosensors can be housed inside a hermetic package. Furthermore, not all biosensors require long-term (years) protection that needs expensive, difficult-to-make hermetic packages. Thus, non-hermetic packages are used frequently.

The polymeric encapsulants are used extensively in the packaging of microsensors because of their low cost, low packaging temperature, and ease of work, although they all have finite permeability to water vapor. The polymeric packages can protect the sensor for a limited time period, from days to months. Frequently, this period is sufficient for some biosensors. In some cases, additional thin passivation layers of silicon nitride, titanium, or carbon may be used to provide additional moisture barriers.

The permeability of vapor through a material can be determined by measuring the amount of vapor, m, (in g or cm^3) that diffuses through an area, A, and thickness, L, under a vapor pressure difference, Δp, in a time, t. The relationship is

$$m = P \frac{At \, \Delta p}{L} \qquad (52)$$

where $P = (mL)/At(\Delta p) = m/(A/L)t \, \Delta p$ is the permeability coefficient of the material. Various units may be used for m, h, t, and Δp, therefore, P may be expressed in (g/cm-s-Torr) or (cm^2/s-cm-Hg) or (kg-m/s-N).[61] Equation 52 can be used to calculate or estimate the rate of diffusion of vapor across a sealing material into a sealed cavity. If the vapor pressure outside the cavity is p_0 and at time zero the pressure inside is p_i, the time, t, it takes for the vapor pressure inside the cavity to rise from p_i to p_2 can be estimated by:

$$t = \frac{VL}{PART} \ln\left(\frac{p_0 - p_i}{p_0 - p_2}\right) \qquad (53)$$

where V is the volume of the cavity; L, the diffusion path length of the seal; P, the permeability coefficient in (g/cm-s-Torr); A, the diffusion area of the seal; R, the gas constant = 3465 (Torr-cm^3/kgH$_2$O); and T, absolute temperature in K.[57]

In selecting materials for packaging to protect sensors from moisture, the adhesion of the packaging material to the substrate is equally, if not more important, than permeability. The interface between the sensor and the package material should have good adhesive bonding to prevent condensation at the interface that develops into a conductive pathway due to the hydrophilic tendency of the encapsulated parts. If there are voids in the material, the permeability will be increased and diffused water vapor can condense in these voids, electrolytic currents can develop that can cause the partial lift-off of the packaging material and damage to the package.

In order to have a good adhesion bond, the parts to be encapsulated should have properly cleaned surfaces. Organic contaminants should be moved, and the level of oxide formation should be controlled. All particulates should be eliminated. Usual methods include solvent rinsing, vapor degreasing, and ultrasonic cold cleaning in fluorocarbon solvents, such as Freon^R TF and Freon^R TMS. A vacuum bake-out prior to bonding is recommended whenever possible, if it does not damage the sensor. Adhesion may also be promoted by various chemical agents or by plasma and ion-beam etching.

The interfacial stresses should be reduced to prevent damage to adhesion bonding. The stress may be induced by (1) dimensional mismatch between the packaging material and substrate due to temperature changes, swelling, shrinkage during curing, etc., and (2) externally applied forces acting upon the interface. A layer of low-elastic-modulus material may be used as the intermediate layer by allowing strain relief at the bonding surface. Silicone rubber is used extensively in sensor packaging for strain relief.

The corrosion potential of the substrate material is another important factor to be considered. Biological fluids are strong electrolytes that may cause corrosions and large leakage currents. If the adhesion bond should fail, allowing moisture to come into contact with a potentially corrosive metal, rapid breakdown of the bond will occur. The adhesive failure will propagate until the stress is great enough to lift the package. Therefore, any galvanic couple formed by dissimilar metal is undesirable. The self-passivation property of titanium and tantalum is desirable. The noble metals, unfortunately, do not possess bonding properties equivalent to other metals.

For mechanical protection, the main considerations are:

1. Ridigity: The package must possess adequate mechanical strength and remain stable throughout the life of the sensor for proper operation and to ensure against damage during handling.
2. Weight, size, and shape of the package for convenience in handling and operation.
3. Structure that lends itself to repair and replacement of disposable components or expensive parts.

For some devices, such as semiconductor-based sensors, the ambient light and heat may alter the operation of the device and produce excessive error. Some form of optical and thermal shielding may be required. The large difference in temperature coefficients between parts may generate damaging stress during the operation and assembly process. The thermal properties of materials used in packaging and sensor structure need to be examined.

The chemical and biological considerations in the protection of the sensor are determined by specific biosensor operation. The degradation of packaging material and sensor parts under the operation and storage environments must be considered.

Besides the protection of sensors, the protection of environment (especially one's body) from biosensors can be grouped under:

1. body reaction to the sensor,
2. toxic products from sensor operation,
3. stabilization.

The material used for packaging and construction of biosensors should not react with the body or environment to generate undesirable responses. Besides selection of materials, the mechanical design of the package, the surface condition, the attachment structure, weight, size, etc., are important considerations.

The metallic corrosion and polymeric degradation may produce undesirable products from the sensor. This is an important consideration. An understanding of the behavior of materials that are used is needed in the selection of packaging. Pertinent literature should be consulted.

In particular applications, biosensors should be sterilized to remove living organisms and bacteria from the sensor. Cool sterilization with ethylene oxide followed by ample outgassing is simple and is frequently used. (Thorough outgassing is necessary to eliminate the toxic residue of ethylene glycol or ethylene chlorohydrin.) Wet-steam and dry-heat sterilizations are also used frequently if the sensor can withstand the temperature, pressure, and the moisture. Nuclear radiation and sporicide solutions, which must be removed completely after sterilization may also be used, with proper precautions, in some cases.

9.5.3 Materials and Properties

Glass, ceramic and metal are materials considered to be impermeable to vapors at conventional thicknesses of 10 micrometers or larger. No other materials are suited for hermetic seals. However, biosensors often do not require hermetic package. Polymeric packages are used frequently for ease of handling and because they are less expensive. Figure 17 illustrates the effectiveness of sealant materials as a moisture barrier in a specific geometry.[57] For organic materials,

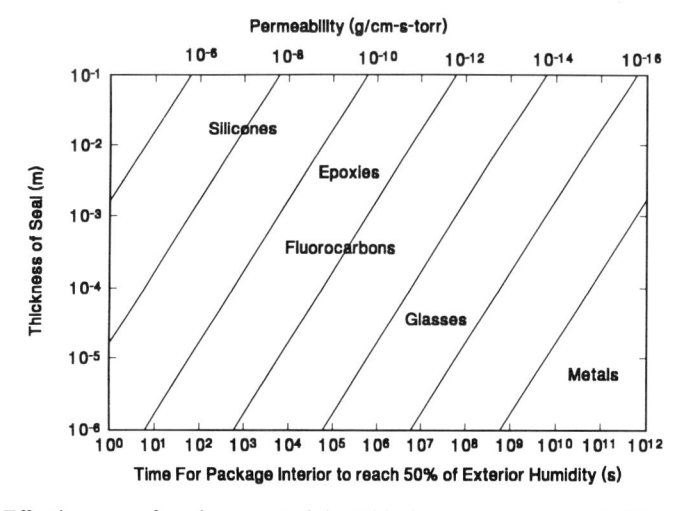

Fig. 17 Effectiveness of sealant materials. This is a comparison of different sealing materials in a specific geometry package.[57]

such as epoxies, silicones, and fluorocarbons, the moisture would penetrate a package having a thickness of a millimeter in minutes to days. A brief summary of common packaging materials for sensors is given below.

Silicone Rubbers (Silastics). They are excellent for potting, encapsulating, and sealing for short-term use. Medical-grade silicon rubbers are biocompatible, flexible, easily applied, autoclavable, and show excellent adhesive characteristics to most substrates. The drawbacks are the tendency to swell in aqueous solution and they are difficult to repair. Dow-Corning manufactures a line of medical grade silicon rubbers—Medical A adhesive, Silastic 382, MDX-4-4210, and nonmedical grade R-4-3117, RTV 3140. The success of the sealing is very much dependent upon the elimination of air bubbles and entrapment air. The techniques used include: vacuum and heat outgassing, cold storage after mixing, and centrifugal spinning.[56]

Polyurethanes. They offer good humidty and chemical resistance, high dielectric strength, good flex-life, and fine tensile properties near room temperature. They are attacked, swollen or dissolved by halogenated solvents and some polar solvents (polar solvents have build in dipole moments).

Epoxies. These two-component systems exhibit good mechanical strength and hardness, but are poor ion barriers and absorb moisture readily. Depending on the filler used, the curing shrinkage, and the thermal, electrical, and mechanical properties will change. The two resins should be accurately mixed in a low-humidity, nitrogen atmosphere to prevent reaction of the curent agent with CO_2 and moisture.

Fluorocarbons. The most well known are polytetrafluoroethylene (Teflon®) and fluorinated ethylene propylene. They have the desirable electrical characteristics of low dielectric constant and low dissipation factor but have poor adhesion and low friction, poor stiffness, and are susceptible to creep.

Acrylics. They are good electrical insulators, hard, rigid, and tough. They have small shrinkage during cure, and are frequently used as conformal coating materials. However, they have poor solvent resistance.

Parylene. This thermal plastic polymer can be produced as a thin, uniform, pinhole-free film by vapor-phase polymerization. They have good insulating properties. Parylene C has a low permeability to moisture and gases, but has poor adhesion. The coating will fail if the film is scratched or cracked.

Polyimides. They are stable over a wide temperature range and have good electrical and mechanical properties. Polyimides are cured at elevated temperature (200–450°C) and are used in the electronics industry as insulators, inter-metal dielectrics, and flexible substrates.

Glasses. Glasses with various electrical and thermal properties are used for packaging sensors and electronics by localized heating, or low-energy laser. Low-temperature glass frits also have been used as sealing materials. The matching of thermal-expansion coefficients is the major consideration. Glasses are much stronger in compression than in tension. The useful strength of a glass is but a fraction of the published intrinsic strength because of surface imperfections that produce localized stress concentrations.

Ceramics. Ceramics are chemically inert, brittle, and have low fracture toughness. They are good electrical insulators and excellent barriers for moisture, but require high heat sealing. The glass–ceramic Macor (by Corning) is a machinable biocompatible ceramic that may have promising applications in packaging when sealing techniques are developed.

Metals. Metals can be welded to form hermetic packages using resistance, electron beam, and laser-microwelding processes. Metals can be welded to ceramics and glasses to form feed-throughs and hermetic packages. Titanium offers light weight, excellent corrosion resistance in aqueous salt solution and is one of the preferred metals for biomedical packages. It forms protective oxide film below 593°C, but above that temperature, a non-productive scale will form rapidly that can shorten the life of the package. Many other metals may be used for the packaging of sensors. Consult materials handbooks and metal handbooks for additional information.

9.5.4 Packaging Techniques and Examples

For electrical protection of leads, many techniques have been developed. Figure 18a shows the structure of an ISFET. The long leads, from drain and source to bonding pads, are diffused paths under oxide and nitride. The pads and the surrounding areas are passivated with silicon rubber or epoxy. Sometimes, SiN_x films are deposited on the top of the polymer to reduce the moisture permeability.

Figure 18b shows the back contact of an ISFET, where the front-side is protected by silicon dioxide and nitride, the leads are connected to the front from the back contacts etched through the silicon substrate. The ISFET chip may be further packaged in a molded shell.

Figure 18c shows the buried feed throughs that connect the sensor to the electronic circuits that process the signal and may telemeter the information from inside the glass capsule to external units.

Figure 18d shows a multi-wafer silicon package that protects the leads under the bonded silicon wafer and provides a well in which the bio-recognition membrane can be formed.[62] Neighboring wells can have different membranes, so that multi-component sensors and differential measurements can be made.

Figure 18e shows a surface-micromachined chamber where the sensor surface may be protected from excessive fluid flow and large protein poisoning.

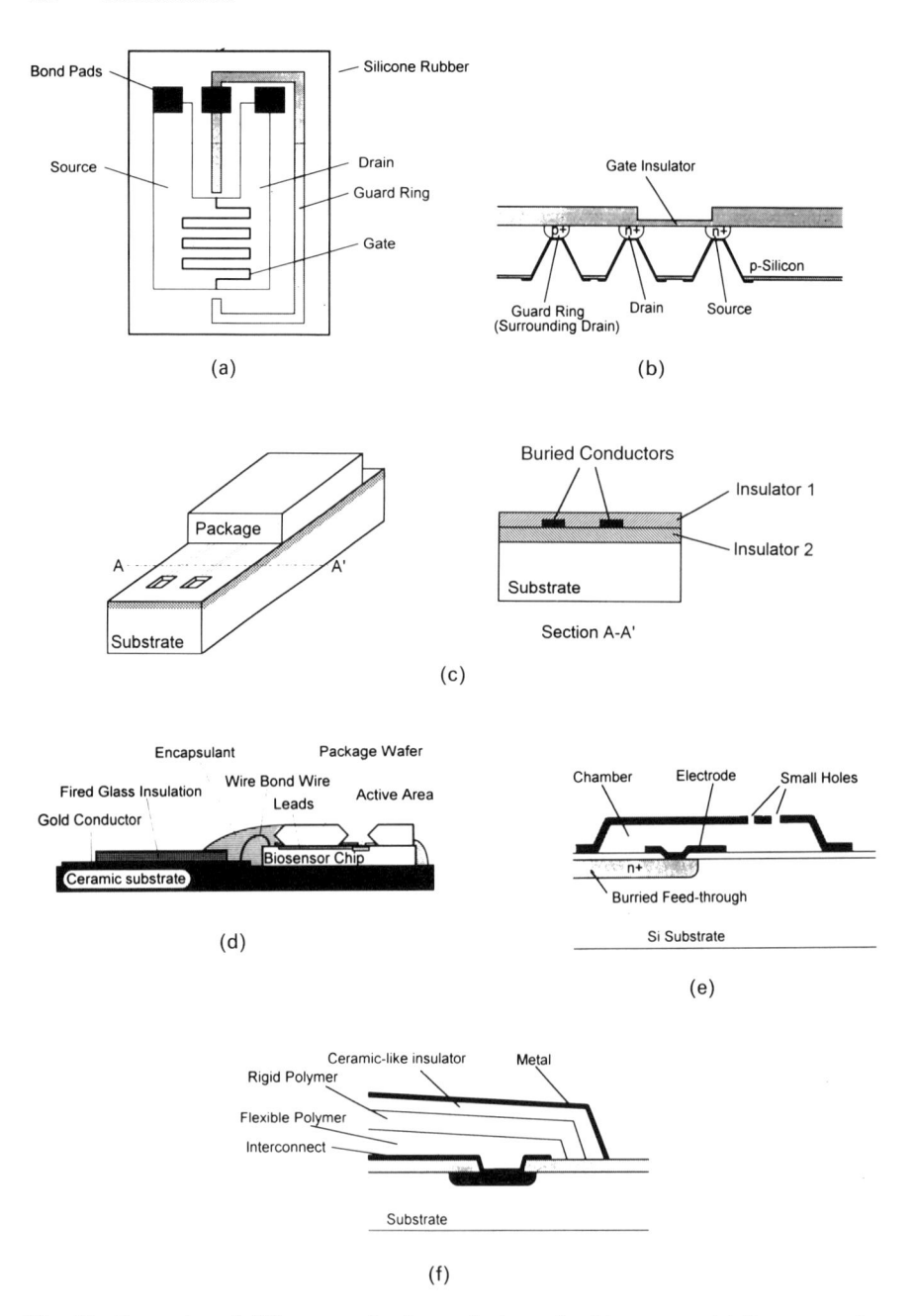

Fig. 18 Examples of different packaging techniques for biosensors. (a) Structure of a pH ISFET sensor. (b) Back-side contacts to an ISFET. (c) Buried feed-throughs. (d) A multi-wafer ENFET package that protects the leads under the bonded silicon wafer and provides a well in which different biomembranes can be formed. (e) Surface-micromachined chamber for electrochemical electrodes. (f) Multi-layer protection package of biosensors.

Furthermore, the product of sensing (pH reversible change) may be confined in the chamber to reduce the effect on the host body tissues.

Figure 18f shows multilayer protection of biosensors. Each layer serves a specific function; together the group satisfies the total specifications or requirements.

For mechanical protection, one has to select materials and structures so that these mechanical considerations are taken into account. Besides ridigity and mechanical strength, the package shape, minimum corner radius and surface smoothness are sometimes important, particularly when the device is in contact with the body or tissues.

Polymeric material may be mixed with additives such as carbon, TiO_2 powders, and glass frits to change its optical, thermal, and shrinkage properties to meet the package requirements.

Each biosensor has its special requirements, features, and operating environment. Therefore, each may need special design and packaging. There is no general rule for packaging of biosensors at this time. A compromise between conflicting requirements usually has to be made in selecting materials, structures, and techniques used in the packaging of sensors.

9.6 SUMMARY AND FUTURE TRENDS

9.6.1 Summary of the Chapter

This chapter presented an overview of biosensors. A biosensor was defined as a biological-recognition membrane in intimate contact with a traditional transducer. The biological active material in the recognition membrane selectively detects the analyte through a reaction or specific adsorption, and the transducer converts the result of the detection into a usable signal, usually electrical or optical. The high selectivity and sensitivity of biosensors is due to the shape-specific recognition of biological-recognition materials. A wide variety of biological recognition materials and transduction principles are possible because of the wide diversity of life.

The operational characteristics of biosensors depend on both the properties of and reactions occurring within the bio-recognition layer and the operation characteristics of the transducer. The details of the operation of the different transducers used in biosensors are covered in other chapters of this book, so this chapter concentrated on the biological-recognition membrane. Possible bio-recognition elements include molecules (e.g., antibodies, antigens, enzymes, DNA, RNA), organelles, cells, tissues, and whole organisms. The concept of biomimetic structures was presented as one of the new and exciting ways to take advantage of the biochemistry and chemical amplification that occurs in cells. To use these bio-recognition elements, four methods of immobilization were discussed: membrane confinement, matrix entrapment, adsorption, and covalent binding. Since the recognition reaction occurs in a membrane, the mass transport of the analyte, co-reactants, interference, and products determine the response time of the sensors. The loading effects (i.e., the amount) of the biological recognition material in the membrane on the sensitivity was discussed. The importance of matching the transduction principle to the bio-recognition reaction was emphasized, and the different transduction principles were presented, using devices and measurements techniques as the foundation of the discussion.

The functions and characteristics of the often neglected part of sensor design, the packaging, were covered next. A high selectivity and high sensitivity sensor is nearly useless if it cannot be packaged to survive in the environment. The design of the packaging is an integral part of the sensor design, and should be considered at the beginning of the design process. First the mechanical, electrical and chemical requirements on the biosensor packaging were presented. Many of the biosensor applications have the added requirement of biocompatibility. The relevant properties of some commonly used packaging materials were presented. It was emphasized that the only way to get a true hermetic seal is to use glasses, ceramics or metals as the sealing and package materials. However, some biosensor applications exist that do not require true hermetic seals.

9.6.2 Future Trends

The field of biosensors is still in its early development stage. There are common problems that need further research into the materials, biochemistry, sensor technology, sensor principles and basic understanding of the behavior of biosensors. A summary of the challenging problems given below may be useful in selecting future directions of study.

The stability of biosensors in field applications is possibly the most significant problem facing biosensors now. The causes of instability include: (1) the degradation of the recognition membrane due to denaturing, leakage, and poisoning; (2) the alteration of the surface of the active area due to protein clogging, mechanical damage, and chemical or biological poisoning; (3) instability of the transducer, and (4) the development of leaks, corrosion, and degradation of the packaging.

The usefulness of a sensor depends on its sensitivity and selectivity. Here, sensitivity means the detection limit set by the noise level of the device, which includes the interferences and drift over the time needed to record data during the measurement. Higher sensitivity can be obtained by increasing the efficiency of the transduction and by reducing the noise. The random noise can be reduced by averaging over time or over redundant sensors whenever possible. By integrating several different biosensors on a single substrate and using pattern-recognition techniques, including neural networks, multiple analyte concentrations can be measured simultaneously, with the potential of higher sensitivity and lower noise than for individual sensors.

The operating environment of biosensors usually contains many materials competing for the reaction sites of the recognition membranes, which can cause interferences. High selectivity is necessary for good sensitivity when the interfering materials cannot be eliminated. Selectivity can be improved by better recognition membrane materials and by the analysis of measured data from redundant sensors.

One of the desirable and significant applications of biosensors is to measure the parameters on-line, as a part of an implant unit. These sensor–actuator implants would make it possible to produce therapeutic instruments that could correct body malfunctions. If *in vivo* glucose sensors can be stable over long periods, diabetics can be treated with implanted, closed-loop insulin pumps. Many similar thereapeutic systems are waiting for *in vivo* sensors. Solid-state sensor–actuator systems are advantageous for implant systems to reduce the size and weight. Successful *in vivo* systems would require reliability of the device and package, long sensor lifetime, built-in self-test, and built-in self-calibration.

A part of the future developments on biosensors will be to meet the challenges outlined above. Besides these, genetic-engineering may soon provide biosensor elements that can be tailored to meet the specific requirements of particular biosensors. The rapid growth in the understanding of the biochemistry of cells, genes, viral agents, hormones, enzymes, antibodies, etc., will greatly improve the understanding of how biosensing elements actually work, allowing the sensor engineer to design and build higher performance sensors. Biosensor design and fabrication will go from being an art to being a technology in the near future.

The engineering technology used to fabricate present day biosensors includes the bulk and surface-micromachining techniques discussed in Chapter 2. The potential of these technologies is not yet fully utilized. The advance of micro-actuators and micro-electro-mechanical systems (MEMS) has opened another wide field of possibility. It is now practical to consider research and fabrication of microsystems that integrate sensors, logic and control circuits, actuators, and mechanical components in micro-scale in the same manufacturing cycle.

ACKNOWLEDGMENTS

The authors would like to thank Dr. Chung-Chiun Liu, Director of the Electronics Design Center, and Dr. James Zull of the Department of Biology at Case Western Reserve University for their critical reading of this chapter and their helpful insights.

PROBLEMS

1. Derive Eq. 6 and show that the Michaelis constant is given by Eq. 7. K_M is the ratio of the rate of formation and the rate of consumption of the enzyme–substrate complex, *ES*. Plot the rate of product formation (v) versus the substrate concentration $[S]$ using the asymptotes.

466   BIOSENSORS

2. The catalytic activity of enzymes is dependent on the pH, temperature, etc. To model the pH effects, assume that enzyme exists in the three states of oxidation (EH_2, EH^- and E^{2-}) and only one of them, EH^-, is catalytically active. The Michaelis–Menton reaction becomes

$$S + EH^- \underset{k_{-1}}{\overset{k_w}{\longleftrightarrow}} EH^-S \overset{k_2}{\longrightarrow} P + EH^-.$$

The dissociation reactions for the three oxidation states of the enzyme are given by

$$EH_2 \leftrightarrow EH^- + H^+ \qquad K_{es1} = \frac{[EH^-]_0[H^+]}{[EH_2]_0}$$

and

$$EH^- \leftrightarrow E^{2-} + H^+ \qquad K_{es2} = \frac{[E^{2-}]_0[H^+]}{[EH^-]_0}.$$

Assume that the total amount of enzyme in the system is fixed $[E]_{tot} = [EH_2]_0 + [EH^-]_0 + [E^{2-}]_0$.

a) Derive the equation for the rate of product formation (the modified Michaelis–Menton kinetics) in terms of the total enzyme concentration, $[E]_{tot}$, the hydrogen-ion concentration, $[H^+]$, and the rate constants, K_M, K_{es1}, and K_{es2}. (Hint: $[EH^-]_0 = [EH^-] + [EH^-S]$.)

b) For a fixed substrate concentration, at what pH value is the rate of product formation maximum?

c) Plot v_{max} versus pH for $pK_{es1} = 6$ and $pK_{es2} = 8$.

3. In Section 9.3.3, the effects of the enzyme loading in the sensor membrane on the output response of a biosensor were discussed with respect to a transducer whose output was proportional to the concentration of the product at the surface of the transducer. Equation 25 shows the dependence of the product concentration on the substrate concentration at the membrane surface. Now, the sensor is placed in a well-stirred solution with mass transport constant K_D, derive a relationship for the sensor output as a function of the substrate concentration in the bulk solution.

4. A piezoelectric biosensor is made from an AT-cut quartz crystal which has an area of one square centimeter and a resonant frequency, after the antibody is immobilized, of 10 MHz. A frequency counter with a 0.1 Hz resolution is used to make the frequency measurements.

a) If there are 10^{12} site/cm^2 antibodies immobilized on the surface of the crystal, and the average diameter of the antibody is 10 nm, what is the fractional surface coverage?

b) Assume the antigen has a molecular mass of 10 kg/mol. What is the minimum number of molecules that can be detected?

c) During the incubation period, along with the specific adsorption of the antigen, proteins from the solution also adsorb on the surface. If after washing, roughly 10 ng/cm^2 of protein remain, what is the detection limit if it the signal-to-noise ratio is required to be 5?

d) If the sensor is used repeatedly and only 90% of the bound antigens can be removed by the cleaning procedure, how many times can the sensor be used before it loses 75% of its original sensitivity?

5. Consider a thermal-detection scheme where a large volume analate solution is allowed to flow through a calorimeter, as described in Section 9.4.3, so that the signal comes to a steady-stage value during measurement. First, the solution flows through a heat exchanger to set its initial temperature. Next, the solution flows through the column containing the immobilized enzyme. At the end of the column, the temperature of the solution is measured again. Assume that (1) the analate solution is dilute enough so that all the substrate is reacted, (2) the heat capacity of the calorimeter system is dominated by that of water (4.18 J/gm°C, at 20°C), (3) the effects of the flow of the solution on the measurement are negligible, and (4) the effective volume of the measurement region is 1 cm^3. The immobilized enzyme is glucose oxidase, and the analate is glucose, thus the enthalpy of the reaction is 80 kJ/mol.

a) Calculate the smallest concentration that can be detected if the thermistor can detect a change in temperature of 0.001°C.

b) If the friction and turbulence due to the flow produce 0.01 joules of heat, what is the detection limit if a signal-to-noise ratio of 5 is required?

6. Consider a mediated amperometric enzyme electrode that follows the kinetics as described in Eqs. 35 and 36. If the electrode is operating in the kinetic-limited mode, and the membrane thickness is 10 μm, from the data given below, find the rate constants K_M, k_2 and k_e.

TABLE P6a Flux to the Electrode for a Constant Mediator Concentration [M] = 0.001 M as a Function of Substrate Concentration, [S], for Different Enzyme Concentrations [E]

	Flux (mol/cm^2-s^1)				
[S] (M)	[E] = 0.001 M	[E] = 0.002 M	[E] = 0.003 M	[E] = 0.004 M	[E] = 0.005 M
1.000×10^3	9.615×10^{10}	1.923×10^9	2.885×10^9	3.846×10^9	4.808×10^9
3.000×10^3	9.740×10^{10}	1.948×10^9	2.922×10^9	3.896×10^9	4.870×10^9
5.000×10^3	9.766×10^{10}	1.953×10^9	2.930×10^9	3.906×10^9	4.883×10^9
7.000×10^3	9.776×10^{10}	1.955×10^9	2.933×10^9	3.911×10^9	4.888×10^9
9.000×10^3	9.783×10^{10}	1.956×10^9	2.935×10^9	3.913×10^9	4.891×10^9

TABLE P6b Flux to the Electrode for a Constant Enzyme Concentration $[E] = 0.001$ M and Different Mediator Concentrations, $[M]$, as a Function of Substrate Concentration

	Flux (mol/cm²-s)			
$[S]$ (M)	$[M]=0.002$ M	$[M]=0.003$ M	$[M]=0.004$ M	$[M]=0.005$ M
1.000×10^3	1.852×10^9	2.679×10^9	3.448×10^9	4.167×10^9
3.000×10^3	1.899×10^9	2.778×10^9	3.614×10^9	4.412×10^9
5.000×10^3	1.908×10^9	2.798×10^9	3.650×10^9	4.464×10^9
7.000×10^3	1.913×10^9	2.807×10^9	3.665×10^9	4.500×10^9
9.000×10^3	1.914×10^9	2.812×10^9	3.674×10^9	4.487×10^9

7. A glass package is sealed to a silicon substrate by a layer of Hysol epoxy, as shown in Fig. P1. Assume that the glass and silicon are nonpermeable to moisture and gases, and the permeability coefficient of the epoxy is 10^{-11} (g/cm-s-Torr). Find the leakage rate in cm³/s at $t = 0$, if the package is under 25°C, 760 Torr absolute pressure, and the pressure inside package is zero. (*Hint*: convert the permeability from (g/cm-s-Torr) to (cm³/cm-s-Torr) first.)

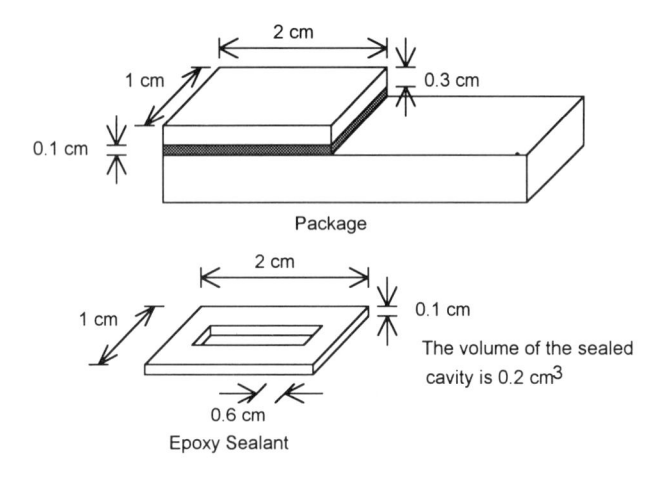

Fig. P1 The package and seal geometry for Problems 7, 8, and 9.

8. For the package in Problem 7, the package is used in an environment where the atmosphere outside is saturated with moisture. If the relative humidity reaches 50% inside the package, the electronics will malfunction. Find the lifetime of the package.

9. Calculate and plot the vapor pressure inside (in % of outside pressure) versus time curve for the package in Fig. P1.

REFERENCES

1. A. F. P. Turner, "Preface," *Biosensors: Fundamentals and Applications*, A. F. P. Turner, I. Karube and G. S. Wilson, Eds., Oxford Science Publications, Oxford, 1987, p. v.
2. L. C. Clark and C. Lyons, "Electrode system for the continuous monitoring in cardiovascular surgery," *Ann. NY Acad. Sci.* **102**, 29–45 (1962).
3. M. F. Chaplin and C. Bucke, "Fundamentals of enzyme kinetics," *Enzyme Technology*, Cambridge University Press, Cambridge, 1990, pp. 1–39.
4. H. Stieve, "Sensors of biological organisms—biological transducers," *Sens. Actuators* **4**, 689–704 (1983).
5. A. Dautry-Vasart and H. F. Lodish, "How receptors bring proteins and particles into cells," *Sci. Am.* **250**, pp. 52–8 (1984).
6. H. Schmidt, W. Schuhmann, F. W. Scheller, and F. Schubert, "Specific features of biosensors," *Sensors: A Comprehensive Survey, Volume 3: Chemical and Biochemical Sensors, Part II*, W. Gopel, *et al.*, Eds., VCH, New York, 1991, pp. 717–817.
7. J. E. Rothman and J. Lenard, "Membrane asymmetry," *Science* **195**, 743–853 (1977).
8. J. B. Davenport, "Physical chemistry of lipids," *Biochemistry and Methodology of Lipids*, A. R. Johnson and J. B. Davenport, Eds., Wiley–Interscience, New York, 1971, pp. 47–83.
9. A. Arya, U. J. Kruell, M. Thompson, and H. E. Wong, "Langmuir–Blodgett deposition of lipid films on hydrogen as a basis for biosensor development," *Anal. Chim. Acta* **173**, pp. 331–6 (1985).
10. U. J. Krull and M. Thompson, "The lipid membrane as selective chemical transducers," *IEEE Trans. Electron Devices* **ED-32**, 1180–4 (1985).
11. M. Sugawara, M. Kataoka, K. Odashima, and Y. Umezawa, "Biomimetic ion-channel sensors based on host-guess molecular recognition in Langmuir-Blodgett membrane assemblies," *Thin Solids Films* **180**, 129–33 (1989).
12. H. Bader, K. Dorn, B. Hupfer, and H. Ringsdorf, "Polymeric monolayers and lioposomes as models for biomembranes," *Adv. Polymer Sci.* **64**, 1–63 (1985).
13. A. F. P. Turner, I. Karube, and G. S. Wilson, Eds., *Biosensors: Fundamentals and Applications*, Oxford Science Publications, Oxford, 1987.
14. L. J. Blum and P. R. Coulet, Eds., *Biosensor Principles and Applications*, Marcel Dekker, New York, 1991.
15. W. Gopel *et al.*, Eds., *Sensors: A Comprehensive Survey, Volume 2: Chemical and Biochemical Sensors, Part I*, VCH, New York, 1992.
16. W. Gopel, *et al.*, Eds., *Sensors: A Comprehensive Survey, Volume 3: Chemical and Biochemical Sensors, Part II*, VCH, New York, 1992.
17. A. F. P. Turner, Ed., *Advances in Biosensors: A Research Annual*, Vol. 1, JAI Press, London, 1991. (The first in an annual series).
18. A. E. G. Cass, Ed., *Biosensors: A Practical Approach*, Oxford University Press, Oxford, 1990.
19. W. Hartmeier, "Methods of immobilization," *Immobilized Biocatalysts: An Introduction*, Springer-Verlag, Berlin, 1986, pp. 22–50.
20. M. F. Chaplin and C. Bucke, "Preparation and kinetics of immobilized enzymes," *Enzyme Technology*, Cambridge University Press, Cambridge, 1990, pp. 80–137.

21. J. Klein and F. Wagner, "Methods for the immobilization of microbial cells," in *Applied Biochemistry and Bioengineering, Volume 4: Immobilized Microbial Cells,* I. Chibata, et al., Eds., Academic Press, New York, 1983, pp. 2–51.
22. S. D. Caras, J. Janata, D. Saupe, and K. Schmitt, "pH-based enzyme potentiometric sensors. part 1. theory," *Anal. Chem.* **57**, 1985, pp. 1917–20.
23. R. M. Olsen, *Essentials of Engineering Fluid Flow,* International Textbook Company, Scranton, 1966.
24. M. J. Eddowes, "Theoretical methods for analyzing biosensor performance," in *Biosensors: A Practical Approach,* A. E. G. Cass, Ed., Oxford University Press, Oxford, 1990, pp. 211–63.
25. J. Janata, "Chemical selectivity of field-effect transistors," *Sens. Actuators* **12**, 121–8 (1987).
26. S. Collins and J. Janata, "A critical evaluation of the mechanism of potential response of antigen polymer membranes to the corresponding antiserum," *Anal. Chim. Acta* **136**, 93–9 (1982).
27. P. Bergveld, "Future applications of ISFETs," *Sens. Actuators* **B4**, 125–33 (1991).
28. A. J. Bard and L. R. Faulkner, *Electrochemical Methods: Fundamentals and Applications,* Wiley, New York, 1980.
29. J. Janata, *Principles of Chemical Sensors,* Plenum, New York, 1989.
30. J. Kauffmann and G. G. Guilbault, "Potentiometric enzyme electrodes," in *Biosensor Principles and Applications,* L. J. Blum and P. R. Coulet, Eds., Marcel Dekker, New York, 1991, pp. 63–82.
31. S. S. Kuan and G. G. Guilbault, "Ion-selective electrodes and biosensors based on ISEs," in *Biosensors: Fundamentals and Applications,* A. F. P. Turner et al., Oxford Science Publications, Oxford, pp. 135–52 (1987).
32. G. F. Blackburn, "Chemically sensitive field effect transistors," in *Biosensors: Fundamentals and Applications,* A. F. P. Turner et al., Oxford Science Publications, Oxford, 1987, pp. 481–571.
33. T. Kuriyama and J. Kimura, "FET-based biosensors," in *Biosensor Principles and Applications,* L. J. Blum and P. R. Coulet, Eds., Marcel Dekker, New York, 1991, pp. 139–62.
34. M. J. Eddowes, "Response of an enzyme-modified pH sensitive ion-selective device; analytical solution for the response in the presence of pH buffer," *Sens. Actuators* **11**, 265–74 (1987).
35. G. S. Wilson, "Fundamentals of amperometric sensors," in *Biosensors: Fundamentals and Applications,* A. F. P. Turner, et al., Eds., Oxford Science Publications, Oxford, 1987, pp. 165–79.
36. P. N. Bartlett, "Conducting organic salt electrodes," in *Biosensors: A Practical Approach,* A. E. G. Cass, Ed., Oxford University Press, Oxford, 1990, pp. 47–95.
37. W. J. Albery and D. H. Craston, "Amperometric enzyme electrodes: theory and experiment," in *Biosensors: Fundamentals and Applications,* A. F. P. Turner, et al., Oxford Science Publications, Oxford, 1987, pp. 180–210.
38. M. F. Cardosi and A. F. P. Turner, "The realization of electron transfer from biological molecules to electrodes, in *Biosensors: Fundamentals and Applications,* A. F. P. Turner, et al., Oxford Science Publications, Oxford, 1987, pp. 257–75.
39. G. Bardeletti, F. Sechaud, and P. R. Coulet, "Amperometric enzyme electrodes for

substrate and enzyme activity determinations," in *Biosensor Principles and Applications,* L. J. Blum and P. R. Coulet, Eds., Marcel Dekker, New York, 1991, pp. 7–45.

40. P. N. Bartlett, "The use of electrochemical methods in the study of modified electrodes," in *Biosensors: Fundamentals and Applications,* A. E. G. Cass, Ed., Oxford Science Publications, Oxford, 1987, pp. 211–246.

41. H. P. Bennetto, J. Box, G. M. Delany, J. R. Mason, S. D. Roller, J. L. Stirling, and C. F. Thurston, "Redox-mediated electrochemicsty of whole organisms: from fuel cells to biosensors," in *Biosensors: Fundamentals and Applicatins,* A. F. P. Turner *et al.,* Eds., Oxford Science Publications, Oxford, 1987, pp. 291–314.

42. D. B. Kell and C. L. Davey, "Conductimetric and impedimetric devices," in *Biosensors: A Practical Approach,* A. E. G. Cass, Eds., Oxford University Press, Oxford, 1990, pp. 125–54.

43. W. R. Seitz, "Chemical sensors based on fiber optics," *Anal. Chem.* **56**, 16A–34A (1984).

44. W. R. Seitz, "Optical sensors based on immobilized reagents," in *Biosensors: Fundamentals and Applications,* A. F. P. Turner, *et al.,* Eds., Oxford Science Publications, Oxford, 1987, pp. 599–616.

45. B. P. H. Schaffer and O. S. Wolfbeis, "Chemically mediated fiberoptic biosensors," in *Biosensor Principles and Applications,* L. J. Blum and P. R. Coulet, Eds., Marcel Dekker, New York, 1991, pp. 163–94.

46. M. A. Arnold, "Fluorophore- and chromophore-based fiber optic biosensors," in *Biosensor Principles and Applications,* L. J. Blum and P. R. Coulet, Eds., Marcel Dekker, New York, 1991, pp. 195–211.

47. V. S. Danilov and A. D. Ismailov, "Bacterial luciferase as a biosensor of biologically active compounds," in *Applied Biosensors,* D. L. Wise, Ed., Butterworths, Boston, 1989, pp. 39–78.

48. F. McCarpa, "Potential applications of bioluminescence and chemiluminescence in biosensors," in *Biosensors: Fundamentals and Applications,* A. F. P. Turner, *et al.,* Eds., Oxford Science Publications, Oxford, 1987, pp. 617–37.

49. B. Danielsson and K. Mosback, "Theory and application of calorimetric sensors," in *Biosensors: Fundamentals and Applications,* A. F. P. Turner, *et al.,* Eds., Oxford Science Publications, Oxford, 1987, pp. 575–95.

50. B. Danielsson, "Enzyme thermistor devices," in *Biosensor Principles and Applications,* L. J. Blum and P. R. Coulet, Eds., Marcel Dekker, New York, 1991, pp. 83–105.

51. J. H. T. Luong and G. G. Guilbault, "Analytical applications of piezoelectric crystal biosensors," in *Biosensor Principles and Applications,* L. J. Blum and P. R. Coulet, Eds., Marcel Dekker, New York, 1991, pp. 107–38.

52. J. Janata, "Mass sensors," *Principles of Chemical Sensors,* Plenum Press, New York, 1989, pp. 55–80.

53. M. S. Nieuwenhuizen and A. Venema, "Mass sensitive devices," in *Sensors: A Comprehensive Survey, Volume 2: Chemical and Biochemical Sensors, Part I,* W. Gopel *et al.,* Eds., VCH, New York, 1991, pp. 647–60.

54. R. J. Oliveira and S. F. Silver, "Immunoassay with coated piezoelectric crystals," *U.S. Patent* 4,242,096 (1980).

55. T. K. Rice, "Sandwich immunoassay using piezoelectric oscillator," *U.S. Patent* 4,314,821 (1982).
56. W. H. Ko and T. Spear, "Packaging materials and techniques for implantable instruments," *IEEE Eng. Med. Biol. Mag.* pp. 22–38 (March 1983).
57. R. K. Traeger, "Nonhermeticity of polymeric lid sealant," *IEEE Trans. Parts, Hybrids and Packaging* **PHP-13**, 147–52 (1977).
58. J. W. Boretos, *Concise Guide to Biomedical Polymers*, Thomas, Springfield, IL, 1973.
59. D. F. Williams, Ed., *Biocompatibility of Clinical Implant Materials*, CRC Press, Boca Raton, FL, 1981.
60. *JEDEC Standard* No. 22-A-109, Electronic Industries Association, Washington DC, USA (June 1988).
61. *Engineering Materials Handbook*, Vol. 3, ASM International, 1990 (ISBN-0-87170-281-9)
62. A. S. Dewa, "A silicon based enzyme biosensor utilizing Langmuir–Blodgett film immobilization," *Ph.D. Dissertation*, Case Western Reserve University, Cleveland OH, 1993.

10 Integrated Sensors

K. NAJAFI and K. D. WISE
University of Michigan
Ann Arbor, MI, USA

N. NAJAFI
IBM Microelectronics Division
Essex Junction, VT, USA

10.1 INTRODUCTION

10.1.1 Overview

Microelectronic systems have gone through many fundamental changes due to the rapid progress achieved in microelectronics; it is feasible now to realize many control/instrumentation system components as single monolithic or hybrid modules. The introduction of the microprocessor in the early 1970s revolutionized the design and use of control/instrumentation systems by allowing system operation to be defined in software, thus permitting a substantial increase in signal-processing and user-interface features. Analog elements (e.g., analog-to-digital converters) have also been improved substantially to satisfy high-speed and high-accuracy requirements;[1] however, progress in the development of sensors with which these electronic systems can interface has been slow.[2] Figure 1 shows the overall system architecture of a measurement/control system. External physical and chemical parameters are measured and converted into an electrical format using an array of sensors. The sensed data are collected, processed, and digitized using in-module (integrated or hybrid) circuitry, and are transmitted over a digital bus to a host controller. The host controller uses this information to make appropriate decisions, and feeds control information back to the external environment through an array of actuators. Such systems are increasingly needed in many applications, including

---

Semiconductor Sensors, Edited by S. M. Sze.
ISBN 0-471-54609-7 © 1994 John Wiley & Sons, Inc.

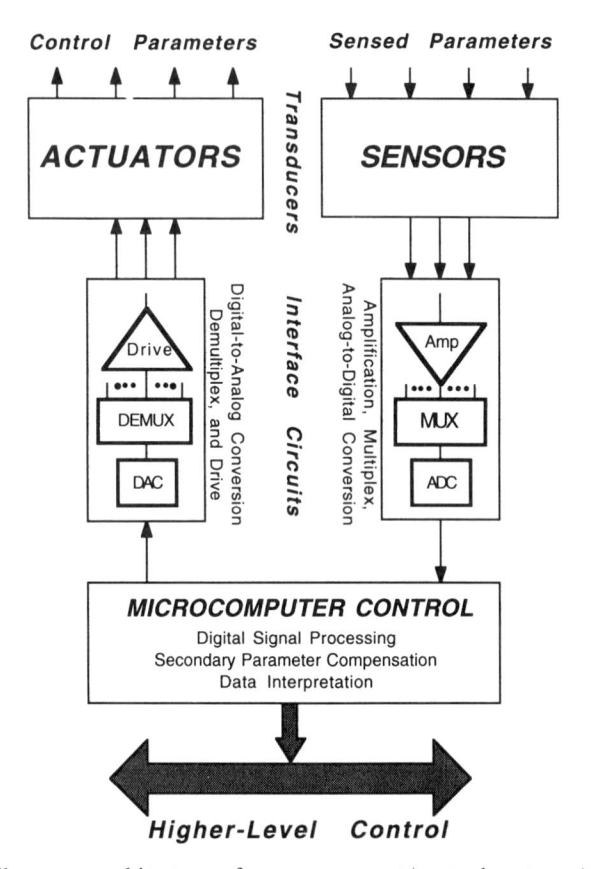

Fig. 1 Overall system architecture of a measurement/control system. (After Ref. 26)

automotive, health care, manufacturing, environmental monitoring, industrial processing, avionics, and defense.

In sensing systems, the sensors are the determining factor of system accuracy and represent the most critical system element. However, sensors currently represent the weakest link in the development of most emerging and next-generation instrumentation, data-acquisition, and control systems. They are often unreliable, rarely offer adequate accuracy, sometimes cost more than the controller, and provide few, if any, fault-tolerance or fault-detection capabilities. The lack of such features results in expensive system maintenance and repair costs. For example, as far as long-term reliability and fail-safe operation are concerned, sensors have been identified as the weakest link in today's automotive control systems.[3] The development of solid-state microsensors has helped improve some performance characteristics, in particular in terms of reproducibility, size, cost, and accuracy. However, the full potential of solid-state microsensors and microactuators has not been realized yet.

In the past few years "integrated sensors" that monolithically combine the sensor structure and at least some portion of the interface and signal-processing electronics on the same substrate have begun to emerge. By emerging microsensors and circuits, integrated "smart" sensors can go far beyond simple sensors, providing features such as standard interfaces, self-testing, fault-tolerance, and digital compensation to extend overall system accuracy, dynamic range, and reliability in ways not otherwise practical, while reducing the capital and ownership costs. Another aspect that has often been neglected in considering the advantages of integrated sensors is that a variety of new sensors, which were not practical in the past, may now be feasible. There is a wide unexplored range of possibilities for combining digital processing power with current solid-state technology to produce novel sensors and instrumentation techniques. However, the full potential of integrated sensors cannot be achieved simply by replacing conventional sensors.[2,4]

The most important aspect of any sensing system is the performance achieved by the entire system, rather than that of a single component, such as a sensor. Indeed, users desire a high-performance sensing system, not an isolated smart sensor. Furthermore, many features of smart sensors are only meaningful when considered in the broader context of the overall sensing system. The system structure must change as well to accommodate the full capacity of smart sensors.

This chapter will review the state of the art in the development of integrated sensors that include a variety of on-chip signal processing functions on the same substrate as the sensor itself. We will first review the history of sensor development. Section 2 will discuss the architecture and design of bus-origanized sensing systems and integrated sensors. Section 3 will present a review of basic circuit functions needed in integrated sensors, and Section 4 will discuss the different fabrication techniques required in the implementation of integrated sensors. Section 5 presents two examples of integrated sensors illustrating their features and advantages. Finally Section 6 will present a view of future trends in the development of integrated sensors and will examine areas requiring further research and development.

10.1.2 Solid-State Sensor Evolution

Sensors have slowly evolved through at least four generations of interface.[5,6] First-generation devices had essentially no electronics in them and produced the end effect with virtually no signal processing, as in a bimetal strip. Second-generation devices contained amplification and perhaps some temperature compensation. Typically, all of the electronics was remote from the sensor. The data was analog and data flow was one-way from the sensor to the display. By the third-generation devices (the state of many current devices) some amplification and signal buffering was occurring in-module using discrete or hybrid electronics, as shown in Fig. 2. The sensor operated into a remote signal-processing package consisting of an analog-to-digital converter (ADC) and a microcomputer. The communication link from the sensor was one-way, high-level analog. Many of today's automotive sensors are third-generation devices. The fourth-generation sensor is realized using a higher level of

476 INTEGRATED SENSORS

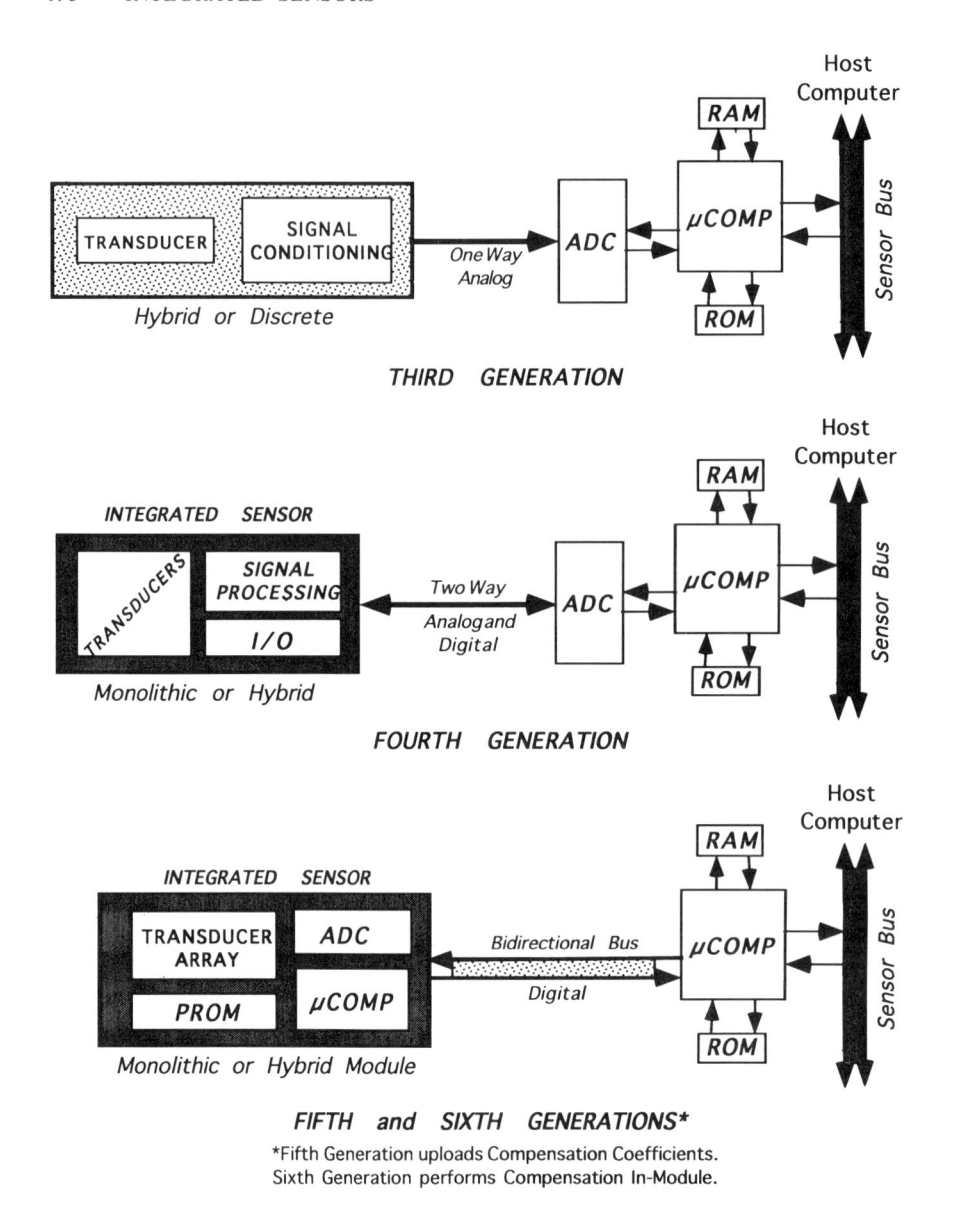

Fig. 2 Evolution of sensor interface from third to sixth generations. (After Refs. 6, 9)

integration, with some or all of the sensor electronics monolithically integrated on the sensor chip itself. Fourth-generation devices are addressable from the processor and feature both control inputs and (usually) processed analog outputs. The communication link is two-way using a digital address and an analog high-level voltage or time-analog pulse-rate-modulated output for data.

Compensation may be at the sensor or may be performed remotely using a separate output from the interfering parameter (usually temperature).

Many emerging sensing systems require high-performance sensor features that are only attainable at a system level. For such applications, fifth/sixth-generation sensors, which are system components (not simple devices), are under development. As shown in Fig. 2, in the fifth-generation sensors the data conversion to digital is performed at the sensor itself, so that the link to the host controller is over a digital bi-directional sensor bus. The sensor is self-testing and addressable. Data compensation may still be remote (i.e., fifth generation) or local (i.e., sixth generation). Calibration is via a PROM (rather than laser-trimmed resistors), which is programmed during a sensor factory test. The entire analog signal path is tested as a unit, reducing overall testing cost. It is also desirable to calibrate the sensor and perhaps the electronics *in situ* upon demand, especially if sensor characteristics, such as gain and offset, drift over time. In order to perform auto-calibration remotely, however, a known input signal needs to be generated and the sensor should be capable of being controlled in a closed loop. This can be achieved using some of the techniques developed recently for microactuators, including actuation and control of micromachined elements using electrostatic forces. These techniques are already being employed to calibrate accelerometers[7] and microflowmeters,[8] and it is exprected that many future sensors, fabricated using either bulk- or surface-micromachining techniques, will incorporate some type of force feedback for auto-calibration and self-testing. A comprehensive discussion of different protocols used in the past for different sensor generations is provided in Ref. 6.

Although the designation of these various generations is somewhat arbitrary, the trend to implement more electronics at the sensing node is clear and is significant. In addition to the increasing level of electronics at the sensing node, the above discussion also points to the fact that many future sensing systems will include a large number of distributed sensors and to the importance of bus-organized systems. In the next section we will review the architecture of these bus organized sensing systems, the architecture of a parallel and a serial standard bus developed at the University of Michigan, the organization and interface requirements for sensing nodes, and several different standard sensor interfaces developed for various applications.

10.2 SYSTEM ORGANIZATION AND FUNCTIONS

10.2.1 Bus-Organized Sensing Systems

Figure 3 shows the system architecture of a high-performance distributed sensing system, that consists of four principal parts: a host computer, which performs overall system control; a bus structure, which transmits node addresses, host commands, and data to and from the sensing nodes; a microprocessor-driven sensing node, which interfaces with its sensors and the host controller, interprets

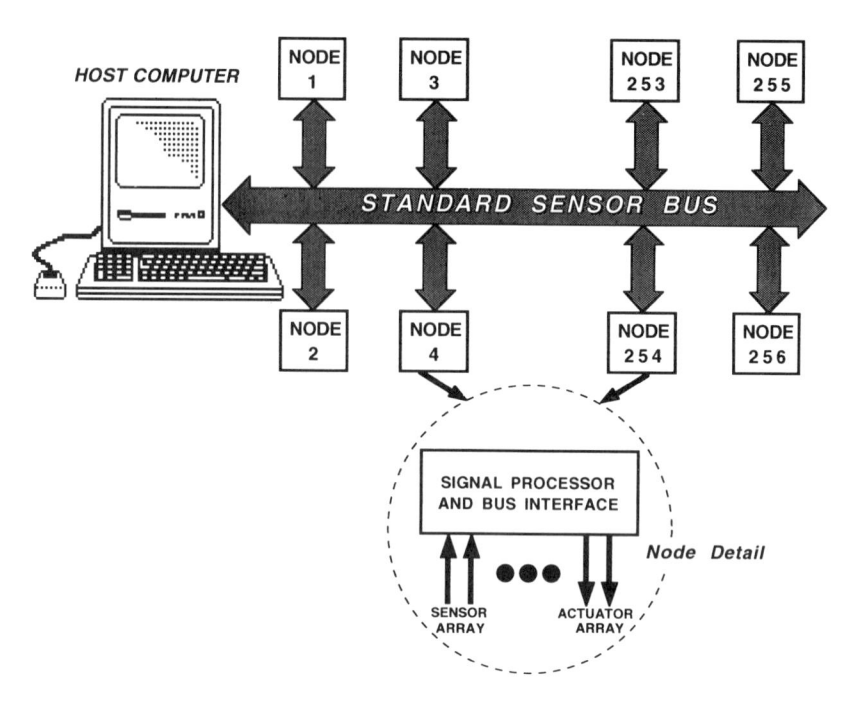

Fig. 3   Overview of a smart sensing system. (After Refs. 6, 9)

and executes commands, and provides the requested information to the host controller; and, finally, a sensor front-end, which contains the sensors along with the necessary circuitry.[6] In this section, we will review and discuss the design and architecture of each of the parts in this system and will present several possible designs and configurations for each part. It should be noted that these examples are used to demonstrate the issues presented, and that other designs can also be developed for different applications.

The host computer is the main controller of the system and usually is the only node allowed to initiate messages. Whenever a node detects its address on the bus, it is responsible for executing and responding to the incoming command. Once the node has executed the command, it continuously monitors the bus until it detects its address again. In the meantime, however, the node also continuously and autonomously samples its sensors and stores the sensor data in RAM so that when the host asks for this data it is readily available. This organization makes sensor interrogations appear much like memory accesses, yielding very fast response times to the host.

The distributed sensing nodes are self-testing, addressable, programmable, compatible with a standard digital bus, capable of handling commands sent by the host controller, offer 12 bits of accuracy using internally-stored compensation coefficients, and are capable of operating up to 32 sensors each. It should be noted, however, that in the present form of this node structure, there is no

10.2 SYSTEM ORGANIZATION AND FUNCTIONS

interpretation of the sensor data by the node and hence no need for *a priori* knowledge of the larger system. The node is smart in terms of its ability to process data and respond in complex ways to the host, but ignorant in terms of the host system and its functions. However, as the complexity of the node electronics evolves, some interpretive ability could be downloaded to the nodes as system optimization dictates.

It is desirable to make the host computer as "generic" as possible. Most current data-acquisition/instrumentation controllers have a great deal of analog electronics for interfacing with their environment (i.e., sensors and actuators). They tend to be custom-designed for a specific range of applications and their control programs are usually not portable. The cost of the electronics required to interface to the system periphery is usually much more than that of the more standard computing hardware of the host comoputer. Due to the cost of this electronics and the associated interconnect problems, the number and the complexity of the sensors and actuators that such systems can operate is limited. Furthermore, although the controller-interface electronics sometimes offers attractive features, such as 16-bit A/D conversion, the final data is normally not very accurate since the analog sensor outputs, which are usually low in amplitude, have to interface via relatively long leads in a noisy environment, resulting in a low signal-to-noise ratio.

To overcome the interconnection problems between the sensors and the controller, a bus-organized system is needed. The sensor output is converted to a digital representation by the sensing nodes, to preserve the signal-to-noise ratio at its node level. This is highly important because of the very low-level outputs that sensors usually provide and because of the high accuracy required by many emerging systems. The converted data is transmitted to the host via a standardized digital bus. As a result, there is no need for extra analog hardware at the host for converting and manipulating low-level sensor oututs. Thus, the only extra hardware imposed on the host computer is that of a simple digital bus driver.[9]

Figure 4 shows an example of a parallel bus developed for communication between the nodes and the host at the University of Michigan. This Michigan Parallel Standard (MPS)[6,9] bus contains 16 lines, including eight bi-directional data lines (DO-D7), a parity line, four control lines for synchronizing message transfers, and three power lines; this bus demonstrates the basic elements needed in a bus-organized sensing system. The control lines are designated according to their primary functions as host handshake (HHS), node handshake (NHS), node bus request (NBR), and special purpose (SP). Message transfer is initiated by the host when it addresses a specific node to request or send data. The first byte (an address) is decoded by all nodes on application of HHS. The addressed node acknowledges by pulsing NHS high and sets NBR high, locking out other nodes for subsequent bus data. Subsequent message bytes are entered and acknowledged using HHS and NHS, respectively, as indicated in Fig. 4. Messages are of variable length, with the bus released when NBR drops low. The SP line can be used as a system interrupt if an event requires immediate

480 INTEGRATED SENSORS

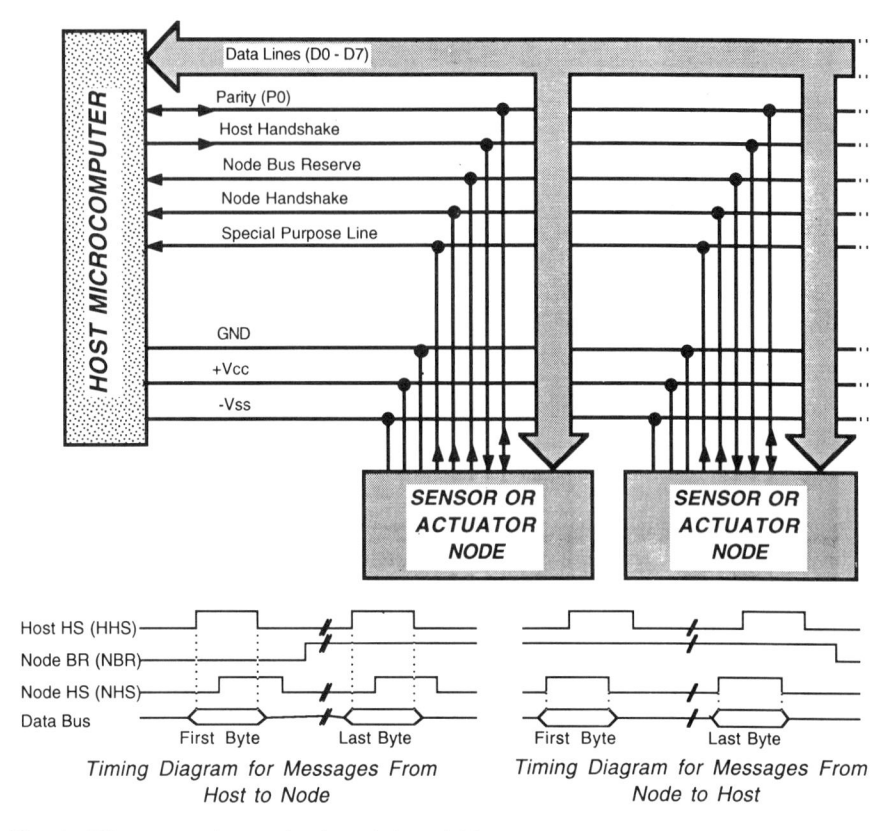

Fig. 4 Diagram and organization of the Michigan Parallel Bus Standard (MPS). (After Refs. 6, 9)

action by the host or it can be used to signal the insertion of a new sensor if there is a replacement or addition to the system. Thus, while the host normally initiates all messages, the nodes can do so under special circumstances where their local capabilities for signal interpretation are sufficient.

For applications where the number of bus lines and/or the bus noise immunity are of paramount importance, a serial bus has also been developed[6,9] as shown in Fig. 5. This Michigan Serial Standard (MSS) bus contains 4 lines: one for data and three for power. Figure 5 also shows the message structure for message transfers to and from the host on the MSS data line. The communication is half-duplex, capable of operating in both directions but not simultaneously. Messages are of variable length using 8-bit words. Message transfer is initiated by the host when it addresses a specific node to request or send data. The node then acknowledges and, in the case of a read command, returns the requested data. Messages are transferred word by word. Figure 5 shows the format of each word. The idle state of the data line is high. Sensor nodes (or the host) begin receiving a word whenever there is a start bit on the data line, that is,

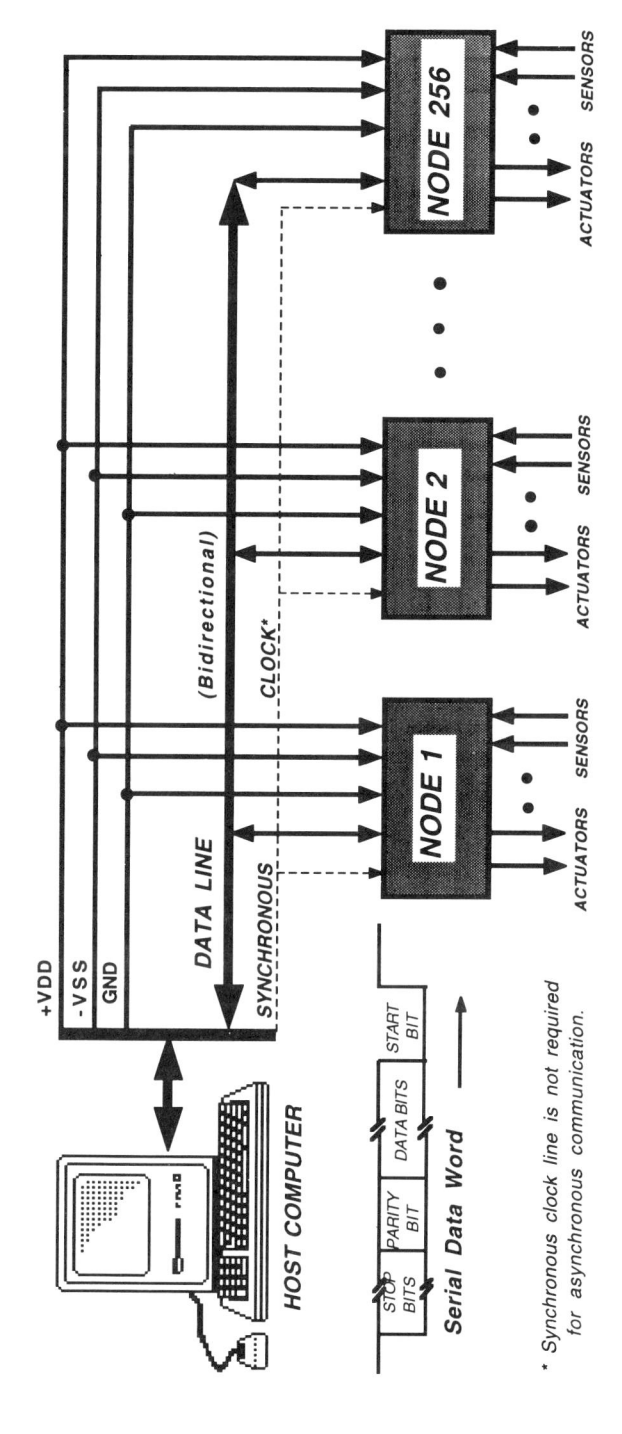

Fig. 5 Diagram and organization of the Michigan Serial Bus Standard (MSS). (After Refs. 6, 9)

whenever the output level of the data line changes from the idle state (high) to a low level. In other words, each start bit triggers the receiver (the host or nodes) to obtain and sample the transmitted word. The receiver sampling rate should be the same as the transmission rate. Alternatively, bits can be detected via edge transitions and verified through timing checks when transmission rates are slower. Since there can be several (up to 256) nodes in the system, address words and data words should be distinguishable by the nodes. There are two ways to implement this: adding an address/data bit to each word or coding the message. Parity as well as timing checks are used to guard against noise. In the event of a parity error, the host requests retransmission, and in the event of repeated errors, the node can be tested by the host, and a particular sensor, or an entire node, can be removed from service. The goal is to be able to detect sensor drift prior to catastrophic failure so that timely replacements can be made and equipment downtime can be avoided. While nodes could initiate messages in this system, they typically do not and there is no real interrupt capability as there is in the parallel structure. An alternative to setting clock timing in each node is the use of an additional external clock line for system synchronization as indicated in Fig. 5.

Since in normal operation the nodes respond to commands from the host but do not intiate messages, there are no contention problems in this hierarchical control structure. In unusual situations where node-initiated messages do occur, any contention problems are detected by node circuitry, which verifies the bus voltage during node transmissions. Retransmissions are then performed subject to a queuing delay proportional to the node address. Parity errors or other problems detected at the host can also lead to polling of the nodes to resolve problems. Given the relatively slow response times of most process variables (milliseconds to fractions of seconds) and the fact that preprocessed information (not raw data) is being transmitted, bus speed is not expected to be a critical issue in determining performance.

It is useful to compare the parallel MPS and serial MSS bus organizations with other present communication standards. The closest equivalent to the parallel bus is probably the IEEE-488 (HP-IB) structure. This bus was designed for communication among several instrumentation systems. Like MPS, it employs eight bi-directional digital data lines; however, there is no parity line in the IEEE-488 structure, and it can handle only 15 nodes as opposed to the 256 in MPS. In addition, power lines are not part of the IEEE-488 structure, whereas they are included as part of MPS. Power specification is particularly important for distributed sensing systems because stable voltage references must be generated within (or supplied to) each of the various nodes. Excluding the power and parity lines of MPS, IEEE-488 bus requires four more lines than MPS, and its bus protocols are more complex, resulting in more sophisticated node hardware and slower communication. With regard to serial buses, MSS functionality is similar to a subset (half-duplex two-wire) of the RS-232 and RS-449 standards. MSS employs a single data line, however, and its protocols are custom-designed for sensing systems with centralized controls. The J1850

10.2 SYSTEM ORGANIZATION AND FUNCTIONS

bus, designed for automotive applications, is addressable and provides a half-duplex serial communication with the use of one data line, as in MSS. Its structure and protocols are, however, different from MSS. J1850 is designed for multi-node systems with no central controller. Its complexity (structure, realization, and protocol) is accordingly much higher than MSS, which provides higher communication speed and simpler hardware and software implementations. Moreover, MSS includes power as part of its specifications.

In order to make the sensing system user friendly, a command library is provided that includes subroutines to execute a set of standard commands (from the host contoller to the smart sensing nodes). As a result, the entire sensing system, including bus, communication, and node structure, is transparent to the user. Introducing a command set is an important step towards the standardization of integrated sensing systems. Application/control programs are written based on this command set, and hence, softward portability completely depends on the wide acceptance of the command set. However, the fact that there are many types of sensors and actuators generates many obstacles to the introduction of one comprehensive command set. As a result, it is a better strategy to divide the command set into two segments: a global set of commands that are shared by most sensors;[9] and a number of custom command sets, one for each specific type of sensor (e.g., mass flow controllers). Each custom command set, in reality, should be a secondary standard created and introduced by manufacturers of that specific sensor. A proposed basic command set for the sensing system is shown[9] in Table 1. The first three commands can also be "broadcast" to all nodes at the same time. To accomplish this broadcasting, there is a dedicated address in the system to which all sensing nodes respond by executing the command associated with this braodcast address without acknowledging the command (a routine that is usually performed).[9] The VLSI sensing-node realization will be discussed in detail next.

TABLE 1 A Proposed Command Set for Smart Sensing Nodes (After Ref. 9)

Command no.	Description
1	Reset
2	Emergency
3	Synchronize
4	Send data of all sensors
5	Send data of one sensor
6	Send identification information of all transducers
7	Send identification information of one transducer
8	Change an actuator output
9	Change the sensity node address
10	Change the sampling list

10.2.2 Sensing Node Organization and Interface Standards

Figure 6 shows the representation of a sensing node in a hybrid "fifth-generation" sensor design.[2,5,6,9] The node is partitioned here as three chips: a sensor chip, containing limited front-end interface electronics; a signal processing/interface chip, including amplification, data conversion, and a microprocessor-based micro-controller which interfaces to the external sensor bus through a standard communication interface; and a PROM which contains the node identification information, including the node address, as well as information pertaining to the sensor compensation techniques. The PROM can be either commercial or custom-designed for this application. While in future applications the entire node might appear as a single monolithic chip using merged process technologies, at the present time a multi-chip hybrid implementation appears more reasonable.

To implement a high-volume, low-cost sensing node, the front-end sensor chip must present a standardized interface to the more generic VLSI interface chip. Only after such a standard interface is utilized is it possible to fabricate generic VLSI interface chips for the high-volume market required to justify their development and production. Since different applications have different

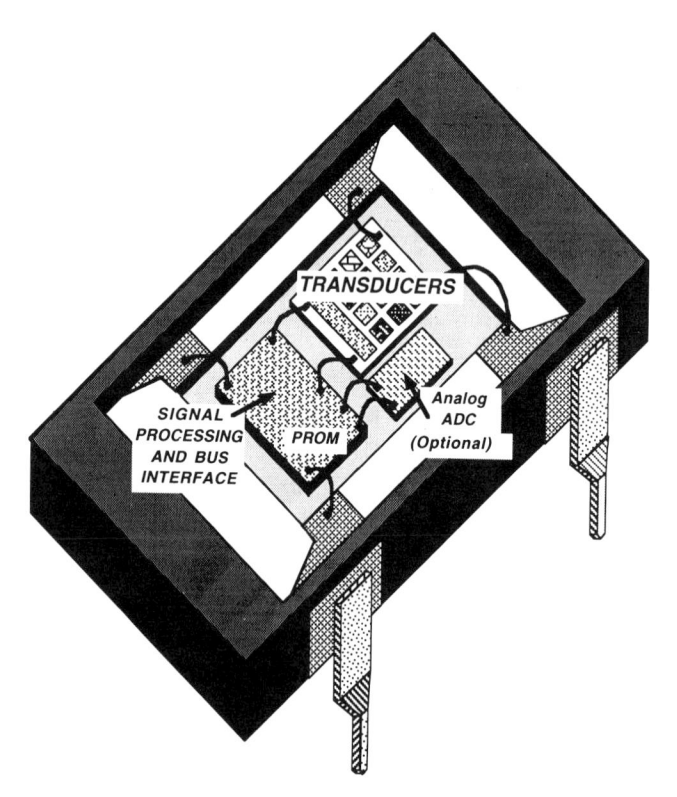

Fig. 6 Schematic representation of a "fifth-generation" hybrid sensing node. (After Refs. 6, 9)

10.2 SYSTEM ORGANIZATION AND FUNCTIONS

sensor-interface requirements, it is difficult to develop a single universally accepted interface standard that can meet all these requirements. Some applications may require a very small number of output leads, while others may require a digital or frequency output. It is, however, possible to define a number of interface standards each of which can satisfy the needs of a different application. Seven different front-end interface standards with lead counts from eight to two for different applications are introduced comprehensively in Ref. 9, and will be briefly discussed below. These interface standards are used to illustrate the signal transmission and interface issues faced in connecting sensors to more sophisticated generic VLSI chips.

The eight-line front-end interface, its command structure, and its timing diagram are shown in Fig. 7. A total of eight interconnect lines are assumed for this interface: three for power (VDD, VSS, and GND), clock input (CLK), serial code input (CIN), strobe input (STR), sensor data output (DO), and sensor-data-valid output (DV). Communication starts when STR is activated.

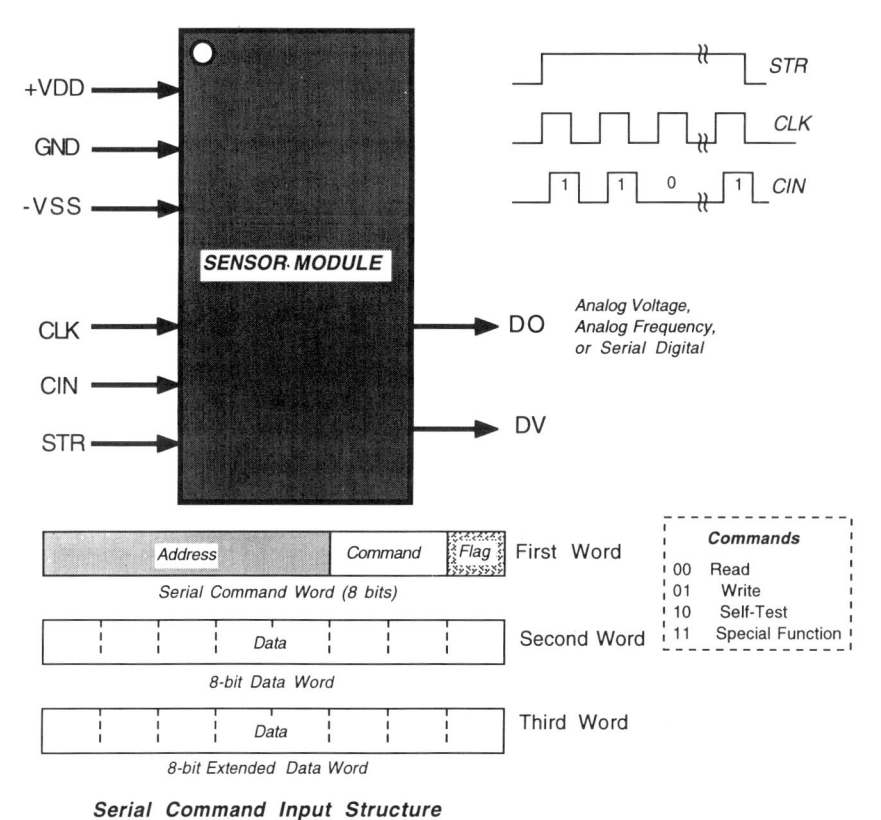

Fig. 7 Standardized eight-line sensor front-end interface and its command structure. (After Refs. 6, 9)

The serial input information (i.e., commands and, if necessary, data) is then clocked into the input register of the sensor chip on CIN. After the serial information is transmitted, STR is deactivated to finish the communication and to initiate the conversion/data output. The sensor chip provides the sensor data output via DO and indicates the integrity of the output by activating DV. As shown in Fig. 7, a command word has three sections: a five-bit sensor address, two bits specifying the command to be executed, and flag bit (which is used solely in the write command). This command structure is the same for all proposed front-end standards from two to eight lines. There are four different input commands: write, read, self-test, and special function. By using the write command, control inputs can be given to front-end microactuators and/or programmable circuit blocks, e.g., DACs (digital-to-analog converters and digital registers). The flag bit specifies the length of the data to be written (one or two bytes). The read command causes the addressed sensor to output its data on the DO line, with DV signaling data valid/ready in the case of on-chip data conversion. It is assumed that output data can be in voltage-amplitude, pulse-rate (frequency), or serial-digital formats. The DV line can also serve as a command acknowledgment. In applications where sensor self-testing is required, the self-test command can be utilized to enable appropriate front-end circuitry. The special-function command is accommodated to provide required design flexibility, and customization of individual applications can be achieved via this command.

There are a total of 32 addressable front-end sensors, any of which can be written, read, given special functions to operate, and remotely tested. While this front-end I/O structure is general and thought to be generic to many future sensors, it nonetheless requires only a modest amount of circuitry on the sensor chip. For most piezoresistive and capacitive sensors for variables such as pressure, temperature, and flow, the required front-end circuitry consists of an input register and latch, an amplifier or oscillator or data converter, and associated circuitry for command decoding and test. For resonant sensors, additional drive/detect circuits might be required. This eight-line front-end standard has been demonstrated in two examples of the sensor chips: a high-performance monolithic microflow sensor,[8] and a multi-element gas analyzer[10] to be discussed later.

The eight-line standard can be reduced to a seven-line implementation for applications that are more sensitive to lead count. As shown in Fig. 8, a two-level clock is employed in order to merge CLK and STR lines together. The clock levels are +VDD and −VSS. Two consecutive GND-to-VSS clocks indicate the start of the word (shown as SSTR in Fig. 8), while a single GND-to-VSS clock informs the end of the word (shown as FSTR in Fig. 8). After the starting strobe, the serial code is entered by the GND-to-VDD clocks the same as the eight-line standard.

The six-line front-end interface and its timing diagram is shown in Fig. 8, where CLK, STR, and DV lines are merged as one bi-directional line. The communication procedure is the same as the seven-line approach except that

10.2 SYSTEM ORGANIZATION AND FUNCTIONS   487

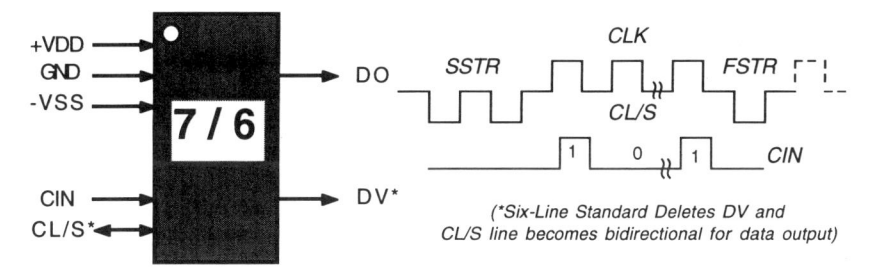

Fig. 8 Seven/six-line sensor front-end interfaces and their timing diagram. (After Ref. 9)

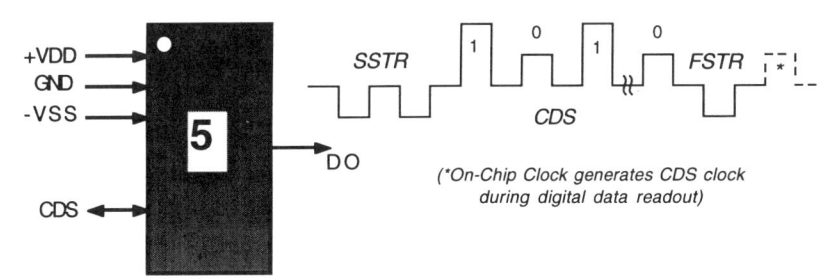

Fig. 9 Five-line sensor front-end interface and its timing diagram. (After Ref. 9)

after a command is transmitted, the communication direction of this bi-directional line is changed, and the sensor chip sends one clock cycle to indicate that the data on DO line is valid. A simpler six-line interface, however, can be achieved by utilizing the seven-line standard without the DV line for applications where a data valid signal is not required.

To reduce the number of the front-end lines to five, it is necessary to use a three-level clock: VSS, VDD/2, and VDD, as shown in Fig. 9. The CLK, CIN, STR, and DV lines are merged together. Communication starts by sending a strobe signal, that is, two consecutive GND-to-VSS clocks. In order to superimpose serial data on the clock, two positive clock outputs are utilized: VDD and VDD/2. After activating the strobe signal, any transition from ground to either VDD or VDD/2 is a clock cycle; a transition to VDD indicates a code with logic "1", while a transition to VDD/2 represents a code with logic "0". The end of the transmitting serial codes, that is, the end of the strobe, is identified by one GND-to-VSS clock. Then, the communication direction is changed, and the sensor chip sends one GND-to-VDD/2 clock to announce that the data on the DO line is valid.

By adding circuitry to the front-end to allow signaling over the supplies and running data over clock, it is possible to define front-end interfaces which collapse the number of required I/O lines from the five to as few as two. Figure 10 depicts the four-line and three-line front-end interfaces and their timing diagrams. The three-line front-end configuration is the same as the four-line interface except that just one supply (i.e., VDD) is used instead of two (i.e., VDD and VSS). Extra circuitry might be utilized in the sensor chip to generate two internal voltage supplies from this one input supply in situations where a negative reference is needed. The strobe signal is superimposed on the VDD line. Two consecutive clocks on the VDD line indicate the start of the word (shown as SSTR in Fig. 10). A single clock on this line signals the end of the word (shown as FSTR in Fig. 10). Serial data are transferred the same way as the five-line approach so that any transition from ground to either VDD or VDD/2 identifies a clock cycle, where a transition to VDD indicates a "1", and a transition to VDD/2 represents a "0". Note that in this interface, unlike the others with eight to six lines, there is no DO line, and the sensor output is superimposed on the C/R line. After the serial command is received, the communication direction is changed, and the sensor chip sends one GND-to-VDD/2 clock to announce that the data on the DO line will be valid. Then, the sensor chip uses the C/R line as an output line to send its output signal which can be analog, pulse-coded (frequency), or serial-digital. The reason that the VDD line, not the C/R line, is used for strobe is that, otherwise, after a read command there would be contention between the sensor chip, which drives the C/R line to send its output, and the VLSI chip, which is trying to use this line to strobe the sensor chip to input the next command.

A two-line front-end interface has been used for areas such as biomedicine, where lead counts can be critical[11,12] and there the sensor chip and the node may be physically separated. The structure and timing diagram of the two-line standard is shown in Fig. 11. All communication and data lines are superimposed on the VDD line with two extra voltage levels (here, these levels are called V1 and V2), while the ground line serves as both ground and voltage and common reference. The strobe signal is identified by a "long" VDD-to-V2 clock pulse; the sensor chip should be capable of distinguishing this long clock pulse from normal clock pulses used for transmitting data. The serial communication is performed the same way as the four-line standard (i.e., any transition from VDD to either V1 or V2 identifies a clock cycle, where a transition to V2 indicates a logic "1", and a transition to V1 represents a logic "0"). The major fact that

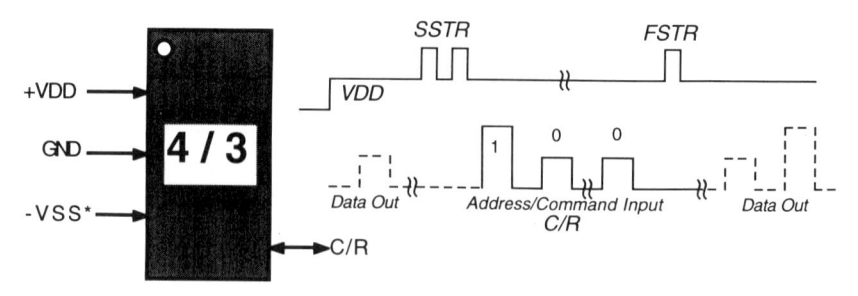

(*There is no VSS line in the three-line interface)

Fig. 10 Four-line and three-line sensor front-end interfaces. (After Ref. 9)

10.2 SYSTEM ORGANIZATION AND FUNCTIONS

distinguishes this standard from the others is that the input data is frequency-coded in the form of high-level current spikes on either the VDD line or the ground line. Generation and detection of these high-level current spikes are relatively easy tasks; however, extra circuitry is required since there is a trade-off between the standard lead count and the simplicity of the front-end circuitry. A simpler form of this standard has been used in a microcatheter.[11,12] In this application there is no command, and a one-level clock, superimposed on the VDD line, controls a multiplexer where the output of the multiplexer is a frequency-coded stream of high-level current spikes. Thus, many multiplexed sensors are monitored utilizing the two power lines.

These front-end interface circuits are typically integrated on the same substrate as the sensor, and feed their signals to a more generic VLSI chip. For example, the eight-line front-end standard can be connected to a VLSI control chip whose block diagram is shown[9,10] in Fig. 12. In this design, it is assumed that the sensor chip produces either a pulse-coded (i.e., frequency-mode) or

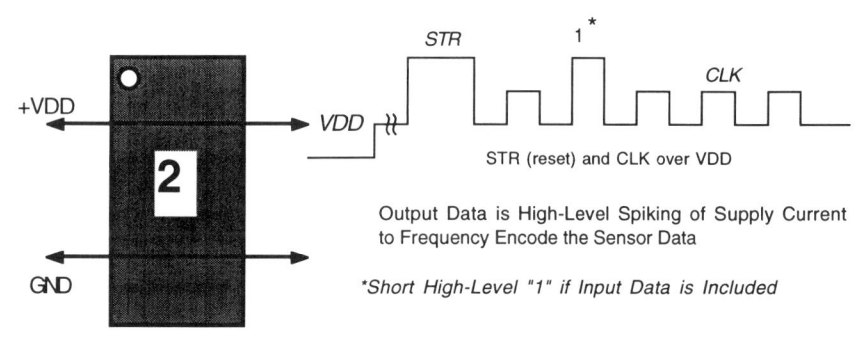

Fig. 11 Two-line sensor front-end interface and its timing diagram. (After Ref. 9)

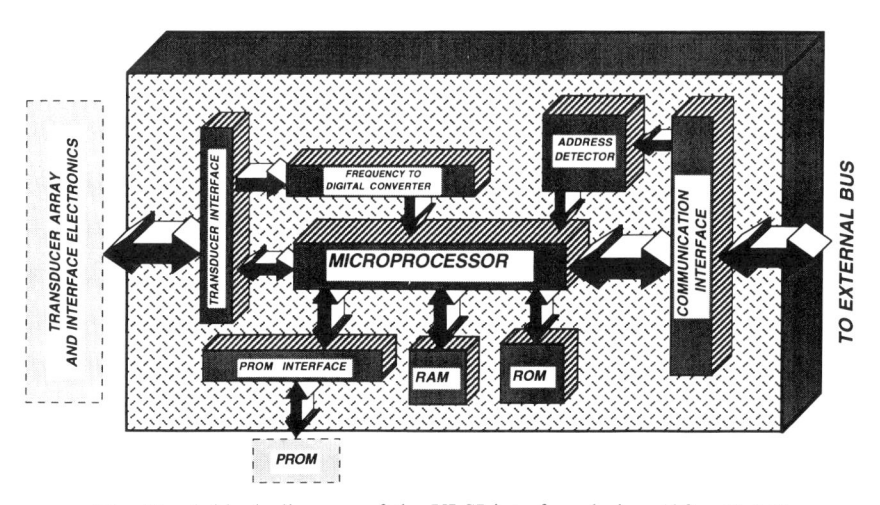

Fig. 12 A block diagram of the VLSI interface design. (After Ref. 9)

digital output. A custom-designed 12-bit frequency-to-digital converter produces a digital representation of the frequency input. Thus, the VLSI signal-processing chip is designed with a digital process, resulting in a much more compact layout, even with an expanded on-chip memory, on chip PROM, and a parallel interface.

An ADC, if necessary, would be located on an extra (optional) analog chip. It should be mentioned, however, that most microsensors with analog output have preamplifiers on their chips, and, hence, by adding a small amount of circuitry (e.g., a Schmitt-trigger oscillator) to the sensor chip it is possible to convert the analog signals to pulse-coded (frequency) signals compatible with this VLSI design.

A RAM stores the digitized sensor data, and a 512×8-bit ROM stores the control-program code for the microprocessor. The external PROM is serially interfaced with the signal-processing chip to reduce the number of I/O pads. With the control-program code for the sensing node stored in its ROM, the PROM accesses are needed only in node initialization and when uploading the compensation coefficients to the host. Therefore, the speed penalty associated with serial PROM communication is not critical here. In an optimized implementation of this node design, both ROM and PROM might be merged into a single EPROM on the processor chip itself with a parallel on-chip data path. The communication interface is configured from the Michigan Serial Standard (MSS) bus.[6]

A custom-designed microprocessor is the main controller of the VLSI interface chip. The microprocessor is designed to be as simple as possible. Since 12-bit accuracy and a 12-bit address bus (4K memory) are required, a 12-bit microprocessor with a 8 MHz clock and only one internal bus is needed. The microprocessor can handle 22 instructions with three different addressing modes and can accept one maskable interrupt. Although the microprocessor has a 12-bit data path, its memory is 8-bits wide so that it can interface with commercially available PROMs, which are configured in 8-bit words. The implementation of this generic VLSI chip is practical using today's technology. For example, using a conventional 1.2 μm double-metal single-poly CMOS technology, this chip requires a total area of only ≈ 12 mm^2.

The previous discussion shows that it is practical to develop integrated sensors with some custom-designed interfaces that can be connected to generic VLSI processing and control chips. The computational and processing power provided by these integrated-circuit chips can significantly enhance the functionality of the sensor and transform it from a passive component to a system peripheral that can communicate in complex ways with the controller. As mentioned before, one of the most promising features of integrated sensors is the versality and functionality they offer in terms of self-test, calibration, and compensations features. These features are illustrated below and in examples at the end of this section.

10.2.3 Sensor Characterization and Compensation

Digital compensation is one of the most attractive and powerful features offered by integrated sensors. It should be performed in a way that is transparent to the user, either at the sensor node ("sixth-generation" sensors) or at the host controller ("fifth-generation" sensors). The great majority of problems in nonideal sensor behavior demand only modest data-processing power, and, thus, can be readily compensated in software. In contrast, the analog compensation of such nonideal behavior is usually complicated, requires custom-designed circuitry and results in higher cost. It should be noted that the entire sensor analog-signal path can also be digitally compensated with better performance and higher reliability than with analog compensation.

As an example, offset and nonlinearity compensation calculations can be performed by using a polynomial equation, where compensated data is deduced from a polynomial with uncompensated data as its variable. Here, sensor characterization is achieved by specifying the coefficients of this specific polynomial. Nonlinearity of a variety of commercial and custom capacitive pressure sensors has been improved by more than an order of magnitude using this approach. Figure 13 shows the uncompensated and compensated data of a commercial piezoresistive pressure sensor (Motorola MPX-100A). Notice that a relatively simple digital compensation algorithm, that is a fifth-order, one-dimensional polynomial equation, improves the sensor performance considerably.

Cross-performance sensitivity is probably the most common sensor problem and is more difficult to deal with. Sensors usually respond to other conventional variables, especially temperature, in addition to their primary sensing parameter. A major consideration of sensor design has always been maximizing the desired sensitivity and minimizing the undesired sensitivity. It has been shown[13,14] that utilizing two-dimensional polynomial equations to compensate for the temperature dependence of pressure sensors is very effective. In the same way, it is expected that multi-dimensional polynomial equations should be effective for the compensation of higher-order cross-parameters sensitivies.

There are two major problem areas associated with using digital compensation: the cost of generating the proper compensation coefficients, and the computational power and time required to perform the actual compensation. In the principal method for generating the digital compensation coefficients, sensors are characterized individually, and each one has different coefficients. It is highly important to have an efficient characterization system to improve both device performance and final device cost. This characterization process can, fortunately, be automated to perform the measurement, calculation, and storage of appropriate coefficients for thousands of sensors in one run, resulting in mass sensor characterization.[9]

The second problem area with regard to using digital compensation is associated with the required computational power and time. To analyze the use of digital compensation objectively, measures for both digital compensation complexity and overall system performance are required. Here, sensor sampling rate is considered an overall system performance measure, while the degree of the compensation polynomial is considered an identifier for compensation complexity. Generally speaking, sensor sampling rate depends on many parameters, including the number of sensors in the system, the number of sensing

492 INTEGRATED SENSORS

nodes, the number of sensors per node, the host computer speed, the sensor node speed (or microprocessor clock frequency), the bus speed, the digital compensation complexity, and the degree of oversampling (i.e., averaging sensor data to minimize noise). In most applications, sampling rate is limited by the sensor node speed when a small amount of digital compensation is applied, while with complex compensation the sampling rate is limited by the host computer. To reduce the overall noise in the system, it is also possible to oversample the sensor data.

The example shown in Fig. 13 illustrates the fact that although the sensor structure and design may not provide optimum performance, various digital signal-processing techniques can be used to improve the accuracy. These features, of course, can be achieved only if the sensor response is stable over time. Therefore, one of the most important goals in the design of semiconductor sensors is stability. As long as this is achieved, digital compensation and signal processing can be utilized to overcome problems such as nonlinearity, offset, and cross-parameter sensitivity. If the sensor response drifts over time, a new set of polynomial coefficients have to be generated and used in any data compensation routines. This will require calibration of the sensor, which normally has to be done in a controlled environment. Sensor calibration requires the application of a reference signal of known quantity to the sensor and measuring the sensor output. It is extremely desirable to be able to perform this calibration without having to remove the sensor from its environment.

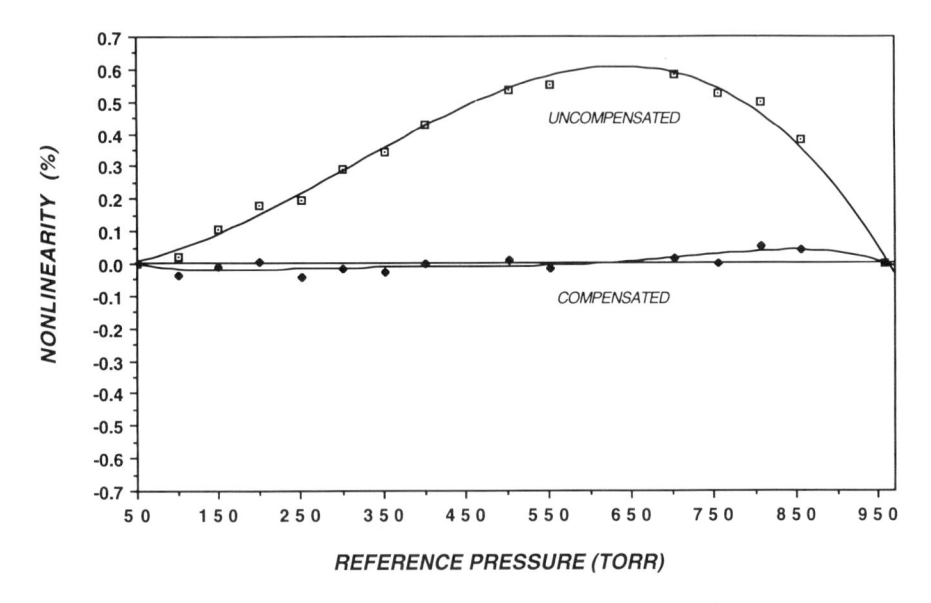

Fig. 13 Compensated and uncompensated data nonlinearity of a commercial piezoresistive pressure sensor at 21.5 degrees centigrade. (After Ref. 13)

Some emerging solid-state integrated sensors have some self-testing and calibration capability, although much more progress and developement is needed it this area.

10.3 INTERFACE ELECTRONICS

One of the most important functions of sensor interface electronics is to convert typical sensor signals into a format that is more compatible with external electronic systems that control the sensing system. Most semiconductor sensors exhibit electrical performance characteristics or possess terminal characteristics that are not compatible with the overall electronic system performance and interface requirements. The electrical signals generated are generally low in amplitude, or the sensor generates changes in its elemental value, for example, capacitance or resistance changes in response to applied external parameters, that are only a very small percentage of the nominal values. Many sensors possess interface impedances at the required signal frequencies that are very high and need to be buffered. Furthermore, it is extremely desirable to multiplex the outputs of many sensors into a single output line to reduce the number of output leads. In addition to these characteristics, sensors sometimes exhibit drift in their response over time, temperature, and other secondary parameters, and could suffer from other problems such as nonlinearity and offset due to the transduction technique employed. Previous chapters have discussed some of these characteristics in terms of the different types of sensors. This section presents a review of typical circuit blocks used in integrated sensors.

10.3.1 Standard Circuit Blocks

In this section we will review the basic functions of interface circuitry that are required to enable semiconductor sensors to interface with higher-level control electronics.

Amplifiers. Amplification of sensor signals remains one of the most common and important functions in many applications where the amplitude of the signal is low. Amplification of these signals at the sensor site, before transmission to the outside world, not only enhances the overall signal-to-noise ratio, but also allows for the full utilization of the dynamic range of an ADC for those sensors that incorporate an ADC in the sensor module. For many integrated sensors, this amplification can be achieved by using MOS and bipolar amplifiers that require only a minimal amount of circuitry with nominal gain, bandwidth, and performance specifications. CMOS amplifiers are the most suitable since they provide high gain and high input impedance through a relatively simple and compact circuit, and are compatible with the integration of high-density digital circuitry and sensors on the same chip.[15,16]

Figure 14 shows the circuit diagram of a typical two-stage CMOS operational

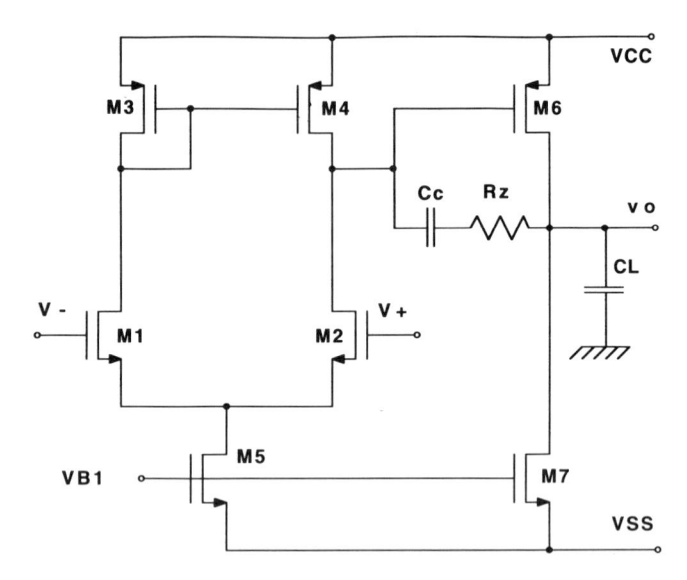

Fig. 14 Circuit diagram of a two-stage CMOS operational amplifier.

amplifier. The input stage consists of a source-coupled differential pair that can provide the high differential gain and low common-mode gain needed to reject common-mode signals. This input stage also performs differential to single-ended conversion without the need for an additional stage through the active current-mirror load configuration employed. Additional gain is provided by the second gain stage consisting of transistors M6 and M7. The overall differential voltage gain of the amplifier is:[15,16]

$$A_{vd} = \frac{g_{m2}}{g_{ds2} + g_{ds4}} \times \frac{g_{m6}}{g_{ds6} + g_{ds7}} \quad (1)$$

$$g_{mi} = \sqrt{4 k_i I_{Di}}$$

$$I_{Di} = k_i (V_{GSi} - V_{Ti})^2$$

$$k_i = \tfrac{1}{2} \mu C_{ox}(W/L)_i$$

$$g_{dsi} = \lambda_i I_{Di}$$

where I_{Di} is the DC bias current through the i-th transistor, V_{Ti} is the threshold voltage of the i-th transistor, μ is the surface mobility of electrons or holes, C_{ox} is the gate oxide capacitance per unit area, W and L are the channel width and length of the transistor respectively, λ_i's are the channel length modulation parameters determined by the particular CMOS process used to implement the amplifier, g_{mi}'s are the transconductances, g_{dsi}'s are the channel conductances, and V_{GSi} is the gate-to-source voltage of the i-th transistor.[15,16]

It is evident that the gain of the amplifier can be set through proper selection of the DC bias current, transistor dimensions, and CMOS process parameters. In addition to the incremental voltage gain of the amplifier, the gain-bandwidth product is another important parameter that can affect the amplifier performance. For a CMOS amplifier, it is generally given by (assuming low output load capacitances):[15,16]

$$\text{Gain-bandwidth} = \frac{g_{m2}}{C_c} \qquad (2)$$

where C_c is the Miller compensation used around the second gain stage of the amplifier to ensure stability of the amplifier response at high frequencies. Trade-offs between gain and bandwidth can easily be made for the amplifier. It is worth noting that for most sensor applications the required signal bandwidth is typically in the tens of kilohertz range, which is much lower than that needed in most other applications. This means that relatively high gains can be achieved with small operating currents in small areas. In addition, for these applications one needs to intentionally limit the upper cut-off frequency of the amplifier to reduce total input-referred noise. This can be done by limiting the bandwidth of the amplifier by increasing the value of the Miller compensation capacitor.

CMOS amplifiers are now easily capable of providing open-loop gains of 90 dB, are fast, can be compensated for their offset using various techniques involving charge storage and cancellation,[15,16] and consume little power and die area. These amplifiers are 3 to 5 times smaller than bipolar amplifiers, making it possible to include tens of per-channel amplifiers on a single chip. In addition, the input impedance of CMOS amplifiers is very high, which makes them ideal for sensor applications. As seen from Fig. 14, these amplifiers require only a few transistors and are easy to design using a variety of CAD (computer-aided design) tools.

In addition to the standard amplifier configuration shown in Fig. 14, a variety of other topologies have been developed to satisfy the needs of different applications. In applications where very high gain is needed, cascode stages can be utilized to reduce the output conductance, g_{ds}, provided by the transistor active loads, thus increasing the voltage gain appreciably. Using cascode stages, however, reduces the output voltage range and the input common-mode range of the operational amplifier. This can be partially overcome using a folded cascode configuration.[16] Both of these approaches result in larger area and higher power consumption. These amplifiers can also provide a higher gain-bandwidth product, although for most sensor applications the bandwidth provided by the standard amplifier topology is quite sufficient.

Another important performance parameter for operational amplifiers used in sensor interfaces is the total input-referred noise. As mentioned before, most sensor signals are very low in amplitude, and it is desirable to lower the total input noise through appropriate circuit and process design techniques. The total input-referred noise voltage of the CMOS amplifier shown in Fig. 14 is dominated by the noise of the input transistors, and primarily consists of thermal and $1/f$ noise.[15] The noise contribution from the load transistors M3 and M4 can be ignored because of their small width-to-length ratios. Therefore, the total input-referred noise voltage can be approximated by:

$$v_{eq}^2 = 4kT\left(\frac{2}{3g_m}\right)\Delta f + \frac{K\Delta f}{C_{ox}WLf} \tag{3}$$

where the first term represents thermal noise and the second term represents $1/f$ noise. K is the flicker noise ($1/f$) coefficient that is a function of a particular fabrication process, Δf is the bandwidth of the amplitifier, T is the absolute temperature, k is the Boltzmann constant, and the other parameters are defined as before. It can be seen that the $1/f$ noise can be minimized by making the dimensions of the input transistors as large as possible, and by making the gate oxide thickness as small as possible, while the thermal noise contribution can be minimized by increasing the transconductance of the input transistors. It should be further noted that p-channel transistors generally provide a lower level of noise, and are, therefore, preferred as input devices in sensor applications.[17] Amplifiers that utilize bipolar or JFET transistors have a lower noise level than MOS amplifiers. Nonetheless, various circuit and fabrication techniques can be used to minimize the noise in CMOS amplifiers[15,18] without having to use these other technologies that may complicate the overall process.

Signal filtering can also be incorporated in these amplifiers. For multiplexed multisensor systems, filtering of the signal is often required to prevent problems such as aliasing that can introduce high-frequency noise into the signal passband thus override the low-amplitude sensed signal. Signal filtering can also improve the signal-to-noise ratio of the sensor as it filters the out-of-band high-frequency noise that may be introduced by the electronic devices themselves. A variety of low-pass, bandpass, and high-pass filter circuits are available using switched-capacitor circuit techniques or more conventional resistor–capacitor techniques.[19]

Buffers. In addition to signal amplification, impedance transformation is also often required for both resistive and capacitive sensors. A low impedance output from the sensor is required not only to ensure maximum signal transfer to the next stage, but also to drive output leads and reduce the susceptibility of the sensed signal to environmental noise. Typical CMOS operational amplifiers similar to those discussed can be easily used for this purpose. Often the amplification and impedance transformation can be achieved simultaneously in such circuits. For many applications it is possible to simply use source followers that require only a few transistors and consume very little area and power.[16] For capacitive-based sensors, one has to ensure that the input capacitance of the buffer is as small as possible to minimize capacitive loading. Several approaches have been implemented using various source-follower configurations, and will not be discussed here.[20]

Multiplexing. Reducing the number of output leads is another important design consideration for a majority of sensors. Data multiplexing can not only reduce the amount of circuitry required for the sensor (circuit blocks such as the ADC can be shared among several sensors), but also reduces the number of external leads by either multiplexing several sensed signals onto a single output lead, or by superimposing clock and control signals over the power lines.[11,12] Low-noise analog multiplexing is an especially important function in sensing arrays and systems that require the simultaneous measurement of many signals to accurately extract the parameter of interest. In addition, reducing the number of leads that are interconnected to a sensor package is very important in simplifying the packaging, reducing the cost, and improving the long-term reliability of the system in which the sensor is integrated. Automotive applications of sensors have found this to be of paramount importance and it is expected that almost all sensors fabricated for this area will have particular attention paid to lead reduction.[9]

In addition to these primary functions, the interface signal-processing circuitry is increasingly being used to perform functions such as self-testing of the analog circuit blocks and offset cancellation that will permit full utilization of the dynamic range of the ADC.[21] It should be emphasized again that signal amplification, filtering, and multiplexing of several data channels onto a single data line also reduce packaging problems and inherently improve the reliability of the device by enhancing the signal-to-noise ratio and by eliminating the need for too many external leads.

Two points should be mentioned with regard to the sensor interface circuitry. First, although the design and layout of many of these circuit blocks is still custom, the progress made in the last few years in the CAD of analog integrated circuits and the availability of cell designs for analog functions have greatly simplified the implementation of many of the functions discussed above. Silicon compilers are now available for a variety of analog functions, which relieve the sensor designer from performing detailed circuit and layout designs.[22] Many of these CAD tools have reduced design time and improved reliability. Therefore, it is now possible to design a semi-custom analog integrated circuit for a particular sensor and fully test its functionality without committing the design to silicon. Second, the integration of many of these circuits with the sensor has become less complicated as integrated sensors have evolved towards greater compatibility with integrated-circuit fabrication processes, as will be discussed later.

10.3.2 Interface Circuits for Resistive Sensors

Many semiconductor sensors convert the physical parameter being measured into a resistance change by utilizing various elements, including piezoresistors used in pressure and acceleration sensors, magnetoresistors, and Hall bridges. The most common interface circuit used to measure the resistance change is the traditional differencing amplifier shown[23] in Fig. 15. Many resistive sensors

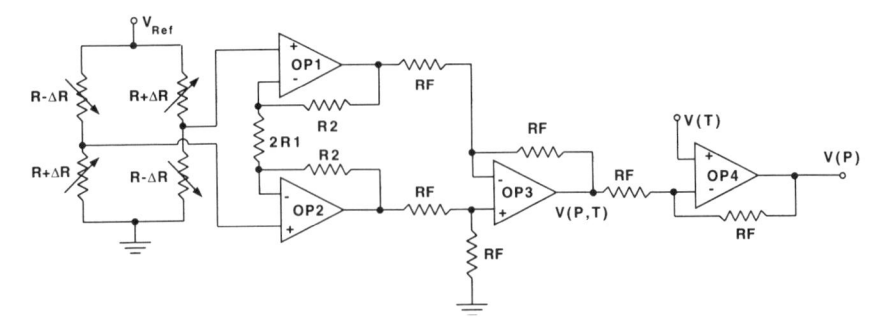

Fig. 15 Circuit diagram of a differencing amplifier used in the measurement of resistance changes produced in resistive sensors. (After Ref. 26)

are configured so that the variable sensor resistance is part of a resistance bridge. The bridge is usually balanced (i.e., the resistances in the opposite legs of the bridge are nominally equal) and is biased by a voltage reference. As the resistance of the elements in the bridge changes by an amount ΔR, an output voltage is generated across the output terminals of the bridge. This voltage change can be measured and amplified using the differencing amplifier. The output voltage of the complete circuit is shown to be equal to:[23]

$$V_{out} = \frac{\Delta R}{R} \frac{R_2}{R_1} V_{Ref}. \qquad (4)$$

The circuit shown in Fig. 15 avoids taking current from the bridge and produces an overall gain that is equal to R_2/R_1. Amplifier OP3 acts as a voltage follower, although it can be designed to provide additional gain if necessary. The output of this amplifier is a function of both the parameter being measured, and of temperature. Temperature sensitivity stems from both the sensor itself, and from the interface electronics and the resistive elements in the sensor. Response to temperature can be compensated for by providing a temperature-dependent signal input through amplifier OP4. Note that the output voltage provided by the bridge is a linear function of the reference voltage. If resistance changes are very small, the value of this reference voltage can be increased in order to increase the amplitude of the signal received by the amplifier. This will, however, increase the current in the bridge and the power dissipation. Another point of interest in the design of the bridge is its linearity. It is often desirable to achieve perfect linearity between the output voltage and the resistance elements used in the bridge. This can be done by ensuring that the four resistances in the bridge are all equal under no-load conditions. Usually, this is hard to achieve due to the unavoidable variations in the fabrication process, which necessitates trimming of the individual bridge resistors prior to packaging. This is achieved by laser trimming techniques to the point where the bridge is prefectly balanced and produces zero output voltage.

10.3.3 Interface Circuits for Capacitive Sensors

Although resistive sensors have been widely used in the past, recently, capacitive sensors have begun to emerge in many areas. Capacitive sensors are attractive because they possess a number of important features desirable in many integrated systems. They generally exhibit a much lower temperature sensitivity than their resistive counterparts because of the elimination of the temperature-sensitive resistor elements in their structure. They also provide an overall higher sensitivity and resolution, and consume much less power, which makes them suitable for many low-power applications such as biomedical and instrumentation systems.

For these capacitive sensors, several possible read-out approaches exist as shown in Fig. 16. The most common approaches use the capacitor as the timing element in an oscillator so that the output frequency is a function of the

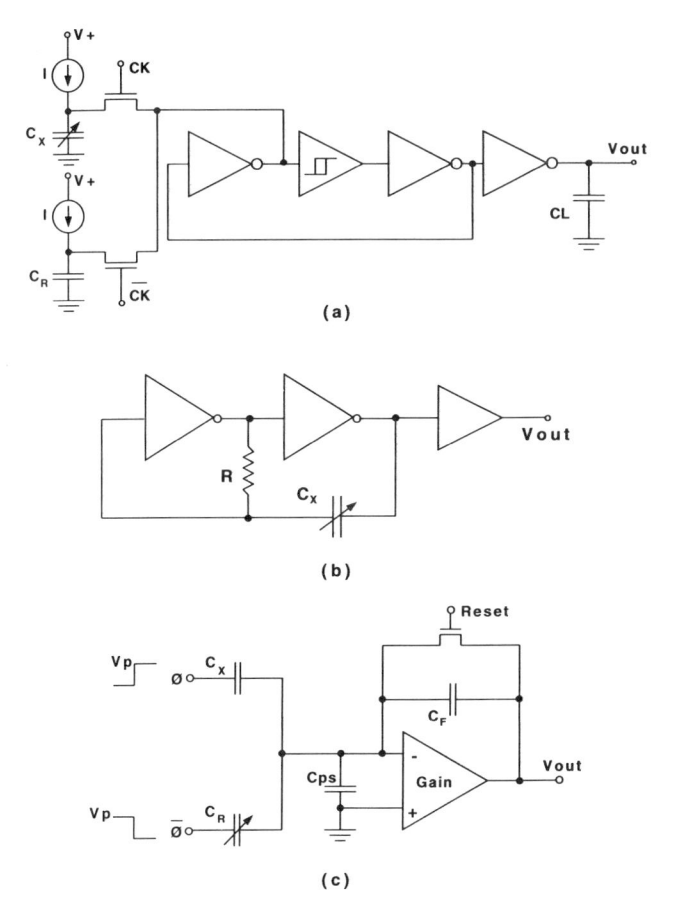

Fig. 16 Circuit diagrams of three interface circuits for measuring capacitance change. (After Ref. 26)

capacitance and, hence, of the parameter being monitored. Figures 16a and b show a Schmitt-trigger oscillator[24,25] and an *RC* oscillator design.[26,27] The Schmitt-trigger oscillator design uses a high-threshold buffer to set the timing, driving the capacitor with a current source to ramp the input voltage. The hysteresis provided by the Schmitt trigger causes the closed-loop ring oscillator to oscillate at a frequency that is a function of the threshold voltages of the trigger, of the amplitude of the current source, and of the capacitance provided by the variable capacitor C_X. This approach can be sensitive to both power supply voltage and to temperature. However, since many capacitive sensors have very low temperature coefficients, the output frequency can be differentiated against an on-chip reference capacitor to subtract the influence of temperature from the offset and to reduce the slope error to a percentage of value.

The *RC*-oscillator approach reduces power-supply sensitivity, but requires both plates of the capacitor to be off ground, which may not be desirable in many applications. Nonetheless, temperature compensation is still possible. Note that for both of these oscillator techniques, the effective read-out speed is slow. For example, for a frequency of 1 MHz and a precision of 8 bits, a compensated read-out will take $\approx$ 1 ms. This is fast enough for most applications, but could be marginal in highly multiplexed systems.

A considerably faster read-out approach, and one that can potentially provide a much higher resolution than other techniques, is the switched-capacitor technique shown[28-30] in Fig. 16c. Here the sensor capacitance is pumped against a reference capacitor, C_R, to which it is nominally matched. The difference charge, which is proportional to $C_X - C_R$ is integrated to produce a voltage pulse whose amplitude is proportional to the capacitance difference divided by the feedback capacitance, as shown below:

$$V_{out} = V_p \frac{C_X - C_R}{C_F}. \quad (5)$$

It is evident that the output signal is insensitive to the gain of the integrator (so long as the gain is reasonably high) and to input parasitic capacitance, C_{ps}, since the op-amp (operational amplifier) maintains a virtual ground across the two terminals of this capacitor. Signal read-out in this technique takes less than 20 μs, and is governed by the settling time of the op-amp. Input offset variations and reset charge injected through the MOS transistor switches can be removed using several techniques such as correlated double sampling,[31] and charge redistribution and digital correction techniques.[32,33] In this latter technique, charge redistribution and sensing is used to measure the difference between the unknown capacitance and the reference capacitance, as discussed above, using digital techniques. It is thus possible to eliminate the effects of initial mismatch, charge injection due to the MOS switches, and kT/c noise, and obtain measurement resolutions down to $30-50 \times 10^{-12}$ F at speeds approaching several tens of kHz.[32,33]

In addition to these capacitance sensing circuits, a number of other circuits have been developed for interfacing to capacitive sensors that are used to measure physical parameters such as pressure and acceleration. Most of these sensors utilize a microstructure that moves in response to the parameter being measured. The movement of the microstructure then produces a capacitance change. The circuits discussed above all directly measure the capacitance change between the moving microstructure and a second fixed capacitor plate, and do not interfere with the movement of the microstructure. A disadvantage associated with this movement is that the output of the sensor is a nonlinear function of the parameter being measured, since the capacitance value is inversely proportional to the separation between the two plates of the capacitor.

Another technique that can be used to measure the parameter of interest involves the use of force feedback to the sensing microstructure to keep its position fixed.[34-36] The amplitude of the signal used to generate sufficient force to achieve this is then a function of the parameter of interest. For most current semiconductor sensors the force is generated electrostatically. An implementation of this force feedback circuit is shown[36] in Fig. 17. The moveable microstructure MS forms one plate of two variable capacitors C1 and C2, while two fixed plates MP form the other two plates of these capacitors. Plates MP are driven with drive signals VD and $\overline{VD}$, and the signal induced on plate MS is buffered by a high-impedance buffer. When MS is in the neutral position, we have C1 = C2 and no signal is generated on MS. As MS moves in response to the parameter of interest, the charge induced on it is nonzero and a signal is created that is proprotional to the differential change in capacitance. This signal is filtered and amplified to generate the feedback signal V_0 which is fed back to MS through an isolation buffer stage, to relocate plate MS to its neutral position. Other implementations of the same basic circuit have been demonstrated,[34,35] employing the same idea.

This force-balancing technique offers several features. First, since the microstructure is effectively stationary this technique offers improved linearity over the capacitance-sensing technique. Second, it can offer very high precisions at moderate bandwidths. Third, because of its higher precision, the technique can be used for high-sensitivity applications such as sensitive accelerometers and pressure sensors. Note that this technique provides self-testing and calibration capability using the same circuit blocks used for sensing. A possible

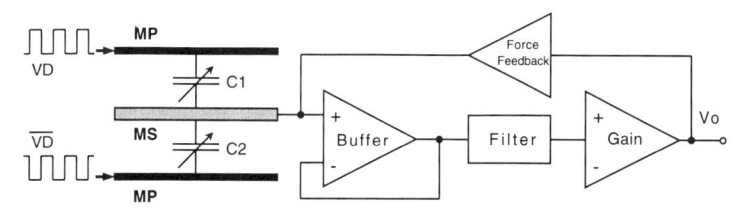

Fig. 17 Application of force feedback read-out in solid-state sensors.

disadvantage of this technique is its more elaborate electronic and interface circuitry, and possibly higher voltage supplies required to generate sufficiently high forces to implement the force feedback.

These circuit blocks represent the most common and useful interface circuits used in semiconductor sensors. There are, however, a variety of other custom-designed circuit blocks that have been developed for specific applications that offer more optimized characteristics.

10.4 FABRICATION TECHNIQUES

One of the main features of integrated sensors is the capability to integrate on-chip interface and control circuitry with the sensor itself. As discussed previously, integrated sensors can provide many desired functions in sensing systems that require data acquisition from many sensors with high precision and high speed. On-chip circuitry can be used to improve signal-to-noise ratio, to buffer signals, and to achieve desirable features such as self-testing, compensation, and auto-calibration. The level of circuit integration, and the appopriate partitioning of the system to achieve proper balance between overall sensor performance, packaging requirements, testing techniques, and cost, have to be determined for individual applications. Semiconductor sensors that are more compatible with circuit integration are obviously preferred. The fabrication process used for the sensor should provide minimal disturbance to the process used for the electronics, and should, ideally, be performed either before or after the circuit process is completed. The choice of the sensor technology is often forced by the required performance characteristics of the sensor. In this section we will first review different standard circuit-fabrication technologies, and will then discuss integrated sensor technologies that have been developed to fabricate a number of different devices.

10.4.1 Review of Circuit-Fabrication Technologies

To select an appropriate sensor on-chip technology, a number of requirements have to be satisfied. These include high packing density (small die area), low power dissipation, functional versatility, since the on-chip circuitry requires the integration of both analog and digital electronics on the same substrate, fabrication simplicity (high yield), and drive capability. Historically, integrated circuits have been fabricated using three main technologies: (1) bipolar junction transistor (BJT); (2) metal-oxide semiconductor (MOS), and (3) complementary MOS (CMOS).

Bipolar ICs have typically been used in very high-speed and high-drive applications where large loads need to be driven at fairly high speeds. They, therefore, consume large amounts of power and area. In the digital-circuit arena, bipolar technologies have been mostly replaced by MOS technologies because bipolar devices require complicated processing and have low packing density.

10.4 FABRICATION TECHNIQUES 503

In the analog regime, however, many bipolar circuits and functions are being used because of the inherently higher gain and speed of bipolar analog circuits and because of the higher uniformity achieved with bipolar transistors. However, bipolar technologies do not lend themselves easily to the implementation of a number of analog functions that are critical to circuits required in sensor applications. The first important requirement is the need for high input impedance interfaces for a variety of sensors, which is usually difficult to achieve with bipolar devices unless area and power-consuming circuit techniques are employed. Second, bipolar device structures do not lend themselves easily to analog multiplexing and cannot be used as bilateral switches. To overcome some of the shortcomings of the BJT, other technologies such as integrated injection logic[37] and bipolar-CMOS (BiCMOS)[38] have been developed. BiCMOS technologies have become more popular because they combine the high packing density/low power dissipation of MOS technology and the high drive/gain capability of bipolar devices. They, however, have a more complex fabrication process, which restricts their application to very high-performance applications (e.g., very high-speed VLSI circuits).

MOS circuit technologies, on the other hand, offer very high packing density and low power dissipation which makes them very suitable for many applications. In addition, their very simple fabrication process results in high yield and increased reliability. They best satisfy the requirements for integrated microsensors. MOS devices can be easily used as bilateral analog switches (or transmission gates), which is useful for making a small multiplexer. Although they have a higher $1/f$ noise than BJT devices, the noise level is still lower than the minimum required level for many sensor applications. The two most frequently used MOS technologies nowadays are enhancement-depletion n-channel MOS (ED-NMOS), and CMOS. ED-NMOS circuit technology offers a very simple fabrication process and moderate device performance levels. The fabrication process requires only five masking steps and standard wafers without the need for epitaxial layers. Operational amplifiers and a variety of digital and analog functions have been designed using this technology and are used in a number of integrated sensors.[39-41] In spite of its fabrication simplicity, ED-NMOS digital circuits are not desirable for low-power applications because of the static power required in their operation. In addition, the maximum gain achievable in ED-NMOS amplifiers is low due to the lack of a complementary device pair. Therefore, the use of this technology in integrated sensors has been limited to applications where fabrication simplicity is an important requirement.

Because of the abovementioned characteristics, it is believed that CMOS is currently best suited as the circuit technology of choice in most integrated sensors and actuators. It provides a higher gain per stage than NMOS circuits (due to the presence of both p-channel and n-channel transistors), is capable of implementing digital as well as high-performance analog circuits, has a low-power dissipation, offers good packing density and high speed, and has a mature fabrication technology that is reliable and has high yield. The fabrication technology, however, is more complicated since two different device types

(p-channel and n-channel) have to be fabricated on the same substrate. A single-well CMOS process requires a minimum of about nine masking steps, and demands the accurate control of many high-temperature steps to produce uniform and reproducible threshold voltage values. A number of sensors that have been developed incorporate CMOS circuitry.[42-45]

Although standard CMOS technology satisfies most of the circuit requirements for integrated sensors, the implementation of some functions still requires bipolar devices. These functions include voltage references, voltage regulators, low-voltage current sources and sinks, and precision current sources. Voltage regulation is of special interest because, as integrated sensors incorporate more circuit functions, it will become increasingly desirable for them to handle all their power regulation and referencing locally. This is even more desirable for implantable biomedical sensors that operate using radiofrequency telemetry.[46,47] Based on these observations, a BiCMOS process technology that offers moderate bipolar device performance in terms of device speed and power handling capabilities will be required for many future integrated sensors and micromechanical systems. The use of parasitic bipolar devices and modified BiCMOS technologies to satisfy this need is perhaps the easiest and least expensive approach. It should be noted that BiCMOS technology is becoming more popular in the IC industry because of its higher drive capability and improved speed, and several foundries offer BiCMOS as a standard technology. In summary, all the IC technologies discussed have matured to a level such that any one of them is capable of meeting the needs of many integrated-sensor applications. The choice, as mentioned previously, is often dictated by the particular sensor technology adopted for the sensor itself.

10.4.2 Custom Technologies and Requirements

The two main sensor-fabrication technologies are bulk and surface micromachining (see Chapter 2 for references), both of which can produce microsensors that are superior in many respects to their discrete counterparts. Several techniques have been developed that allow the fabrication of integrated sensors in either of these technologies.

In bulk-micromachining techniques, the sensor is fabricated from the same silicon wafer that eventually houses the circuitry. This requires that the silicon substrate be of an impurity type and concentration compatible with integrated circuits, while allowing micromachining steps to be incorporated for the precise fabrication of the required microstructures. One of the techniques that has been utilized to satisfy these requirements utilizes an electrochemical etch-stop to create microstructures such as diaphragms. Figure 18a shows the cross-section of a typical integrated sensor implemented using this technology.[43] The microstructure in this process is usually formed from single-crystal silicon material of one type (n- or p-type) grown epitaxially over a standard silicon substrate of opposite impurity type (p- or n-type). The epi layer is grown to the proper thickness dictated by the sensor requirements, and has the

10.4 FABRICATION TECHNIQUES

appropriate impurity concentration required for the fabrication of active devices. The circuit process is then performed on the silicon wafer, using either bipolar or MOS fabrication technologies, to implement the desired functions. Following circuit integration, the sensor microstructure is formed using an electrochemical etch-stop. The sensor microstructure is formed by etching through the back of the silicon substrate through an opening. If a voltage is applied across the p–n junction formed between the epi layer and the substrate, the substrate etch will come to an abrupt end when the junction is exposed to the etch solution because the applied potential will result in the growth of a thin silicon dioxide film that will end the etching of the silicon. Thus, one can achieve the desired thickness for the microstructure, without using nonstandard process steps. Several investigators have fabricated semiconductor sensors using this technology.[43] One of the shortcomings of electrochemical etch-stops is the requirement for electrical contacts to the epi layer and substrate to achieve the appropriate potential difference, across the entire silicon wafer, needed for the etch-stop. Resistive voltage drops in the epi layer, in particular, can cause large nonuniformities in the thickness of the final microstructure, unless special care is taken to avoid them.[48] In addition, junction leakage current can cause a premature etch-stop unless care is taken in applying the voltage across the junction. Both of these require the use of potentiostats and reference electrodes to precisely monitor the voltage and current in the p–n junction.

Another custom bulk-silicon micromachining technology developed for integrated sensors, relies on deep boron diffusion and boron etch-stop techniques to fabricate the microstructure, while it uses lightly-doped silicon to house the needed on-chip circuitry.[49] This process also allows the fabrication of bipolar transistors, available in a standard CMOS technology, that can be used for implementing special functions that require them.[47] Figure 18b shows the cross-section of an integrated sensor fabricated using this process. In this process, boron etch-stops are formed at the beginning of the fabrication sequence to avoid exposure of the circuit to the high thermal processing required for the deep boron diffusion, followed by a nearly unaltered circuit process, in this case CMOS. The final silicon etch removes the undoped silicon and stops at the highly-doped areas, thus, completing the process. With such a sequence, minimal change is required in normal circuit processing. In the case of a deep boron sensor diffusion (for a diaphragm or other microstructures), the p-well can be formed simultaneously with the boron etch-stop diffusion. The CMOS circuitry is integrated on a n-epi layer that is grown over a p-substrate. Deep boron diffusion from the front-side of the wafer allows the use of boron etch-stop techniques to form beams and diaphragms, thus, eliminating electrochemical etch-stops between the n-epi layer and the p-substrate, resulting in a much simpler process. The process utilizes a total of ten masking steps and results in high-performance CMOS devices and circuitry. The boron diffuses through the n-epi layer and connects to the underlying p-substrate creating isolated n-epi regions that will house the circuitry. These p^+-diffused regions will protect against lateral etching of the sensor during the ethylene-diamine pyrocatechol

(a)

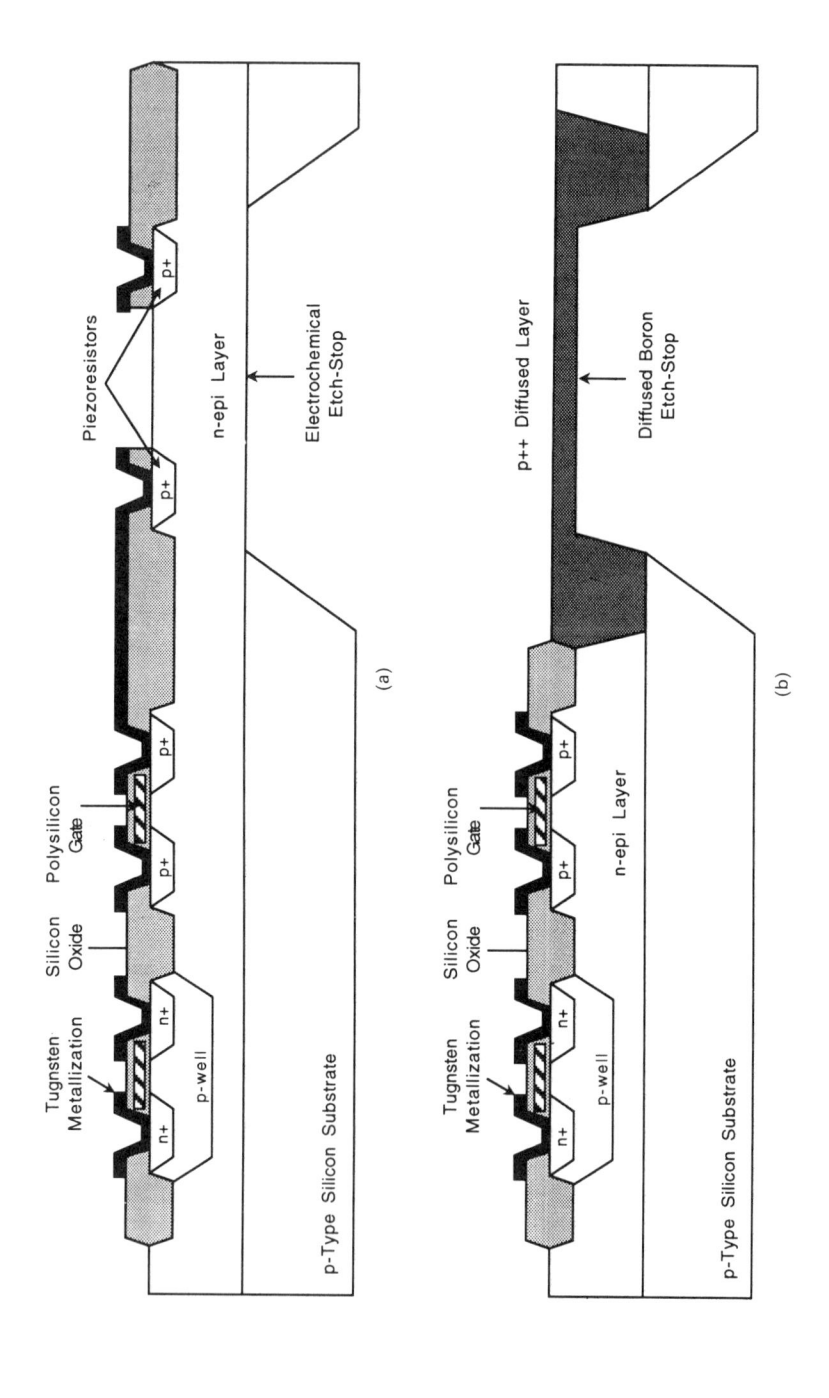

(b)

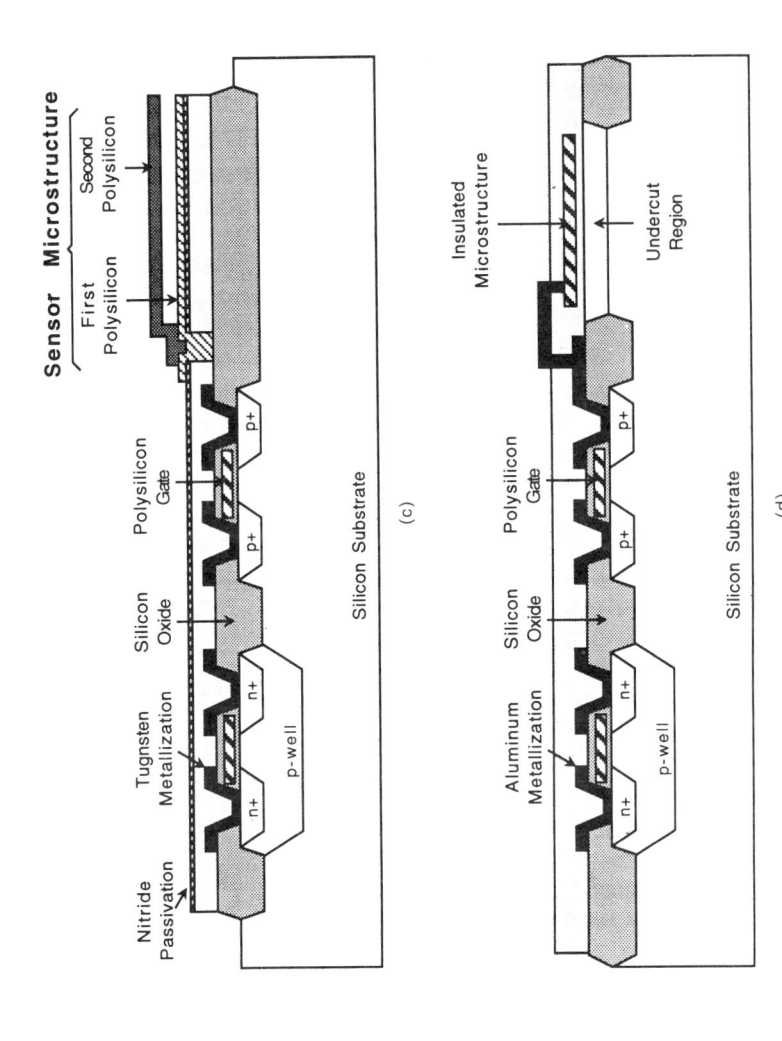

Fig. 18 Custom technologies for solid-state integrated sensors: (a) cross-sectional view of integrated sensors fabricated using p–n-junction electrochemical etch-stop; (b) bulk-silicon integrated sensors fabricated using boron etch-stop; (c) surface micromachined integrated sensors; and (d) CMOS foundry-compatible integrated-sensor fabrication technology.

(EDP) etching process. These diffused regions also provide electrical isolation between n-epi regions, and allow the fabrication of isolated-collector, vertical npn bipolar transistors. These isolated-collector npn bipolar transistors have been shown[47] to have a current gain in the range of 100 to 150. The CMOS technology also utilizes two polysilicon layers to implement on-chip integrated capacitors, which are often required in switched-capacitor and analog circuits. In addition, it allows the use of refractory interconnections instead of the conventional aluminum. Refractory interconnects are desirable in many sensor/actuator applications since they can be further insulated on top with low-pressure chemical-vapor-deposited (LPCVD) thin films allowing the entire sensor/circuit structure to be isolated from destructive and often corrosive external environments. In effect, LPCVD nitride and oxide films provide a long-sterm, non-hermetic package for the circuit while still allowing the sensor to interface with the external world for parameter measurement.

In surface-micromachining technologies, the sensor microstructure is fabricated using deposited thin films, as discussed in Chapter 2, while the circuit is fabricated in the supporting silicon substrate, as shown[20] in Fig. 18c. Typically in surface-micromachined integrated sensors, the sensor processing is performed at the end of the circuit process, in contrast to bulk micromachining. The circuit process is usually carried through the metallization step, at which point the sensor process begins. Since several high-temperature steps have to be performed to complete the sensor process, the metallization material used in the circuit should be capable of tolerating these steps. Therefore, refractory metal silicides are usually employed. A process developed at the University of California, Berkeley, uses a combination of titanium nitride and titanium silicide to form a high-quality contact region, and a layer of CVD tungsten to form the metal interconnects.[20] After the circuit process is completed, the entire wafer is covered with a low-temperature oxide and silicon nitride passivation layer to protect the electronics from the etchants used in the sensor-fabrication process. Sensor fabrication utilizes two layers of low-pressure CVD polysilicon that are patterned in a standard surface-micromachining technology, with a silicon dioxide sacrificial layer between them. The sacrificial layer is finally etched in hydrofluoric acid to finish sensor processing and release the microstructures formed to implement the sensor. Note that the circuitry is protected from the HF etch using the nitride passivation layer deposited previously. This process has been used to develop an integrated accelerometer,[20] and a modification of the process has been used by Analog Devices, Inc. to produce a commercially available accelerometer.[36]

A second approach to fabricating surface-micromachined integrated sensors, uses a completely standard CMOS technology available from semiconductor-fabrication foundries.[50,51] In this process, wafers that have been through a standard CMOS process sequence are processed further to release microstructures formed from the thin films available in this process (such as silicon dioxide and polysilicon), as shown in Fig. 18d. The sensor microstructure, typically polysilicon, is released by undercutting sacrificial layers, such as silicon dioxide, using a standard surface-micromachining step. It is also possible to etch a portion of the underlying silicon substrate to release and form more sophisticated microstructures. An attractive feature of this process is that it does not require any custom sensor processing to be performed on the wafer, and that the only additional processing step required is the final release etch. The price one pays for this simplicity, on the other hand, is the lack of any control over the material properties of the microstructure material and therefore, a limitation in the range of quality of the devices that can be implemented. A number of integrated sensors have been demonstrated using this technology, including flow sensors and infrared detectors.[50,51]

As is evident, the main difference between integrated bulk- and surface-micromachined sensors is that in the former the circuit-fabrication steps are performed after the main sensor-fabrication steps are completed, while in the latter the circuit process is completed before the sensor is fabricated. It should also be noted that in surface-micromachined integrated sensors, careful attention has to be paid to minimizing the thermal budget needed by the sensor process to reduce the affects of high-temperature steps on circuit performance and device characteristics. While each of these technologies have their advantages and disadvantages, the final selection criteria are dictated by a combination of the sensor performance, the availability of processing capabilities at the fabrication site, and the cost of the overall fabrication process. The cost of the overall process, in turn, is a function of the overall sensor size, the overall circuit area, the degree with which processing steps can be shared between the two, testing cost, and packaging cost. Yield modeling and calculation should be performed before a particular technology is chosen, and cost estimation based on the above parameters should be made.

10.4.3 Economic Considerations

Integration of on-chip signal circuitry with the sensor structure is to a large extent dependent on economic considerations. Circuit integration can only be justified when either the overall cost of the larger system housing the sensor can be reduced, or when the overall performance characteristics of the sensor are significantly improved. In many applications, it is sufficient to package the sensor and its interface and signal-processing electronics in a single module using any of a number of hybrid technologies. The choice of on-chip versus off-chip circuitry, is dictated mostly by how the overall yield, the packaging cost, performance, and the overall system cost are affected.

An important economic factor for integrated sensors with on-chip circuitry is the yield of both the sensor and the circuitry. Any yield analysis has to take into consideration three specific areas: the yield of the sensor; the yield of circuits integrated with sensors; and finally the affect of circuit integration on the overall yield. For most semiconductor sensors it is expected that the yield will be dominated by gross defects, or parametric processing problems, as opposed to the situation in integrated circuits where randon defects dominate the yield.[52]

This is due to a number of factors. First, most semiconductor sensors have much larger feature sizes than typical integrated circuits and are, therefore, less affected by random defects. Second, the majority of semiconductor sensors have a totally different device structure and physics, which in most cases are simpler and less sensitive to random defects. Finally, operation of many sensors is less dependent on electronic behavior and more on mechanical behavior. Therefore, it is expected that if one sensor on a silicon wafer is inoperative, other sensors in the vicinity of the defective sensor will also be bad, which is a distinctive feature of gross defects. This can happen for a number of reasons including bad diffusion, bad silicon-glass bond, nonuniform intrinsic stress, etc.

For integrated circuits, the yield is dominated by random defects such as pinholes, photolithographic defects, and leakage defects.[52] However, if the circuit features are kept rather conservative, many of these defects will not grossly degrade the overall yield. As sensors decrease in area, however, one should try to minimize the circuit area as well. The yield of integrated circuits is strongly dependent on the area of the circuit. Therefore, no matter how big the sensor with which the circuit interfaces, the circuit yield is still determined by the circuit area, the defect density, and circuit sensitivities. The overall yield of the sensor-circuit chip will also depend on the number of additional processing and masking steps that have to be performed, which results in a reduction in the yield. Therefore, in order to determine the yield of a sensor with on-chip circuitry exactly, one should construct a yield model for the specific process. The overall yield and reliability of a sensor cannot be determined unless packaging issues are also considered.

One of the most important problems associated with sensors is their packaging. The package for each type of sensor is different and has different requirements because it should transfer the variable being measured to the sensor, while it protects the sensor from harmful environmental effects. The sensor package not only affects the quality of a sensor, but it also influences its final cost. In some cases, the lack of a proper package can prohibit the longer-term use of the sensor.[12,39] Therefore, no sensor can be designed without a consideration of the final package. Depending on the application and the type of sensor, on-chip circuitry can potentially help reduce packaging problems. These problems very often deal with transmission of recorded electrical signals to the outside world, and the number of output connections needed. Typically these signals are low-amplitude, high-impedance signals which are easily affected by external noise. For multisensor devices, the number of output leads can also be a limiting factor. By integrating on-chip circuitry, it is possible to amplify and buffer the recorded signals before transmission to the outside world and relieve the package from having to protect otherwise weak signals. This is even more important for the long-term reliability of the sensor. For multisensor systems, multiplexing of several data channels into a single data line not only relieves packaging and encapsulation requirements but also improves the lifetime and yield of the final product by minimizing the number of leads and bonds, which are potential points of failure. Although integration of on-chip circuitry may be advantageous for packaging of the sensor, one has to provide encapsulation for the circuitry itself, which very often has different packaging requirements than the sensor.

Integration of on-chip circuitry along with the sensor, in most cases, increases the cost due to the additional processing and design effort. However, the final cost of manufacturing a reliable sensor is not always increased. First, integration of on-chip circuitry can reduce packaging problems and the number of wire bonds, which reduces the packaging cost and increases long-term reliability. Second, by incorporating circuitry along with the sensor, it is possible to reduce the number of components and the cost of assembling these components on a circuit board. This means that any sensor-system interface has to be divided into an optimum number of components. In order to do this, one should be able to predict the yield and cost of each component and their combined effects. Third, on-chip circuitry can improve overall system performance that otherwise has to be obtained through external processing circuitry, which results in higher system cost. In summary, for many sensory systems the cost of adding on-chip circuitry to the sensor is often offset by reduced tolerances and system functions in other portions of the system.

The cost of integration of on-chip circuitry on the sensor substrate also depends on the area and size of the sensor. As the area of the sensor increases, the cost of on-chip circuitry increases since the number of chips per wafer decreases. Therefore, the sensor and circuit should be processed separately so that maximum yield is obtained for the sensor. In this situation, too, models for the yield and cost of the sensor and circuit are required to determine the optimum separation point. The issues discussed above have different implications for different sensors. While integration of on-chip circuitry can be beneficial to one sensor, it might not be cost effective for another. It is certain that as new semiconductor sensor technologies develop and as new high-volume markets begin to open for integrated sensors, it will become more cost effective to mass produce many different integrated sensors.[36]

10.5 EXAMPLES OF INTEGRATED SENSORS

During the past few years we have seen great progress in the development of many integrated sensors using different technologies. In this section we will present two examples of integrated sensors that have been fabricated recently using bulk-silicon-micromachining techniques. The first example, integrated multichannel recording/stimulating microprobes, is an excellent example of sensors that benefit immensely from on-chip interface and signal-processing circuitry because of the small amplitude of the sensor signals and the need for a minimum number of output leads. The second example, a multi-element gas analyzer system, desribes an approach to the implementation of multi-sensing systems where bidirectional communication with the sensor through a standard digital bus and self-testing capability in the sensor are of particular interest.

10.5.1 Multichannel Recording/Stimulating Microprobes

An important application area for solid-state sensors and actuators is in health care, biomedicine, and in biological research. These applications require the development of miniature, high-performance stable sensors and actuators to accurately and reliably interface with the biological environment. One area of interest has been in recording electrical activity from the nervous system, and in delivering electrical signals to the nervous or muscular tissue for selective activation of paralyzed muscles. The use of microelectrodes to record extracellular biopotentials generated electrochemically within individual neurons has been one of the principal techniques for studying the central nervous system at the cellular level. Long-term recording of electrical discharges from several neurons simultaneously will help us gain a greater understanding of the neural systems at the circuit level. Multielectrodes can also be used to simulate a variety of biological tissue by selective injection of charge through appropriately designed electrodes. For both of these types of electrodes, on-chip circuitry is required to amplify recorded signals, to multiplex input and output data lines to reduce the number of output leads, and to buffer interconnect lines to minimize problems due to leakage.

Figure 19 shows the structure of a micromachined silicon microprobe developed for both electrical recording and stimulation of the central nervous system.[39,45,53] The micromachined silicon substrate supports an array of conductors that interconnect recording/stimulating sites located at the tip of the probe, and on-chip signal-processing electronics housed on the back-end of the probe substrate. The silicon substrate is formed using deep boron diffusion and boron etch-stop techniques. Boron doping is also used around the perimeter

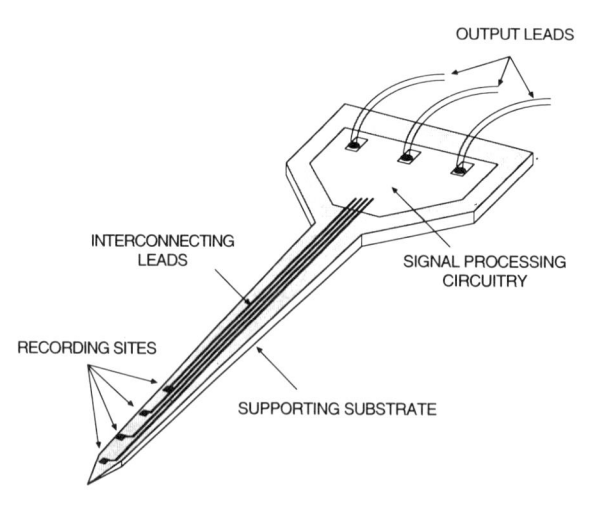

Fig. 19 A multichannel multiplexes intracortical recording electrode array. (After Ref. 45)

10.5 EXAMPLES OF INTEGRATED SENSORS

of the rear of the probe for dimensional control of the probe, while a silicon well of normal impurity concentration is retained for circuit fabrication.

Figure 20 shows the block diagram of the on-chip circuitry that is integrated on the back of the probe substrate. The on-chip circuitry consists of ten per-channel pre-amplifiers followed by an analog multiplexer and a broadband output filter to drive the external data line.[39] In order to allow the external regeneration of the on-chip clock, a synchronization pulse is inserted once each frame as an eleventh channel. The external electronics strips off these pulses, regnerates the sample clock, and demultiplexes the neural signals for external recording and processing equipment. The on-chip electronics is also capable of self-testing the electrode impedance levels on demand. When testing is desired, the power supply is pulsed from its normal value of 5 to 8 V, enabling the test-waveform generation circuitry. This circuit generates a 50 mV, 1 kHz signal that is capacitively coupled to all the recording electrodes. The resulting induced electrode voltages are amplified and multiplexed in the normal way, providing an indication of the electrode impedance levels and of the functionality of the on-chip electronics. After testing is performed, the circuitry can be returned to its normal mode by turning the power supply off and then back on again at a 5 V level. The design for the on-chip circuitry requires only three output leads for power, ground, and data while maintaining self-test capability.

A second-generation 32-channel microprobe has also been developed[45] that allows the user to select 8-out-of-32 channels to extend the functionality of the

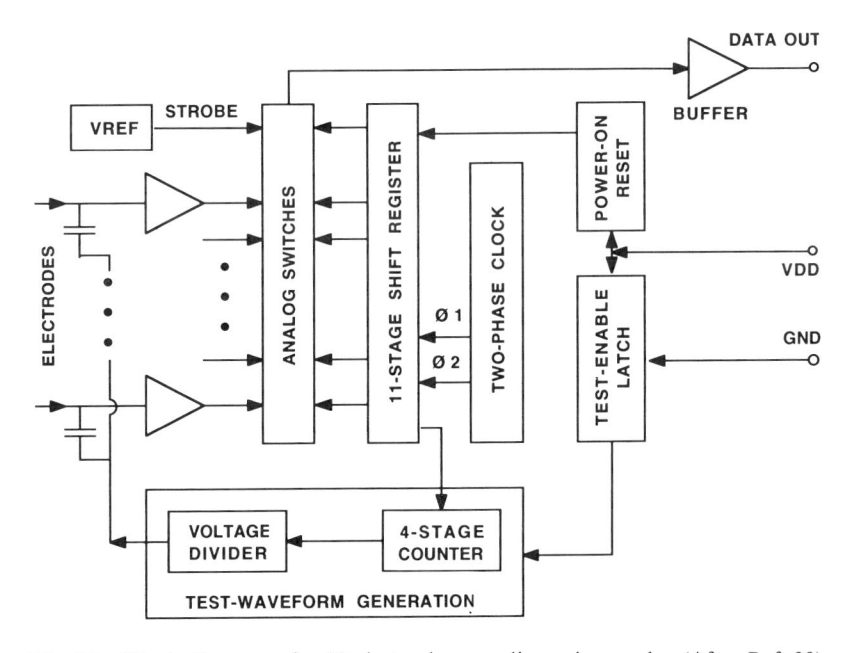

Fig. 20 Block diagram of a 10-electrode recording microprobe. (After Ref. 39)

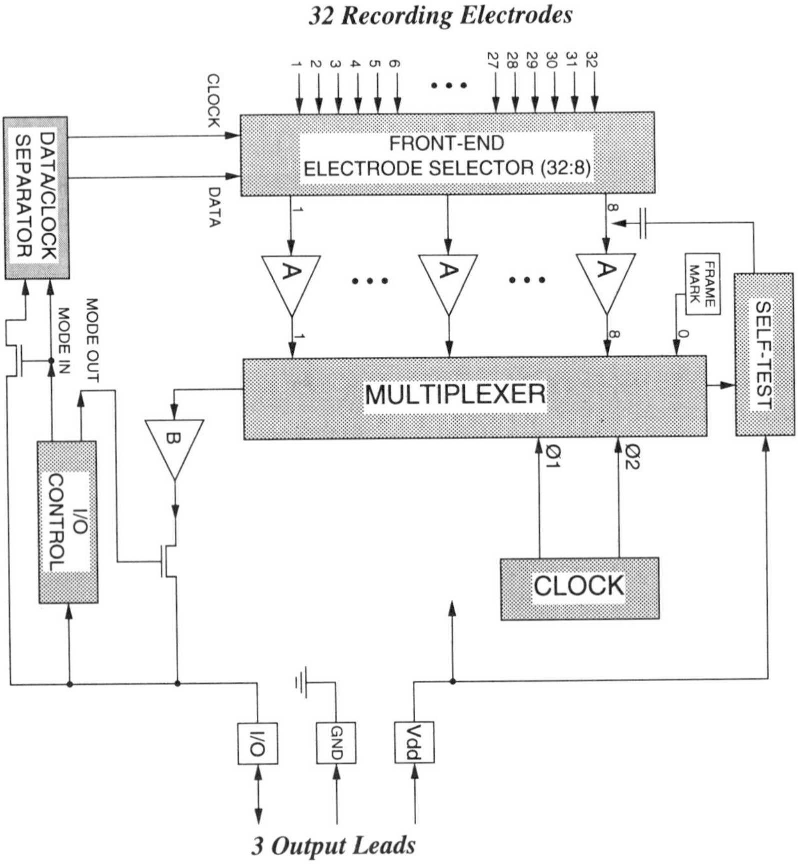

Fig. 21 Block diagram of the on-chip circuitry designed for an electronically configurable 32-electrode recording microprobe. (After Ref. 49)

on-chip circuitry, as shown in Fig. 21. An input channel selector is used to select eight active recording sites from among 32 sites located on the probe. This approach allows the channels selected for processing to be optimized for maximum electrode-cell coupling and effectively implements electronic site positioning. This enhanced capability is desirable for many next-generation solid-state sensors where user-sensor interaction is needed. The eight selected sites feed their recorded data to per-channel amplifiers whose outputs are subsequently multiplexed onto a single data line that is connected to the outside world. The amplifiers are designed with low-pass filters implemented using a new diode-capacitor approach. These filters are required to remove the DC input drift generated by the recording site–tissue interface, and the amplifier offset voltage. This will permit the amplifiers to be designed for maximum AC gain while avoiding saturation due to these DC offsets. Channel selection is

10.5 EXAMPLES OF INTEGRATED SENSORS

obtained over the same lead normally used for multiplexed data output from the probe. A 5 V pulse applied externally to this lead selects its mode. In the output mode, the eight signal samples and frame marker are time-multiplexed to the external circuitry, while in the input mode, binary input data and clock signals are superimposed on the data lead. On-chip circuitry separates these signals and clocks the input channel selector to enter the desired mode. On-chip self-test capability is also maintained in this second-generation probe.[45]

Figure 22 shows optical and SEM views of the multi-electrode recording microprobes. The probe shank is 15 μm thick, about 3 mm long, and tapers to a width less than 20 μm at the tip of the probe. The circuit is fabricated using a standard 3 μm p-well CMOS process using the technology discussed in Section 4, consumes an area of about 1.5 mm^2, and has a power requirement of less than 2 mW from a single 5 V supply. The CMOS process is performed following the deep boron diffusion step needed to define the probe shank areas using boron etch-stop techniques. Following the CMOS process, the entire circuitry is insulated on top using deposited LPCVD silicon dioxide and nitride dielectrics. This requires the use of refractory metal silicides such as tungsten silicide to produce low-resistance contacts to the silicon regions.

In these recording microprobes, the primary functions of the on-chip circuitry are signal amplification and multiplexing. In stimulating electrode applications, the primary need is to reduce the number of interconnections between the external world and the implanted device. This is required for improved long-term reliability of the implant, for reducing the tethering forces exerted on the implanted probe due to the interconnect leads, and for simplifying packaging and assembly.

Electrical stimulation for neural activation is usually achieved by using microelectrodes to deliver small currents into the tissue. Stimulating probes for many prosthetic applications should be able to deliver constant current pulses into tissue through multiple sites precisely spaced on a substrate. The block diagram of a 16-channel active stimulating microprobe is shown[53] in Fig. 23. This device has a DAC for every channel, allowing all sixteen channels to be actively stimulating at any given time, with individual control over the current amplitudes. The circuit receives 16-bit data words that are shifted into the data pad using a 4 MHz clock. The timing diagram for the data word is also shown in Fig. 23. Four bits represent the site address, two are mode bits, eight set the current amplitude, and two are used for strobing and status. In the normal mode of operation, the specified current is set up on the addressed channel and remains on until the channel is accessed again to alter it. Other channels are set up sequentially to source or sink current as desired. At a clock rate of 4 MHz, the time resolution between any two changes in the probe current pattern is 4 μs, which is effectively simultaneous to the tissue. A second mode of operation grounds the associated stimulus channel allowing the DAC currents to be calibrated externally without stimulating the tissue. A third mode of operation allows the analog voltage developed in response to stimulus current to be observed externally to check site integrity at any time. Finally, a fourth

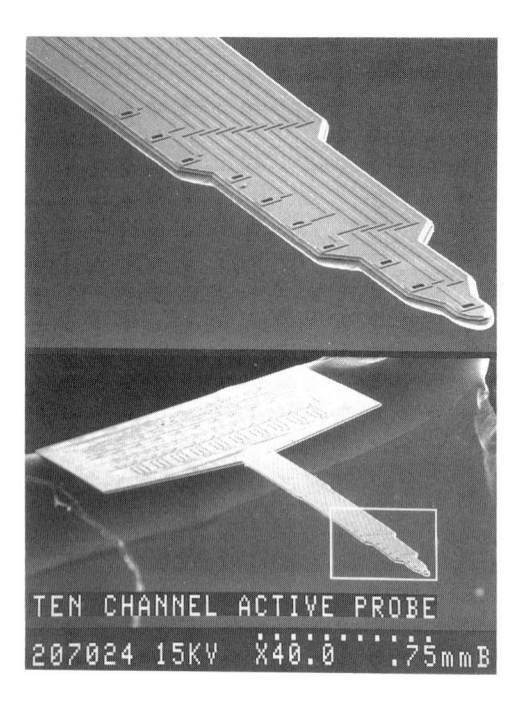

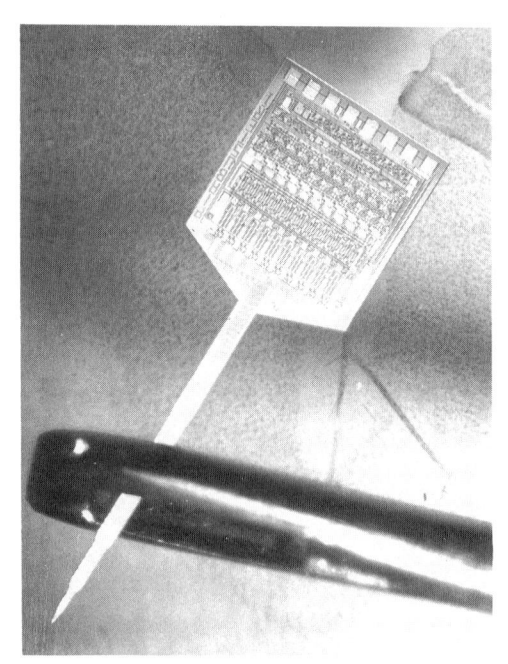

Fig. 22 Optical and SEM photographs of multichannel recording microprobes. (After Refs. 39, 49)

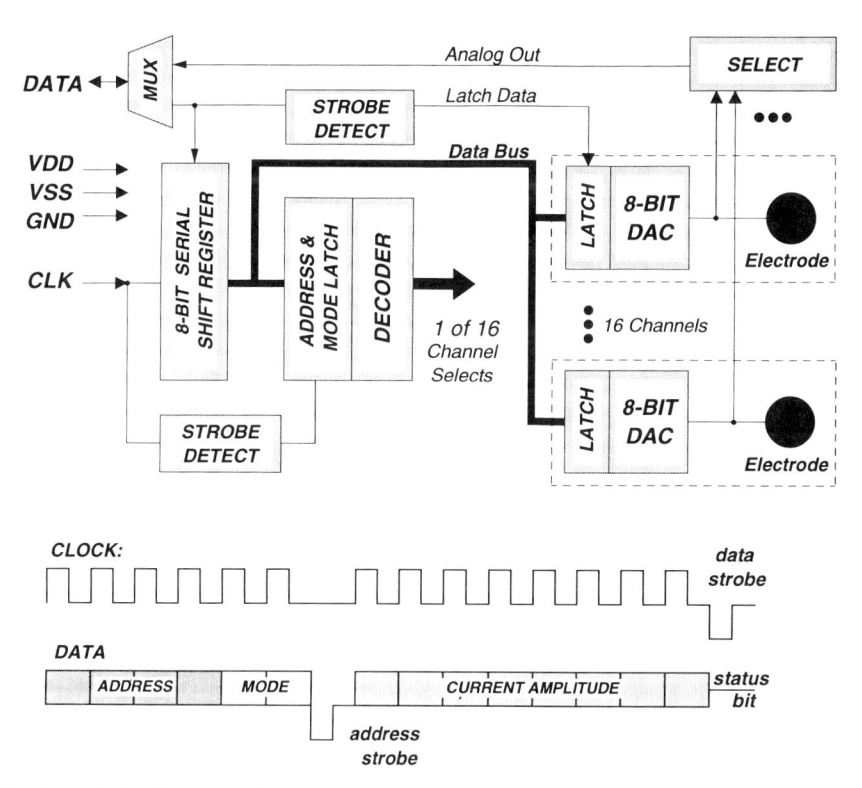

Fig. 23 Block diagram of a 16-channel stimulating microprobe used in the stimulation of the central systems. (After Ref. 53)

mode of operation biases the electrode above ground between pulses to increase the charge delivery of the irridium oxide electrode sites.

Figure 24 shows a photograph of a fabricated stimulating microprobe. This probe consists of two shanks each of which supports eight stimulating sites. Each shank is 3 mm long and 135 μm wide, and the shanks are 500 μm apart. The site areas are 1000 μm^2 and are spaced 250 μm apart. The circuitry is fabricated using a 3 μm p-well single-metal single-poly bulk CMOS process, occupies an area of 4.4 mm × 2.5 mm, contains about 7100 transistors, consumes 80 μW of power when idle, operates from ±5 V supplies, and operates at clock rates as high as 10 MHz. The output currents are linear to 8 bits with sourcing and sinking currents matched within ±$\frac{1}{2}$ least significant bit (LSB), ensuring adequate charge balancing at the stimulating sites.[53]

These examples of integrated sensors clearly demonstrate the advantages and features of on-chip circuitry, the role it plays in improving signal-to-noise ratio and in reducing the number of I/O leads, which will eventually simplify packaging requirements for these sensors that have to operate in the harsh biological environment. In addition, on-chip circuitry enhances functionality in

518 INTEGRATED SENSORS

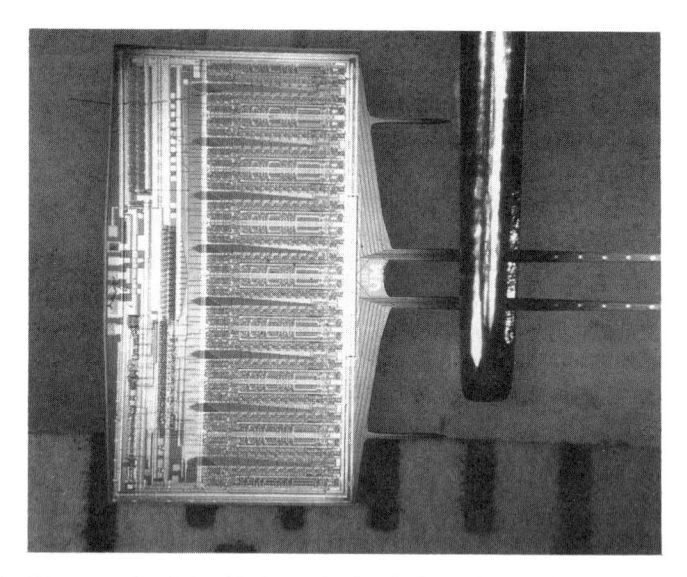

Fig. 24 Photograph of the 16-channel stimulating microprobe. (After Ref. 53)

terms of testability and programmability that enable the user to operate the sensor on demand. The above examples also represent applications where the user interfaces with one sensor at a time. In many sensing systems, however, an important requirement is to interface with many sensors through a bi-directional digitally addressable bus. The next example demonstrates one such application.

10.5.2 A Multi-Element Smart Gas Analyzer

Gas sensors are in high demand for many applications including automotive, medical, and process control, and consequently, a significant amount of research has been concentrated on them during the past decade. However, substantial performance problems remain unresolved, the most important of which involve slow response, low sensitivity, poor selectivity, long-term drift, and high input power. Recently, a conductivity-type gas sensor based on the use of ultra-thin metal films on selectively-formed dielectric windows has been reported.[54,55] Early results indicate that such devices have great potential and are capable of overcoming many of the current problems. For instance, fast detection of sub-ppm oxygen levels in CF_4 has been demonstrated with negligible drift and hysteresis.[55]

The cross-section and top view of the basic structure of a new single-element gas detector are shown in Fig. 25, and a photograph of a dual-detector chip is shown[10] in Fig. 26. This sensor is a conductivity-type gas sensor based on the use of ultra-thin metal films on selectively-formed dielectric windows[54,55] and relies on the dielectric window to provide high thermal isolation between the thin sensing film and the silicon-chip rim, minimizing the amount of input

10.5 EXAMPLES OF INTEGRATED SENSORS 519

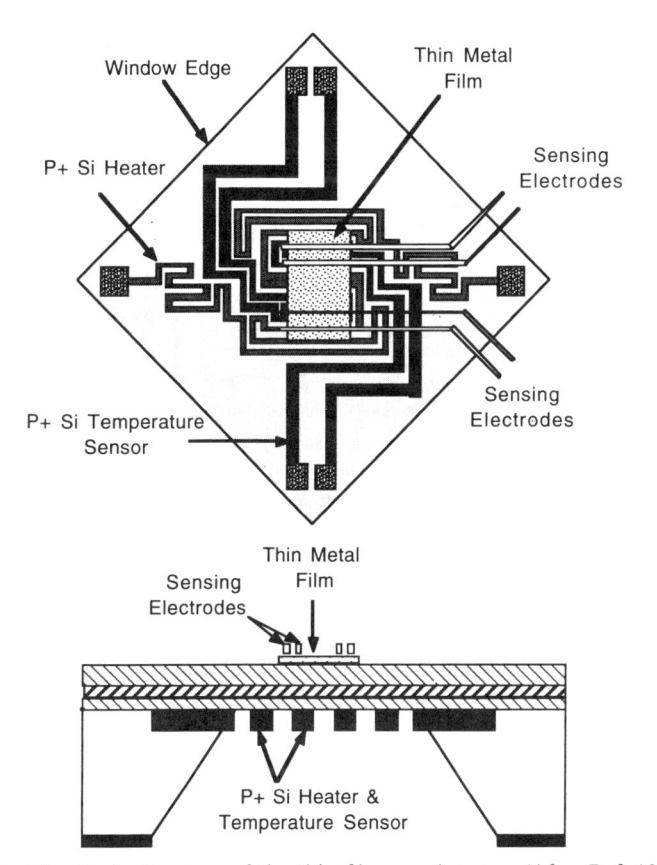

Fig. 25 Basic structure of the thin-film gas detector. (After Ref. 10)

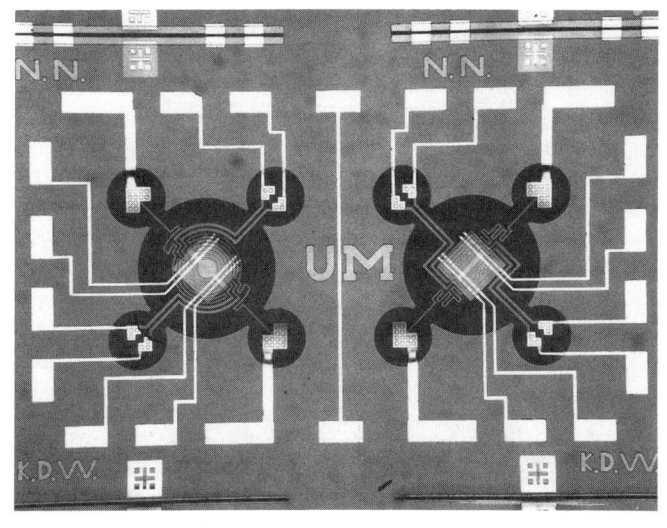

Fig. 26 Photograph of a dual-detector gas sensing chip fabricated using bulk micromaching techniques. (After Ref. 10)

power required by the detector for a given operating temperature. Such detectors can produce window temperatures in excess of 1300°C, well above those required for most gas-sensing applications (i.e., 250–500°C). The conductance of the thin film depends on the types and quantities of gases present, gas adsorption, window temperature, and possible surface reactions. The response of gas is measured as an adsorption-induced thin-film resistance change. Here surface adsorption dominates over bulk effects in determining film conductivity and response time. Surface adsorption is normally reversible by applying higher temperatures.

Thin films are highly sensitive to some gases and inert to some others. Therefore, a multi-element gas analyzer, having different thin-film detectors, has been developed to both overcome the drawbacks of the original gas detectors and to improve selectivity and specificity in analyzing gaseous mixtures. Figure

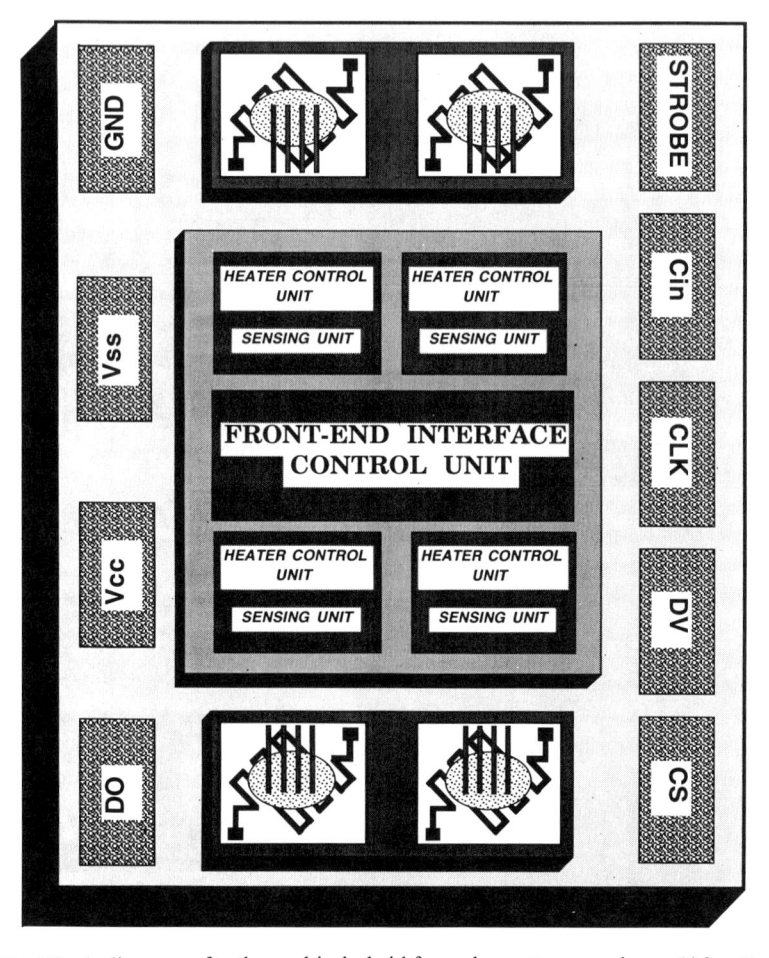

Fig. 27 Block diagram of a three-chip hybrid four-element gas analyzer. (After Ref. 10)

10.5 EXAMPLES OF INTEGRATED SENSORS

27 shows the block diagram of the four-element hybrid gas analyzer which consists of three chips: two dual-window gas detectors and a control sensor chip. Figure 28 shows a photograph of one of the hybrid analyzers. For normal operation only eight pins are used (required for the front-end standard interface). However, for applications that demand higher window temperatures extra pins for supplying power directly to the gas-detector heaters are required.

The control chip communicates over the eight-line standardized interface discussed before to either an embedded VLSI microcontroller (forming a smart sensing node) or to a remote processor, allowing the analyzer to be used as a smart peripheral. Utilizing this front-end standard provides a very sophisticated and intelligent gas analyzer with only eight lines. Indeed, in this four-element gas analyzer, 12 sensors can be read, 12 DACs/actuators can be programmed, and 6 internal nodes are available for self-testing. Furthermore, it is possible to realize an eight-element hybrid package with the same eight lines, by utilizing two control chips and four dual-element gas detectors.

The control sensor chip is capable of controlling four gas-detecting elements where each element operates independently of all others and all elements can be monitored and programmed simultaneously. This control chip allows the

Fig. 28 Photograph of a multi-chip hybrid gas analysis module. (After Ref. 10)

temperature of each window to be set from ambient to 1000°C with eight bits of resolution. In order to investigate and characterize a broad range of thin detecting films, the film conductance measurement is accomplished in one of the 255 (eight bits) programmable ranges for full-scale resistances from 5 Ω to 1.3 MΩ over a frequency range from DC to 2 MHz. The control chip consists of over 4000 transistors and is fabricated using a 2 μm p-well double-poly, double-metal CMOS process in a die size of 4.4 mm × 6.6 mm.

Figure 29 shows the block diagram of the control chip and its connection to just one of the four corresponding gas detectors. The chip consists of three principal circuit blocks: the control/interface unit, the heater controller, and the conductivity (AC and DC) sensing circuitry. The control/interface unit (common among all four detectors) is responsible for communicating with the user (over the eight-line front-end interface) and for receiving, interpreting, and executing commands sent by the user. It also generates proper control signals

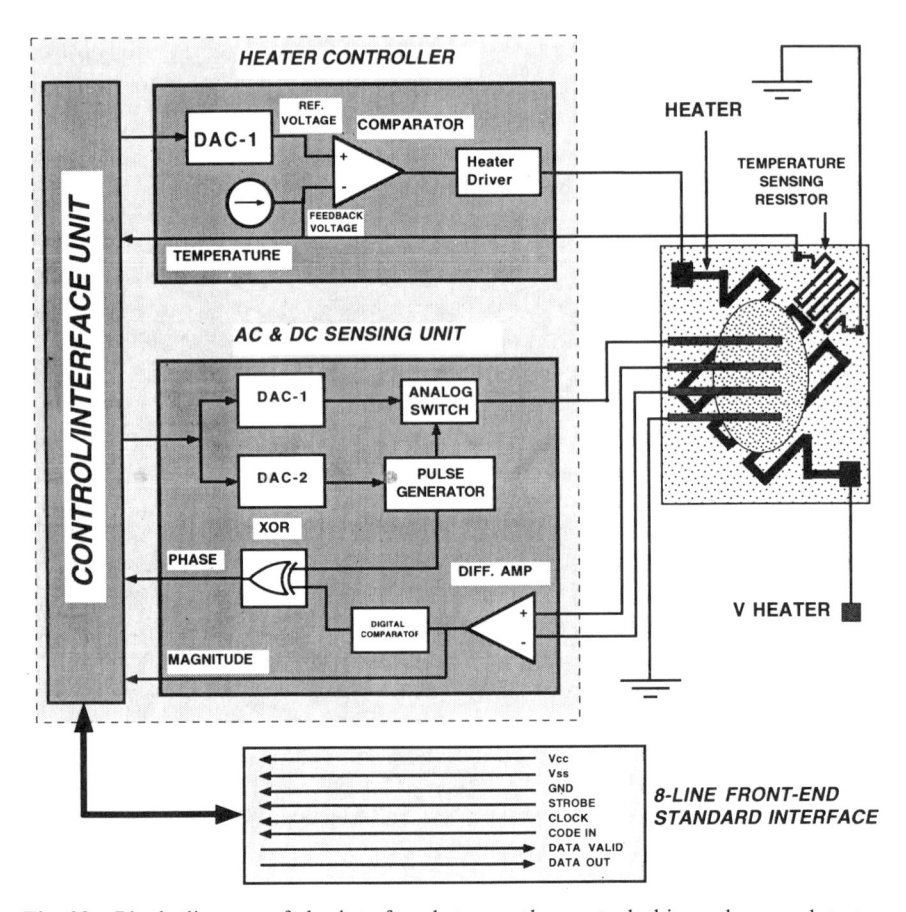

Fig. 29 Block diagram of the interface between the control chip and a gas detector. (After Ref. 10)

for the other units. There are three programmable eight-bit cumstom-designed DACs for each of the four elements.

The desired heater temperature set point is entered serially and latched in a DAC, which generates a reference voltage for a MOS comparator that drives the heater. Temperature feedback is derived from the sensing resistor on the window and is used to limit the comparator drive at the appropriate level. This window temperature can also be read externally. The heater controller maintains a stable temperature at the set point, which is challenging, because the energy required for the heater is a function of many parameters, including gas pressure and its associated thermal conductivity. Simulations have shown that the window temperature is stable at a temperature of 250°C with less than 0.5% variations.

The conductivity sensing unit measures both AC and DC using four sensing electrodes on the thin film. The two outer electrodes are used to inject current into the thin film, and the two inner probes are employed to measure the voltage drop across the inner part of the thin film. Two eight-bit DACs can be programmed to select the appropriate current level and frequency of the measurement (eight bits each). Magnitude and phase information is sensed and fed off-chip, again using the eight-line interface. The chip has been designed to operate over 255 programmable ranges with films having resistances (full scale) from 5 Ω to 1.3 MΩ, allowing it to be used with a wide variety of different film types.

10.6 SUMMARY AND FUTURE TRENDS

We have observed tremendous growth and progress in the development of a variety of semiconductor sensors during the past two decades. This has been made possible by the development of silicon micromachining techniques and the utilization of a wide array of technologies already developed for the fabrication of integrated circuits. Semiconductor sensors are potentially cheaper, offer higher performance and reliability, and are smaller in size than their discrete counterparts. They are being used in a number of applications areas, including instrumentation, transportation, health care, industrial processing and manufacturing, avionics and defense, and consumer appliances. The full potentials of these sensors can be realized only when they are used together with signal processing and control circuits.

Integrated sensors take full advantage of the worlds of semiconductor sensors and integrated circuits by combining the best features they have to offer. Sensor signals can be amplified, multiplexed, and buffered before transmission to microprocessor-based control systems. The microprocessor-based system operates on these signals and offers a standard digital data stream to the user making the entire sensing module behave like a smart-system peripheral rather than a passive component. The system designer can manipulate the sensing module remotely to achieve complex functions such as self-testing, calibration, and data compensation. This added flexibility and versatility will, in turn, reduce system overhead in terms of cost, configuration, and software requirements, and will enhance overall stability, reliability, and performance.

Many new technologies remain to be developed before integrated smart sensors can be widely accepted and utilized in different applications. First, new system architectures that can take full advantage of the features of integrated sensors need to be developed. These integrated or distributed systems should be capable of receiving sensor data in a standard format, and should be capable of processing and interpreting the data to minimize the responsibilities imposed on the final user. To achieve this, standard signal-processing electronics, communication interfaces, data processing, and interpretation techniques should be developed. Second, new, truly compatible sensor-circuit fabrication technologies should be developed to maximize the overall yield to keep the cost down. It is believed that CMOS and BiCMOS circuit technologies will be increasingly used in bulk and surface-micromachined integrated sensors. Third, new packaging technologies need to be developed to improve sensor reliability and performance, and to keep the unit cost down. The total cost of most semiconductor sensors is dictated mainly by testing, calibration, and packaging steps. Wafer-level packaging techniques, hybrid, and multi-chip module technologies that can accommodate a variety of different sensor structures are badly needed. Batch wafer-to-wafer process technologies, microstructures, and assembly techniques are needed which will permit stacked three-dimensional systems to be formed using a set of robust construction primitives, including support pillars, lead columns, flexible multi-lead interconnect cables, mechanical elements, and signal-processing platforms. We believe that unless breakthroughs are made in these areas, integrated sensors will have a difficult time getting into new applications and markets. Fourth, computer-aided design tools need to be developed for sensor design and simulation. This is a very important aspect of the development of integrated sensors since one can check the functionality of both the sensor and the circuit before committing resources to fabrication. These tools should include databases incorporating process/material data, gateways to commercial finite-element code, and links to solid-state process/device/circuit simulators. Fifth, standardization of sensing systems and their various components is badly needed and should be pursued at all levels. Standard circuit modules and interfaces need to be developed to avoid design duplications and reduce the overhead when a new sensor is to be developed. This will inevitably require closer cooperation between sensor designers, circuit designers, system designers, and process engineers. Sixth, much improved understanding, characterization, and control over the mechanical properties of the films used in the fabrication of microstructures is badly needed. Such issues as the control of stress and strain in various thin films and the removal of stress using a variety of annealing techniques will have to be resolved for both existing and new microfabrication technologies. It is with this resulting knowledge base that we can hope to design, simulate, and successfully fabricate new and more complicated microsystems that will perform as intended. Seventh, semiconductor sensors for the measurement of a number of physical parameters have been successfully developed, however, semiconductor chemical sensors are not yet highly developed. These sensors offer a number of challenges to sensor designers especially in terms of stability and reproducibility, and require further research to develop new sensing materials and transduction techniques. Finally, we have not adequately addressed the integration of optical sources and detectors as part of overall microsystem development beyond limited efforts at fiber-optic repeater systems. Recent developments in the fabrication of light-emitting diodes using porous silicon are very encouraging and hold promise for eventual monolithic integration of optical sources with silicon-based sensors.

From a business perspective, important and practical new applications for semiconductor sensors and sensing systems must be defined in new emerging industries, while in more established industries we must overcome the inertia and very real costs associated with moving to this new technology. While the main commercial markets for semiconductor sensors up to now have required the development of individual sensor devices, such as pressure sensors and accelerometers used in automobiles, it is believed that integrated sensors have great potential application in many emerging instrumentation and control systems that require the accurate measurement of a number of physical and chemical parameters such as pressure, acceleration, vibration, gas concentration, humidity, position, force, temperature, and strain. Thus, while we have made substantial progress in some areas of semiconductor integrated sensors, the challenges before us are broader still and must be the focus for major efforts in the future. Solving these problems will require the focusing of multidisciplinary teams from engineering, chemistry, physics, business, and the life sciences, both in universities and in industry.

ACKNOWLEDGMENTS

The material presented in this chapter has been based on the collective efforts of a large number of individuals all around the world who have, over the years, worked on the development of solid-state sensors. In particular, a major effort in the research and development of solid-state integrated sensors has occurred here at the University of Michigan. A great many of the doctoral students, as well as the technical staff in the Solid-State Electronics Laboratory, have been involved in the completion of various projects discussed here. Special thanks go to our sponsors who have provided us with the encouragement and means to carry out research programs. The authors gratefully acknowledge: Drs. T. Hambrecht and W. J. Heetderks of the National Institutes of Health for their support of our biomedical sensor projects, Dr. G. Hazelrigg of the National Science Foundation; many individuals at the Semiconductor Research Corporation, and Drs. Lance Glasser and Ken Gabriel of the Advanced Research Projects Agency.

PROBLEMS

1. For the resistive bridge circuit shown in Fig. 15 show that the output voltage is given by the expression provided in Eq. 1.

2. For the switched-capacitor interface circuit shown in Fig. 16 prove that the output voltage is as given by the expression provided in the figure, i.e.

$$V_{out} = V_p \frac{C_X - C_R}{C_F}$$

3. The expression given in Problem 2 assumes that the op-amp has an infinite open-loop gain and that the parasitic capacitance, C_{ps}, is zero. Derive an exact expression assuming that the op-amp has an open-loop gain of A and that the parasitic capacitance is nonzero.

4. Draw the block diagram of the circuit that can implement the 3-line standard interface. Note that you can use circuit blocks such as op-amps, comparators, and other digital circuits.

5. Draw the block diagram of the circuit that can implement the 2-line standard interface.

6. The 2-line standard interface circuit produces output current pulses superimposed on the power supply. Draw the block diagram of the external circuit that can extract these current pulses from the power supply line while providing the supply voltage needed.

7. The final cross-section of the bulk-silicon micromachined-CMOS process is shown in Fig. 18b. Develop a process sequence that shows the individual process steps that need to be carried out to implement the device shown in the figure. You should write the sequence showing the major process steps for both the sensor and the CMOS circuit.

8. Discuss the advantages and disadvantages of both surface- and bulk-micromachined sensor-circuit processes presented in Fig. 18.

9. This problem deals with the design of a simple CMOS op-amp. Assume that we have a fabrication technology that provides the following process parameters:

$$V_{Tn} = 0.8 \qquad V_{Tp} = -0.8 \text{ V}$$

$$C_{ox} = 9 \times 10^{-8} \text{ F/cm}^2$$

$$\lambda_p = \lambda_n = 0.01$$

$$\mu_n = 600 \text{ cm}^2/\text{V-s} \qquad \mu_p = 300 \text{ cm}^2/\text{V-s}$$

Design an op-amp that provides an open-loop gain of 80 dB, and a gain-bandwidth product of 2 MHz, dissipates an average power of 2 mW from ± 5 V power supplies, has a slew rate of 20 V/μs, drives a load capacitance of 2 pF, and uses transistors with channel lengths of 6 μm. You should come up with width dimensions for all transistors.

10. Explain how the Schmitt-oscillator interface circuit for capacitive sensors works. Draw timing diagrams for the output voltage signal and the voltage signal at the input of the Schmitt trigger.

REFERENCES

1. K. D. Wise, "Solid-state microsensors," *Sensors*, March (1988).
2. K. D. Wise and N. Najafi, "The coming opportunities in microsensor systems (invited paper)," *Dig. Int. Conf. on Solid-State Sensors and Actuators, Transducers '91, San Francisco, CA, June, 1991*, p. 2.
3. R. P. Knockeart and R. E. Sulouff, "Integrated micromachined silicon: vehicle sensors of the 1990s," *Dig. Int. Congress on Transportation Electronics, New York, 1988*.
4. K. Najafi, "Smart sensors," *J. Micromech. Microeng.* **1**, 86 (1991).
5. N. Najafi, K. W. Clayton, W. Baer, K. Najafi, and K. D. Wise, "An architecture and interface for VLSI sensors," *Dig. IEEE Solid-State Sensor and Actuator Workshop, Hilton Head, SC, June, 1988*, p. 76.
6. N. Najafi and K. D. Wise, "An organization and interface for sensor-driven semiconductor process control systems," *IEEE Trans. Semiconductor Manufacturing*, November, 230 (1990).
7. D. W. deBruin, H. V. Allen, and S. C. Terry, "Second-order effects in self-testable accelerometers," *Dig. IEEE Solid-State Sensors and Actuators Workshop, Hilton Head, SC, June 1990*, p. 149.
8. S. T. Cho and K. D. Wise, "A high-performance microflowmeter with built-in self test," *Sens. Actuators* **A36**, 47 (1993).
9. N. Najafi, A generic smart sensity system utilizing a multi-element gas analyzer, *Ph.D. Dissertation*, The University of Michigan, 1990.
10. N. Najafi, and K. D. Wise, "An integrated multi-element ultra-thin-film gas analyzer," *Dig. IEEE Solid-State Sensors and Actuators Workshop, Hilton Head, SC, June 1992*, p. 19.
11. H. L. Chau, and K. D. Wise, "An ultraminiature solid-state pressure sensor for a cardiovascular catheter," *IEEE Trans. Electron Devices*, **ED-35**, 2355 (1988).
12. J. Ji, S. T. Cho, Y. Zhang, K. Najafi, and K. D. Wise, "An ultraminiature CMOS pressure sensor for a multiplexed cardiovascular catheter," *IEEE Trans. Electron Devices* **ED-39**, 2260 (1992).
13. S. B. Crary, W. Baer, J. C. Cowles, and K. D. Wise, "Digital compensation of high-performance silicon pressure transducers," *Dig. 5th Int. Conf. on Solid-State Sensors and Actuators, Montreux, Switzerland, June 1989*, p. 85.
14. S. B. Crary, W. G. Baer, J. C. Cowles, and K. D. Wise, "Digital compensation of high-performance silicon pressure transducers," *Sens. Actuators* **A21**, 70 (1990).
15. P. R. Gray and R. G. Meyer, *Analysis and Design of Analog Integrated Circuits*, Wiley, New York, 1993.

16. P. E. Allen and D. R. Holberg, *CMOS Analog Circuit Design,* Holt, Reinhart and Winston, New York, 1987.
17. J. C. Bertails, "Low-frequency noise considerations for MOS amplifiers design," *IEEE J. Solid-State Circuits,* **SC-14**, 773 (1979).
18. K. C. Hsieh, P. R. Gray, D. Senderowicz, and D. G. Messerschmitt, "A low-noise chopper-stabilized differential switched-capacitor filtering technique," *IEEE J. Solid-State Circuits* **SC-16**, 708 (1981).
19. R. Gregorian, K. W. Martin, and G. C. Temes, "Switched capacitor circuit design," *Proc. IEEE* **71**, 941 (1983).
20. W. Yun, R. T. Howe, and P. R. Gray, "Surface micromachined, digitally force-balanced accelerometer with integrated CMOS detection circuitry," *Dig. IEEE Solid-State Sensor and Actuator Workshop, Hilton Head, SC, June 1992,* p. 126.
21. J. E. Brignell, "Sensors in distributed instrumentation systems," *Sens. Actuators* **10**, 249 (1986).
22. M. G. R. Degrauwe, O. Nys, E. Dijkstra, J. Rijmenants, *et al.*, "IDAC: an interactive design tool for analog CMOS circuits," *IEEE J. Solid-State Circuits* **SC-22**, 1106 (1987).
23. E. O. Doebelin, *Measurement Systems, Application and Design,* McGraw-Hill, New York, 1983.
24. C. S. Sander, J. W. Knutti, and J. D. Meindl, "A monolithic capacitive pressure sensor with pulse period output," *IEEE Trans. Electron Devices* **ED-27**, 307 (1980).
25. A. Kjensmo, A. Hanneborg, J. Gakkestad, and H. Von Der Lippe, "A CMOS front-end circuit for a capacitive pressure sensor," *Sens. Actuators* **A21–A23**, 102 (1990).
26. K. D. Wise and K. Najafi, "VLSI sensors in medicine," *VLSI in Medicine* N. G. Einspruch and R. D. Gold, Eds., Academic Press, San Diego, CA, 1989.
27. A. B. Grebene, *Bipolar and MOS Analog Integrated Circuit Design,* Wiley, New York, 1984.
28. Y. E. Park and K. D. Wise, "An MOS switched-capacitor readout amplifier for capacitive pressure sensors," *IEEE Proc. Custom Integrated Circuit Conf., May 1983,* p. 380.
29. K. Chun and K. D. Wise, "A high-performance silicon tactile imager based on a capacitive cell," *IEEE Trans. Electron Devices* **ED-32**, 1196 (1985).
30. K. Suzuki, K. Najafi, and K. D. Wise, "A 1024-element high-performance silicon tactile imager," *IEEE Trans. Electron Devices* **ED-37**, 1852 (1990).
31. M. H. White, D. R. Lampe, F. C. Blaha, and I. A. Mack, "Characterization of surface channel CCD image arrays at low light levels," *IEEE J. Solid-State Circuits* **SC-9**, 1 (1974).
32. J. T. Kung, H. S. Lee, and R. T. Howe, "A digital readout technique for capacitive sensor applications," *IEEE J. Solid-State Circuits* **23**, 972 (1988).
33. J. T. Kung and H. S. Lee, "An integrated air-gap capacitor pressure sensor and digital readout with sub-100 attofarad resolution," *Proc. IEEE/ASME J. Micro Electro Mechanical Systems* **1**, 121 (1992).
34. F. Rudolf, A. Jornod, J. Bergqvist, and H. Leuthold, "Precision accelerometers with μg resolution," *Sens. Actuators* **A21–A23**, 297 (1990).

35. H. Leuthold and F. Rudolf, "An ASIC for high-resolution capacitive microaccelerometers," *Sens. Actuators* **A21–A23**, 278 (1990).
36. *Monolithic Accelerometers with Signal Conditioning, ADXL50*, Application Note, Analog Devices.
37. K. Hart and A. Slob, "Integrated injection logic: a new approach to LSI," *IEEE J. Solid-State Circuits* **SC-7**, 346 (1972).
38. M. Kubo, I. Masuda, K. Miyata, and K. Ogiue, "Perspective on BiCMOS VLSI," *IEEE J. Solid-State Circuits* **SC-23**, 5 (1988).
39. K. Najafi and K. D. Wise, "An implantable multielectrode recording array with on-chip signal processing," *IEEE J. Solid-State Circuits* **SC-21**, 1035 (1986).
40. R. T. Howe and R. S. Muller, "Integrated resonant-microbridge vapor sensor," *Des. Int. Electron Devices Meeting (IEDM), December 1984*, p. 213.
41. M. W. Putty, S. C. Chang, R. T. Howe, A. L. Robinson, and K. D. Wise, "Process integration for active polysilicon resonant microstructures," *Sens. Actuators* **20**, 143 (1989).
42. S. Sugiyama, M. Takigawa, and I. Igarashi, "Integrated piezoresistive pressure sensor with both voltage and frequency output," *Sens. Actuators* **4**, 113 (1983).
43. T. Ishihara, K. Suzuki, S. Suwazono, M. Hirata, and H. Tanigawa, "CMOS integrated silicon pressure sensor," *IEEE J. Solid-State Circuits* **SC-22**, 151 (1987).
44. E. Yoon and K. D. Wise, "A multi-element monolithic mass flowmeter with on-chip CMOS readout electronics," *Tech. Dig. IEEE Solid-State Sensor and Actuator Workshop, Hilton Head, SC, June 1990*, p. 161.
45. J. Ji and K. D. Wise, "An implantable CMOS circuit interface for multiplexed microelectrode recording arrays," *IEEE J. Solid-State Circuits* **SC-27**, 433 (1992).
46. T. Akin, B. Ziaie, and K. Najafi, "RF telemetry powering and control of hermetically sealed integrated sensors and actuators," *Dig. IEEE Workshop on Solid-State Sensors and Actuators, Hilton Head, SC, 1990*, p. 145.
47. B. Ziaie, Y. Gianchandani, and K. Najafi, "A high-current IrOx thin-film neuromuscular microstimulator," *Proc. Int. Solid-State Sensor and Actuator Conf., San Francisco, CA, June 1991*, p. 124.
48. S. S. Wang, V. M. McNeil, and M. A. Schmidt, "An etch-stop utilizing selective etching of n-type silicon by pulsed potential anodization," *Proc. IEEE/ASME J. Micro Electro Mechanical Systems* **1**, 187 (1992).
49. J. Ji and K. D. Wise, "An implantable CMOS analog processor for multiplexed microelectrode recording arrays," *Dig. IEEE Solid-State Sensor and Actuator Workshop, Hilton Head, SC, June 1990*, p. 107.
50. M. Parameswaran, H. P. Baltes, and A. M. Robinson, "Polysilicon microbridge fabrication using standard CMOS technology," *Dig. IEEE Solid-State Sensor and Actuator Workshop, Hilton Head, SC, June 1988*, p. 148.
51. M. Parameswaran, R. Chung, M. Gaitan, R. B. Johnson, and M. Syrzycki, "Commercial CMOS fabricated integrated dynamic thermal scene simulator," *Dig. Int. Electron Devices Meeting (IEDM 91), Washington DC, December 1991*, p. 753.
52. C. H. Stapper, F. M. Amstrong, and K. Saji, "Integrated circuit yield statistics," *Proc. IEEE* **71**, 453 (1983).
53. S. J. Tanghe and K. D. Wise, "A 16-channel CMOS neural stimulating array," *IEEE J. Solid-State Circuits* **SC-27**, 1819 (1992).

54. C. L. Johnson, N. Najafi, K. D. Wise, and J. W. Schwank, "Detection of semiconductor process gas impurities using an integrated ultra-thin-film detector," *Dig. Technical Papers, SRC TECHCON '90, San Jose, October 1990*, p. 417.
55. C. L. Johnson, K. D. Wise, and J. W. Schwank, "A thin-film gas detector for semiconductor process gases," *Dig. Int. Electron Devices Meeting, Washington DC, December 1988*, p. 662.

APPENDIX A
List of Symbols

Symbol	Description	Units
a	Lattice constant	Å
B	Magnetic induction	Wb/m^2
c	Speed of light in vacuum	cm/s
C	Capacitance	F
D	Electric displacement	C/cm^2
D	Diffusion coefficient	cm^2/s
E	Energy	eV
E_F	Fermi energy level	eV
E_g	Energy bandgap	eV
$\mathscr{E}$	Electric field	V/cm
$\mathscr{E}_m$	Maximum field	V/cm
f	Frequency	Hz
h	Planck's constant	J-s
$h\nu$	Photon energy	eV
I	Current	A
J	Current density	A/cm^2
k	Boltzmann constant	J/K
kT	Thermal energy	eV
L	Length	cm or μm
m_0	Electron rest mass	kg
m^*	Effective mass	kg
$\bar{n}$	Refractive index	
n	Density of free electrons	cm^{-3}
n_i	Intrinsic density	cm^{-3}
N	Doping concentration	cm^{-3}
N_A	Acceptor impurity density	cm^{-3}
N_D	Donor impurity density	cm^{-3}
p	Density of free holes	cm^{-3}
P	Pressure	N/m^2
q	Magnitude of electronic charge	C

APPENDIX A: LIST OF SYMBOLS

Symbol	Description	Units
Q_{it}	Interface-trap density	charges/cm^2
R	Resistance	Ω
t	Time	s
T	Absolute temperature	K
v	Carrier velocity	cm/s
v_s	Saturation velocity	cm/s
V	Voltage	V
V_{bi}	Built-in potential	V
V_B	Breakdown voltage	V
W	Thickness	cm or μm
x	x direction	
∇T	Temperature gradient	K/cm
ε_0	Permittivity in vacuum	F/cm
ε_s	Semiconductor permittivity	F/cm
ε_i	Insulator permittivity	F/cm
$\varepsilon_s/\varepsilon_0$ or $\varepsilon_i/\varepsilon_0$	Dielectric constant	
τ	Lifetime or decay time	s
θ	Angle	rad
λ	Wavelength	μm or Å
ν	Frequency of light	Hz
μ_0	Permeability in vacuum	H/cm
μ_n	Electron mobility	cm^2/V-s
μ_p	Hole mobility	cm^2/V-s
ρ	Resistivity	Ω-cm
ϕ	Barrier height	V
ϕ_m	Metal work function	V
ω	Angular frequency ($2\pi f$ or $2\pi v$)	Hz
Ω	Ohm	Ω

APPENDIX B
International System of Units

Quantity	Unit	Abbreviation	Units
Length	meter	m	
Mass	kilogram	kg	
Time	second	s	
Temperature	kelvin	K	
Current	ampere	A	
Frequency	hertz	Hz	s^{-1}
Force	newton	N	$kg\text{-}m/s^2$
Pressure	pascal	Pa	N/m^2
Energy	joule	J	N-m
Power	watt	W	J/s
Electric charge	coulomb	C	A-s
Potential	volt	V	J/C
Conductance	siemens	S	A/V
Resistance	ohm	Ω	V/A
Capacitance	farad	F	C/V
Magnetic flux	weber	Wb	V-s
Magnetic induction	tesla	T	Wb/m^2
Inductance	henry	H	Wb/A

APPENDIX C
Physical Constants

Quantity	Symbol	Value
Angstrom unit	Å	1 Å = 10^{-1} nm = 10^{-4} μm
		= 10^{-8} cm = 10^{-10} m
Avogadro constant	N_{AVO}	6.02204×10^{23} mol^{-1}
Bohr radius	a_B	0.52917 Å
Boltzmann constant	k	1.38066×10^{-23} J/K (R/N_{AVO})
Elementary charge	q	1.60218×10^{-19} C
Electron rest mass	m_0	0.91095×10^{-30} kg
Electron volt	eV	1 eV = 1.60218×10^{-19} J
		= 23.053 kcal/mol
Gas constant	R	1.98719 cal/mol-K
Permeability in vacuum	μ_0	1.25663×10^{-8} H/cm ($4\pi \times 10^{-9}$)
Permittivity in vacuum	ε_0	8.85418×10^{-14} F/cm ($1/\mu_0 c^2$)
Planck constant	h	6.62617×10^{-34} J-s
Reduced Planck constant	$\hbar$	1.05458×10^{-34} J-s ($h/2\pi$)
Proton rest mass	M_p	1.67264×10^{-27} kg
Speed of light in vacuum	c	2.99792×10^{10} cm/s
Standard atmosphere		1.01325×10^5 N/m^2
Thermal voltage at 300 K	kT/q	0.0259 V
Wavelength of 1-eV quantum	λ	1.23977 μm

APPENDIX D
Properties of Si and GaAs in 300 K

Properties	Si	GaAs
Atoms/cm^3	5.0×10^{22}	4.42×10^{22}
Atomic weight	28.09	144.63
Breakdown field (V/cm)	$\sim 3 \times 10^5$	$\sim 4 \times 10^5$
Crystal structure	Diamond	Zincblende
Density (g/cm^3)	2.33	5.32
Dielectric constant	11.9	13.1
Effective density of states in conduction band, N_C (cm^{-3})	2.8×10^{19}	4.7×10^{17}
Effective density of states in valence band, N_V (cm^{-3})	1.04×10^{19}	7.0×10^{18}
Effective mass, m^*/m_0:		
Electrons	$m_l^* = 0.98$, $m_t^* 0.19$	0.067
Holes	$m_{lh}^* = 0.16$, $m_{hh}^* = 0.49$	$m_{lh}^* = 0.082$, $m_{hh}^* = 0.45$
Electron affinity (V)	4.05	4.07
Energy bandgap (eV)	1.12	1.424
Heat capacity (J/mol-K)	20.07	47.02
Index of refraction	3.42	3.66
Intrinsic carrier concentration (cm^{-3})	1.08×10^{10}	2.1×10^6
Intrinsic Debye length (μm)	24	2250
Intrinsic resistivity (Ω-cm)	2.3×10^5	10^8
Lattice constant (Å)	5.4309	5.6533
Linear coefficient of thermal expansion, $\Delta L/L \Delta T$ (°C^{-1})	4.2×10^{-6}	6.86×10^{-6}
Melting point (°C)	1412	1238
Minority-carrier lifetime(s)	2.5×10^{-3}	$\sim 10^{-8}$
Mobility (drift) (cm^2/V-s):		
μ_n (electrons)	1500	8500
μ_p (holes)	450	400
Optical phonon energy (eV)	0.063	0.035
Phonon mean free path λ_0 (Å):		
Electrons	76	58
Holes	55	
Poisson's ratio, $\langle 100 \rangle$ orientation	0.28	0.31

APPENDIX D: PROPERTIES OF Si AND GaAs IN 300 K

Properties	Si	GaAs
Specific heat (J/g-°C)	0.7	0.35
Thermal conductivity (W/cm-°C)	1.5	0.46
Thermal diffusivity (cm^2/s)	0.9	0.44
Vapor pressure (Pa)	1 at 1650°C	100 at 1050°C
	10^{-6} at 900°C	1 at 900°C
Young's modulus (GPa)	130	85.5

INDEX

Absolute humidity, 367
Absorption, 275
   coefficient, 276, 277, 310
Absorptivity, 338, 340
Acceleration sensors, 58, 126, 176
Accelerometer, 17, 105, 113, 116, 135, 143, 153, 157, 162, 187, 191, 192, 197, 477, 525
   devices, 115
Accumulation layer, 387
Acidic etchant, 49
Acoustic measurand, 5
Acoustic plate mode (APM), 97
   device, 112
Acoustic resonator, 105
Acoustic sensing, 110
Acoustic sensor, 97, 104, 109, 113, 497
Acoustic streaming, 143
Acoustic wave, 99, 117
   gas sensor, 124
Acoustic wavelength, 119
Acoustic-wave sensors, 356
Acrylics, 460
Activation energy, 48, 394
Active suspension control, 197
Actuation, 4
Actuator, 42, 105, 109, 479
Additive process, 80
Adsorption, 425, 463
   -FET (ADFET), 406, 409
Affinity reactions, 454
Ag/AgCl electrode, 408
Airbag, 18, 197
   actuator, 127
   deployment, 17
Air conditioning, 17
Air/fuel (A/F) ratio, 398
AlAs/GaAs heterojunction, 228
Alcohols, 383, 401
AlGaAs/GaAs heterojunction, 225, 228, 245
Alignment and exposure, 27
Alkaline etchants, 46
Alkanes, 402
AlN, 99, 105, 106, 108, 117
Al_2O_3, 108, 395, 408
α-Amino acid, 423
Alpha particles, 273
AlSb, 313
Aluminum (Al), 21, 49, 56, 65, 66, 107, 313, 314, 353, 409
   nitride, 107

   oxide, 43
Ambient conditions allowed, 7
Ammonia (NH_3), 61, 124, 451
   gas, 440
Ammonium hydroxide, 49
Amorphous glass, 42
Amorphous silicon, 56, 308
Amperometric biosensors, 441
Amperometric detection, 445
Analog output, 9
Analog–receptor reaction, 448
Analog signal, 3
Analog-to-digital converter, 475
Anemometers, 358, 359
Anisotropic etchants, 29
Anisotropic etching, 42, 43, 49, 80, 113, 155, 366
Anisotropic fabrication, 178
Anisotropic scattering factor, 230
Anodic bonding, 51, 160
Anodic bonds, 50
ANSYS, 345
Antibody, 418, 419, 463
   antigen binding, 416
   antigen reaction, 454
Antigen, 418, 419, 463
Antilock brake system, 197
Antimony, 353
Argon (Ar), 22, 156, 365
Arsenic, 302, 303
   oxide, 391
Artificial bilayer liquid membranes, 422
AT-cut quartz crystal resonator, 111
Atmospheric CVD, 57
Atomic energy, 3
Atomic weight, 535
Attenuation coefficient, 98
Au, 48, 304
   film, 118
$AuCl_3$, 304
Auger-electron production, 276
Auger lifetimes, 300
Au/Si system, 54
Automotive applications, 497
Automotive sensors, 475
Automotive systems, 17
Avalanche photodiode, 293, 295, 296
Avionics, 523
Axial heat flow, 343

Back-etched SOI, 59

537

538 INDEX

Back-side etch-stop, 113
Bacteria, 98, 459
Bandgap, 276
Bandwidth, 118
Batch process, 12
BCl_3, 65
Beam-type sensors, 177
Bending moments, 188
Bessel functions, 349
Beta particle, 273
B_2H_6, 57
BiCMOS, 227, 503, 504, 524
Bi_2S_3, 313
Bi-directional line, 486
Bifunctional reagent, 425
Bilateral analog switches, 503
Bimetallic warping, 52
 effect, 34
Bimetal strip, 475
Binary code, 9
Binding reaction rate, 419
Bio-affinity recognition, 416
Biochemical analysis, 143
Biological chemicals, 417
Biological diagnostic system, 17
Biological measurands, 5
Biological-recognition membrane, 421, 463
 processes, 421, 437
Bioluminescence based biosensor, 451
Biomaterials, 417
Bio-metabolic recognition, 416
Biomimetic structures, 421, 463
Bioreactor, 427
 membrane, 428
Bio-recognition processes, 415
Biosensor, 383, 415
 sensitivity curve, 432
Bipolar amplifiers, 493
Bipolar IC process, 249, 257
Bipolar IC technology, 233, 254, 261, 305
Bipolar magnetotransistors, 210, 218
Bipolar p-n-p transistors, 237
Bipolar silicon technology, 365, 502
Bipolar transistors, 10, 137, 246, 260, 508
Bismuth, 353, 363
 /antimony thermocouples, 363, 370
 molybdate, 398
 molybdate sensor, 404
 -tin layer, 231
Black body, 339
 radiation, 338
β-nicotinamide adenine
 dinucleotide, 443, 451
Bohr radius, 294
Boltzmann factor, 390
Boltzmann transport equation, 211
Bolometer, 362, 363, 366
Bonding techniques, 159
Boron, 44, 45, 302, 303
 -doped regions, 48
 doping, 50
 etch-stop method, 157

Borosilicate, 52
 glass, 63
Bose–Einstein statistics, 280
Boundary layer, 336
 flow sensor, 359
 wind-flow sensor, 360
Bovine serum albumin, 425
 molecules, 140
Bragg curve, 284
Breakdown field, 535
Breakdown voltage, 263, 295
Bremsstrahlung, 284
Bridgman method, 312, 313
Bridgman techniques, 302
Brillouin zone, 170
Bromine-based plasma, 66
Buffer HF, 66, 496
Bulk-micromachined sensors, 19, 20, 25
Bulk, micromachining, 42
 of silicon, 154
 technique, 504
Bulk-wave delay line, 110
Bulk-wave (BAW) oscillators, 452
Bulk-wave piezoelectric gas (BWPG) detector
 was, 124
Bulk-wave resonator, 112
Bull's-eye pattern, 65
Buried layer, 254
Burstein–Moss shift, 281, 300
Burst noise, 289
Bus-organized system, 479

CAD, 495, 497
Calibration cycles, 9
Calorimetric method, 359
Cameras, 17
Cantilever beam, 116, 127, 174, 196, 342, 347
 accelerometer, 136
 piezoresistive acceleration sensor, 42
Capacitive accelerometers, 186
Capacitive-based sensors, 496
Capacitive semiconductor accelerometers, 192
Capacitive sensors, 13, 153, 185, 486, 499
Capacitive transduction, 194
Carbohydrates, 425
Carbon, 441, 457, 463
 monoxide, 403
Carrier deflection, 212
Carrier-domain magnetometers, 208, 232, 258, 260
Cartesian system, 164
Cascade stages, 495
Catalyst, 384, 394
Cavity, 73
CCD, 332
 camera, 305
CCl_4, 65, 406
CCl_2F_2, 66
$Cd_{1-y}Zn_yTe$, 302
CdSe, 313
CdTe, 277, 299, 304, 312
CdZnTe, 313

Cell membranes, 421
Cellulose acetate, 425
Ceramics, 461
Ceramic superconducting Y-Ba-Cu-O compound material, 209
Ceramic Y-Ba-Cu-O, 228
Cesium hydroxide, 49
CF_4, 64, 518
C_2F_6, 66
CH_4, 125, 391, 402
C_2H_6, 301
CH_3COOH, 49, 154
Charge-coupled devices, 285
Charge density, 104
ChemFET, 404
Chemical amplification, 421
Chemical etching, 136
  process, 11
Chemical measurands, 5
Chemical sensor, 525
Chemical signal, 3
Chemical vapor deposition, 23, 107
Chemiluminescence-based biosensors, 451
CHF_3, 66, 156
Chlorine, 312
Chlorine-based plasma etching, 65
Choline, 49
Chrome, 353
  -gold films, 136
Chromium, 27, 65
  oxide, 391
Chromophore-based biosensors, 451
Circular, horizontal four-layer CDM, 262
Circular, horizontal three-layer CDM, 262
Cl_2, 156, 409
Closed membranes, 349
CMOS, 58, 231, 260, 493, 503, 505, 508, 524
  circuitry, 245, 265
  IC technology, 247, 249, 265
  operational amplifier, 494
  process, 494, 508, 515, 517
  technology, 227, 237, 246, 251, 254, 255
CO, 125, 383, 387, 391, 401, 402, 406
CO-enzyme biosensors, 443
CO_2, 125, 436
  laser, 106
Collagen, 425
Combustion control, 197
Combustible gas, 397
Command library, 483
Compass, 205
Competitive-binding approach, 454
Competitive-binding sensors, 449
Complementary MOS, 502
Compressed powders, 396
Compton effect, 283
Compton scattering, 276, 283
Computer-aided design (CAD), 227, 524
Conduction, 333
Conductivity, 333
  sensing unit, 523
  -type gas sensor, 518

Consumer appliances, 523
Contact aligners, 27
Contactless switching, 206
Contact resistance, 304
Continuity equations, 238
Convection, 335
Conversion phenomena, 6, 7
CoO, 400
Copper, 353
Corbino disc, 217
Corning glass 7740, 52
Co-sputtering, 22
Coulombic processes, 283
Coulombic scattering, 300
Covalent bonding, 426, 463
Crankshaft position sensors, 206
Cr-Ag, 48
Cr-Au pads, 118
Critical-point drying, 70
Cross-performance sensitivity, 491
Cross-shaped Hall devices, 245
Crystalline quartz (SiO_2), 102, 107
Crystal structure, 535
Cu, 48
Current domain magnetometers, 210
Cut-off frequency, 495
Curvature measurements, 35
Custom command sets, 483
CVD films, 56
CVD tungsten, 508
Cyclic fatigue, 33
Cylindrical drift chamber, 318

Damping, 193
DAMS, 246
Dangling bonds, 385
Deep-level transient spectroscopy (DLTS), 300
Deformation detection, 158
Delta rays, 283
Density, 535
  of molecules, 334
  of states, 535
Depleted JFET, 320
Depleted p-channel MOS, 320
Depletion layer, 389
Depletion-mode PMOS device, 128
Deposition, 20
Design considerations, 45
Detection means, 6, 7
Detectivities, 287
Devitryifying glasses, 52
Dew-point sensors, 368
Dew-point temperature, 367
Diamond, 32, 334
Diaphragm pressure sensor, 6, 10, 42
Dichlorosilane ($SiCl_2H_2$), 61
Dielectric constant, 535
Differencing amplifier, 497
Differential amplification magnetic sensor, 23, 237
Differential voltage gain, 494

Diffusion coefficient, 428
Diffusion constant, 393
Digital compensation, 492
Digital output, 9
Digital signal, 3
Dihydroxyalcohol, 66
Dimensional control, 38
Dimethyl sulfoxide, 140
Direct tensile tests, 33
Direct transition, 277
Displacement detection, 158, 206
Dissociation energy, 392
DNA, 143, 463
   recognition, 454
Dopants, 301
D orbital, 385
Double-bridge flow sensor, 359
Double layers, 384
Double-sided aligner, 29
Drift chambers, 324
Drilling, 154
Dry etching, 29
   parameters, 30
Dual-detector chip, 518
Dual-drain MAGFET, 243, 245, 246, 248

Early effect, 355
EDP, 47, 48, 50, 116, 128, 129, 154
   solution, 116
Effective mass, 170, 228, 535
Elastic flexural plate waves (FPWs), 97
Elastic properties, 32
Elastic stiffness, 104
Elastic wave, 97, 98
   devices, 102
   motion, 100
Elasto-electric conversion, 119
Electrical energy, 3
Electrical signal, 3
Electrical-thermal analogies, 342
Electric measurands, 5
Electrochemical control (ECC) Etch-Stop, 113
Electrochemical detection, 437
Electrochemical etching, 43, 116
Electrochemical etch-stops, 50, 190, 504
Electromagnetic radiation, 271, 274
Electromagnetic waves, 99, 274
Electromechanical coupling, 122
Electron affinity, 299, 535
Electron-beam evaporation, 109
Electron diffraction, 303
Electroplating, 75
Electrostatic bonding, 12, 51
Electrostatic pressure, 52
Ellipsometry, 303
Emissivity, 339, 340
End-point linearity, 8
Energy gap, 535
ENFET, 437, 439, 440
Enhancement-depletion n-channel MOS, 503
Enthalpy, 333, 451

Environmental control, 374
Enzyme, 369, 417, 418, 435, 463
   electrode, 417
   field-effect transistor, 428
   immobilization, 440
   kinetics, 445
   loading factor, 434
   reaction, 451
   reaction rate, 433
   substrate interaction, 420
   thermistors, 452
Epitaxial growth, 109
Epitaxial layer, 234
Epoxies, 55, 460
Equilibrium constant, 419
Equivalent-noise charge, 318, 322
Error function, 429
Etching, 29
   profile, 40
   silicon, 64
Etch masks, 46, 49
Etch rate, 43, 65
   ratio, 48
Etch selectivities, 48
Etch-stop film, 40
Ethyl alcohol, 404
Ethylene diamine/pyrocatechol (EDP), 155, 189
Ethylene diamine, 47
   pyrocatechol etching process, 508
Ethylene oxide, 459
Ettinghausen effect, 231
Ettinghausen-Nernst coefficient, 231
Ettinghausen-Nernst effect, 231
Eutectic alloys, 54
Eutectic bonds, 160
Evaluation of sensors, 10
Evanescent field, 450
Evaporation, 21, 57
Excition, 280
Expansion, 535
External quantum yield, 286

Fabrication techniques, 502
Fail-safe operation, 474
Fano factor, 314
Faraday constant, 438
Faraday rotation, 208
Fast electrons, 271, 273, 275, 283, 309
Fast-etching rate, 47
F atoms, 64
Fe, 304
Feedback amplifier, 127
Fe_2O_3, 399, 400
Fermi distribution function, 171
Fermi energy, 390
   control, 395
Ferrocene, 443
Ferromagnetic materials, 208
FET, 389
   -based sensor, 383
   devices, 404
Fiber optics, 285, 298, 449, 450

INDEX   541

Fick's laws, 428
Fick's second law, 433
Field-effect transistor, 318, 383, 406
Fifth-generation sensors, 477, 484, 491
Film microstructure, 32, 38
Finite-element method, 189
Flash-evaporation, 22
Flat bands, 386
Flexible beams, 33
Flexural plate device, 112
Flexural-plate-wave (FPW) sensors, 111
Flexural rigidity, 139
Flicker noise, 252, 288, 496
Floating membrane, 342, 345
  structure, 365
Float-zone growth, 312
Float-zone technique, 310
Flow injection analysis (FIA) system, 430
Flow sensors, 153, 157, 176, 187, 193, 341,
  348, 357, 358, 509
Fluid density, 140
Fluorocarbons, 460
Fluorophore based biosensor, 451
Flux-gate magnetometer, 208, 264
Focused ion-beam technology, 228, 234
Force-balancing technique, 501
Four-element hybrid gas analyzer, 521
Fourier series method, 431
FPW device, 141, 142
FPW sensor, 137, 140
FPW transducer, 138
Fracture, 33
Frequency-modulated output, 9
Frequency output, 9
Frequency shift, 111
Frequency synthesizer, 120
Front-end interface circuits, 489
Full scale, 7
Fully depleted junction charge-couple devices,
  320
Fusion bonding, 50, 53
FWHM, 318, 322

GaAs, 15, 78, 104, 107, 117, 122, 207, 221, 225,
  226, 227, 259, 277, 297, 302, 313, 535
  Hall device, 234
  IC technology, 234
Gage factor, 161
Galavanomagnetic effects, 205, 206, 210, 211,
  212, 228
Gallium arsenide (GaAs), 42, 43, 391
γ-Amino propylsilane, 55
Gamma function, 229
Gamma photons, 272
Gamma-radiation detectors, 310, 312, 313
Ga_2O_3, 391
GasFET, 405
  sensor, 409
Gas injection, 197
Gas-phase detections, 455
Gas sensors, 105, 125, 518
Gas-type sensors, 364
Ge, 259, 277

Gelatin, 425
Generation-recombination rate, 238
Geomagnetic field, 206, 207
Germanium (Ge), 161, 168, 172, 283, 297, 298,
  305, 324
Gibbs free energy, 438
Glass, 43, 461
  bonding, 50
  flint, 52
  -frit seals, 160
G-line mask aligner, 78
Global positioning system, 127
Global set of commands, 483
Glucose, 370, 373, 440, 443
  oxidase, 417, 451, 452
  sensor, 449, 465
Glutardialdehyde, 425
Glycerol, 66
Gold, 21, 49, 54, 107, 114, 303, 313, 353, 400,
  441
  mask, 136
  -silicon eutectic bond, 12
Grain boundaries, 399
Gravimetric acoustic sensors, 111
Gravimetric effects, 98
Gravimetric sensing, 140
Gravimetric sensitivity factor, 111, 139
Gravitational energy, 3
Guckel rings, 37

H_2, 25, 125, 383, 384, 391, 401, 402, 404
Hall angle, 214, 216, 219
Hall bridges, 497
Hall coefficient, 214, 216, 217, 229, 230
Hall cross, 234
Hall devices, 208, 218, 222
Hall effect, 205, 238
  sensor, 206, 221, 222, 226, 231, 241
Hall elements, 208, 228
Hall field, 213, 214, 217, 238
Hall IC chip, 234, 264
Hall MAGFET, 241, 242
Hall magnetic sensor, 221
Hall measurements, 303
Hall mobility, 212, 225, 230, 242, 245
Hall mode, 239
Hall plate, 211, 218, 221, 223, 225, 226, 232,
  240, 242
Hall scattering coefficient, 245
Hall scattering factor, 242
Hall switches, 234
Hall voltage, 214, 220, 226, 230, 231, 233, 238,
  242, 246, 257, 288
Halogenated hydrocarbons, 401
HBr, 66
HCl, 304, 406
H_2/CO mixture, 403
He, 22
Health care, 523
Heat, 332
  capacity, 452, 535
  flow, 331, 342
  transfer, 332, 340, 350

542 INDEX

Heat (*Continued*)
 -transfer coefficient, 334, 337, 365
Heavy-charged particles, 271, 273, 275, 283
Helium, 365
 nuclei, 273
Helmholtz double layer, 407
Helmholtz plane, 446
Heterojunction avalanche photodiodes, 293, 297
Hexacyanoferrate, 443
Hexamethyldichlorosilazane, 25
Hexamethyldisilazane, 55, 64
HF, 49, 59, 70, 304
 solutions, 157
Hg, 300, 301, 303
HgCdTe, 298, 324
 infared sensor, 298
 system, 298
HgI_2, 312
HgTe, 277, 299
High-energy oxide implantation, 59
High-energy photodiodes, 308
High g-level acceleration, 134
 packaging, 131
 Si accelerometers, 129
High-Q SAW resonator, 137
Hillocks, 49
HNO_3, 49, 407
H_2O_2, 78, 406
Homogeneous heating, 350
Hooge noise parameter, 14
Hooge parameter, 226
Hooke's law, 101, 188
Horizontal Hall plates, 238
Host computer, 478
Host controller, 477
Hot-wire system, 193
H_2PO_4, 67
H_2S, 125, 401, 403, 406, 409
Human senses, 1, 367, 373
Human transferrin sensor, 454
Hybrid amplifier, 137
Hydrazine, 49, 155, 189
Hydrocarbons, 383
Hydrogen, 124, 294, 298, 333, 383, 403
 bond, 425
 chloride, 124
 ion, 436
Hydrophobic forces, 425
Hysteresis, 8

Ice, 335
IC technology, 353
IDT, 119, 120, 126
IEEE-488 (HP-IB) structure, 482
IGFET, 439
IMFET, 437
Immobilization, 423
Immunochemical field-effect transistor, 437
Impact ionization, 277, 281
Impedometric detection, 445
InAs, 216, 226, 298

Incident irradiance, 276
Index of refraction, 59, 277, 535
Indirect absorption, 280
Indium (In), 303, 304
Industrial processing, 523
Infrared radiation, 340
 sensor, 231, 340, 341, 357, 362, 373, 509
In_2O_3, 393, 400, 403
Inorganic spin-on glassers, 20
InSb, 216, 225, 226, 298
Insertion loss, 111, 118
In situ stress, 36
Instrumentation, 523
Insulin pumps, 465
Integrated accelerometer, 508
Integrated bimetal actuator, 357
Integrated bulk Hall devices, 232
Integrated circuit, 12
 fabrication, 12
Integrated-circuit technology, 18
Integrated Hall plates, 232
Integrated Hall sensors, 231, 234
Integrated microbeam accelerometer, 127
Integrated optics, 124
Integrated resistors, 351
Integrated sensors, 2, 14, 197, 517
Integrated silicon magnetic sensors, 237
Integrated watt-hour meters, 206
Interdigital transducer, 99, 117, 356
Interface circuits, 497, 499, 510
Interface electronics, 493
Interface standards, 484
Intrinsic carrier concentration, 535
Intrinsic Debye length, 535
Intrinsic resistivity, 535
Inversion layer, 241, 389, 405
Inverted-pyramid topography, 294
In vivo sensors, 465
Iodine, 302
Ion-beam deposition, 109
Ion-beam milling, 303
Ion-channel sensors, 422
Ionic crystal, 385
Ionic forces, 425
Ion implantation, 12, 233, 234
Ion-selective electrodes, 439
IR detectors, 109
IrOx, 408
ISFET, 406, 408, 428, 436, 437, 438, 439
 sensor, 409
Isopropyl alcohol, 48
Isothermal Nernst effect, 231
Isotropic etching, 29
Isotropic etch-stop, 156

J1850 bus, 483
JFET, 306, 320, 321, 496
Johnson–Nyquist noise, 222
Josephson junctions, 209
Junction-isolated Hall plates, 233

KCl, 408

INDEX 543

Kinetic effects, 140
Kinetic viscosity, 336, 337
Kirchhoff's law, 338
Knee frequency, 289
KOH, 48, 50, 59, 113, 116, 138, 154
K-shell, 276
K-space, 170

$LaCrO_3$, 400
Lactate oxidase, 443
Lambert's law, 340
Lamb waves, 101
Laminar flow, 335
Landau levels, 304
Langmuir–Blodgett (LB) technique, 422
Laplace equation, 238
Lasers, 450
    ablation, 109
    recrystallization, 158
    trimming, 234
Lateral magnetotransistors, 232, 249, 265
Lateral MT, 246
Latex films, 64
Lattice constant, 535
Lattice-oxygen mobility, 398
LCD displays, 57
Lead-zirconate-titanate (PZT), 99
Li, 395
$Li_2B_4O_7$, 104
Life-off, 29
Lifetime, 288
LIGA, 75
Light-emitting diodes, 451
$LiNbP_3$, 102, 103, 104, 127
Linearity, 8
Liquid-phase chemical reaction, 43
Liquid-phase epitaxy, 302
Liquid-type sensors, 364
$LiTaO_3$, 104
Lithographic transfer errors, 39
Lithography, 25
    step, 12
Lithium-doped ZnO films, 107
Lithium-drifted silicon detectors, 310
Lithium niobate ($LiNbO_3$), 102, 103, 125
Load-deflection measurement, 33
Loading effect, 64
Longitudinal piezoresistance coefficient, 178
Longitudinal wave, 100
Lorentz angle, 214, 217
Lorentz-current, 243
    deflection, 205, 214, 238, 240, 249, 250
    force, 205, 207, 208, 211, 213, 217, 248, 249, 260
    mode, 220, 239, 240
    operation, 246
Lowest-order antisymmetric mode, 102
Low-pressure thermal CVD, 23
    etching, 47
    glass bonding, 52
    oxide, 70
LPCVD, 35, 57, 63, 73, 508, 515
    process, 138

Luciferase, 451
Luciferin, 451
Lyophilized drug, 143

Madelung model, 385
Magnetically controlled oscillator, 245
Magnetic detector, 254
Magnetic energy, 3
Magnetic field effect transistor, 232, 240
Magnetic heterojunction devices, 245
Magnetic induction, 211, 240
    noise, 223
Magnetic measurands, 5
Magnetic semiconductors, 304
Magnetic sensors, 17, 206
    modeling, 238
Magnetic signal, 3
Magnetic storage media, 207
Magnetite, 205
Magnetoconcentration, 238, 246
    effect, 205, 217, 218
Magnetodiodes, 208, 210, 218, 232, 258, 259
Magnetometer, 206
Magneto-operational amplifier (MOP), 254
Magneto-optic effects, 208
Magneto-optic sensors, 208
Magnetoresistance, 205, 207, 214
    effect, 212, 215, 216
Magnetoresistive devices, 208
Magnetoresistive (MR) sensors, 210
Magnetoresistive (MR) switching, 208
Magnetostriction, 208
Magnetotransistor, 208, 246, 256, 264, 497
Magnetron, 106, 107
    sputtering, 106, 109
Manufacturing, 523
Many-valley model, 171
Mask making, 26
Mass balance equation, 434
Mass detection, 452
Mass energy, 3
Mass-flow measurement, 359
Mass sensitivities, 112
Mass transport, 23, 427, 445
Matrix entrapment, 425, 463
Matrix immobilization, 432
Measurand, 6, 13, 25, 98, 185, 196, 207, 287
    range, 7, 9
Measurement systems, 3
Mechanical energy, 3
Mechanical measurands, 5
Mechanical sensors, 34, 185
Mechanical signal, 3
Mechanical warping, 35
Mediator molecules, 441
Melting point, 535
Membrane, 196, 418
    confinement, 424, 463
    deflection, 38
    loading effect, 432
    stress, 184
    type sensor, 176

Merged epitaxial lateral overgrowth, 59
MESFET, 137
Message transfer, 479
Metabolic sensor, 431
Metal-can packages, 34
Metal cation, 387
Metallic seals, 50
Metal-organic CVD, 107, 302
Metal oxide, 52
  gas sensor, 393
  sensor, 383
Metal-oxide semiconductor (MOS), 502
Metals, 461
Metal-semiconductor diodes, 293
Methacrylates, 425
Methyl-methacrylate, 25
Michaelis constant, 420
Michaelis–Menton kinetics, 420, 433, 443, 444
Micro-actuators, 357, 465
Micro-calorimeters, 38, 357, 369, 373
Microcomputer, 475
Microcontroller, 372
Micro-electro-mechanical (MEM) process, 128
Microelectromechanical sensing system (MEMS), 18
Micro-electro-mechanical systems, 465
Microelectronics circuits, 1, 2
Microelectronics industry, 18
Microflow meters, 477
Microflow system, 143
Micromachined sensors, 17, 19
Micromachining, 12, 373, 374
  error, 39
  technique, 523
Micromechanical sensor, 197
Micro-mechanical systems, 12
Micromixers, 143
Microphones, 17
Micropumps, 143
Microprocessor, 13, 73, 490
  based system, 523
Microsensors, 2, 143, 474
Microsystem, 197
Miller compensation capacitor, 495
Miller indices, 155
Minority-carrier lifetime, 535
MIS capacitors, 304
Miscibility, 298
Mn, 304
Mobility, 170, 245, 535
  fluctuation, 288
  induction product, 225
MOCVD, 109
Modifier, 3, 4
Modulate/demodulate, 4
Modulating sensor, 4, 357, 373
Modulation-doped AlAs/GaAs superlattice structure, 228, 245
Modulation frequency, 288, 290, 295
Modulation index, 295
Moisture, 456
  barrier, 459

Molar enthalpy, 452
Molding, 75
Molecular beam epitaxy, 107, 227, 245, 302
Molecular energy, 3
Molecular velocity, 334
Molecules, 98
Monolithic SAW devices, 117
MOS, 493
  studies, 299
  transistor, 57, 500
MOSFET, 210, 225, 241, 243, 297, 320, 383
Multichannel recording/stimulating microprobes, 512
Multi-element smart gas analyzer, 518
Multiple sensors, 126
Multiplexing, 497
Multiplication-noise factor, 296
Multisensor systems, 496

Nanoindentation apparatus, 32
NaOH, 113, 116
Nb_2O_5, 400
Nernst voltage, 231
Nernst equation, 408, 438, 439
Nernstian response, 400, 438
Neural activation, 515
Neutron, 271, 273, 275, 285
  transmutation, 312
Newton's force law, 101
NH_3, 125, 401, 406
NH_4F, 66
Nickel, 65, 209, 353
Nickolsky equation, 439
NiCo, 208
Ni-Cr combination, 65
NiFe, 207, 208
$Ni_{81}Fe_{19}$, 208
Night-vision systems, 17
NiO, 400
Nitrogen, 365
N-methylphenazinium, 443
NO, 406
NO_2, 125, 406
Noise, 252, 287
  correlation, 253
  equivalent power, 285, 286
  phenomena, 222
Nondegenerate semiconductors, 212
Nonlinearity compensation, 491
Non-organic bond techniques, 159
Non-volatile memory, 109
n-p-n-p structure, 261
Nuclear collision, 284
Nuclear-decay processes, 274
Nuclear energy, 3
Nuclear magnetic resonance (NMR), 264
  magnetometry, 206
Nuclear-particle radiation, 273
Nucleic acids, 418
Nusselt number, 338, 360
Nyquist noise, 252
Nyquist relation, 223

INDEX 545

O_2, 384, 401
O_2 plasma, 304
<100>, 230
   axes, 162, 164
   direction, 155, 171
   surface, 181
(100) etch fronts, 44
[010] direction, 171
<100> plane, 178
   direction, 155, 158
(110)-silicon, 46
(111) planes, 46
   planes, 44, 178
<111> direction, 44, 155, 166
<111> orientation, 175
<111> slopes, 181
1/f noise, 222, 226, 241, 245, 252, 288, 289, 302, 496, 503
Offset, 9, 35, 234, 242, 491
   ohmic contacts, 304
   voltage, 224
Ohm's law, 228, 286
On-chip amplifier, 134
On-chip circuitry, 511, 513, 517
On-chip feedback, 14
Operating life, 9
Operational amplifier, 220, 254, 503
Optical communications, 298
Optical detection, 447
Optical fiber, 124, 209
Optical interference, 158
Optical interferometer, 209
Optical interferometry, 158
Optical measurands, 5
Optical phonon energy, 535
Optical sensors, 449
Optoelectronic magnetic sensors, 208
Organelles, 418, 463
Organic bonds, 54
Organic contaminants, 54
Organic dyes, 443
Organic films, 64
Organic materials, 20
Organic salt, 443
Organisms, 418
Output format, 9
Overload characteristics, 9
Over range, 9
Oxidation, 23, 63, 443
Oxidized silicon, 107
Oxygen, 451
   adsorption, 389
   ion vacancy, 393
   vacancies, 393
Ozone-based organic etchers, 70

Package, 510
Packaging, 455
Packaging technologies, 524
Pair production, 276
Palladium hydride phase, 406
Paramagnetic materials, 208

Parity, 482
   error, 482
Partial pressure, 397
Parylene, 64, 460
Passive delay line, 110
Patterning techniques, 29
PbI_2, 313
PbS, 298
PbSe, 298
PbTe, 298
Pd, 125, 399, 402, 404
   -gate H_2 sensor, 405
   -gated MOSFET, 406
   -gated sensor, 409
Pd_2Si, 298
PdO, 406
PECVD, 62, 73, 74
   silicon film, 58
Peltier cooling element, 373
Peltier element, 368
Penicillin, 373, 440
Penicillinase-penicillin, 436
Performance/cost ratio, 13, 18
Permalloy, 208
Permanent magnets, 207
Permeability, 73, 207, 274
Permittivity, 274
PH, 451
   ISFET, 440
   value, 441
PH_3, 57
Phase-locked resonator measurements, 120
Phase shift, 110, 209, 290
Phase velocity, 98, 100, 103, 110, 138
Phonon, 332, 334
   detector, 362
   mean free path, 535
Phospholipids, 421, 425
Phosphosilicate, 52
   glass, 63, 66
   glass bond, 52
Photo-assisted MBE, 302
Photoconductive mode, 290
Photoconductors, 285, 451
Photodetection system, 4
Photodiode, 4, 289, 292, 451
Photoelectric effect, 276
Photoelectron, 276
Photolithography, 25
Photomagnetoelectric effect, 6
Photon energy, 274
Photoresist, 25, 70
Photosensitive resists, 20
Phototransistor, 297, 451
Photovoltaic effect, 4
Photovoltaic mode, 289
PHOTOX, 304
Physical constants, 15, 534
Piezoceramic films, 106
Piezoelectric biosensors, 455
Piezoelectric cantilever-beam accelerometer, 128

546 INDEX

Piezoelectric cantilever-beam devices, 127
Piezoelectric capacitor, 114, 134
Piezoelectric coefficient, 43
Piezoelectric coupling, 99, 102, 103, 117, 120
Piezoelectric crystal plates, 97
Piezoelectric effect, 102, 117, 192
Piezoelectricity, 104
Piezoelectric materials, 104
Piezoelectrics, 56
Piezoelectric stiffness, 104
Piezoelectric substrates, 356
Piezoelectric thin films, 105
Piezoelectric transducers, 454
Piezoelectric zinc oxide, 126
Piezo-Hall effect, 230
Piezo-Hall sensors, 158
Piezojunction effect, 158
Piezoresistance, 162, 230, 242
   coefficients, 158, 162, 163, 165, 166, 167, 168, 171, 172, 175, 182, 183, 196
   effect, 160, 161, 171, 191, 196, 224
   factor, 173
Piezoresistive accelerometer, 176, 185, 192, 193, 194, 195
Piezoresistive coefficient, 174
Piezoresistive effect, 10
Piezoresistive material, 162
Piezoresistive membrane sensors, 189
Piezoresistive pressure sensors, 153, 157, 160, 161
Piezoresistive semiconductor accelerometers, 191
Piezoresistive sensors, 153, 161, 166, 174, 175, 486
Piezoresistive strain gages, 158
Piezoresistivity, 153, 160, 162, 174
Piezoresistors, 174, 176, 177, 180, 497
p-i-n diodes, 295, 308
Pinning, 70
Pixel, 305
Planarization, 74
Planar technology, 18
Plasma CVD, 57
Plasma deposition, 24
Plasma-enhanced CVD, 24
Plasma etching, 29, 156
Plasma polymerization, 25
Plate-mode sensors, 103, 105
Plastic fibers, 450
Platinum, 50, 107, 141, 303, 353, 363, 441
   resistors, 351
   wire, 115
PLZT, 106
PMMA, 78
PMOS, 131
p-n junction, 43
p-n-p transistors, 250
P_2O_5, 66
Point-contact rectifier, 10
Poisson equation, 238, 295, 390
Poisson's ratio, 36, 133, 139, 536
Poisson statistics, 314

Pollutant emissions, 144
Polyacrylamides, 425
Polycrystalline piezoresistors, 177
Polycrystalline silicon, 56
Polyimide, 20, 55, 70, 74, 460
   film, 64
Polymer, 55, 362
   film, 125
Polymeric encapsulants, 457
Polymeric material, 463
Polymerization, 77
   reaction, 425
Polymethacrylates, 425
Poly-methyl methacrylate, 77
Polypeptide chains, 424
Polysilicon, 157, 194, 197, 231, 351, 353, 508
   /aluminum thermocouple, 363
Polystyrene, 64
Poly-tetrafluororethylene, 25
Polyurethanes, 460
Polyvinylidene fluoride, 362
Ponyting vector, 274
Popcorn noise, 289
Porous silicon, 359
Positron, 273, 283
Potassium hydroxide (KOH), 48, 155, 189
Potentiometric biosensor, 428, 435, 439
Potentiometric detection, 438
Potentiometric enzyme biosensor, 431
Potentiometric sensor, 437
Power supply, 4
Prandtl number, 337, 338
Pressure sensitivity, 189
Pressure sensor, 6, 17, 57, 105, 145, 153, 157, 176, 187, 197, 525
Promoter, 394
Propane, 383, 404
Protein, 421, 423, 425
   chain, 426
Proximity switches, 206
Pt, 100, 125, 351, 394, 400
PTAT temperature sensors, 370
Pt-H bond, 395
Pt-H groups, 394
Pt-O bonds, 395
PtSi, 298
   infrared detector, 293
Pyrazine, 47
Pyrex glass, 192
Pyrex 7740 borosilicate glass, 160
Pyrocatechol, 47
Pyroelectric, 56
   sensor, 362
Pyruvate oxidase, 443
PZT, 99, 105, 109, 117
   films, 103

Quality factor, 111, 159
Quantum efficiency, 277, 285, 291, 294, 308, 309
Quartz, 42, 43, 78, 104, 125, 127, 192
Quartz crystal, 453

microbalance, 97, 452
Quaternary semiconductor, 294
Quinones, 443

Radial heat flow, 343
Radiant energy, 3
Radiant signal, 3
Radiation, 338
  measurants, 5
  sensors, 271
Radon, 333
RAM, 478
Rapid thermal anneling, 35
Rate vectors, 41
Ratiometric techniques, 38
Rayleigh waves, 453
RC oscillator, 500
RC time constant, 347
Reaction-limited processes, 23
Reactive diode sputtering, 106
Reactive growth, 23
Reactive-ion etching (RIE), 74, 99, 154, 156
Reactive magnetron sputtering, 106
Reactive metal bonding, 54
Reactive surface, 104
Receptor, 418
  -ligand binding, 416
Recombination noise, 296
Recovery time, 9
Refractive index, 274
Refractory-materials, 22
Registration errors, 39
Relative detectivity, 362
Relative humidity, 367
Relativistic velocities, 284
Relativity theory, 3
Relaxation time, 212
Repeatability, 9
Residual stress, 34, 35, 36, 57
Resistive sensors, 497
Resistivity, 10, 59, 62, 294, 362
Resistor heating, 350
Resolution, 9
  errors, 39
Resonant frequency 110, 133, 159
Resonant sensors, 486
Response time, 358
Reynolds number, 336, 338
RF-magnetron-sputtered ZnO, 138
RF sputtering, 106, 109
RIE, 154
Rigidity, 188
RNA, 463
Rocking curve, 108
RuO_2, 400
Rutherford backscattering, 284

Sacrificial etching, 42, 67, 80
  layer, 63, 154, 508
  materials, 67
Salmanella typhimurium DNA sensor, 454
Sampling rate, 492

Sapphire, 302
SAW, 99, 101, 103, 105, 119
  cantilever-beam accelerometer, 135
  communication device, 144
  convolver, 106
  correlator, 106
  delay line, 106, 110, 124, 136
  gas sensor, 124
  interdigital transducer, 138
  resonator, 122
  sensor, 109, 111, 117, 138
  vapor sensor, 122, 453
  velocity, 100, 101
Scattering coefficients, 229, 230
Scattering factor, 212
Scattering parameters, 120, 121, 122
Schmitt trigger, 234
  barrier, 293, 294
  oscillator, 490, 500
Schottky avalanche photodiodes, 29
Schottky barrier, 293, 294
Schottky diode, 298
Schottky equation, 390
Sealing, 72
Seebeck coefficient, 15, 351, 352
Selectivity, 9, 39, 43, 48, 70
Selectivity and uniformity trade-off, 40
Self-generation sensor, 4, 357, 369, 373
Self-heating effect, 344
SEM, 36, 515
Semiconducting metal oxides, 383
Semiconductor devices, 1, 10
Semiconductor drift chamber, 315
Semiconductor magnetic sensors, 208
Sem-insulating GaAs, 234
Sensing node organization, 484
Sensing viscosity, 140
Sensitivity, 9, 221, 245, 248, 251, 257, 262
Sensor, 197
  characterization, 7, 490
  evaluation, 475
  nodes, 480
  sampling rate, 491
Sentire, 1
7740 pyrex glass, 43
Sezawa mode, 119
SF_6, 64, 66, 156
S-gun systems, 106
Shape-specific binding, 416
Shear-mode oscillations, 453
Shear stresses, 162
Sheet resistance, 225
Shockley–Read–Hall lifetime, 300
Shot noise, 252, 287, 295
Si, 15, 116, 117, 128, 225, 226, 227, 259
Si/Al_2O_3 interface, 259
Si-Au cantilever beams, 357
SiC, 308
$SiCl_4$, 65, 156
Side illuminated diode, 293
Sidewall profiles, 65
SiGe alloys, 298

Sigma-delta A/D analog-to-digital converter, 256
Signal amplification, 515
Signal conditioning, 4
 circuitry, 131, 264
Signal filtering, 496
Signal-to-noise ratio, 14, 117, 222, 517
Signal transmission, 4
Silane pyrolisis reaction, 57
Silica sol, 401
Silicon, 1, 10, 42, 52, 53, 56, 80, 125, 138, 153, 161, 168, 169, 173, 176, 207, 212, 225, 227, 230, 282, 283, 297, 302, 334, 340, 345, 353, 385, 460
 aluminum thermopiles, 363, 370
 beams, 186
 color sensors, 305
 compilers, 497
 dioxide, 56, 156, 233, 359
 direct bonding, 160
 fusion bonding (SFB), 160
 magnetic sensors, 208
 membrane, 180, 186, 190
 nitride, 44, 48, 56, 59, 67, 137, 154, 156, 194, 359, 409, 457
 -on-insulator devices, 53
 oxide, 62, 154
 -rich nitride, 67, 77
 -rich nitride films, 70
 -rich silicon nitride, 35
 rubbers, 460
 Schottky diodes, 293
 silicon-dioxide interface, 320
 substrates, 320
 technology, 373
 to-glass bond, 159
SiN, 73
Single-crystal silicon, 58
Single-wafer plasma etchers, 65
SiO_2, 48, 52, 64, 116, 128, 129, 254, 304, 402, 405, 408
 bandgap, 292
Si/SiO_2 interface, 259, 260, 314
Si_3N_4, 64, 408
$Si_xN_yH_2$, 62
16-bit A/D conversion, 479
"Sixth-generation" sensors, 491
Slider-type devices, 33
Slow-wave transmission, 138
Smart sensors, 374
$SnCl_2$, 401
SnO_2, 383, 384, 385, 392, 393, 399, 402, 403
 sensor, 404
SO_2, 125, 406
Sodium hydroxide, 49
Sodium-rich glass, 51
Sodium-rich substrates, 43
SOI diode, 293, 308
Sol-gel methods, 99
Sol-gel techniques, 109
Solid-state sensors, 1
Solubility, 20

SOS (silicon-on-sapphire), 259
 hybrid amplifier, 137
Spark erosion, 154
Spatial distribution, 6
Specific heat, 332, 342, 536
Spectral density, 253
Spectrum, 4
Speed of response, 10
SPICE, 345
Spin casting, 20
Spin-on glass, 63, 160
Slurries, 52
Spiral drift chamber, 320
Sputtering, 22, 57
Sputter S-gun, 107
Squeeze-film effect, 193
SQUID, 209
SSIMT, 254, 257
Stability, 10, 403, 492
ST-cut quartz, 137
Step coverage, 21, 22
Stefan–Boltzmann law, 339
Stimulating microprobe, 517
Stoney equation, 36
Strain, 32, 33, 134
Strain-gage applications, 11
Stray stress, 33, 35, 36, 52, 84, 386
Stress, 32, 188
 detection, 158
 relaxation, 34
Strip detectors, 314, 324
Structural time constant, 358
Substraction process, 80
Substrate current, 252, 254
Suhl effect, 206, 217
Sulfide, 124
 dioxide, 124
Surface acoustic waves (SAW), 97
 delay-line oscillators, 452
 sensor, 1, 13
Surface adsorption, 520
Surface loading, 104
Surface-micromachined sensors, 19, 20
Surface micromachined structures, 334
Surface micromachining, 55, 154, 177, 504
 of silicon, 157
Surface-micromachining technologies, 508
Surface mobility, 21
Surface wave device, 112
Superconducting coils, 207
Superconducting quantum interference device, 264
Superconductor magnetic sensors, 209
Superlattice structures, 294
Supermagnetoresistor, 258, 263
Suppressed sidewall-injection MT, 249
Switched-capacitor, 508
 technique, 500
Symbols, 531
Synthetic polymers, 425

Taguchi sensor, 383, 400, 403

Ta_2O_5, 408
Te, 301, 304
Technological aspects, 6, 7
Telluride, 363
  compounds, 363
Temperature coefficient, 181, 184, 224
  of offset, 82
Temperature gradient, 333
Temperature sensor, 341, 371
Temporal distribution, 6
Tensile stresses, 37
Tension, 104
TeO_2 layer, 304
Terminal-based linearity, 8
Ternary semiconductor, 294
Tetrafluoroethylene, 64, 401
Tetramethylammonium hydroxide, 49
Thermal capacitance, 332, 342, 348
Thermal conductivity, 38, 62, 194, 334, 336, 342, 364, 366, 536
Thermal detection, 451
Thermal diffusivity, 38, 337, 536
Thermal energy, 3
Thermal expansion coefficient, 34, 52
Thermal measurands, 5
Thermal noise, 223, 252, 287, 496
Thermal parameters, 38
Thermal reactive growth, 72
Thermal resistance, 341, 342, 343, 346, 348, 373
Thermal resistivity, 342
Thermal-sensing elements, 350
Thermal sensor, 6, 38, 364, 365, 373
Thermal sheet resistance, 343, 348
Thermal signal, 3, 341
Thermal stability, 43
Thermal strains, 34
Thermal stresses, 34
Thermal structures, 348
Thermal time constant, 348
Thermistors, 369, 451
Thermocouples, 351, 360, 451
Thermoelectric Seebeck effect, 351
Thermomagnetic effects, 230
Thermopile, 341, 346, 348, 535
  -based sensors, 373
  sensor, 362, 363
Thickness shear-mode (TSM) sensor, 97
Thin film, 20, 55
  materials, 56
  sensor, 351, 397, 399
  stoichiometry, 22
Thin metal film magnetic sensors, 208
Three-electrode configuration, 190
Threshold, 9
  voltage, 241, 406, 494
Thyristors, 312
Time constant, 346, 362
Time-of-flight measurement, 359
Timing checks, 482
TiO_2, 392, 393, 399, 400, 403, 463
Tire pressure monitoring, 197

Tissues, 418, 463
Titanium, 77, 457
  nitride, 508
  silicide, 508
Total input-referred noise, 495
Transducer, 2, 4, 17, 341
Transduction, 2, 31, 523
Transfer function, 346, 348
Transistor, 354
  temperature sensors, 355
Transit times, 290
Translator, 105
Transverse piezoresistance coefficient, 166, 178
Transverse voltage gage, 158
Transverse waves, 100
Trap-assisted tunneling, 300
Traveling-heater method, 312
Trenches, 65
Triethanolamine, 125
Triode sputtering, 106
True-rms converters, 357, 365, 373
Two-dimensional polynomial equations, 491
Two-sided processes, 113
Tungsten, 21
Turbulent flow, 335
TV SAW filters, 106
TV VIF filters, 105

UHF SAW oscillator, 137
Ultra-clean chamber, 54
Un-binding reaction rate, 419
Uniaxial state of stress, 165
Uniformity, 22
Unipolar comb-like electrodes, 103
Unit, International System of, 15, 533
Urea, 440
Ureum, 373
UV Lithography, 39

Vacuum sensors, 348, 357, 364
Van der Waal Forces, 53, 425
Vapor pressure, 536
  probes, 251
Vehicle navigation system, 143
Vertical four-layer CDM, 261
Vertical Hall device, 232, 235, 236, 238
Vertical magnetotransistor (MT), 232, 247
Vertical MT, 246
V-groove, 44, 45, 46
Visible-light color sensors, 305
Viscosity, 336
Vitreous glasses, 52
VLSI, 99
  interface chip, 484
  microcontroller, 521
  processing, 134
  signal-processing chip, 490
  technology, 19
VMT, 257

Wafer, 335

Wafer (*Continued*)
    bonding techniques, 42, 374
    lamination process, 12
    to-glass bonding, 80
Warburg impedance, 446
Wave amplitude, 4
Wavelength, 110
Wet chemical etching, 29, 41
Wheatstone bridge, 158, 178, 182, 191
    configuration, 196
Wien's displacement law, 340
WO_3, 125, 393, 400, 403
Wulff construction, 41

Xe, 303
Xenon, 303, 333
X-ray, 276, 277, 283, 324
    detection, 293, 317
    diffraction, 38
    diffraction pattern, 108
    emission, 281, 290
    fluorescence spectroscope, 447

imager, 321
mask, 75, 77

Y-Ba-Cu-O ceramic, 264
Y-Ba-Cu-O films, 258
Young's modulus, 32, 33, 36, 139, 177, 188, 536

Zero-stress resistance, 180
Zinc, 302
    oxide, 103, 144, 192
ZnO, 1, 99, 102, 105, 106, 107, 108, 117, 120, 122, 128, 392, 393
    capacitor, 127
    on-Si, 99
    on-Si device, 118
$ZnO/SiO_2/Si$ overlay transducer, 122
ZnO-SiO_2-Si structure, 119
ZnS, 304
ZnSe, 313
ZnTe, 313
Zone-melting method, 312